T0310062

MOBILE POSITIONING AND TRACKING

MOBILE POSITIONING AND TRACKING

FROM CONVENTIONAL TO COOPERATIVE TECHNIQUES

SECOND EDITION

Simone Frattasi

Sony Mobile Communications, Sweden

Francescantonio Della Rosa

Radiomaze Inc., USA,
Tampere University of Technology, Finland

This edition first published 2017

Registered offices
John Wiley & Sons. Inc., 111 River Street, Hoboken, NJ07030, USA
John Wiley & Sons Ltd, The Atrium, Southern Gate, Chichester, West Sussex, PO19 8SQ, UK

Editorial Office
The Atrium, Southern Gate, Chichester, West Sussex, PO19 8SQ, UK

For details of our global editorial offices, customer services, and more information about Wiley products visit us at www.wiley.com.

Wiley also publishes its books in a variety of electronic formats and by print-on-demand. Some content that appears in standard print versions of this book may not be available in other formats.

Library of Congress Cataloging-in-Publication Data applied for

Hardback: 9781119068815

Cover design by Wiley
Cover image: (Swirl) © DrHitch/Shutterstock; (Abstract Map) © Supphachai Salaeman/Shutterstock

Set in 10/12pt TimesLTStd by SPi Global, Chennai, India
Printed and bound in Malaysia by Vivar Printing Sdn Bhd

10 9 8 7 6 5 4 3 2 1

… to my dear dad, Luigi, and my grandad, Francesco
Simone Frattasi

… to my lovely wife Anna, my mom Emilia, my dad Liberto and my brother Gianluca
Francescantonio Della Rosa

Contents

About the Authors

Simone Frattasi became European Patent Attorney (EPA) in 2016. He received his PhD in wireless and mobile communications from Aalborg University (AAU), Aalborg, Denmark, in 2007, and his MSc degree *cum laude* and his BSc degree in telecommunications engineering from the University of Rome "Tor Vergata", Italy, in 2002 and 2001, respectively. Additionally, he obtained a certificate as a project manager from Act2Learn in 2009 and a certificate as an instructor in the IEEE Leadership Course from the IEEE in 2008.

He is the head of the Patent Section at Sony Mobile Communications, Lund, Sweden, where he was previously employed as a senior patent attorney. From 2011 to 2015, he worked as a patent consultant at Patrade. From 2010 to 2011, he worked as a postdoc in the Center for TeleInFrastruktur (CTIF) at AAU, where he was technical project manager for the FP7 project ASPIRE, proposal coordinator for the SOS-4-HEALTH project (including AAU, Aalborg Hospital, Telenor, Care4All, IctalCare, and G4S) and lecturer for the course "IPR, Patenting and Technology Transfer" for the M.Sc. on innovative communication techniques and entrepreneurship. In 2009, he worked as a patent consultant in Plougmann & Vingtoft. From 2007 to 2008, he worked as a postdoc at AAU, where he was technical project manager for the industrial project LA-TDD, a collaboration with Nokia Siemens Networks (NSN). From 2005 to 2007, he fundraised and worked as a manager for the Danish-funded project COMET at AAU. From 2002 to 2005, he worked as a research assistant at AAU on two FP5 projects (STRIKE and VeRT) and one industrial project (JADE) in collaboration with the Global Standards & Research Team, Samsung Electronics, Korea.

He is author of the first edition of the book *Mobile Positioning and Tracking: from Conventional to Cooperative Techniques* (John Wiley & Sons Inc., June 2010). He is author/co-author of more than 65 publications, including papers published in journals, magazines and proceedings of international conferences, book chapters, encyclopedia papers and technical reports. He is inventor/co-inventor of one US patent and four Danish patent applications. He has served as a reviewer for several technical and IPR journals (including the *Oxford Journal of Intellectual Property Law & Practice*), magazines and international conferences, and as a guest editor for several special issues in various technical journals and magazines. He has been an instructor for a half-day tutorial on wireless location at IEEE PIMRC'07 as well as for seminars and tutorials on IPR at MobileHCI'15 and GWS'14.

He was co-founder of Kyranova Ltd, and co-founder and president of the International Symposium on Applied Sciences in Biomedical and Communication Technologies (ISABEL), the International Conference on Cognitive Radio and Advanced Spectrum Management (CogART) and One2One (Business & Science Match).

He is an editorial board member of the journal *Recent Advances in Communications and Networking Technology* (formerly *Recent Patents in Telecommunications*, Benthamscience Publishers). He has been a board member of IPR Nord, Chairman of the Danish Section of IEEE Graduates of the Last Decade (GOLD) and a member of the IEEE Aerospace & Electronic Systems Society.

His research interests include (but are not restricted to) cooperation in wireless networks, link layer techniques, wireless location, quality of service mechanisms, next-generation wireless services and architectures, user perspectives and sociological dimensions related to the evolution of technology and society.

Francescantonio Della Rosa received an MSc degree in electrical and electronic engineering from Aalborg University, Denmark, a BSc degree in telecommunications engineering from the University of Cassino, Cassino, Italy, and is a PhD candidate at Tampere University of Technology, Tampere, Finland.

He is an accredited business coach for the European Commission at Horizon 2020 SME Instrument, also serving as chairman at Ekin Labs Oy (Finland). Currently he is the managing director of Technological Innovation at Radiomaze Inc (California), funded by Singularity University Labs and selected to solve the Global Grand Challenges for Humanity in the Security sector at NASA Ames Research Park in Mountain View (USA).

He has successfully coached and instructed more than 20 technology-based ventures, turning research ideas and results into products and businesses.

Francescantonio served as IEEE Finland Section Executive Board Officer and as Honorary Jury at CES 2017 for Space and UAV Category. He funded and managed multi-million euro projects focusing on the commercialization of innovative solutions for business, security, IoT, the space industry, big data, artificial intelligence, such as the EU FP7 Multi-technology Positioning Professionals Marie Curie and the European FP7 project GRAMMAR (Galileo Ready Advanced Mass Market Receiver), leading the team who built the first Galileo receiver for the mass market. He has also commercialization research and development results and the Watchdog project, realizing the first home security solution that can detect a human presence based on the radio wave fluctuations in conventional wifi systems available in domestic environments.

He is the winner of many of international research, innovation and business awards as a result of the commercialization of his research results, for example the CES 2015 Innovation Award Honoree, Las Vegas, Best European Startup 2015 in the Smart Spaces category at EIT Digital, Best Technology Transfer Awards Hipeac Network of Excellence and the NOKIA Foundation Award. He also gained an Entrepreneurial Achievement award from Kauffman FastTrac TechVenture, which grants the bearer the right and responsibility to build an "uncommon company", and received the nomination as Young Research Entrepreneur of the Year 2016 in Finland and the Honorary Technical Creativity and Business Award from Tampere City (Finland).

He is co-author and editor of three books and many chapters, patents and scientific publications focusing his research interests of positioning and navigation, GNSS, location-based services, big data, IoT, wireless communications, artificial intelligence, business and innovation.

List of Contributors

Gilberto Berardinelli Aalborg University, Denmark

Tanveer Bhuiyan Aalborg University, Denmark

Guido Bruck Lehrstuhl für KommunikationsTechnik, Universität Duisburg-Essen, Germany

Davide Dardari University of Bologna, Italy

Francescantonio Della Rosa Radiomaze Inc., Cupertino, USA, Tampere University of Technology, Finland

Simone Frattasi Sony Mobile Communications, Sweden

João Figueiras Aalborg University, Denmark

Ismail Guvenc Wireless Access Laboratory, DOCOMO Communications Laboratories, Palo Alto, CA, USA

Peter Jung Lehrstuhl für KommunikationsTechnik, Universität Duisburg-Essen, Germany

Nicola Marchetti CTVR, Trinity College Dublin, Ireland

Jari Nurmi Tampere University of Technology, Finland

Mauro Pelosi Radiomaze Inc, Cupertino, California

Andreas Waadt Lehrstuhl für KommunikationsTechnik, Universität Duisburg-Essen, Germany

Preface

Localization is a research topic that is receiving increasing attention from both academia and industry. Previously considered as vital information for vehicle tracking and military strategy, location information has now been introduced into wireless communication networks. In contrast to dedicated solutions, such as the global positioning system (GPS), that were designed to simply provide positioning information, the new solutions for wireless networks are able to supply the combined benefit of both communication and positioning. As a consequence, the network operator, as well as the service provider and the end user, can profit from such position-enabled communication capabilities. Indeed, while the network operator is able to manage the resources of its network more efficiently, the service provider is able to offer location-based services to the end user, who can fully enjoy such personalized location-dependent services. In particular, it can be found from the literature that location information is being used as a basic requirement for the deployment of new protocols (e.g., routing and clustering), new technologies (e.g., cooperative systems) and new applications (e.g., navigation and location-aware advertising). From the point of view of the industry, the use of location information has been stimulated mainly by applications such as navigation, location-dependent searching and social networking. Since wireless communication networks are nowadays present anywhere and anytime, every location-dependent networking enhancement, service or application can be spread rapidly and used globally.

The above-mentioned trends are a major stimulator for the development of novel solutions for obtaining positioning information in wireless networks. Chapter 1 outlines the motivation behind these solutions and presents potential categories and applications of location-based services (both conventional and network-related).

Chapter 2 introduces the main application areas for positioning, providing an overview of the localization ecosystem and its usability with a look at the main patent trends. Chapter 3 presents the fundamentals of wireless communications for positioning, describing the main radio propagation characteristics of both conventional and cooperative. Chapter 4 presents the fundamentals of positioning, proposing a classification of positioning methods, techniques and main error sources. Chapter 5 describes these various types of data association algorithms, showing the advantages and disadvantages of each. Chapter 6 deals with the fundamentals of tracking, in particular several mobility models (including group-based and socially based models) that are used in the following chapters will be introduced. Chapter 7 considers some advanced techniques (from the realm of signal processing) used to mitigate the errors mentioned in previous chapters, thus trying to enhance the accuracy of the overall location estimation process. Chapter 8 presents the state of the art of satellite-based and terrestrial based positioning systems, spanning the range from outdoor to indoor environments, from wide-area networks to short-range networks, and from orthogonal frequency division multiplexing to ultra-wideband (UWB) technologies. In Chapter 9 we introduce the topic of UWB positioning by describing fundamentals about regulations and positioning approaches for tracking targets. Chapter 10 presents indoor positioning approaches in wireless local area networks by highlighting the effect the environmental impairments and human body signal absorption have on signal strength measurements. Chapter 11 introduces the topic of multi-tag localization by adopting radio frequency identification systems and experimental activities as well. Replicating cooperative human behavior in wireless communications has resulted in a number of emerging research fields. In particular, its application in wireless location has flown in a new breed of techniques that may revolutionize the entire field. Hence, in Chapter 12 we take a tour through the state of the art of what we call "cooperative augmentation systems", that is, mobile positioning systems that exploit the cooperation of users, terminals and networks to boost their location estimation accuracy in both simulated and real environments.

Acknowledgements

The authors would like to thank the direct contributors to the book, namely, João Figueiras, Gilberto Berardinelli, Nicola Marchetti, Andreas Waadt, Guido Bruck, Peter Jung, Tanveer Bhuiyan, Davide Dardari, Jari Nurmi, Mauro Pelosi, Ismail Guvenc, Rasmus Olsen, Hanane Fathi, Basuki Priyanto, and the indirect contributors, who, in one way or another, have been involved in several of the activities that provided the knowhow for writing the book. Finally, the authors would like to thank the Wiley team, who offered them their unceasing help in order to make this book a reality: Sandra Grayson, Tiina Wigley, Preethi Belkese, Teresa Netzler and Stephan Schwindke.

List of Abbreviations

1L-DF	one-level data fusion
2D	two-dimensional
2L-DF	two-level data fusion
3D	three-dimensional
3G	third generation
3GPP	Third Generation Partnership Project
4G	fourth generation
5G	fifth generation
ACPS	Ad-Coop Positioning System
A-GNSS	Assisted Global Navigation Satellite System
A-GPS	Assisted Global Positioning System
ACK	acknowledge
ADC	analog-to-digital converter
AOA	angle of arrival
AP	access point
API	ppplication programming interface
ASP	application service provider
AWGN	additive white Gaussian noise
B2B	business-to-business
B2C	business-to-consumer
BCCH	broadcast control channel
BF	beamforming
BPM	burst position modulation
BPSK	binary phase shift keying
BPZF	band-pass zonal filter
BS	base station
BTS	base transceiver station
CA	collision avoidance
CAS	cooperative augmentation system
CATV	cable television
CD	collision detection
CDF	cumulative distribution function
CDMA	code division multiple access
CEP	circular error probability

CH	cluster head
CID	cell ID
CG	cluster gateway
CIR	carrier-to-interference ratio
CLI	caller location information
CLS	constrained least squares
CM	cluster member
COFDM	coded orthogonal frequency division multiplexing
COMET	Cooperative Mobile Positioning System
coop-EKF	cooperative extended Kalman filter
coop-WNLLS	cooperative weighted nonlinear least squares
CR	cognitive radio
CRC	cyclic redundancy check
CRLB	Cramér–Rao lower bound
CRB	Cramér-Rao bound
CRMSE	cooperative root mean square error
CSMA	carrier sense multiple-access
CS-MNS	clock sampling–mutual network synchronization
CSN	connectivity service network
CTM	current transformation matrix
CTS	clear to send
CW	continuous wave
DAC	digital-to-analog converter
DAA	detection and avoidance
DGPS	Differential Global Positioning System
DL	downlink
DOP	dilution of precision
DP	direct path
DR	dead reckoning
DS	direct sequence
DSSS	direct-sequence spread spectrum
E911	enhanced 9-1-1
eCall	emergency call
ED	energy detector
EDGE	enhanced data rates for GSM evolution
EGNOS	European Geostationary Navigation Overlay Service
EIRP	effective isotropic radiated power
EKF	extended Kalman filter
e.m.	electromagnetic
EM	expectation maximization
EPS	evolved packet system
ERP	equivalent radiated power
EUWB	European ultra-wideband
FBMC	filter bank multicarrier
FCC	Federal Communications Commission
FDD	frequency division duplex

FDMA	frequency division multiple access
FEC	forward error correction
FFT	fast Fourier transform
FHSS	frequency-hopping spread spectrum
FIM	Fisher information matrix
FRP	fixed reference point
FS	fixed station
FT	fixed terminal
G-CRLB	generalized Cramer–Rao lower bound
GAGAN	GPS-aided Geo-augmented Navigation System
GDOP	geometric dilution of precision
GERAN	GSM/EDGE radio access network
GFDM	generalized frequency division multiplexing
GIS	Geographic Information System
GLONASS	Globalnaya Navigationnaya Sputnikovaya Sistema
GP	guard period
GPB	generalized pseudo-Bayesian
GNSS	Global Navigation Satellite System
GPS	Global Positioning System
GSM	Global System for Mobile Communications
GPS-CM	GPS-equipped CM
HazMat	hazardous material
HLOP	hybrid lines of position
HTAP	hybrid TOA/AOA positioning
HTDOA	hybrid TDOA/AOA positioning
IAD	identify and discard
ID	identification
i.i.d.	independent, identically distributed
IEEE	Institute of Electrical and Electronics Engineers
IF	intermediate frequency
IFFT	inverse fast Fourier transform
IMU	inertial mobile unit
IP	Internet Protocol
IPDL	idle period downlink
IPO	interior-point optimization
IR-UWB	impulse radio UWB
IS	idle slot
IT	information technology
KF	Kalman filter
LAC	local area code
LAN	local area network
LBS	location-based service
LCS	location services
LD	laser diode
LDC	low-duty cycle
LEACH	low-energy adaptive clustering hierarchy

LED	light-emitting diode
LLS	linear least squares
LMS	least median of squares
LMU	location measurement unit
LO	local oscillator
LOS	line of sight
LP	local positioning
LRT	likelihood ratio test
LS	least squares
LT	location and tracking
LTE	long-term evolution
LTS	least-trimmed squares
mmW	millimeter wave
MAC	medium access control
MAP	maximum a posteriori
MCC	mobile country code
MF	matched filter
MIMO	multiple-input–multiple-output
MISO	multiple-input–single-output
ML	maximum likelihood
MLE	maximum-likelihood estimator
MMSE	minimum mean square error
MNC	mobile network code
MPP	Mobile Positioning Protocol
MS	mobile station
MSAS	Multifunctional Satellite Augmentation System
MSC	mobile switching center
MSE	mean squared error
MSK	minimum shift keying
MUI	multi-user interference
MUR	multistatic radar
MVNO	mobile virtual network operator
NAV	network allocation vector
NAVSTAR	navigational satellite timing and ranging
NICT	new information and communication technology
NLLS	nonlinear least squares
NLOS	non-line-of-sight
NNSS	Navy Navigation Satellite System
NTP	network time protocol
OBU	on-board unit
OFDM	orthogonal frequency division multiplexing
OFDMA	orthogonal frequency division multiple access
OOB	out of band
OOK	on–off keying
OSI	open system interconnection
OTDOA	observed time difference of arrival

OTDOA-IPDL	observed time difference of arrival–idle period downlink
ppm	part per million
P2P	peer-to-peer
PAM	pulse amplitude modulation
PC	power control
PCR	predictive channel reservation
PDA	personal digital assistant
PDF	probability density function
PDP	power delay profile
PF	particle filter
PHY	physical layer
PLMN	public land mobile network
POI	point of interest
PPM	pulse position modulation
PR	pseudo-random
PRF	position reporting frequency
PRN	pseudo-random number
PSAP	public safety answering point
PSD	power spectral density
PSTN	public switched telephone network
PTP	precision time protocol
QAM	quadrature amplitude modulation
QoS	quality of service
QP	quadratic programming
QPSK	quadrature phase shift keying
QZSS	Quasi-Zenith Satellite System
RBF	recursive Bayesian filtering
RF	radio frequency
RFID	radio frequency identification
RLS	recursive least squares
RMS	root mean square
RMSE	root mean square error
RP	reference point
RRC	root raised cosine
RRM	radio resource management
RSS	received signal strength
RSSI	received signal strength indicator
RT	residual test
RT	response time
RTD	relative time difference
RTLS	real-time locating system
RTS	request to send
RTT	round-trip time
RTS	request to send
RV	random variable
RX	receiver

RXLEV	receiver level
SBAS	Satellite-Based Augmentation System
SBS	serial backward search
SD	spatial diversity
SDK	software development kit
SDMA	space division multiple access
SFN	system frame numbers
SIC	self-interference cancellation
SIMO	single-input–multiple-output
SINR	signal-to-interference-plus-noise ratio
SIR	signal-to-interference ratio
SISO	single-input–single-output
SM	spatial multiplexing
SNR	signal-to-noise ratio
SP	service provider
STBC	space-time block codes
STC	space–time coding
S-V	Saleh-Valenzuela
TA	timing advance
TDMA	time division multiple access
TDOA	time difference of arrival
TH	time-hopping
THSS	time-hopping spread spectrum
TOA	time of arrival
TOF	time of flight
TSF	timing synchronization function
TWR	two-way ranging
TX	transmitter
U-TDOA	uplink–time difference of arrival
UE	user equipment
UFMC	universal filtered multicarrier
UHF	ultra-high frequency
UKF	unscented Kalman filter
UL	uplink
UMTS	Universal Mobile Telecommunications System
UTRAN	UMTS terrestrial radio access network
UWB	ultra-wideband
VCO	voltage-controlled oscillator
WAAS	wide area augmentation system
WGS84	World Geodetic System 1984
WiFi[Wi-Fi]	wireless fidelity
WiMAX	Worldwide Interoperability for Microwave Access
WLAN	wireless local area network
WLS	weighted least squares

WNLLS	weighted nonlinear least squares
WPAN	wireless personal area network
WSN	wireless sensor network
WSR	wireless sensor radar
ZZB	Ziv-Zakai bound

Notations

Symbol	**Description**
x, y, a	scalar
X, Q, R	vector or matrix
α, μ, θ	scalar, vector or matrix
X_{rel} or X^{rel}	X with a descriptive label "rel"
X^{T}	transpose of X
X^{-1}	inverse of X
$\hat{X}$	estimator of X
$\tilde{X}$	$\hat{X} - X$
$\overline{X}$	mean of X
$\lvert X \rvert$	determinant of X
$X^{(i)}$	X with respect to element i
$X^{(i)(j)}$	X with respect to a relation between i and j, both from the same group
$X^{(i)[j]}$	X with respect to a relation between i and j from a different group
X_i	equivalent to $X^{(i)}$
$X_{i,j}$	equivalent to $X^{(i)(j)}$
X_k or $X(k)$	X at discrete time k
$X_{k\mid l}$	X at discrete time k compared with time l
$X_{l:k}$	vector or matrix containing all occurrences of X from discrete time l until k
Δx	difference between two values of x
$\binom{n}{i}$	set of i combinations of a set n
$\mathrm{P}(x)$	probability of x
$\mathrm{f}(x)$	PDF of x
$\mathrm{Norm}(\mu, \sigma)$	normal distribution with mean μ and standard deviation σ
$\mathrm{Unif}(a, b)$	uniform distribution between a and b
$Exp(\overline{\gamma})$	exponential distribution with parameter $\overline{\gamma}$
$\mathrm{Tri}(a, b)$	triangular distribution with lower limit a, upper limit b and center $(a + b)/2$
$\log\ (\bullet)$	logarithm with base 10
$\ln\ (\bullet)$	logarithm base e (neper number)
f	frequency
λ	wavelength
c	speed of light

Symbol	**Description**
p	power
α	constant term in the path loss or received power equation
β	path-loss exponent
t	time
t	time difference
d	distance
d	difference between distances
$E[\bullet]$	expected value
η	noise component
μ	mean value
σ	standard deviation
θ	angle
γ	weight value
i or j	iteration constant
x or X	position coordinate
z or Z	measured value
$\mathcal{H}(\bullet)$	function relating position coordinates and measured values
H	Jacobian of $\mathcal{H}(\bullet)$
$\mathcal{A}(\bullet)$, $\mathcal{F}(\bullet)$	function relating position coordinates at different points in time
A	Jacobian of $\mathcal{A}(\bullet)$
∇	gradient operator
$\mathcal{J}(\bullet)$	cost function in an optimization procedure

This list normalizes the notation throughout the book, however, at selected and well identified parts a different notation may be used for convenience purposes.

1

Introduction

João Figueiras[1], Francescantonio Della Rosa[2,3] and Simone Frattasi[4]
[1]*Aalborg University, Aalborg, Denmark*
[2]*Radiomaze Inc, Cupertino, California, USA*
[3]*Tampere University of Technology, Finland*
[4]*Sony Mobile Communications, Lund, Sweden*

Over the past few decades, wireless communications have become essential in everyone's daily life. The world has become mobile, and continuous access to information has become a requirement. Owing to this necessity for fresh, real-time, first-hand information, devices such as mobile phones, computers, pagers, data cards, sensors and data chips have been entering our lives as typical technology "buddies". For this reason, wireless services have gained popularity, and location information has become useful information in the wireless world. The continuous demand for information has created a huge potential for business opportunities and it has promoted innovation towards the development of new services. As a consequence, the infrastructure that enables communication among wireless devices has been rapidly growing towards higher coverage, higher flexibility and higher interoperability. This reality is so visible that it is nowadays unthinkable to envision our lives without these technologies. Furthermore, with the rapid deployment of wireless communication networks, positioning information has become of great interest. Because of the inherent mobility behavior that characterizes wireless communication users, location information has also become crucial in several circumstances, such as rescues, emergencies and navigation. This position dependency has boosted research, development and business around the topic of positioning mechanisms for wireless communication technologies. The result is a wide variety of integrated and built-in solutions that can combine and interoperate with communication and position information. Thus, this book covers the topic of positioning mechanisms for wireless communication technologies. It explains in detail the services, wireless communication protocols, positioning and tracking algorithms, error mitigation techniques, implementations used in wireless communication systems, and the most recent techniques of cooperative positioning.

Positioning or location can be understood as the unambiguous placement of a certain individual or object with respect to a known reference point. This reference point is often assumed

Mobile Positioning and Tracking: From Conventional to Cooperative Techniques, Second Edition.
Simone Frattasi and Francescantonio Della Rosa.

to be the center of the Earth coordinate system. In practice, the reference point can be any point on the Earth[1] that is known to the system and which all the coordinates can relate to. Although the position itself is obviously a very important source of information, this position must be related to a specific time to be even more useful. In particular, when we consider tracking systems, time information is a key necessity, not only for knowing the position of a certain device at a specific time, but also for inferring higher-order derivatives of the position, that is, speed and acceleration. Thus, "tracking" is a method for estimating, as a function of time, the current position of a specific target. "Navigation" is a tracking solution that aims primarily at using position information in order to help users to move towards a desired destination.

Although enabling position estimation with communication technologies is currently a hot topic, the necessity for determining the position of individuals, groups, animals, vehicles or any type of object is an ancient necessity that can be seen as a basic need. This necessity has been present in human life for many centuries and it is so important that humans and animals have in-built biological mechanisms that permit individuals to localize and orient themselves in many situations. When we consider the actual methods for obtaining position information, the history is long and the systems are numerous. During the early years, a few millennia ago, orientation and positioning were already possible using devices that resemble present-day magnetic compasses, using maps of sea currents and winds, or using celestial navigation techniques. For centuries, celestial navigation, by means of observations of the positions of stars and the Sun, was the most important technique for estimating position information. By knowing, for instance, the position of the Sun or well-known star constellations, people were able to determine their orientation and navigate on the Earth. This technique is so important that it is still used today to provide a rough sense of one's orientation when no other tool is available. Although the mechanisms for orientation based on star position readings and compass readings have been used for at least two millennia, these mechanisms were widely used and improved during the period of sea exploration in the 15th and 16th centuries. This was an important period in the history of positioning systems. Several objects, such as the cross-staff and astrolabe (and, later, the quadrant and the sextant, invented in the 17th and 18th centuries), permitted sea explorers to read the position of the stars and subsequently calculate their own position, often supported by the incomplete maps that were available at that time. Associated with these tools were other techniques for predicting and inferring future positions based on the analysis of past movements. These techniques were often used to navigate when, for instance, the sky was not clear and they were generally complemented by anchor points of known location, such as points on the shore. In the 18th century, the chronometer was invented. This tool, widely used in ship navigation, permitted calculations of position to be connected more efficiently to corresponding timestamps. By the end of the 19th century, wireless communications emerged and, along with this remarkable discovery, the first position solutions based on electromagnetic waves started to be developed. This period marked an important turn not only in the history of positioning systems, but also in the entire history of communication technology. The first positioning system to be invented was the radio direction finder, a device which was able to determine the direction from where radio waves were being generated. The basic concept of this device was to find the null (i.e., the direction which results in the weakest signal) in the signal observed with a directional antenna mounted on a portable support. Only by the mid-20th century were the first radars invented. Ever since, these systems have been used and enhanced, and they are still widely used for several positioning

[1] Or in space, if the system is meant to be used in outer space.

purposes. Radar is a system that transmits radio waves towards a target and is able to read the signals that are received back after they have been reflected from the target itself. From this period onwards, great development in positioning systems has occurred, culminating in the wide variety of systems currently available. One of the most famous systems is the long-range navigation (LORAN) system, proposed in 1940, which was characterized by beacons radiating synchronized signals that were then read by target receivers. The receivers had to be able to measure time differences in the arrival of the signals in order to calculate their positions. Later on, by the 1960s, satellite positioning systems started to be deployed, and space exploration began. Currently, Global Positioning System (GPS) is the best known and most used satellite positioning system. Over the past 30–40 years, several other positioning solutions have been deployed for use in various scenarios, based on various approaches, using infrared, ultrasound, image processing or electromagnetic waves.

In parallel with the deployment of positioning and communication systems during the last decade, research has evolved in the direction of integrating communication and positioning into a single system. Some examples are the current standards for the Third Generation Partnership Project (3GPP) and ultrawideband (UWB), which include information about positioning mechanisms in the communication specifications. The combination of these functionalities has leveraged new services, namely location-based services (LBSs). In contrast to the earlier dedicated solutions that were designed to simply provide positioning, the new solutions for wireless networks are able to provide the combined benefit of both communication and positioning. As a consequence, the whole network, as well as the end user and the service providers, can benefit from position-enabled communication capabilities: the network operator can manage all the network's resources in a more efficient way, the service provider is able to deliver new services based on the user's position, and the user can enjoy new personalized, location-dependent services. Many of the services proposed in research documents assume location information as a basic requirement for deployment of new protocols (e.g., routing and clustering), new technologies (e.g., cooperative systems) and new applications (e.g., navigation and location-aware advertising). Furthermore, in the industrial environment, location information has also been widely used in applications such as navigation, location-dependent searching and social networks. Since wireless communication networks are nowadays almost present anywhere at any time, any new location-dependent networking enhancement, service or application can be spread rapidly and used globally.

The location information mechanisms in wireless communication technologies can be classified into positioning systems and positioning solutions, as shown in Figure 1.1. A positioning system concerns the hardware and software necessary to measure the properties of wireless links and to subsequently process those measurements in order to estimate the user's position. Positioning systems are typically integrations of entire systems into communication technologies, such as the GPS integration into mobile phones and cellular base stations (BSs). In contrast, a positioning solution is typically a software-based implementation, where indirect measurements of the user's position are obtained in an opportunistic fashion from the adaptation of mechanisms existing in communication technologies. The enablement of location information in current wireless communication networks can be done either by integrating a positioning system into the network or by implementing a positioning solution that extracts location information, thus exploiting the potential of the network. By integrating a positioning system into the network, it is usually possible to obtain better accuracy than what is obtained by a direct implementation in the network. The disadvantage is that integrating

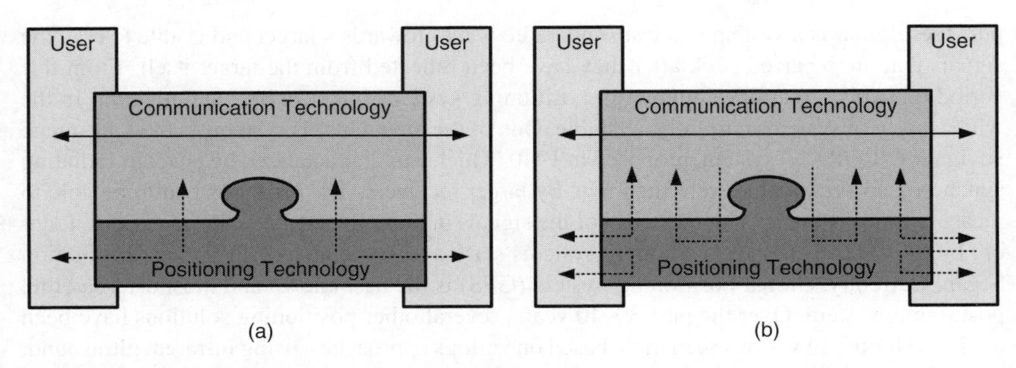

Figure 1.1 Schematic representation of a positioning system (a) and a positioning solution (b). The solid arrows represent the flow of communication and the dashed arrows represent the flow of position information. In (a), it is possible to see that position is always generated in a positioning technology. In contrast, a positioning solution is often calculated based on opportunistic information obtained from communication technology.

additional hardware necessarily implies additional costs, higher power consumption and higher complexity. For this reason, a direct software-based positioning implementation that opportunistically exploits the existing hardware and extracts location information is commonly preferred. Typical positioning solutions make use of mechanisms present in most of the current advanced wireless technologies, such as power control and synchronization schemes. The requirement is that the positioning solution is able to perform measurements of the wireless channel that can be used as indirect observations of the user's position. Examples of such measurements are the received signal strength and the propagation delay of signals.

The possible scenarios and conditions where location information is needed are numerous, and no localization system is able to perform localization under all conditions. For this reason, several localization systems and solutions are available and many others are still under research. For instance, the traditional satellite-based GPS represents a solution with worldwide coverage; however, it is not usable in some specific situations, such as in dense urban areas, indoors, underwater and underground. For these cases, many other systems are being developed or are even already available on the market.

The book is organized as follows. Chapter 2 introduces location-based services as the main driver for wireless positioning. Chapter 3 then introduces the basics of wireless communications. The foundations of positioning are presented in Chapter 4 and are extended to more complex data-processing algorithms in Chapter 5. Tracking algorithms are explained in Chapter 6. As none of the previous chapters reviews the actual preprocessing of raw data, Chapter 7 introduces some algorithms for mitigating propagation effects in wireless positioning. The systems and solutions for wireless positioning currently existing in the market are described in Chapter 8, while Chapter 9 presents the topic of ultra-wideband positioning and tracking. Wireless local area network (WLAN) approaches in indoor positioning are presented and discussed in Chapter 10. In Chapter 11 theoretical and experimental approaches are discussed for cooperative multi-tag localization in RFID systems and, finally, some novel cooperative schemes for wireless positioning are introduced in Chapter 12. The sections below give some insight into each of the chapters.

1.1 Application Areas of Positioning (Chapter 2)

Location-based services are introduced in this chapter as a motivational engine for combining communication and positioning information, and then integrating wireless systems and positioning technologies. Owing to the increasing number of services and systems using location and the even larger number of services envisioned for the future, it has become mandatory to have a framework for layering positioning systems in a similar way to what has been done for the open system interconnection (OSI) stack in the communication world. This framework defines several layers, each of which has clear, defined purposes. This framework, apart from other benefits, will allow services to run independently of the overlaid technology. This chapter concerns mostly the applications of LBS and the entities involved in the services. These entities are, for instance, service providers, users, network operators, location infrastructure operators, developers and portals. The applications are categorized according to the service that is being provided: emergency services, automotive applications, medical applications, monitoring, navigation, management and entertainment. Then, the major benefits of combining the two functionalities are pointed out: interoperability and enhancement of both positioning-based communication and communication-based positioning. Typical examples are radio resource management and radio planning.

1.2 Basics of Wireless Communications for Positioning (Chapter 3)

In this chapter, the basics of wireless communications and how they relate to mobile positioning are introduced. First, the radio propagation scenario is described by classifying the various physical phenomena that underlie wireless communications, that is, propagation loss, shadowing and multipath fading. Then, the possibilities and challenges of multiple-antenna techniques are discussed, focusing in particular on the two big families of *spatial diversity* and *spatial multiplexing*, and on the gains achievable with them. The other important family of multiantenna techniques, that is, beamforming, is discussed later in the chapter, when we will treat space division multiple access (SDMA), an access technique that is based on beamforming. The chapter reviews modulation and access techniques, focusing in particular on orthogonal frequency division multiplexing (OFDM) and orthogonal frequency division multiple access (OFDMA), which are the most relevant techniques for future wireless positioning systems. Moving up in the OSI stack, some radio resource management (RRM) techniques are illustrated, namely handoff, interference management, channel reuse, channel reservation and power control, and we illustrate how localization can benefit from them, and vice versa. Finally, we outline some important ideas about two emerging areas within wireless communications: cooperative communications and cognitive radio technology. Their interconnection with mobile positioning closes the chapter.

1.3 Fundamentals of Positioning (Chapter 4)

After reviewing some basics of wireless communications in the preceding chapter, we introduce the basics of wireless positioning. Systems are classified according to several aspects: topology, physical coverage and type of integration into communication technologies. Then, the various types of measurements are explained. The cell ID is identified as the simplest type of measurement to obtain and also the one that results in the least accurate solution. Range measurements such as time of arrival (TOA) and received signal strength (RSS) measurements are introduced as having, apart from the effect of propagation impairment, the potential to provide more accurate positioning information than what is possible with Cell ID measurements. Two alternatives are measurements of the time difference of arrival (TDOA) and the angle of arrival (AOA). These measurements enable several basic positioning techniques that are explained in detail. Proximity-sensing techniques rely on measurements of proximity or coverage. Triangulation techniques rely on trigonometric theory and channel measurements in order to obtain the positions of mobile stations (MSs). Finally, fingerprinting techniques require an extensive precalibration phase that is aimed at collecting measurements of the wireless channels at several predefined positions. Then, while in operation mode, the measurements obtained are correlated with information from the calibration in order to localize the MSs. As none of the positioning systems are error-free, due to channel noise, the chapter also introduces several of the most important types of errors existing in the measurements. These errors are mostly due to propagation effects, geometry and the technology used, and the consequent topology and communication protocols. Owing to the errors, it is necessary to define the metrics of positioning. The most common metrics are the circular error probability (CEP), the dilution of precision (DOP) and the Cramér-Rao lower bound (CRLB).

1.4 Data Fusion and Filtering Techniques (Chapter 5)

Because of the various error sources, it is necessary to implement mechanisms capable of handling this noise. The first approach explained in this chapter concerns the least squares (LS) method. This approach aims at determining the position that minimizes the squared error between the actual measurements from the channel and the expected measurements. In order to execute the algorithm, it is necessary to define a model to approximate the relation between the position coordinates and the measurements. In this context, linear and nonlinear methods are explained. While the linear LS methods present a closed-form solution and a derived method for recursive calculations, the nonlinear LS (NLLS) methods must be solved by numerical optimization algorithms. An alternative to the LS methods is the Bayesian framework. This framework defines a recursive mechanism to calculate the position based on the least mean squared error. Kalman filters (linear and nonlinear), particle filter methods and grid-based methods are illustrated as implementations of the Bayesian framework, each to be used depending on the properties of the system. As commonly happens in wireless positioning, there may be several parameters that are unknown, even though the model relating the measurements and the positions is known or at least can be approximated. For this reason, these parameters are determined by means of either precalibration or on-line combined estimation of parameters and positions. The chapter closes by describing some alternative solutions, such as fingerprinting and time series analysis.

1.5 Fundamentals of Tracking (Chapter 6)

As a first step towards tracking of MSs, this chapter starts by conceptually explaining the complexity introduced into systems when time is considered, that is, when users move. Then, as a further step in complexity, cooperative schemes are also introduced. Mobility models are presented as an essential aspect of tracking applications. The first class of movement models is that of conventional models, which is given by the well-known physical laws of the straight-line movement. Then, several stochastic models for individual mobility and group mobility are presented. While individual mobility models tend to permit unconstrained movement or are constrained only by scenarios or boundaries, group mobility models tend to be constrained by the mobility of other individuals belonging to the same group. The final class of mobility models is that of cooperative or social-based models, which resemble not only group behavior, but also the social behavior of individuals. The latter models provide some of the foundations of cooperative wireless positioning presented in Chapter 12. After presenting the models, the chapter presents a sequence of components commonly present in a tracking solution. The first phase is to treat the raw measurements in order to exclude some possible destructive effects that can be detected at this early stage. The measurements are then used as an input for tracking algorithms; we first present non-maneuverable and then maneuverable algorithms. Finally, at the output of the tracking algorithm there exists an intelligent platform capable of learning users' positions and tracking patterns; two important algorithms are presented in this context.

1.6 Error Mitigation Techniques (Chapter 7)

Non-line-of-sight (NLOS) situations are commonly encountered in modern wireless communication systems in both indoor and outdoor environments. They result in biased time-of-flight estimates, which, if not handled properly in the localization algorithm, may considerably degrade the localization accuracy. In this chapter, common techniques that can be employed for the mitigation of NLOS effects in a wireless positioning system are reviewed for time-based positioning systems. After providing a generic system model for time-based location estimation, we summarize fundamental lower bounds and maximum likelihood (ML) solutions for line of sight (LOS) and NLOS scenarios. Various NLOS mitigation methods are classified into four different categories and reviewed: (1) least-squares estimators, (2) constraint-based estimators, (3) robust estimators and (4) identify-and-discard estimators. Least-squares algorithms improve the localization accuracy by giving less emphasis to NLOS measurements in the mobile location's solution. Constrained-based estimators use NLOS measurements to define certain constraints on the location estimate; under these constraints, they typically evaluate the location of the mobile station by using LOS measurements. Robust estimators make use of the rich body of literature and methods available in robust estimation theory for detecting and discarding outliers in a given set of data (i.e., NLOS measurements are treated as outliers and discarded). Finally, the identify-and-discard estimators aim to accurately identify NLOS measurements and determine the location of the mobile station by using only LOS measurements. The key contributions in the literature associated with each of the above categories are summarized and compared with each other.

1.7 Positioning Systems and Technologies (Chapter 8)

This chapter presents in a concise way the positioning solutions commonly used in the various wireless communication technologies. The first class of systems to be introduced is satellite-based systems. The basic principle relies on measuring the propagation delays that signals are subject to when traveling from the satellites to the users on Earth. Then, cellular positioning solutions are introduced, with Global System for Mobile Communications (GSM) and Universal Mobile Telecommunications System (UMTS) communication systems the main ones considered. While GSM typically relies on methods based on cell identification and range measurements, UMTS relies on differences of range measurements. The next class of systems introduced in this chapter is WLAN systems. These tend to use both cell identification techniques and measurements of received signal strength (RSS). UWB technology is a special case in which time measurements are preferred, as such a wide frequency band permits time measurements to reach a higher accuracy than what is permitted by RSS measurements. Some dedicated solutions are also presented: radio frequency identification (RFID), infrared and ultrasound. The chapter closes by summarizing some hybrid techniques that combine more than one technology in order to achieve better results in terms of position estimation accuracy. The combination of cellular and WLAN positioning with the Assisted Global Navigation Satellite System (A-GNSS), for example the Assisted Global Positioning System (A-GPS), has become of great interest.

1.8 Ultrawideband Positioning and Tracking (Chapter 9)

In previous chapters the use of RSS, TOA, AOA or carrier phase measurements from different technologies was discussed to infer the position of a mobile node. Such measurements have several geometric constraints, allowing the location engine to estimate a node's position according to some methods. Unfortunately, RSS might not be always correlated with position/distance due to the random nature of radio propagation, especially in indoor environments and in NLOS conditions, thus leading to poor localization performance. In contrast, time-based measurements are in general better suited when high-accuracy positioning is required due to the fact that the larger the signal bandwidth the higher will be the resolution of time measurements. This might be enough to justify the adoption of wideband signals in the GPS and the interest in UWB signals in indoor positioning applications. Besides active positioning, UWB technology enables a plethora of novel applications such as multi-static radar for non-collaborative localization, life signs detection systems, and through-wall and underground imaging. In this chapter, the main issues and methods for obtaining high-resolution ranging and localization in UWB technology are presented and discussed, and the most successful applications of UWB, that is, those related to positioning and tracking, are identified.

1.9 Indoor Positioning in WLAN (Chapter 10)

WLAN data represent at the same time the most suitable and the most challenging signals to exploit in mobile positioning. The location estimation might be easily provided in different ways, making it one of the most adopted solutions for localizing and tracking people indoors. For this reason the very exploitation of WLAN signals is a tempting approach for researchers and service providers. When the signal strength of nearby hotspots is the adopted solution, we can define three main categories: cell identifier (Cell-ID), fingerprinting and pathloss-based. In this chapter, in addition to the description of the conventional approaches and characteristics or WLAN positioning, we describe how human-induced errors should not be ignored when performing experimental activities for indoor positioning using RSS. Specifically we emphasize that although environmental and human-induced perturbations generate systematic errors, if correctly accounted for and cognitively exploited it is possible to somehow mitigate the negative effects and to obtain enhancements in terms of positioning accuracy.

1.10 Cooperative Multi-tag Localization in RFID Systems (Chapter 11)

RFID is a wireless technology used to identify objects. It is mountable or attachable to a product, animal or a human being, enabling the transfer of its stored data using radio waves. In its simplest architecture it comprises RFID tags and RFID readers. RFID readers broadcast queries to tags in their wireless transmission ranges to obtain information stored in the tags such as their unique serial identification number, enabling contactless identification, even for RFID-tagged objects in motion and/or in NLOS with respect to the RFID reader. There are three types of tags: passive, semi-passive or active. Existing RFID localization mechanisms in literature assume that the object contains only a single tag. In order to increase the localization accuracy, multiple tags can be deployed on a target. Consequently, the RSS information obtained from the tags may be combined with the known information about the physical distance between the tags showing that cooperative multi-tag localization does exhibit higher localization accuracy over single-tag localization.

1.11 Cooperative Mobile Positioning (Chapter 12)

Based on the knowledge acquired in the previous chapters, this chapter introduces cooperative schemes for wireless positioning. As a concept borrowed from the context of robotics, these schemes are described within that context and examined further within the context of sensor networks. As a natural requirement of cooperative schemes, nodes are required to cluster so that cooperative links are known within the cluster members and also within the entire network. With this in mind, wireless mobile networks are considered from the cooperative point of view. The estimation algorithms, previously introduced as based on individual positioning methods, in particular the LS and Kalman filter approaches, are augmented in order to consider several users and the influence between them. These algorithms are used as the core estimation process in the Cooperative Mobile Positioning System (COMET) framework, a cooperative infrastructure which provides location information on a group of users by taking into account the user-to-user communications and the proximity relations among them. The architecture is based on the concept of clustering the users and then gathering the measurements from all the links between users and between users and BSs in order to jointly estimate the positions. The various mechanisms and the entire COMET concept are analyzed based on simulations.

2

Application Areas of Positioning

Simone Frattasi
Sony Mobile Communications, Lund, Sweden

2.1 Introduction

In this chapter, we start by presenting the Location Stack, a framework created by Jeffrey Hightower (Hightower et al. 2002), which offers an overview of an exemplary localization system with all its elements and interrelations, similar to the seven-layer ISO/OSI model used in the field of communication systems. Then, we continue by introducing the first application area of positioning and thereby the formal definitions of a location-based service (LBS), which is basically considered as the intersection of new information and communication technologies (NICTs), the Internet and geographic information systems (GISs). Further, we illustrate the "LBS ecosystem", which is an efficient parallel taken from biology, to show the many actors rotating around this business; in particular, these actors will be split into *operational* and *non-operational* actors. The chapter follows this by presenting taxonomies of LBSs, which are divided into the *service model*, *application*, *market segment* and *content*. We will focus on the second of these classes and define ten application categories within LBSs. Finally, we will proceed to the second application area of positioning, where location information is exploited to enhance network performance rather than to provide new services to the user. In this context, we will consider radio network planning and RRM issues closely.

The rest of the chapter is organized as follows: Section 2.2 presents the Location Stack, Section 2.3 introduces LBSs and Section 2.4 describes location-based network optimization issues. Finally, some concluding remarks are made in Section 2.6.

2.2 Localization Framework

Before we describe the wide set of applications related to positioning, it is necessary to give a general picture of the various elements involved in a localization system. Hightower et al. (2002), by means of a survey of the literature, were able to extrapolate five design rules and

Mobile Positioning and Tracking: From Conventional to Cooperative Techniques, Second Edition.
Simone Frattasi and Francescantonio Della Rosa.

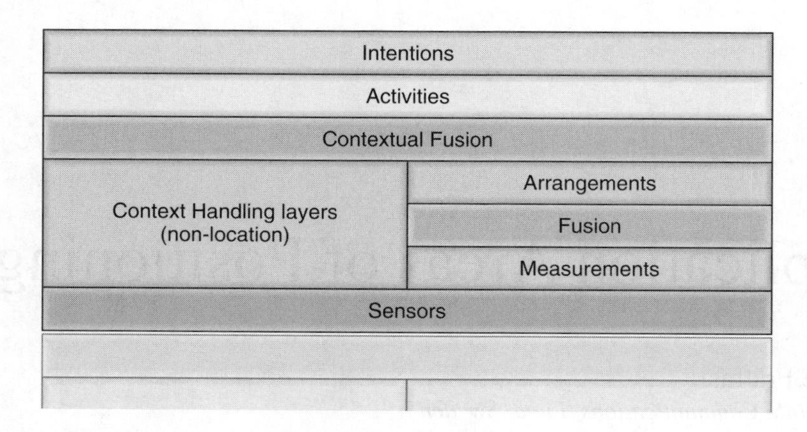

Figure 2.1 The Location Stack (Hightower et al. 2002).

make an abstraction of a localization system; practically, this resulted in a seven-layer model similar to the ISO/OSI model for communication systems (Figure 2.1).

The contents and actions of each of the layers in the seven-layer model for localization systems can be briefly described as follows:

- **Sensors** (Layer 1) represents the available sensors in a localization system (e.g., infrared badges, barcode scanners, cameras, pressure mats, thumbprint readers and keyboard login systems) and provides as output a stream of raw data (e.g., IDs, blob pixels, binary states and login events).
- **Measurements** (Layer 2) represents the algorithms that turn this data flow into its canonical form (e.g., proximities, distances and angles). Along with that, it also provides a model of the uncertainty that is characteristic of the sensor that generated the information.
- **Fusion** (Layer 3) represents the data fusion algorithms (see Chapter 5), which combine the retrieved data from all of the available sensors to calculate the coordinates of the target objects or persons (see Chapter 4).
- **Arrangements** (Layer 4) represents the algorithms that interrelate the positions of these targets, for example by converting their coordinates according to a relative coordinate system (e.g., from the position of a person in a room to the position of that person with respect to the loudspeakers in that room).
- **Contextual fusion** (Layer 5) represents the system that merges the pure location information with other contextual information (e.g., information ranging from personal data to temperature and light level).
- **Activities** (Layer 6) represents the system that, thanks to the contextual information retrieved in Layer 5, recognizes the current activities of the targets (e.g., whether all family members are asleep).
- **Intentions** (Layer 7) represents the system that, based on current activities and predefined user plans, takes decisions about the actions to be undertaken (e.g., increasing or decreasing the temperature in the bedrooms by a few degrees).

While Layers 1–4 will be mostly covered in later chapters, Layers 5–7, which can be seen as elements of the application layer, will be touched upon here.

2.3 Location-based Services

An LBS, in the broadest sense, is any service that extends spatial information processing or GIS capabilities to end users via the Internet and/or wireless networks (Koeppel 2002). However, unlike conventional GIS applications, which are focused on geographic data for land management and planning, LBS applications can provide both the connectivity and the context required to dynamically link an individual's location to context-sensitive information about the current surroundings. This enables a high level of personalization, which facilitates an ability to "put each user at the center of his/her own universe" (Kivera 2002). Some more formal definitions of an LBS are the following:

- **LBS definition 1.** A wireless IP service that uses geographic information to serve a mobile user or any application service that exploits the position of a mobile terminal (OGC 2003).
- **LBS definition 2.** An information service accessible with a mobile device through a mobile network, utilizing an ability to make use of the location of the mobile device itself (Virrantaus et al. 2001).
- **LBS definition 3.** A network-based service that integrates a derived estimate of a mobile device's location or position with other contingent information so as to provide added value to the user (Green et al. 2000).

The above definitions practically describe an LBS as an intersection of NICTs, the Internet and a GIS (Brimicombe 2002; Shiode et al. 2004) (Figure 2.2).

2.3.1 LBS Ecosystem

Inspired by biology, we apply the metaphor of an "ecosystem" to refer to an LBS as a whole composed of several different actors, their roles and their interacting synergism, which forms a

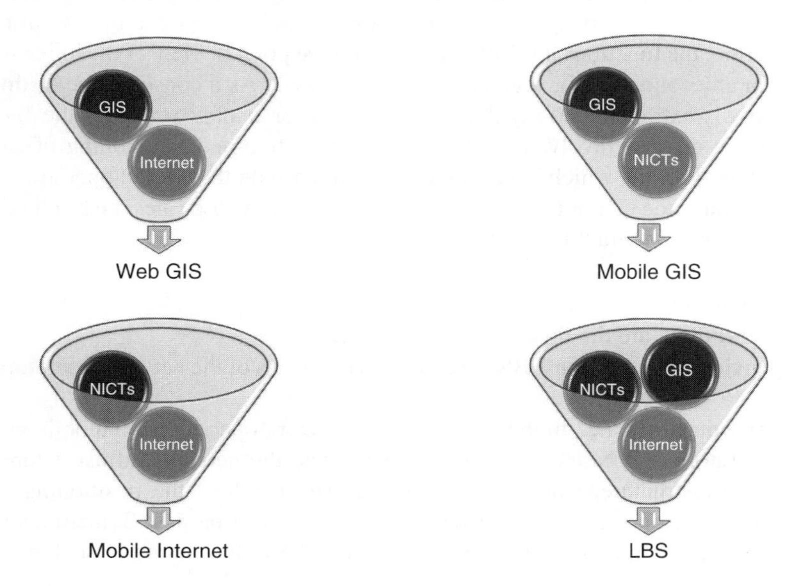

Figure 2.2 An LBS as an intersection of technologies.

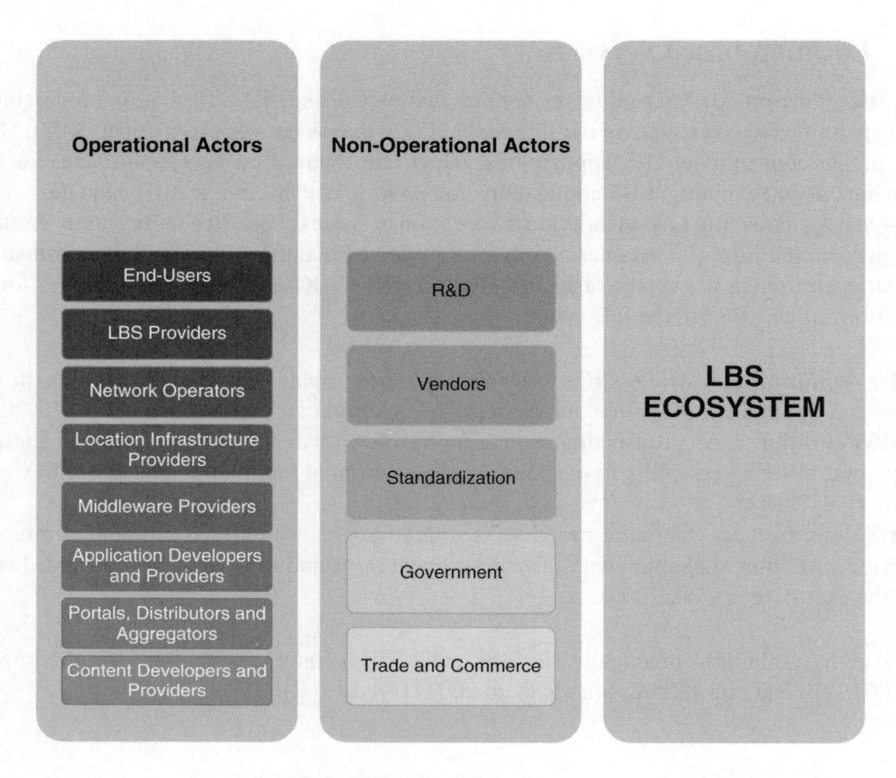

Figure 2.3 The LBS ecosystem.

complex web of interdependencies. As defined in Küpper (2005), an actor is "an autonomous entity (e.g., a person, a company or an organization), which adopts one or several roles that characterize either the functions it fulfills from a technical point of view or the impacts it exerts on LBSs from an economical or regulatory point of view". As a consequence of this definition, we can easily extrapolate to say that two main categories of actors exist: the *operational* actors, which are directly involved in the practical operation and functioning of LBSs, and the *nonoperational* actors, which have an indirect influence on the overall operation of LBSs. Figure 2.3 illustrates the components of the LBS ecosystem, whose roles are briefly described below (Küpper 2005; Steinfield 2004):

- **Operational actors**
  - **End users:** these are the final recipients of LBSs.
  - **LBS providers:** these offer LBSs to end users by means of the network operators' infrastructure.
  - **Network operators**, i.e., mobile and wireless network operators and mobile virtual network operators (MVNOs): these own and manage the network infrastructure and the location infrastructure, from which they retrieve or collect the positioning data, and perform billing operations and provide services to third parties. Sometimes they also interface directly to end users by offering them LBSs, thus acting in such a context as LBS providers.

  - **Location infrastructure providers:** these sell mobile location centers and other hardware and software to network operators.
  - **Middleware providers:** these offer tools that facilitate the use of the various applications made available by application providers to LBS providers.
  - **Application developers and providers:** these create applications that utilize the content gathered by portals, distributors and aggregators, and supply them to LBS providers.
  - **Portals, distributors and aggregators:** these gather the various content offered by the content providers and make it available to application developers.
  - **Content developers and providers:** these consist of GIS application service providers and producers of geographically referenced information – from the location of streets and buildings to points of interest and real-time information such as news, weather and traffic, and mapping services.
- **Nonoperational actors**
  - **R&D:** this embraces universities, research centers, and companies with internal research and development units. This category continuously brings forward and patents novel technological solutions, and thereby dictates the pace at which LBSs progress.
  - **Vendors:** these include manufacturers of handsets and of positioning and network-related equipment. By offering tools to enable and interact with LBSs, this category has a clear impact on network operators and end users.
  - **Standardization:** this encompasses the committees that take care of defining the design principles of LBSs in terms of interfaces, protocols and application programming interfaces (APIs); this process influences a range of actors in the operational chain, from application developers to LBS providers. Note that it is usually the presence of a standard that dictates the success of any given technology since many proprietary solutions, which by definition do not share a common platform, would emerge otherwise; this makes communication between the various actors involved in the LBS ecosystem difficult and reduces the chance to provide end users with seamless, ubiquitous services.
  - **Government:** this comprises the governmental units responsible for legislative issues related to LBSs. This category legislates on regulatory (nontechnical) matters, for example the establishment of rules to be followed when location data is handled, by defining the boundaries between privacy and lawful interception, which affects operational actors such as network operators and LBS providers.
  - **Trade and commerce:** this consists of the companies that make direct or indirect use of LBSs to empower their internal efficiency, their final products and their marketing strategies. On the basis of feedback from this category, developers are required to adjust their applications.

2.3.2 Taxonomies

LBSs can be classified according to the following approaches (Montalvo et al. 2003):

- **Service model.** This approach categorizes a service according to Nepper et al. (2008): (1) the direction of session initialization, (2) the relationship between user and target (i.e., the target may be the same user, other users, an object, etc.), (3) the number of targets and (4) the type of target. The first criterion results in a distinction between *reactive/pull services*

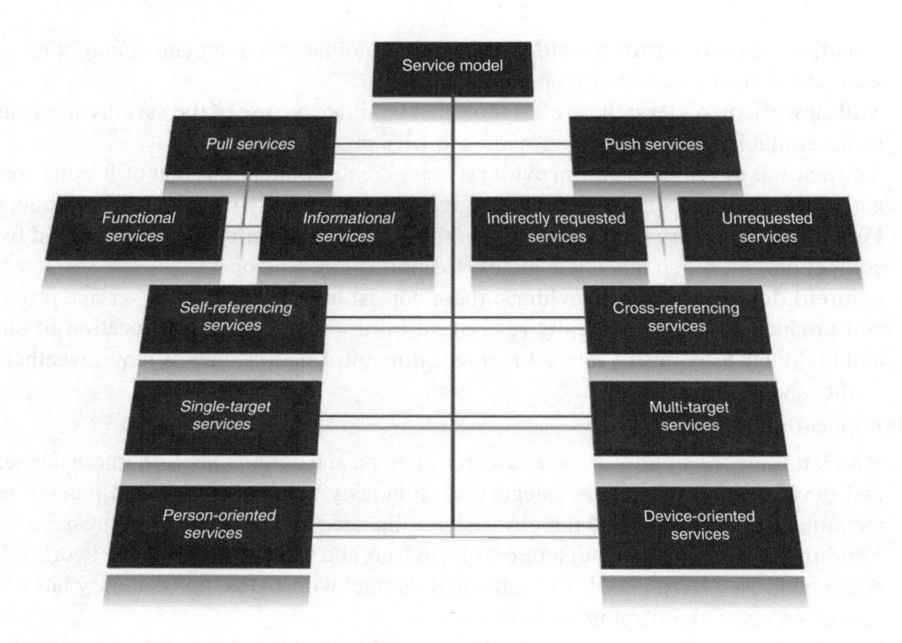

Figure 2.4 Service model.

and *proactive/push services* (Figure 2.4). While the former are triggered by the user, who reactively "pulls" location-based information from the network (e.g., by asking for a reasonably priced café nearby), the latter are triggered autonomously by the infrastructure, which proactively "pushes" location-based information towards the user, owing to the expiry of a timer or the occurrence of a geographically co-located event (e.g., a user in the surroundings of a café which offers a 20% discount at lunchtime). Note that some services, such as the Friend Finder, embrace both pull and push functionalities (Schiller and Voisard 2004). A further separation can be done for pull services into *functional services* (e.g., ordering a taxi or an ambulance just by pressing a button on a device) and *informational services* (e.g., searching for a Chinese restaurant nearby) (Virrantaus et al. 2001). For push services, we can make a further separation into *indirectly requested services* (e.g., a local news service subscription) and *unrequested services* (e.g., a hurricane warning message) (Steininger et al. 2006). The second criterion results in a distinction between *self-referencing services*, where the user and the target are the same entity, and *cross-referencing services*, where the user and the target(s) are separate entities (e.g., a user who is looking for a date in their proximity). The third criterion results in a distinction between *single-target services* (e.g., where a user can visualize the position of a shopping mall on a map) and *multitarget services* (e.g., where a user can visualize on a map the positions of all users matching the first user's personal profile). Finally, the fourth criterion results in a distinction between *person-oriented services* and *device-oriented services* (Schiller and Voisard 2004). The first class comprises services that exploit the target user's position to enhance or enable certain applications, and are controlled by the users themselves (e.g., personal navigation). The second one, in contrast, embraces services that make use of the target device's position for actions that are external and thereby not controlled by the user of that device (e.g., tracking of criminals).

- **Application.** This approach categorizes a service according to its main target application. Specifically, we have derived ten categories (see Section 2.3.2.1): *emergency, safety and security, tracking and tracing, traffic telematics (including vehicle navigation), personal navigation, management and logistics, billing, commerce, enquiry and information, leisure and entertainment (including community and gaming)*, and *supplementary*.
- **Market segment.** This approach categorizes a service according to its main target market (Niedzwiadek 2002), that is, the consumer or business public sector market.
- **Content.** This approach categorizes a service according to the type of information provided (Van de Kar 2004): positions, events, distributions, assets, service points, routes, context, directories and transaction sites.

2.3.2.1 Application Categories

The categorization below has been done by collecting together all of the different and even discordant inputs present in the literature and trying to make them converge as much as possible into a consistent view (Figure 2.5). In particular, the services listed in each category are not intended to form a definitive list, as new services are continuously emerging. Many of them are evidently related to more than one category. However, we have assigned each service to the category that it is most directly linked with. Finally, note that while one can find *infotainment services* (information services + entertainment services) as a stand-alone group in the literature under either "information" or "entertainment", we have decided to spread them between these two categories, specifically between what we have identified as *enquiry and information* (Section 2.3.2.1.8), and *leisure and entertainment* (Section 2.3.2.1.9).

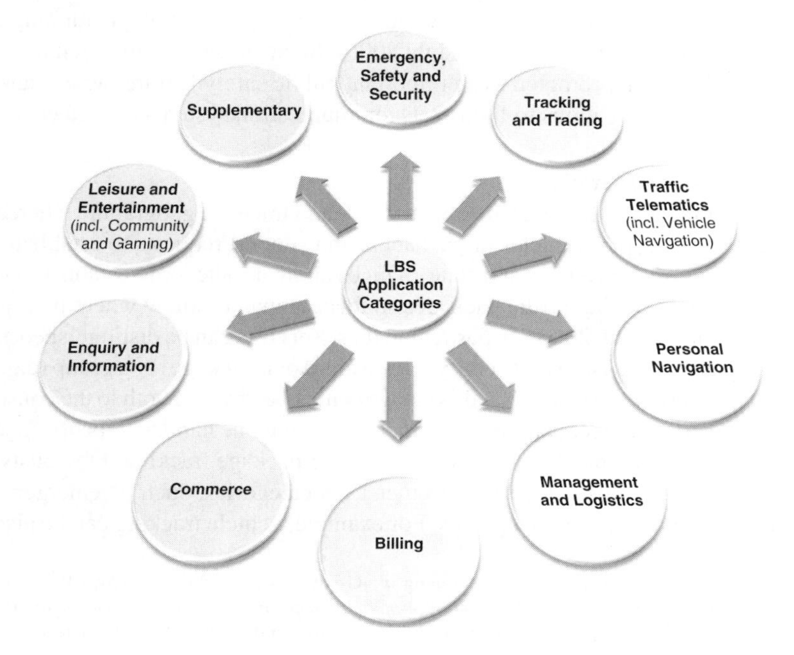

Figure 2.5 LBS application categories.

2.3.2.1.1 *Emergency, Safety and Security*[1]

The most immediate application of LBSs is the enablement of emergency services, that is, services that help users in critical situations, where their safety and security are under threat. Indeed, there are occasions on which we are not able to inform the emergency center of our location (e.g., when we have been involved in a serious traffic accident), we are not able to reveal it (e.g., if we are currently being subjected to a criminal or terrorist attack) or we simply ignore it (e.g., when our car suddenly breaks down). Note also that in many cases, under stressful or shocking situations, callers do not always act calmly and provide the information strictly necessary to allow the service center to immediately dispatch the emergency units needed (e.g., police, firefighters or paramedics). In addition, if users/devices (e.g., mobile phone or car) are able to signal their location, the emergency can be treated much more efficiently, which translates into precious time saved (and consequently more lives saved), higher user satisfaction, etc. While emergency call services are usually regulated by public organizations and therefore are made available to mobile phone users without requiring any pre-subscription, other types of service, such as automotive assistance, are managed by private companies. Table 2.1 shows examples of emergency, safety and security services, which we have assessed in terms of accuracy, environment, business type, revenue potential and related categories.

Practically, these services have driven the development and deployment of positioning technologies. Indeed, geolocation has gained considerable attention over the past decade, particularly since the Federal Communications Commission (FCC) in 1996 passed a mandate requiring that cellular providers in the USA generate accurate location estimates for E911 services (FCC 2001; FCC-14-13 2014) (see Table 2.2[2]). A similar mandate was extended to the EU in 2003, where mobile positioning is considered an even more critical issue owing to the continually increasing number of mobile-originated E112 calls (EUROPA 2003). Nevertheless, the EU will most likely follow the USA and Japan in requiring high positioning accuracy when Galileo is fully operational (Berg-Insight 2006). In the meantime, research in the field of wireless location has been promoted as an important public safety feature, which can also add many other potential applications to future telecommunication systems (Sayed et al. 2005a).

2.3.2.1.2 *Tracking and Tracing*

This category embraces services that are able to track and trace the geographic whereabouts of mobile entities (e.g., people, vehicles or packages) and support requests to establish the location of these entities, their progress and state changes along a route, or to establish prospective future locations (e.g., tracking postal packages so that companies know where their goods are at any time) (Steininger et al. 2006). In particular, these services can be distinguished according to their tracking target, that is, a single entity (e.g., a child or a package) or a group (e.g., a workforce); thus, it is naturally derivable that this category may be directed both to the consumer and to the corporate market. Indeed, applications include, on the one hand, pet, people and vehicle tracking, and, on the other hand, fleet, asset, goods and package tracking. Obviously, there is an intertwining between this category and other LBS categories, such as emergency, safety and security, and management and logistics. For example, vehicle tracking can be also applied

[1] S. Frattasi, M. Monti, "Cooperative mobile positioning in 4G wireless networks", *Cognitive Wireless Networks: Concepts, Methodologies and Visions Inspiring the Age of Enlightenment of Wireless Communications*, Springer, pp. 213–233, September, 2007. Reproduced in part with kind permission of © 2007 Springer Science and Business Media.

[2] For the definition of CEP67, CEP95, and network-based and mobile-based positioning, see Chapter 4.

Table 2.1 Examples of emergency, safety and security services. (Reproduced with kind permission of © 2007 Springer Science and Business Media).

Service	Accuracy[a]	Environment	Business type	Revenue potential	Brief description	Related categories
Emergency call	High	Indoor/ outdoor[b]	business-to-consumer (B2C)	High	This service allows one to quickly respond to an emergency situation by dispatching police, medical personnel, etc. to the location at which the call was originated (TruePosition 2009).	Tracking and tracing
Automotive assistance	Medium	Outdoor	B2B/B2C	Medium to high	This service allows one to automatically send help directly to the location of a damaged or inoperable vehicle. Sophisticated location-enhanced automotive assistance services offer useful on-the-road services, such as providing distances and directions to gas stations, automobile repair services or motels (AAA 2009).	Tracking and tracing
Emergency alert	Low	Outdoor	B2C	Low	This service allows one to deliver emergency information such as information about inclement weather, road conditions, heavy traffic or any life-threatening incident that is relevant to the user's location (Telenity 2009).	Tracking and tracing
Personal medical alert	High	Indoor/ outdoor	B2C	High	This service allows a person who requires medical attention to simply press a button on a device (or type a preset number into a mobile phone) to alert an emergency response service, so that an ambulance can be quickly sent to the appropriate location. This saves precious moments in the response time and provides location information even when the person is unable to speak. This service is ideal for senior citizens and subscribers with life-threatening health issues (TruePosition 2009).	Tracking and tracing
Family monitoring	High	Indoor/ outdoor	B2C	Medium	This service allows one not only to locate or monitor a child or elder, but also to provide an alert message if the person moves outside a predetermined area; this is commonly referred to as *geo-fencing* (TruePosition 2009).	Tracking and tracing

[a] Accuracy: high, from 0 m to 50–75 m; medium, from 50–75 m to 250–300 m; low, from 250–300 m onward.
[b] "Outdoor" means urban, suburban and rural.

Table 2.2 FCC requirements.

Specification	Network-based	Mobile-based
CEP67	FCC-N 1: 0.1 km	FCC-M 1: 0.05 km
CEP95	FCC-N 2: 0.3 km	FCC-M 2: 0.15 km

to dispatch the nearest ambulance to a given emergency call. A similar application can help companies to manage their workforce (e.g., salespeople and repair engineers) more efficiently to provide their customers with accurate arrival times of personnel (Steininger et al. 2006).

2.3.2.1.3 Traffic Telematics (Including Vehicle Navigation)

The area of traffic telematics is aimed at supporting drivers with a manifold set of services related to their vehicles: vehicle navigation, diagnosis of malfunctions, dissemination of location-based information and warning messages (e.g., traffic, weather and road conditions), automatic configuration of appliances, and added features in the vehicle (Küpper 2005). An emerging research topic within this category is intervehicle communications, which usually rely on short-range wireless technologies (e.g., ZigBee, Wi-Fi or Bluetooth) to establish ad hoc communications between two vehicles without the need for a centralized control. This allows users to interact directly with each other and exchange useful information such as warning messages about accidents, local traffic situations or the position of filling stations (Küpper 2005).

The most popular application in traffic telematics is undeniably vehicle navigation. Whereas for the last decade this service has been enabled by expensive on-board units installed in vehicles, recently there has been growing integration of GPS receivers in portable devices, which can easily be plugged into cars. Basically, whatever the device in use, the user just needs to set the location of the final destination and the terminal is able to locate the current position, calculate the route based on preferences (e.g., the shortest, fastest, most economical or most scenic route) and feed this information back in the most suitable way (i.e., according to the capabilities of the device and stated preferences). In fact, this information can be delivered in text format, on a map or with the help of automatic voice guidance, or by a combination of these. If the terminal is able to connect to a cellular network, up-to-date traffic telematics information related to the vehicle navigation (e.g., traffic jams and road works) can also be received; on the basis of this information, the device can also recommend alternative routes. Finally, vehicle navigation can be combined with a number of services related to parking slots, ranging from registering and charging to automatic guidance and exchange of parking slots among drivers (Küpper 2005).

2.3.2.1.4 Personal Navigation

While traffic telematics embraces vehicle navigation, personal navigation is in its own category. Specifically, this category has emerged from the increasing pervasiveness of mobile devices in our lives (from personal digital assistants (PDAs) to mobile phones), and their continuous and progressive enrichment of integrated technologies (from various wireless communication to positioning technologies); in this context, mobile positioning will soon be enabled in both outdoor and indoor environments. Indeed, users are not only concerned about their navigation while driving, but also while walking, including in indoor environments (e.g., shopping malls and offices). Practically, one main difference with respect to traditional vehicle navigation is that users get a precalculated route via their mobile network operator, as road databases are in many cases not available on the device. Moreover, not being restricted to roads, personal navigation may possibly feature positioning algorithms different from those employed in vehicle navigation, and the visualization of maps may also be different.

2.3.2.1.5 Management and Logistics

The services considered under this category encompass any type of management, from facility to environmental management. The inclusion in such management processes of "location intelligence" is a key asset for enabling higher efficiency and reducing possible risks. For instance, in the case of a complex construction project, such smartness allows easy management of work crews as they move from location to location. As outlined in Graphisoft (2007), "Usual scheduling systems are activity-based (good enough for building an airplane engine in a factory), which implies that the user manually takes into account locations and makes subjective determinations on the duration of a certain activity. The addition of location as an extra dimension provides enormous benefits, since it allows users to compress schedules without increasing risks by balancing the amount of float throughout the project and, therefore, reducing workflow variability." However, there are still open issues that may reduce or delay the provision of such benefits. In the case of facility management, for example, inspectors often find it hard to acquire and manipulate location data about facilities when the facilities do not have any distinctive location marks to be referenced. Although they are equipped with GPS-enabled devices, this may still happen if the location identifier the facility is housed inside a building (Tejavanija et al. 2003). Similar considerations and advantages, deriving from the opportunity to locate and track a workforce, apply also to the case of fleets (e.g., freight, public transportation and emergency vehicles). For instance, the operator of a taxi dispatch center may have all the positions of their company's vehicles displayed on a map. When an incoming customer needs to be served, the operator just informs the target taxi (possibly the closest to the caller) of its new destination and the user of the expected arrival time. Finally, it is important to mention that knowing the location of a product and being able to track it can also serve to support logistics. Indeed, this makes it possible to support faster transportation, different transportation modes and the development of fallback scenarios in case of failures (Küpper 2005).

2.3.2.1.6 Billing

This category can, in one instance, be related to mobile users, who can be charged by their wireless carriers on the basis of their current location. Forms of facilitation and service differentiation such as predefining specific zones (e.g., "home zone", "work zone" and "premium price zone") with predetermined rates can be set up (e.g., calls that originate within the user's home zone may be charged as wired calls). In this context, it is also possible to define areas in which callers are allowed to reach the customer and areas in which calls must be blocked. For example, the user may want to receive certain calls in the home zone, but not at work, and perhaps receive only urgent calls when traveling or on vacation (MobileIN 2009). Finally, we can also expect that operators, in order to maximize their revenue, will offer special types of services in agreement with other service providers. For example, as Kenney Jacob foresees, we can think about an advertisement in front of a Coffee Day shop, which says (Jacob 2007) "Enjoy free Hutch to Hutch Calls from your mobile along with your Coffee". This obviously also reminds us of the fact that location-sensitive billing is often found in combination with the purchasing of goods and services, and therefore is an enabling feature of mobile commerce applications (see Section 2.3.2.1.7).

Billing, that is, charging services in connection with the user's vehicle, can also be related to traffic telematics. Road tolling, which concerns the payment of tolls related to the use of roads, highways, tunnels, bridges, etc., is the most obvious example of this type. In general, such payments have not yet been completely automated; indeed, most of the time, we stop at a barrier and pay our toll to an officer or a machine according to the distance covered. Since such an approach increases traffic congestion, some countries have introduced a subscription that is valid for a certain time period and permits the driver to circulate freely. However, even this solution has the shortcoming that drivers are not billed for the distance that they have really covered (Küpper 2005). Currently, the only alternatives for avoiding such scenarios and enforcing payment automatically are the following: (1) vehicles carry an on-board unit, which, thanks to a transponder, is able to exchange data over the air with the toll barrier, and (2) control stations take pictures of passing vehicles and electronically analyze the license plate number by image recognition. Unfortunately, each country in the past has developed and deployed its own proprietary system and therefore incompatibility issues have emerged when national boundaries are crossed. This is one of the issues that has to be tackled within a framework of globalization as there is a strong need to create international efforts to harmonize such a fragmented picture. Finally, it is important to underline that road tolling is only one of many services that fall under the broader umbrella of *remote tolling*. The latter services can be listed as follows (Antonini et al. 2004): (1) pay-per-use services (e.g., automotive assistance), (2) location-based information services (e.g., traffic information and routes), (3) tracing, tracking and emergency management services for the transportation of hazardous material (HazMat) and heavy goods, (4) access control services (e.g., enforcement of laws against users who enter restricted zones such as historic areas without permission), (5) parking services (see Section 2.3.2.1.3), and (6) pay-per-use insurance, where users pay for the insurance of their vehicles on the basis of the percentage of time the vehicle is used, the type of environment in which the driver normally uses the vehicle, and on-board information that can affect the final insurance cost.

2.3.2.1.7 Commerce

In this category we consider all applications related to commerce, from marketing to the purchasing of goods and services. Obviously, location information opens up new sale scenarios and provides marketers, and thereby companies, with great potential to increase their revenue by intercepting new customers and the potential to empower their customer relationships by providing higher customer satisfaction. This can be achieved by means of targeted advertisements, which are activated and directed to potential buyers according to the buyer's personal profile, buying pattern in the past and proximity to the company's premises. For instance, users passing a shop could be attracted inside by an offer popping up on their mobile phone, where this message could refer to discounts, coupons, allowances and gifts, and could be presented in the form of full-color images, short videos or audio files. For larger stores, such as superstores and hyperstores, customers' navigation through the store could also be supported (Ad2Hand 2009). Naturally, this type of advertisement is applicable not only to users walking by, but also to people traveling by public transport or driving. However, in the latter case it must be guaranteed for safety reasons that the user will not be distracted by such messages (Küpper 2005).

A further point to consider regarding such services is related to the avoidance of spam. In this respect, Donald Spector, Chairman of the Innovation Fund LLC, says (IF 2009): "If you call someone and offer them a 10,000 dollars promotional bonus from GM while they are standing in a Honda showroom, they will probably not get mad. We are not going to be bugging people with 25 cent offers. Shopping malls, restaurants, entertainment venues and the travel industry will be able to give consumers great bargains on items that otherwise would lose value. Tickets for a baseball game that is not sold out can be directed to a passing car making the customer an 'offer he can't refuse' ". " 'Remember,' he adds, 'we already know who likes baseball!' ". From this we can derive how important it is to filter the information according to the users' preferences, such that no one will be "bugged" by a massive load of unsolicited messages, which in the long run would lead to a rapidly declining interest of the customer in this type of service. In addition, it must also be possible to cancel a subscription either permanently or temporarily.

2.3.2.1.8 Enquiry and Information

These services target the distribution of all of the pieces of information that are relevant to the user's context. This is, so far, the most widespread type of LBS and can embrace both users who live in a certain area and users who are there for the first time (e.g., for business or leisure purposes). Location-sensitive information concerns guided tours, transportation services and points of interest (e.g., nearby places of historical and touristic interest, hotels, restaurants, coffee shops, shopping malls, cinemas, pubs, discos, theaters, museums, parks, pharmacies, hospitals, gas stations, ATM counters and retailers). Services that provide points of interest are practically an extension of the well-known Yellow Pages that show entries only of local relevance. In particular, they may be combined with navigation facilities for guiding the user to a point of interest or even inside a point of interest (e.g., in a guided tour of a museum) (Schiller and Voisard 2004). Finally, we also recognize under this category services related to the distribution of information closer to the realm of traffic telematics, such as information about traffic jams, parking slots, and road and weather conditions.

2.3.2.1.9 Leisure and Entertainment (Including Community and Gaming)

Community services underlie the gathering of people in either a physical or a virtual place around some common baseline, which might relate to keeping in touch with family and friends, reactivating connections with old friends and acquaintances, or creating discussion groups on specific topics (from cooking to traveling to eroticism). Usually, each user creates their own profile (including their alias, sex, age, domicile, hobbies, work, marital status, etc.), which is used for discovering people with the same background, interests and hobbies, and communication among the members of the community is enabled by means of chatrooms, whiteboards, etc. Recently, there has been an explosion of such services – see Figure 2.6 (Ziv and Mulloth 2006) – thanks to the increasing and widespread use of the Internet, which consequently has facilitated a tendency towards meeting in "virtual places" and therefore creating new social links or empowering existing links regardless of the real physical distances between people. The inclusion of location information in such community services has launched a new paradigm: from people to people irrespective of geographical place to "people-to-people-in-geographical-places" (Jones and Grandhi 2005). Dodgeball, for example, is one of these services, where location information is used to meet nearby friends, friends of friends or even possible crushes (Dodgeball 2009). In the context of community services, privacy is a highly sensitive matter, as the level of information that we are willing to

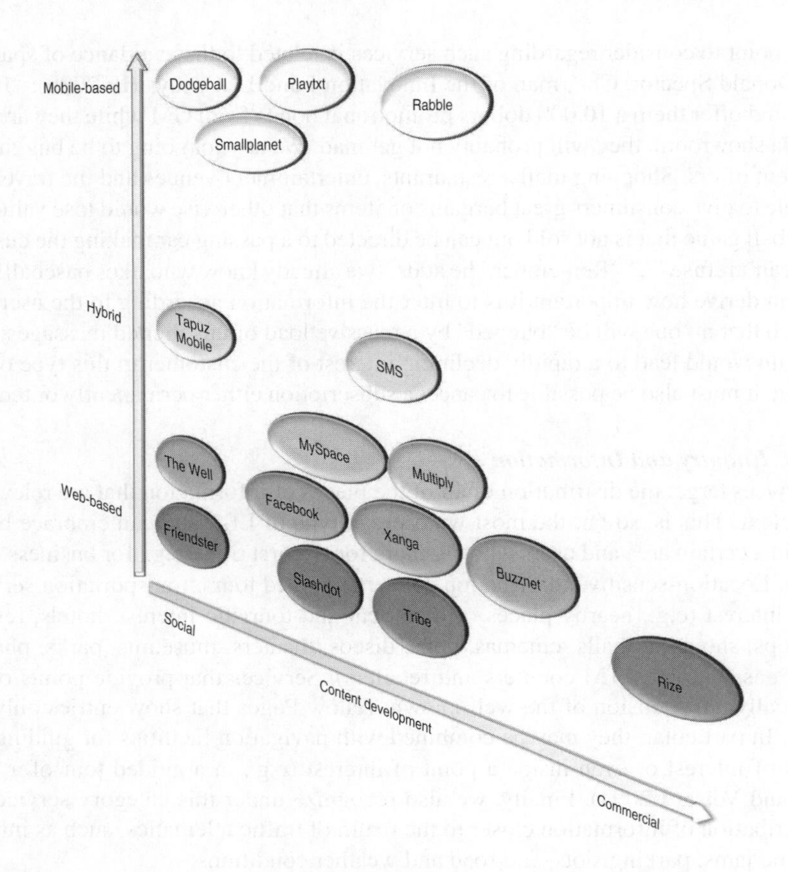

Figure 2.6 The social network matrix (partial list) (Ziv and Mulloth 2006).

make accessible to others depends strictly on the level of social connection that we have with them. That is why, on the one hand, it is important to enable "permission request" messages for every enquiry we receive and, on the other hand, to explicitly declare in our profile the level of detail at which we want to give out information regarding our location. For instance, we can expect that for services such as a family finder or friend finder, the exact position of the user may be freely accessible to both parties, whereas for proximity-matching services or dating services, releasing the location of the user within a certain "privacy range" would be more suitable.

As a primary form of entertainment, at least for youngsters, gaming has evolved tremendously in recent decades, where the trend has moved from a single-player or multiplayer game on a single PC to real-time multiplayer sessions where several computers are connected together in a LAN and to massive multiplayer online games with hundreds or thousands of people connected to the Internet. Recently, owing to the introduction of powerful consoles such as the Play Station Portable (PSP), gaming has gone portable. The next advance in this field will be the introduction of location awareness, which allows users to play in a real environment instead of a virtual one. For example, a shop that is willing to promote its products may sponsor a treasure hunt in a specific location, where the "treasure" is a 29-inch flat-screen TV. The first player reaching that location can claim the prize. Finally, it is important to mention that the next level is foreseen to be the introduction of *augmented reality*, which will allow users to mix reality and virtuality in the game. This technology will further blur the line between what is real and what is computer-generated by enhancing what we see, hear, feel and smell (Schiller and Voisard 2004). This could also create new prospects for worlds that are currently totally virtual, such as Second Life (Second Life 2009).

2.3.2.1.10 Supplementary

Usually, "value-added services" refer to services which enrich the panorama of the user experience beyond basic speech or data services. In practice, many of the services presented above, mostly in the category "enquiry and information", can be included in this category. That is why we prefer to entitle this section "Supplementary", so that we could include all services that fall outside of the previous categories. Practically, the main supplementary services, which can be extended to LBSs, are the following:

1. Call forwarding (or selective routing): this service is usually activated when incoming calls addressed to a user's mobile phone are derouted to a co-located fixed telephone.
2. Coverage information intimation: this service can alert users when they are moving out of coverage or out of a service area (e.g., "high-speed data services end after 2 km") (Chaudhari 2007).
3. Voting: this service can gather votes (e.g., in relation to politics, various events or products) that may have strong local significance. If a regional political election is approaching, for example, political parties can obtain strategic surveys by gathering electronic votes from people on the spot, people who are really informed about local problems.

2.3.2.2 Quality of Service Parameters and Requirements

The quality of service (QoS) parameters are associated with LBSs are listed below (3GPP 2015; Machaj et al. 2012):

- **Localization accuracy:** This parameter can be divided into *horizontal accuracy* and *vertical accuracy* (3GPP 2015). The horizontal accuracy may range from tens of meters (e.g., for navigation services) to kilometers (e.g., fleet management). In general, the majority of attractive LBSs are enabled with a horizontal accuracy between 25 and 200 m (3GPP 2015) (e.g., see Table 2.1). The vertical accuracy may be defined in terms of absolute height/depth or relative height/depth to ground level and may range from about 3 m (e.g., to resolve within one floor of a building) to hundreds of meters (3GPP 2015).
- **Response time (RT):** This parameter is defined as the time period that elapses between the request (e.g., of position determination) and the response (e.g., the position estimate). A trade-off between the localization accuracy and the response time is normally necessary. In particular, the following options are available (3GPP 2015): (a) "no delay": the response time has precedence over the localization accuracy; the system shall deliver either the initial/first position estimate or the last-known/estimated position. (b) "low delay": the response time still has precedence over the localization accuracy, but the system, given the delay requirement, can utilize the longer time to calculate a better position estimate. and (c)

"delay tolerant": the localization accuracy has precedence over the response time so the system can even delay the delivery of the response until the localization accuracy requirement is fulfilled.
- **Position reporting frequency (PRF):** This parameter indicates how frequently a new position estimation is required. Since a high PRF may cause network overload, a trade-off is normally necessary (Machaj et al. 2012).

Tables 2.3 and 2.4 show the QoS requirements for exemplary LBSs in terms of the above-mentioned parameters for pedestrians and vehicles, respectively, since, due to the different speeds, the QoS requirements may change (Machaj et al. 2012).

2.3.3 *Context Awareness*[3]

The continuous development in wireless and wired technologies opens up new and innovative services and applications, and allows the latter to adapt to the ever-changing environment in which they are used. The context in which an application or a service is used by a user is highly relevant for offering functionalities to the user as well as for the user experience. For example, in order to efficiently notify a user in a very noisy environment, the notification mode could change from sound to vibration, or the sound could be transferred to external

Table 2.3 QoS for pedestrians (Dhar and Varshney 2011; Machaj et al. 2012; 3GPP 2015).

LBS	Accuracy (m)		RT (s)	PRF (s)
	Horizontal	Vertical		
Emergency	0–20	0–10	0–1	0–14.5
Tracking	0–50	0–10	0–3	0–36
Personal navigation	0–50	0–10	0–5	0–72
Billing	0–300	0–10	0–5	0–216
Information	0–200	0–10	0–5	0–144

Table 2.4 QoS for vehicles (Dhar and Varshney 2011; Machaj et al. 2012; 3GPP 2015).

LBS	Accuracy (m)		RT (s)	PRF (s)	
	Horizontal	Vertical		Urban	Suburban
Emergency	0–20	0–10	0–1	0–1.4	0–0.9
Tracking	0–200	0–10	0–3	0–14.4	0–9
Vehicle navigation	0–100	0–10	0–5	0–21.6	0–3.5
Billing	0–500	0–10	0–5	0–36	0–22.5
Information	0–500	0–10	0–5	0–36	0–22.5

[3] Rasmus Olsen, Aalborg University, Denmark

loudspeakers. This is an example of application adaptation based on the contextual situation. Another example is adapting screen lighting to the current ambient light (LG Electronics Inc. 2008; Microsoft Corp. 2015).

Context adaptation and awareness is based on *context information*, which is an extremely wide term. Context information has been defined by many people, but one of the most quoted is Dey (2000), who defines it as: "Any information that can be used to characterize the situation of an entity. An entity is a person, place, or object that is considered relevant to the interaction between a user and an application, including the user and application themselves". This definition is very powerful, in the sense that it literally enables any information to be potential context information. *User location* and *time of the day* are two of the most obvious types of context information, as, in fact, a user's need for applications and services greatly depends on those two information elements. Nonetheless, as the definition indicates, any type of information can be regarded as context information and can be potentially useful for an application or a service.

In general, the idea of using context awareness in the provision of an application or a service has existed in the field of computer science for a couple of decades (Schilit et al. 1994). The issue is that it is not necessarily known in advance what information may or may not be relevant, therefore much research has been focused on the management of generic information exchanges. For example, a European project called MAGNET Beyond, a dedicated framework for the management of context information in so-called personal networks, has been developed (Sanchez et al. 2006). The concept is shown in Figure 2.7. The idea is that any application or service that requires any context information can send requests or subscriptions to a *context management framework*, which will then handle issues like discovery of where information should be obtained, secure access and control of the information, and optimization of data flows, for example in case multiple applications try to access the same information. As a consequence, all this information can be shared, instead of increasing the traffic in the network.

The abovementioned concept thus alleviates the programmer's need to worry about discovery mechanisms, which would require solving various issues. By plug-and-play of sub-components in the framework, the *context agent* offers simple interfaces to sensors or any

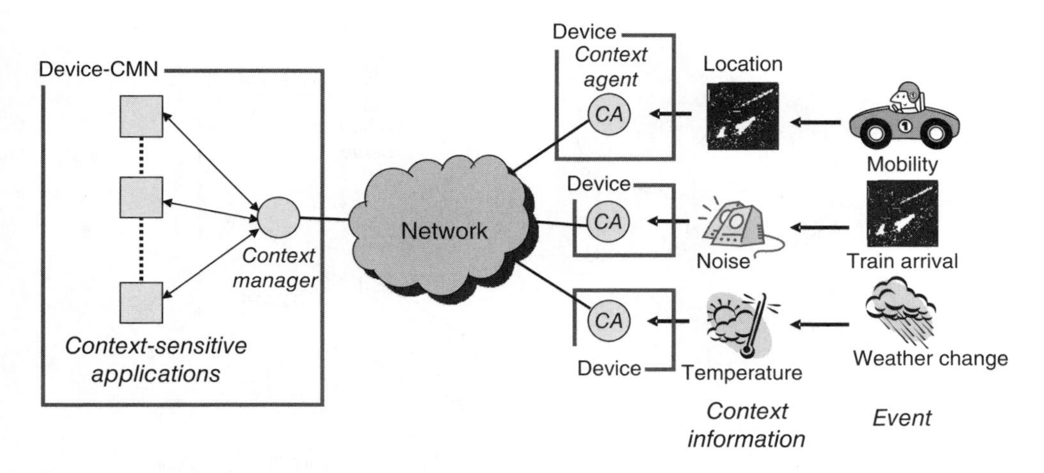

Figure 2.7 Context management and distribution of context information to enable context-aware applications.

other source of information and maps the native data formats (e.g., raw binary measurements from an A/D converter or a proprietary data format) into a common data format, which is then easily exchanged between the agents. Access control can be carried out via a request/response pattern or via a subscription based on events in the present environment or by means of a specific time period.

This concept is particularly effective for devices that do not possess the capability of producing the needed context information but need to use shared resources. Nevertheless, the concept faces several challenges, in particular due to the fact that context information is dynamic and changes over time. As a result, inconsistencies and delays occur. The following example from Olsen et al. (2010) illustrates this problem in regard to *location information*. Consider a system that is able to estimate the position of mobile users by sending RSSI measurements to a positioning server from where the clients can request and obtain their positions (see Figure 2.8). On a request from the client to the server, an estimation process is initiated, which collects enough sample values for the positioning server to provide a proper estimate of the client's position. The result is then sent back to the client to be used by a context-aware application. Note that if the abovementioned context management framework is used, multiple applications will lead to a single request since the framework is caching and reusing the position estimate within a certain time elapse from the initial request. A problem then arises when high localization accuracy is needed, since more samples need to be collected, which results in a longer estimation time. During this longer estimation time, the client may have moved from his initial position, therefore when the estimation process is finally concluded, the position estimate sent to the client is inaccurate, thus becoming useless or even problematic to use.

In Olsen et al. (2010) several models were developed to analyze and understand the link between the reliability of the obtained location information and the processes involved, in particular the delay metrics and the client's velocity were investigated. The main finding was that it is possible to create models based on trade-offs between the accuracy and reliability of location information. For example, if the abovementioned context management framework is used, if accurate location information is required the application has to accept that the information

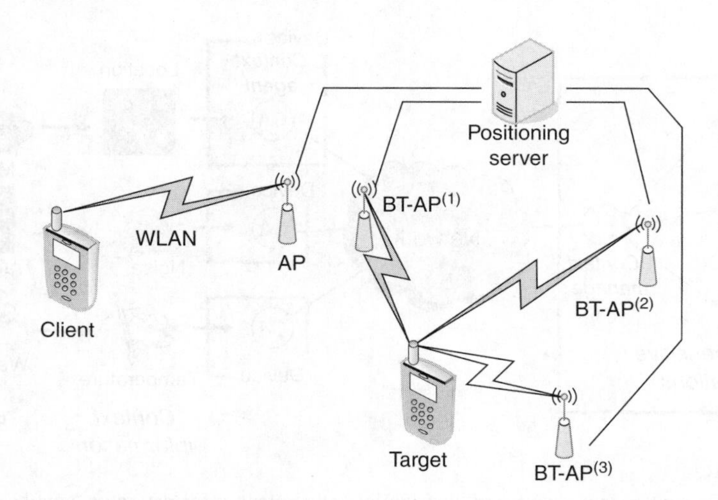

Figure 2.8 Positioning system based on a centralized estimation of the client's position.

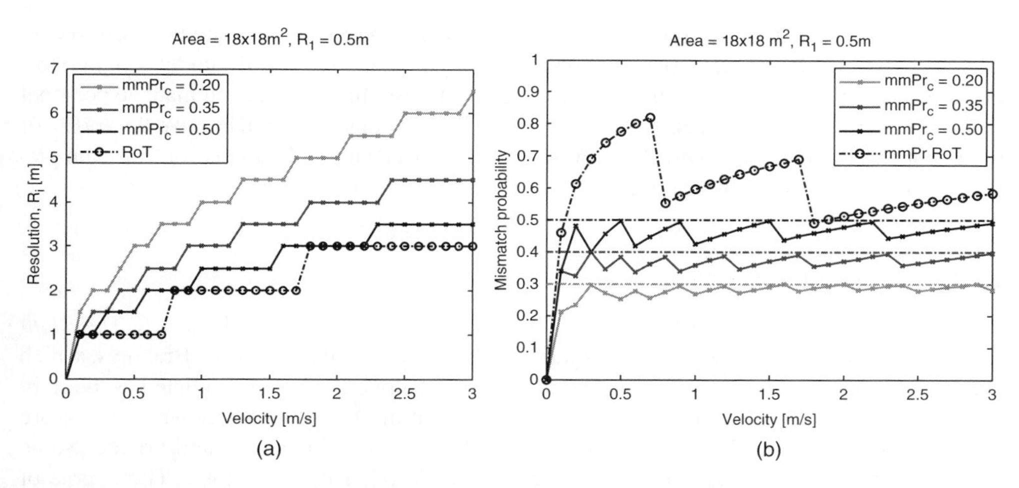

Figure 2.9 Trade-off between accuracy and reliability of the information as a function of velocity. (a) Localization accuracy vs. velocity; (b) Information reliability vs. velocity.

may not be totally reliable and that the faster the client moves the less reliable the information becomes. This can effectively be exploited by the context management framework to provide a certain level of reliability of the context information by automatically adapting the localization accuracy to preset reliability requirements. The context management framework may, for instance, be given three levels of reliability requirements (0.20, 0.35 and 0.50; see Figure 2.9), which may be achieved by adapting the localization accuracy (see Figure 2.9a). As is noticeable from Figure 2.9, the faster the client moves, the less accurate the system should be in order to provide reliable information.

This principle of providing reliable context information is generalized in Bøgsted et al. (2010), where several studies about different access mechanisms are analyzed. In general, the conclusion is that randomness in context information is useful for the context management framework's ability to provide reliable context information. This, in the end, is useful for the user, since reliable context information means reliable behavior of context-aware systems and applications. Reliability is also a highly desirable goal for developers, since their applications and services then behave as expected by adapting to the context.

2.3.4 Privacy[4]

Wireless communication has enabled LBSs to provide users with time and space relevant services to an unprecedented extent. LBSs involve a certain tracking of the users' location to provide geographically relevant and timely services, and possibly even services targeted based on user preferences or interest. However, this may present threats to the users' privacy in that the users of LBSs are susceptible to profiling to provide a related LBS or for a completely different purpose, such as advertising. Privacy protects the right of the people to control what happens to their personal information. The European Commission recently proposed a

[4] Hanane Fathi, Copenhagen Patents (COPA), Denmark

comprehensive reform of personal data protection rules in the EU that will come into force on 24 May 2016. Under EU law, personal data can only be gathered legally under strict conditions, for a legitimate purpose. Furthermore, organizations which collect and manage personal data are responsible for protecting it from misuse. The data protection Directive 95/46/EC of the European Parliament provides a legal framework specifying obligations for, for example, LBS providers and the rights of users.

2.3.4.1 What is Privacy?

Privacy is often referred to as the "right to be let alone" in reference to the article *The Right to Privacy* published in 1890 by the US jurists Samuel D. Warren and Louis Brandeis, which sketched the legal concept of privacy. The right to be let alone can be reformulated as the right to have control of your own personal data in the present information society where LBSs are one type of service that is built on data that can be considered personal and private. Some personal data is private and some is less private, while other data is public. The degree of privacy can be considered unique for each user subjectively. We shall, however, try to abstract from this herein and identify privacy threats and possible countermeasures objectively.

Major privacy threats can be categorized as either a Big brother threat or a Little brother threat (Bellovin 2008), both of which may take advantage of the underlying tracking and profiling taking place in the provision of LBSs. "Big brother", as defined by George Orwell in his novel *Nineteen Eighty-Four*, describes a society where everybody is under surveillance of the main authority. Little brother is the corporate equivalent. It concerns companies that aim to make financial gains by tracking or profiling users and selling personal data. Such tracking and profiling can be done using cookies, third-party cookies, cooperation with third-party advertisement sites and web registration. There are many financial benefits in profiling users to sell targeted advertisements online or offline, and provide targeted services (location-based or not).

Profiling a user may be defined as storing and/or analyzing a user behavior, given that the user is identified or has given personally identifying information. Profiling a user may comprise aggregating personal information with, for example, behavioral information.

2.3.4.2 How does Privacy Apply to LBSs?

To provide LBSs, a user is geographically tracked. The user may consider private the combination of his location and some of his identity (such as login, pseudo, or name). The combination of the two can be considered a privacy violation when used by a third party without explicit consent. A recent MIT study (de Montjoye et al. 2013) showed that four spatio-temporal points, approximate places and times, are sufficient to uniquely identify 95% of 1.5 million people in a mobility database. To this extent, a few locations correlated with time are sufficient to reveal an identity. Thus privacy may become a concern if such time-location data are available to unauthorized parties.

2.3.4.3 Privacy-enhancing Techniques

Several privacy-enhancing techniques to protect user's privacy when using LBSs have been proposed.

2.3.4.3.1 Anonymous Schemes

For protecting users' privacy, anonymous schemes can be efficient if they do not overload communications. Using anonymity in LBSs may be one way to go, but anonymous authentication schemes and anonymous signatures should be used to prevent the abuse of anonymity in performing illegal activities by allowing a server to be authenticated by a user and authenticating the fact that the user is part of a legitimate group.

Anonymizing LBSs is one approach to providing privacy. In an anonymous LBS, the user is anonymously authenticated using, for example, an anonymous password-authenticated key exchange (APAKE) protocol with the LBS server. The anonymous authentication can be performed thanks to an extension of APAKE protocol that allows a user, based on knowledge of a password, to authenticate the LBS server and obtain LBS services while the LBS server can only authenticate the fact that the user is one of the qualified members for LBS services without knowing the user identity. An efficient anonymous PAKE protocol has been proposed by Fathi et al. (2008) for performing anonymous authenticated key establishment, and by Fathi et al. (2010) for credential retrieval.

Alternatively, the user may use APAKE (Fathi et al. 2010) to retrieve credentials from a credential server, and the credential retrieved thus permits the user to generate ring signatures for the packets to be sent to the LBS server providing anonymous services that can consequently control the legitimacy of the packet without disclosing the sender's identity.

The anonymous PAKE protocol aims to provide private key retrieval, semantic security of session keys as well as anonymity against a passive server who follows the protocol honestly, but is curious about identity of a client involved with the protocol.

2.3.4.3.2 Private Information Retrieval

Another approach is to apply private information retrieval techniques to the LBS. Private information retrieval (Chor and Gilboa 1997) is a step towards full privacy protection. It permits a user to retrieve a piece of information from a database without the database knowing which item has been retrieved. It can be an efficient tool against Little brother if it allows the retrieved content to be stored in a database. In the case of LBSs, retrieving information for the LBS from an LBS provider database can be performed in a privacy preserving manner using private information retrieval techniques.

2.3.4.4 Going Forward

For LBSs to further expand in a privacy-preserving manner, privacy must be considered in the design, that is, at the start of the LBS development process. By doing this, there is a higher likelihood of privacy being ensured and the efficiency of the LBS being maintained.

A very promising concept related to privacy is that of data minimization. Data minimization refers to the idea that a company will only retain personal information it actually needs and only for the amount of time that it is needed (FCC Staff 2012). Privacy vulnerabilities are thereby minimized because even in the event of a disclosure, the amount of data at stake has been minimized. However, for various purposes of law enforcement (such as police investigations), there is an interest in retaining more location information for longer periods of time.

There is a need for a balance between privacy preserving and full disclosure, and such a balance may be achieved using privacy-enhancing techniques such as anonymous password-authenticated key exchange schemes or private information retrieval techniques.

2.4 Location-based Network Optimization

In contrast to LBSs, where the user is the main recipient (either passive or active) of the services and the target is mainly to provide appealing applications that benefit from the notion of location, here we are concerned with the optimization of some network functions by making use of this additional information. In practice, this process is transparent to the user and targets mainly the QoS guarantee provided by the network. In particular, while current mechanisms for network optimization are usually reactive, that is, they counteract current, short-term situations with the help of analysis of measurements, location-based intelligence permits the operator in most cases to make proactive decisions, which help to prevent the network suffering from poor performance.

2.4.1 Radio Network Planning

Radio network planning is a process that has the scope to provide ubiquitous coverage to mobile users while conforming to their expected QoS requirements (e.g., a low percentage of dropped calls or good indoor coverage). This process evolves in the following three phases (Chaudhari 2007): *dimensioning*, *detailed planning* and *optimization*. Starting from specifications of coverage, capacity, QoS demands and target service areas (e.g., higher and lower-data-rate services), and plausible radio propagation models, the first phase is characterized by an initial shaping of the network itself, which concerns the number of base stations (BSs) or access points (APs), their sites and their configurations. The second phase concerns more advanced planning of coverage, capacity and frequencies based on an analysis of real-time propagation data (usually gathered during a walk or a drive test) and then on the use of planning tools (e.g., see SEAMCAT at www.ero.dk), which permits one to check for interference and to define frequency reuse patterns, antenna settings, etc. Finally, the third phase consists of a continuous optimization process, which is active during the operation of the network and allows one to dynamically refine the configuration of the parameters initially set during the dimensioning and detailed-planning phases (e.g., transmission powers). In this context, it is obvious how location data can be very useful for identifying possible problems in the network (e.g., areas of poor coverage), for tuning and validating propagation models with real-time data and for creating alternative network plans in the case of traffic congestion (CELLO 2003). During sporting events, concerts or exhibitions, for example, cellular operators could insert additional capacity by adjusting the antenna settings or by using movable BSs (Lathi 2000). In general, location intelligence is an additional tool that can also be exploited prior to radio network planning to perform accurate market and demographic analyses, which will give indications about where and how to expand the network in order to get the maximum return from the initial operator investment (MapInfo 2009).

2.4.2 Radio Resource Management

The aim of radio resource management (RRM) is to utilize the available, often rather limited, radio resources as efficiently as possible. Here, "efficiency" refers to maximizing the traffic load while satisfying the QoS requirements of as many mobile users as possible and overcoming the inherent difficulties of radio wave propagation in the environment (Hasu 2007). This section provides an overview of several RRM techniques (for details, refer to Chapter 3) that can be optimized by introducing the notion of location awareness.

2.4.2.1 Beamforming

Beamforming is a signal-processing technique that permits the directionality of a radiation pattern to be controlled in order to increase the sensitivity of a transmitting/receiving antenna array in the direction of a desired signal while decreasing it in the direction of interference and noise. Basically, this result is obtained by spatially filtering the radio signals transmitted/received by an antenna array by assigning a weight to each antenna element according to the sensitivity pattern that we want to achieve. Owing to phase and gain differences between signals at different antennas, any given set of weights produces constructive interference in some directions and destructive interference in other directions. If the weights are constant and predefined, we refer to "fixed beamforming" or "switched-beam smart antennas"; otherwise, if the weights are selected adaptively on the basis of current measurements, we refer to "adaptive beamforming" or "adaptive-array smart antennas". In the latter case, location information can be used to provide adaptive coverage by identifying areas where there is a sudden need to provide higher capacity (e.g., in a stadium that is hosting the finals of the Olympic Games).

2.4.2.2 Power Control

The problem that power control (PC) deals with is how to achieve the minimum transmission power level for the elements of the network while guaranteeing the desired QoS for each user, usually expressed in terms of a target signal-tointerference-plus-noise ratio (SINR) for each user terminal. In particular, when PC is applied to the downlink, the aim is to minimize multicell interference by balancing the transmission powers at the BSs or APs; when it is applied to the uplink, the aim is to minimize the in-band interference and the energy consumption by balancing the transmission powers at the user terminals. Usually, PC algorithms rely solely on signal strength measurements, either obtained at the BS or AP or reported back by the user terminal. The use of location awareness could greatly enhance their functioning. For example, as Chaudhari (2007) describes, "If a user is in a 'bad zone', the base station can keep its transmission power high even when the mobile is reporting a good signal strength. This will avoid 'spikes' in power level similar to the 'ping-pong' effect in handovers."

2.4.2.3 Packet Scheduling

Packet scheduling is an RRM function that is controlled in downlink by the BS or AP and in uplink by the mobile station. The primary aim of a packet scheduler is to exploit the available physical resources (i.e., time, frequency, space and code) in such a way as to maximize the overall capacity while meeting the QoS requirements (usually defined in terms of a set of predefined parameters or priority levels) for the various links active for each user. Advanced packet-scheduling algorithms use important contingent information such as measures of instantaneous channel conditions (in terms of SINR) to allocate the available resources. For example, in OFDMA systems (see Chapter 3), the packet scheduler at the BS or the AP may assign a certain subchannel to the terminal that is currently experiencing the best channel conditions, which can therefore use the highest modulation and coding scheme to transmit on that band. However, while this approach maximizes the overall capacity of the system, it results in an unfair distribution of resources among the users, as terminals closer to the BS or AP will be served more often than those at the cell edge. A proportional fair metric, which considers not only the current achievable rates but also the history of the rates achieved in the past by the users, has therefore been introduced in the literature to ensure a more democratic distribution of resources (Wengerter et al. 2005). However, in general, the aforementioned scheduling mechanisms can be categorized under the label of "reactive" mechanisms, since they are based on current information about the user. As outlined by Chaudhari (2007), "Sometimes the base station can allocate a significant chunk of bandwidth to a user at the expense of others only to find out that that terminal soon reports back bad channel conditions or, even worse, that the channel conditions of that terminal swing from one extreme to another, thereby driving the packet scheduler in a 'hunt mode', where it constantly changes its forward link transmission parameters." In contrast, "proactive" mechanisms, which feed back information not only about channel conditions but also about the current location and speed of the terminal, can be used in combination with database topological data to predict the future channel conditions and thereby perform a more effective scheduling choice.

2.4.2.4 Handover

Handover is the process that enables the mobility of a user within a network. Basically, it can be either intracell, when the user is moving from one sector to another in the same cell, or intercell, when the user is moving out of their home BS coverage and has to be handed over to a neighboring BS. In the latter case, if both outgoing and incoming connections are maintained continuously, the handover is refer to as "soft" to differentiate from a "hard" handover, where there is a temporary disruption of the ongoing communication. Usually, the criteria on which the handover decision is based are signal strength measurements (if these are provided by the user terminal, the handover is referred to as mobile-assisted), a list of neighbors, the traffic distribution, network capacity, bandwidth and some others (Chaudhari 2007). However, the signal strength might give unreliable indications, owing to its location-dependent variability. For example, if a user is on the borderline between several cells, the terminal will report back measurements that will often trigger a handover, thus creating the so-called "ping-pong" effect. In contrast, if those measurements are combined with location information, the aforementioned effect can be avoided by delaying the handover to a more appropriate moment. Further benefits deriving from such a combination are found in the following applications (Chaudhari 2007):

1. *Interfrequency handover*, when a user is moving to an area where a different carrier frequency has to be used.
2. *Vertical handover*, when a user is moving to an area where there is a different type of network in use (e.g., from UMTS to WLAN).
3. *High-speed microhandovers*: if a user is moving at high speed to an area where there are many microcells (overlapping with the current macrocell), it is not convenient to continuously perform handovers but is better to keep with the same macro-BS.
4. *Handover prioritization*: if the core network has to handle many handover requests coming from many BSs in the network, direction and speed information can be used to prioritize the more time-critical handovers.

Finally, note that other RRM functions are involved when handover is performed, and therefore they can be location-aided:

- **Cell selection.** This is the operation that associates a user with a certain cell out of a set of possible candidates. Usually, the metric used to take this decision is based on signal strength measurements performed between the terminal and the BSs or APs. If cell selection is done with the aid of location information, situations in which certain users suffer from poor QoS can be avoided. For example, if a user is moving towards an area covered by two APs with respect to which the user's terminal is experiencing slightly different channel conditions, an estimation of the user's direction and speed might indicate the cell with lower signal quality as the best choice for serving the user over a certain time period (Chaudhari 2007).
- **Admission control and load control.** These are processes that prevent a network from being overloaded, thus helping to avoid dropped calls, falling QoS, etc. In particular, while admission control aims to avoid congestion by regulating the total number of users who have permission to access the network, load control has to counteract overload situations that the admission control was not able to prevent (e.g., an overload due to a sudden increase in the interference level). If location-related information is available on the network side, these functions can be helped in their task by predicting future congestion. For example, speed and direction could indicate that a user is moving towards a crowded stadium, and therefore admission control and load control will kick in to accommodate the incoming user under the jurisdiction of a BS which is less congested, thus also taking part in performing a load-balancing action.

2.5 Patent Trends

Typically, the result of a patent search offers many powerful business-related indications in respect to, amongst others, deriving the actual interest and the geographical interest for a technology by looking at the evolution of patent filings during the years and the number of patent filings per country/region, respectively (Frattasi and Nielsen 2015). Herein, a patent search has been carried out in order to derive the trends for the services mentioned in Sections 2.3 and 2.4.[5]

Figure 2.10 shows the number of patent families generated within the last 20 years. It is inferable from the trend shown in Figure 2.10 that the number of inventions related to the services mentioned in Sections 2.3 and 2.4 got traction around 1998 and has continuously increased thereafter.

Figure 2.11 shows the geographical distribution of the total number of patent families generated within the last 20 years. It is inferable from the trend shown in Figure 2.11 that the USA, Europe, China, South Korea and Japan are the top five countries in terms of number of generated inventions and/or market appeal.

[5] The numerical results shown in this section are indicative because they are dependent on the keywords and the patent classes used in the search. Also, since patent applications normally remain secret for a period of 18 months from their filing date, the actual numbers may be higher.

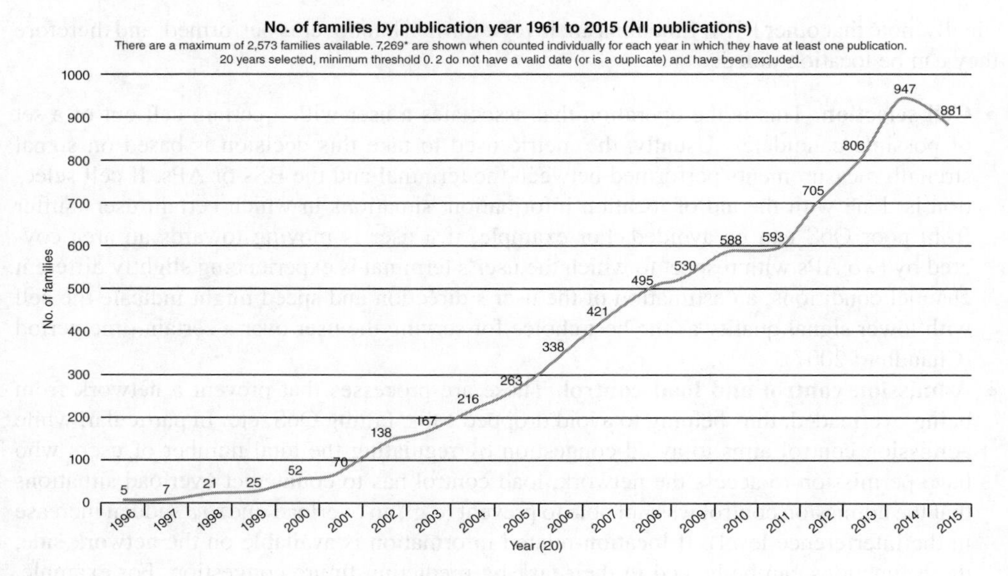

Figure 2.10 Patenting trend within the last 20 years.

Figure 2.12 shows the distribution of the total number of patent families generated within the last 20 years in terms of assignees. It is inferable from the trend shown in Figure 2.12 that Qualcomm, Microsoft, Google, Samsung and Intel are the top five companies in terms of number of generated and/or acquired inventions.

Figures 2.13–2.16 show the trends specifically for location-based gaming and community services (see Section 2.3.2.1.9). It is inferable from the trends shown in Figures 2.13–2.16 that: (1) the number of inventions related to location-based gaming services got traction around 2006 and has slowly increased thereafter, (2) Qualcomm, AT&T, Disney, Google and Nokia are the top five companies in terms of number of generated and/or acquired inventions within location-based gaming services, (3) the number of inventions related to location-based community services got traction around 2006 and has constantly and strongly increased thereafter, (4) Google, Facebook, Yahoo, Nokia and Microsoft are the top five companies in terms of number of generated and/or acquired inventions within location-based community services, and (5) the number of inventions generated within location-based community services is substantially higher that the number of inventions related to location-based gaming services.

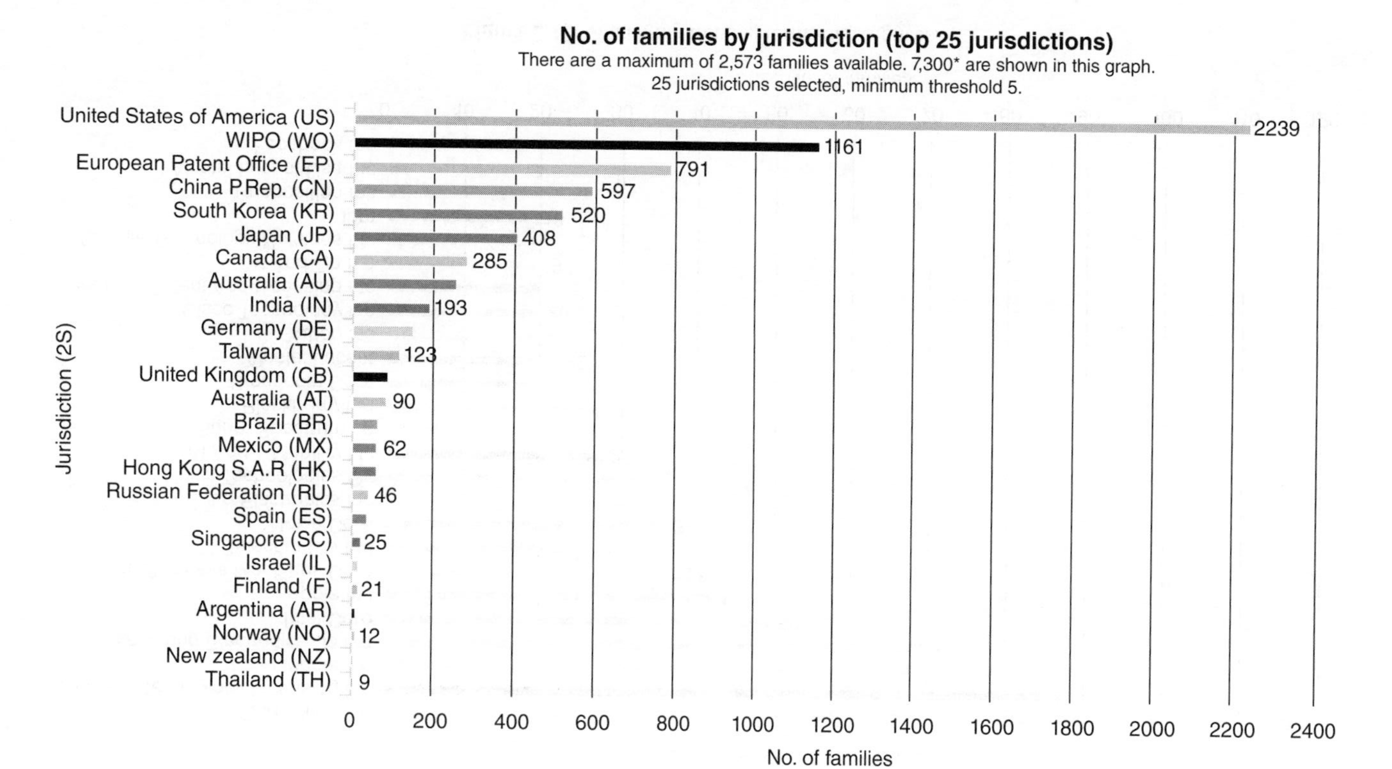

Figure 2.11 Patenting trend in terms of geography.

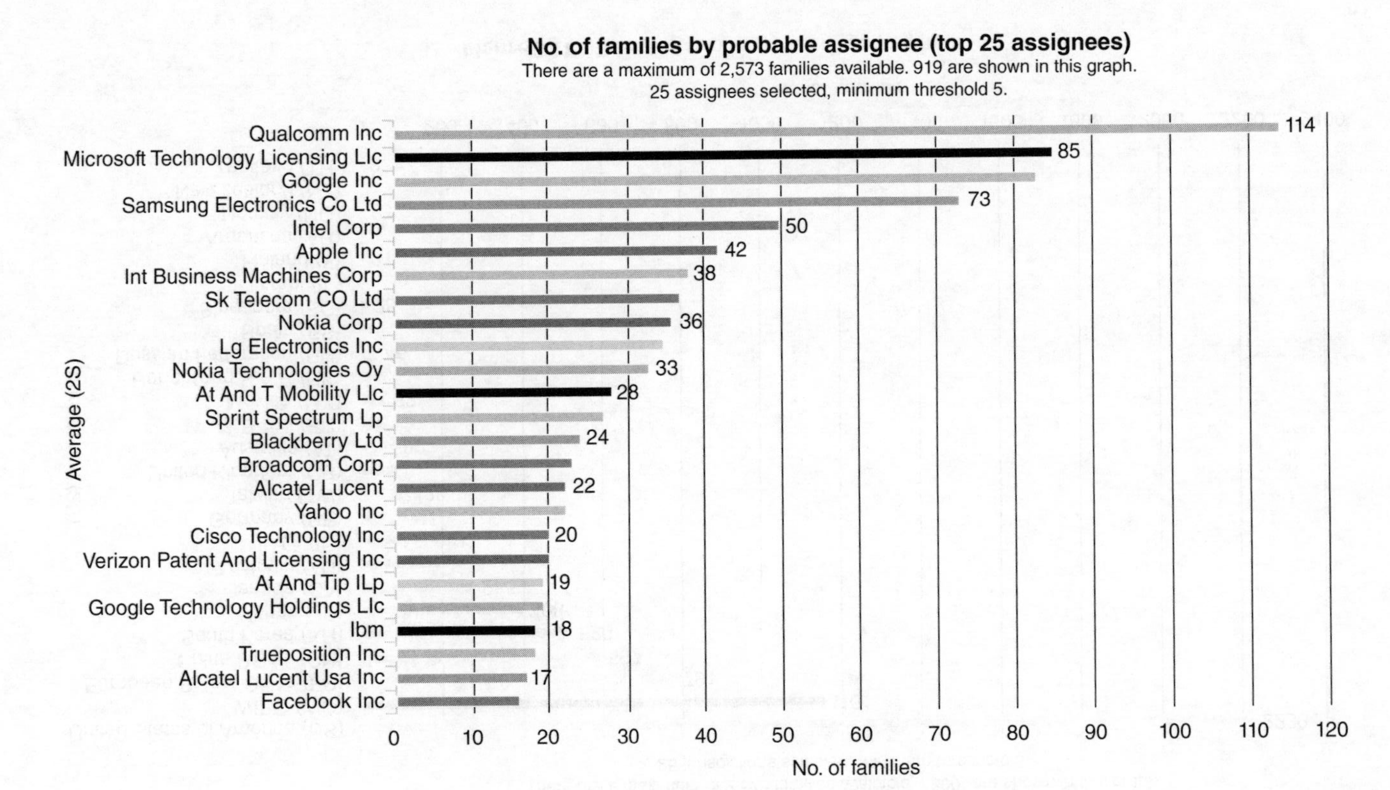

Figure 2.12 Patenting trend in terms of assignees.

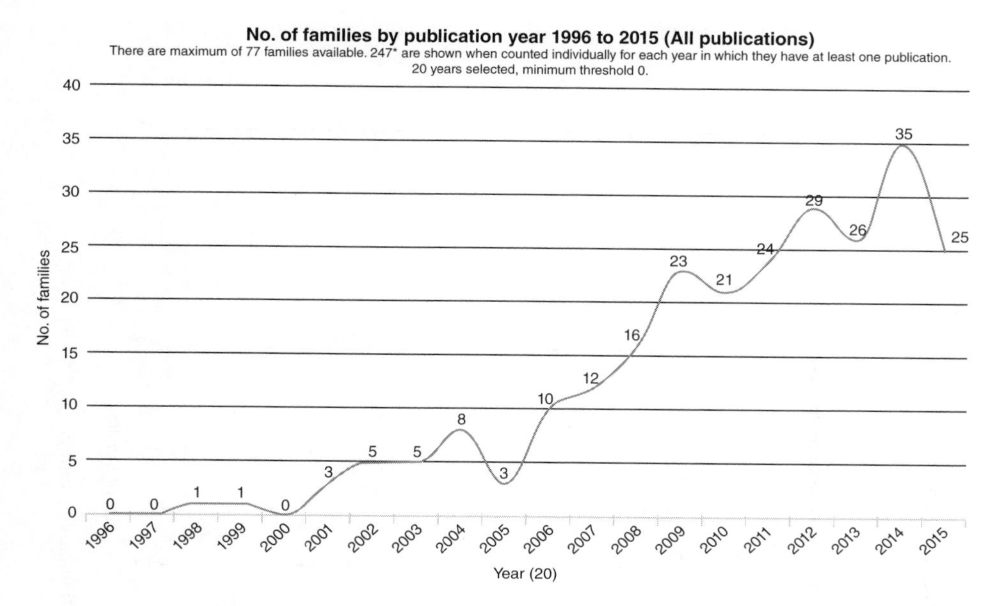

Figure 2.13 Patenting trend within location-based gaming.

2.6 Conclusions

In this chapter we have given an overview of the application areas of positioning, namely LBSs and location-based optimization. According to one of the definitions given in the chapter, an LBS is "an information service accessible with a mobile device through a mobile network, utilizing an ability to make use of the location of the mobile device itself" (Virrantaus et al. 2001). Intuitively, this opens up the possibility of advanced existing services and a previously unforeseen panorama of new appealing services (e.g., proximity matching and gaming in outdoor scenarios). In particular, some of the "killer applications" envisioned in the chapter could result from the enabling of indoor positioning, which current technologies still struggle to provide. In contrast to LBSs, location-based optimization does not imply any active behavior on the user side; instead, it aims at enhancing network performance in terms of coverage, capacity, energy consumption, etc. Such enhancement is in many cases transparent to the user, who will perceive it in terms of higher QoS.

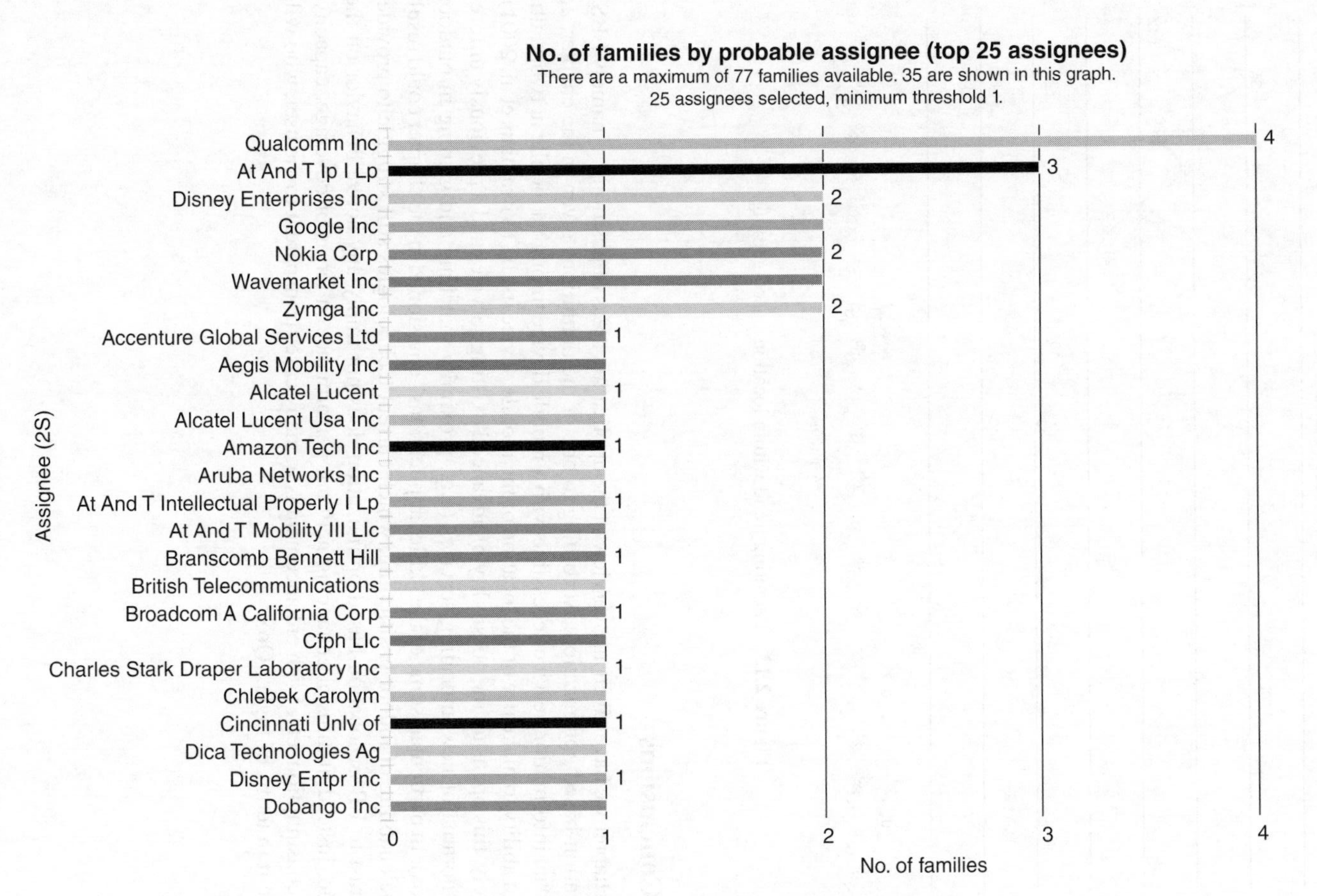

Figure 2.14 Top patent assignees within location-based gaming.

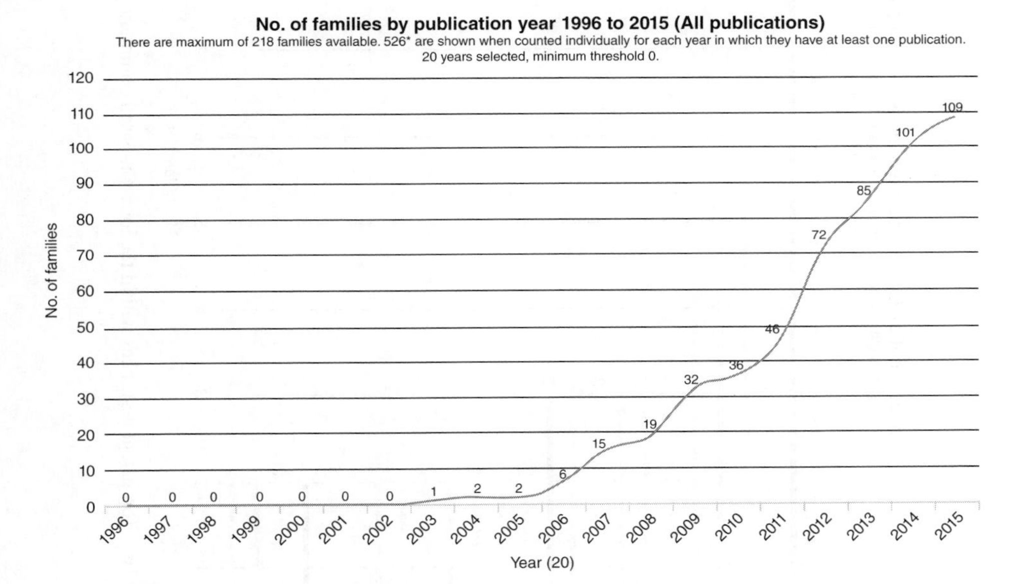

Figure 2.15 Patenting trend within the location-based community.

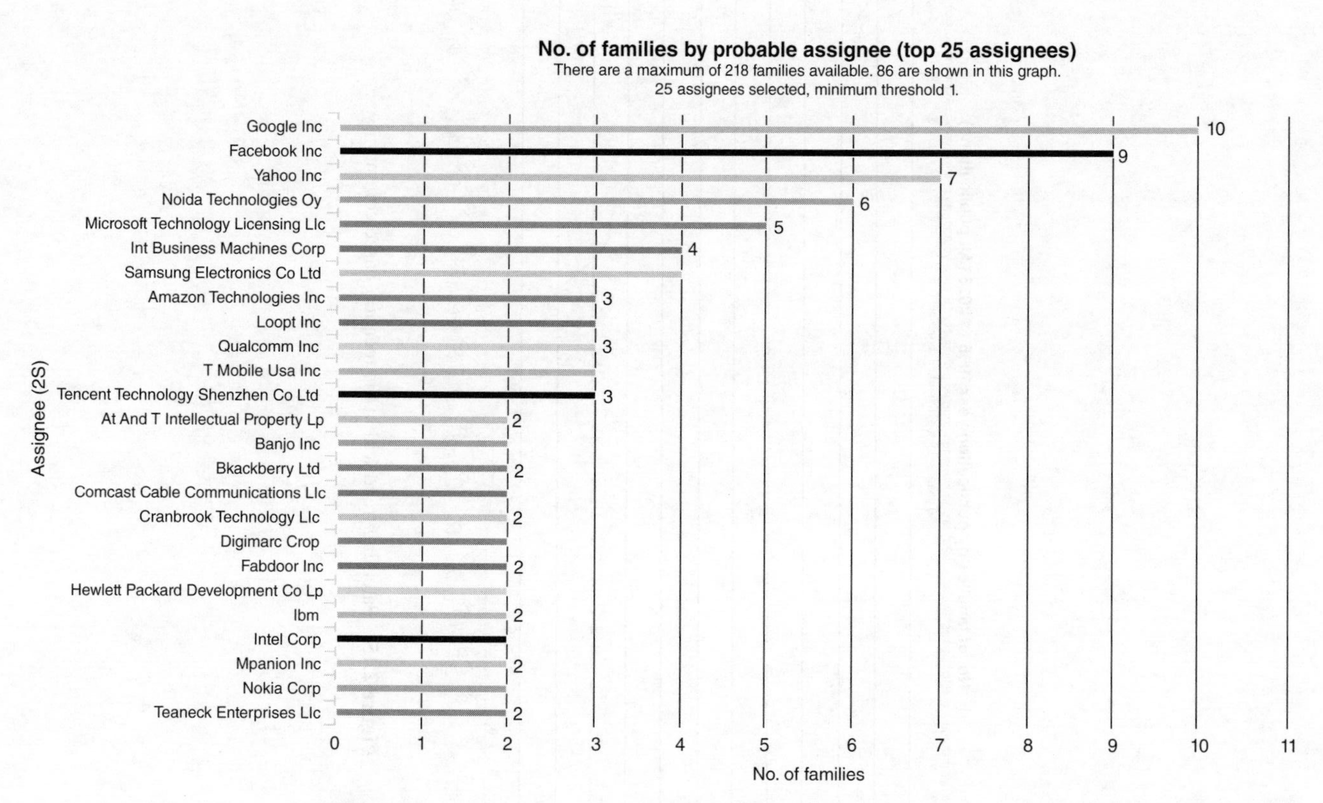

Figure 2.16 Top patent assignees within the location-based community.

3

Basics of Wireless Communications for Positioning

Gilberto Berardinelli[1] and Nicola Marchetti[2]
[1]*Aalborg University, Aalborg, Denmark*
[2]*CTVR, Trinity College Dublin, Ireland*

3.1 Introduction

This chapter introduces the basics of wireless communications with particular focus on their relation to mobile positioning. Section 3.2 describes the principles of radio propagation by classifying the physical phenomena underlying wireless communications in terms of large-scale fading (i.e., path loss and shadowing) and small-scale fading (i.e., multipath propagation). Several positioning algorithms are then briefly reviewed. In Section 3.3, the possibilities and challenges of multiple-antenna techniques are introduced, focusing in particular on *spatial diversity* and *spatial multiplexing* and their achievable gains, namely array gain, diversity gain, multiplexing gain and interference reduction. The other important family of multiantenna techniques, i.e., beamforming, is discussed in connection with space-division multiple access (SDMA). Some considerations of the possibilities offered by multiple-input multiple-output (MIMO) techniques in terms of positioning are provided. Section 3.4 presents the duplexing methods for accommodating two-way transmission over the wireless channel. Next, Section 3.5 overviews modulation and access techniques, with particular focus on multicarrier techniques and access methods such as orthogonal frequency division multiplexing (OFDM) and orthogonal frequency division multiple access (OFDMA). Other techniques such as time division multiple access (TDMA), code division multiple access (CDMA) and carrier sense multiple access (CSMA) are also described. Section 3.6 describes the basics of some radio resource management (RRM) techniques, namely handoff and interference management, channel reuse, channel reservation and power control, and illustrates how localization can benefit from them and vice versa. The principles of over-the-air synchronization are addressed in Section 3.7. In Sections 3.8 and 3.9 we outline

Mobile Positioning and Tracking: From Conventional to Cooperative Techniques, Second Edition.
Simone Frattasi and Francescantonio Della Rosa.

two emerging areas within wireless communications, that is, cooperative communications and cognitive radio technology, and their interconnections with localization. Section 3.10 concludes the chapter.

3.2 Radio Propagation

The propagation of radio signals irradiated by a transmitter is a function of space, frequency and time, and is typically characterized at the receiver side. The received signal is indeed an attenuated and distorted version of the transmitted signal, and such variation can be classified as either large-scale or small-scale fading.

Large-scale fading refers to the signal attenuation over long distances due to the energy dispersion over the wireless channel. Typical prediction models for large-scale fading estimate the mean received signal strength for a transmitter–receiver separation ranging from 4λ to 40λ (Rappaport 1996), where λ is the wavelength of the signal. In that respect, such propagation models are useful for estimating the coverage area of a radio cell. Large-scale fading can, in turn, be classified into path loss and shadowing. Path loss deals with the propagation loss due to the distance between transmitter and receiver, while shadowing describes variations in the average signal strength due to varying environmental clutter at different locations.

Small-scale fading refers to the rapid fluctuations of the received power due to the multipath propagation of the electromagnetic waves in a short period of time or travel distance where the average signal strength remains constant. Small-scale fading includes variations in time and space, as well as frequency-selective fading.

In a generic propagation environment, the signal strength at a particular location depends on both large-scale fading and small-scale fading. As the receiver moves, the instantaneous power of the received signal may rapidly vary, giving rise to small-scale fading. As the distance between the transmitter and the receiver increases, the local average of the received signal power decreases gradually and can be predicted by large-scale fading statistics. The phenomenon of combined large- and small-scale fading is represented in Figure 3.1.

The main factors influencing radio wave propagation and giving rise to the aforementioned fading phenomena are reflection, diffraction and scattering. *Reflection* occurs when the electromagnetic waves impinge upon a surface whose dimension is significantly larger than their

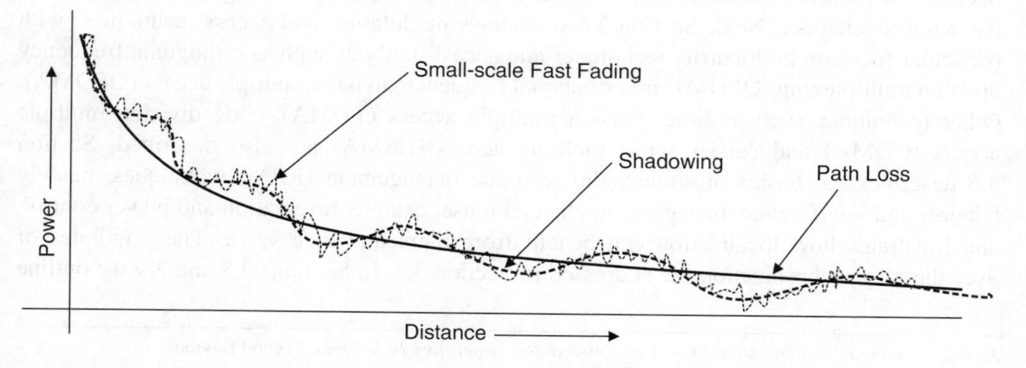

Figure 3.1 Propagation loss (Prasad et al. 2009).

wavelength, and change their direction when returning to the propagation medium. *Diffraction* is due to the effect of sharp edges in the path of the radio waves between the transmitter and the receiver. *Scattering* is caused when the electromagnetic waves encounter objects having significantly smaller dimension than their wavelength.

Most radio propagation models use a combination of empirical and analytical methods for estimating the strength of the received signals. The empirical approach is built on a set of measurements data that are based on the derivation of analytical expressions that reproduce the observed behavior. This has the advantage of implicitly taking into account all of the propagation factors. Typically, the empirical models are derived for specific scenarios and operational conditions, for example carrier frequencies. The validity of an empirical model for frequencies or environments other than those used to derive the model can be established only by additional measurement campaigns (Rappaport 1996).

3.2.1 Path Loss

Predicting the attenuation of a radio signal over large distances is of paramount importance in the design of a wireless communication system since it establishes the conditions and limits for which the communication over the wireless link can take place. We present here an overview of the most common models for path loss prediction currently adopted in engineering.

3.2.1.1 Free-space Path Loss

The free-space path loss refers to the natural energy dispersion of a radio signal irradiated by a source located at a significant distance. In that respect, it subsumes idealistic conditions of unobstructed propagation where phenomena like reflection, absorption and diffraction do not occur. By assuming isotropic antennae at both the transmitter and receiver side, the free-space path loss in dB at a distance d can be expressed by the well-known Friis equation:

$$\alpha_{\text{loss}} = 20\log_{10}\left(\frac{4\pi d}{\lambda}\right) \tag{3.1}$$

Note that such expression is only valid with the assumption that $d >> \lambda$; in the case where $d < \lambda$ the Friis equation may lead to the impossible result of a received signal power higher than the transmitted one.

In the linear domain, the Friis equation leads to a quadratic attenuation law as a function of the distance. In practice, more realistic attenuation exponents are in the order of a power of four of the distance. Though unrealistic, the Friis path loss model represents a useful tool for the estimation of the limit of the radio propagation performance. Further, it is also the basis of empirical models, which are described next.

3.2.1.2 Empirical Models

The complexity of an analytical derivation of realistic path loss models when radio effects such as reflection, absorption and diffraction are taking place has brought researchers to investigate empirical models based on measurement results and targeting specific scenarios. The most common empirical models for path loss prediction are the following:

- **Hata–Okumuru model.** This model was derived from measurements carried out in Tokyo and is intended to predict the path loss in large urban areas where the BS is placed higher than the surrounding rooftops. The path loss in dB can be expressed as:

$$p_{\text{loss}} = 69.55 + 26.16\log_{10}(f_c) - 13.82\log_{10}(h_t) - a(h_r) + (44.9 - 6.55\log_{10}(h_t))\log_{10}d \quad (3.2)$$

where f_c is the carrier frequency, h_t and h_r are the height of the BS and receiver antenna of the mobile station (MS), respectively, and $a(h_r)$ denotes a correction factor for the mobile antenna height based on the size of the coverage area (Rappaport 1996). This model assumes a minimum distance of 1 km and is designed for carrier frequencies between 150 MHz and 1.5 GHz. The Hata–Okumuru model has been widely applied for the network planning of first-generation cellular systems.
- **COST231–Hata model.** This model is an extension of the previous Hata–Okumuru model to carrier frequencies up to 2 GHz. It assumes BS height between 30 m and 300 m, MS height between 1 m and 10 m, and a distance between 1 km and 20 km. In that respect, it is also designed for macro cells (Anon 1991).
- **COST231–Walfish–Hikegami model.** This model targets micro cells or small macro cells and introduces variables related to the height of the buildings, width of roads, building separation and road orientation with respect to the direct path of the desired signal (Lee 1998). In that respect, the model distinguishes between the line-of-sight (LOS) and the non-line-of-sight (NLOS) cases. The model is valid for carrier frequencies between 800 MHz and 2 GHz, BS height between 4 m and 50 m, MS height between 1 m and 3 m, and distance between 20 m and 2 km.
- **Ergec model.** This model introduces the terrain type as a parameter for the path loss prediction (Erceg et al. 1999a). Specifically, three types of terrain are defined: Category A refers to a hilly terrain with moderate tree density, Category B refers to a hilly terrain with light tree density or flat terrain with moderate-to-heavy tree density, Category C refers to a flat terrain with light tree density. Category A has the highest path loss while Category C has the lowest path loss. For a close distance $d_0 = 100$ m, the median path loss (Δp_{loss}) in dB is given by

$$\Delta p_{\text{loss}} = \alpha_{\text{loss}} + 10\beta \log\left(\frac{d}{d_0}\right) + \eta_{\text{shad}}, \quad d > d_0, \quad (3.3)$$

where η_{shad} is a random variable addressed in Section 3.2.2 and

$$\alpha_{\text{loss}} = 20\log\left(\frac{4\pi d_0}{\lambda}\right). \quad (3.4)$$

η_{shad} is the shadowing factor, which follows a log-normal distribution with a typical value around 6 dB, and β is the path loss exponent; β is given by

$$\beta = a - bh_{\text{bs}} + c/h_{\text{bs}}, 10\text{ m} < h_{\text{bs}} < 80\text{ m}, \quad (3.5)$$

where the values of a, b and c depend on the type of terrain (Erceg et al. 1999a) and h_{bs} is the BS antenna height in meters. The model in Equation (3.3) was proposed for a receiver antenna height of 2 m and an operating frequency of 2 GHz. A correction factor for other frequencies and antenna heights can be found in Erceg (2001), and the modified path loss is then

$$\Delta p_{\rm mod} = \Delta p_{\rm loss} + \Delta p_{\rm freq} + \Delta p_{\rm bsh}, \tag{3.6}$$

where $\Delta p_{\rm loss}$ is the path loss given by Equation (3.3), $\Delta p_{\rm freq}$ is a frequency correction term given by $6\log(f/2000)$, where f is the frequency in MHz, and $\Delta p_{\rm bsh}$ is the receiver antenna height correction term given by $-10.8\log(h/2)$, where h is the new receiver antenna height, such that $2\text{ m} < h < 8\text{ m}$.

Considering only the path loss in Equation (3.3), it is possible to define the received power p based on $\Delta p_{\rm loss}$, the transmit power $p_{\rm tx}$ and the transmitter and receiver antenna gains $G_{\rm tx}$ and $G_{\rm rx}$, respectively:

$$p = \alpha - 10\beta\log(d) + \eta_{\rm shad}, \tag{3.7}$$

where

$$\alpha \simeq p_{\rm tx} + G_{\rm tx} + G_{\rm rx} - 20\log\left(\frac{4\pi}{\lambda}\right). \tag{3.8}$$

- **WINNER II models.** The WINNER II models cover a large set of environments found in Europe and North America, including urban macro-cells and micro-cells, indoor office/residential, large indoor halls, as well as outdoor-to-indoor and indoor-to-outdoor scenarios. The models can be applied at frequency bands between 2 and 6 GHz. In that respect, they are often used for the path loss prediction in 802.11-related studies. The path loss in dB can be parametrized as follows:

$$p_{\rm loss} = A\log_{10}(d) + B + C\log_{10}\left(\frac{f_{\rm c}}{c}\right) + X, \tag{3.9}$$

where the parameter A includes the path-loss exponent, parameter B is the intercept, parameter C describes the path loss frequency dependence, and X is an environment-specific term (e.g., wall attenuation in NLOS scenarios). A large set of scenarios are covered by the above equation for different fitting parameters.

3.2.1.3 Ray Tracing

Empirical path loss models provide a useful tool for predicting the propagation performance. However, they are derived statistically from a large set of measurements and they may not correctly cope with specific environments, leading to incorrect estimates. When considering specific sites, the propagation characteristics might be computed by solving the Maxwell equations with the building geometry as boundary conditions. However, the computational complexity of such an approach is cumbersome.

Ray-tracing techniques provide a computationally efficient solution for predicting the signal strength at any given location in specific sites with sufficient accuracy. The ray-tracing model of a mobile radio communication link is based on the physical reflection that an electromagnetic wave undergoes as it strikes the surfaces in a certain environment. The effect of such reflections can be modeled as a combination of a number of directed rays that are emitted by the transmitter. These rays are normal to the surfaces and follow the rule that the angle of incidence is equal to the angle of reflection. The type of surface determines the power and phase of the reflected signal with respect to the incident wave. It also includes the free space loss due to the finite distance between the transmitter and the receiver. The electromagnetic wave received at an observation point can be calculated as the sum of all reflected waves initiated by the transmitter, plus the effect of direct line-of-sight (LOS) between the transmitter and receiver. The reflection and transmission coefficients are necessary for computing each object and arbitrary number of surfaces such as walls, floors, ceilings or partitions, etc. Ray-tracing tools rely on graphical information system (GIS) databases, which enable wireless systems designers to include accurate representations of building and terrain features.

For outdoor propagation prediction, aerial photographs are used in combination with ray-tracing techniques for integrating three-dimensional representations of buildings. In indoor environments, site-specific representation for propagation models is instead typically derived from architectural drawings.

3.2.2 *Shadowing*

The path loss models are distance-dependent and do not capture the varying environmental clutter at different locations having the same transmitter–receiver separation. Such varying clutter may lead to a significant variation of the average receive power with respect to the value predicted by the path-loss model. This random effect is known as *shadowing*.

3.2.2.1 Log-normal Shadowing

Measurements have shown that the receive power at a large number of locations having the same transmitter–receiver separation d can be be approximated as follows Rappaport (1996):

$$P_{\mathrm{r}} = P_{\mathrm{t}} - p_{\mathrm{loss}}(d) - \eta_{\mathrm{shad}} \tag{3.10}$$

where P_{r} and P_{t} denote the receive and transmit power, respectively, and η_{shad} is a Gaussian distributed random variable (in dB) centered on the deterministic average power given by the path-loss model. When converting to the linear domain, the shadowing is modeled as a log-normally distributed random variable. Log-normal shadowing is the most common shadowing model used in computer simulations to provide received power levels at random locations. The standard deviation of the log-normal shadowing model is derived from measurements using linear regression techniques.

3.2.2.2 Auto-correlation Model

The log-normal shadowing assumes that the received power at different locations can be computed on independent realizations of a log-normally distributed random variable. Nonetheless, in practice the shadowing can be spatially correlated between nearby locations. A simple model which takes into account such spatial correlation has been proposed by Gudmundson (1991). The correlation of the shadowing at two points located at distance d can be expressed by the following decreasing correlation function:

$$R(d) = e^{-\frac{d}{d_{\mathrm{c}}}\ln 2} \tag{3.11}$$

where d_c is an environment-specific parameter referred to as the decorrelation length, typically ranging between 5 and 20 m.

3.2.2.3 Cross-correlation Model

Measurements have shown that when an MS is simultaneously receiving/transmitting signals from/to two BSs, the shadowing from the two BSs can be highly correlated depending on the environment. Such cross-correlation can be expressed according to the model proposed in Klingenbrunn (1999):

$$R(\theta) = \begin{cases} 0.8 - |\theta|/150 & \text{if } |\theta| \leq 60^\circ \\ 0.4 & \text{if } |\theta| > 60^\circ \end{cases} \tag{3.12}$$

where θ denotes the angle of arrival difference (AAD) between the two BSs.

3.2.2.4 Distance-dependent Model

The cross-correlation model is a function of the angle of arrival distance between the two signals, but does not capture the impact of the physical distance r between the BSs. Such an effect is instead included by the following model (Klingenbrunn 1999):

$$R(\theta, r) = \begin{cases} f(\theta, X)(0.6 - |\theta|/150) + 0.4 & \text{if } |\theta| \leq 60^\circ \\ 0.4 & \text{if } |\theta| > 60^\circ \end{cases} \tag{3.13}$$

where

$$f(\theta, X) = \begin{cases} 1 - r/X & \text{if } r \leq X \\ 0 & \text{if } r > X \end{cases} \tag{3.14}$$

X denotes the point where the distance-dependent correlation reaches its minimum value, and is typically in the range 6–20 dB.

3.2.3 *Small-scale Fading*

Small-scale fading refers to the rapid fluctuations of the signal amplitude and phase in a short period of time, in which the average receiver power can be considered constant and is dominated by large-scale phenomena. Such fading is generated by the non-coherent combination at the receiver side of multiple delayed versions of the transmit signals due to the presence of surrounding reflective objects such as buildings, hills and trees. The multipath effect is particularly pronounced in urban areas since the height of the mobile antennas is significantly lower than the height of the surrounding reflectors.

Further, the relative motion of the MS with respect to the BS might generate a variation of the multipath delay components over time. This leads to a random frequency modulation of the transmitted signals due to the different Doppler shift of the multipath components.

The variation of the amplitude, phase and delay parameters introduced by the propagation environment can be characterized by a channel impulse response, which can be expressed as follows:

$$h(t, \tau) = \sum_{m=1}^{n_{se}} a_m \delta(t - \tau_m) \exp(-j\theta_m). \tag{3.15}$$

where n_{se} denotes the number of multipath components and a_m, θ_m and τ_m refer to the amplitude, phase and delay of the mth component, respectively. $\delta(\bullet)$ is the Dirac delta function and j is the imaginary unit of complex numbers.

3.2.3.1 Multipath Fading

Multipath propagation gives rise to small-scale fading in the time and frequency domains. The variations in time are due to the presence of multiple propagation paths with different distances between the transmitter and the receiver, while the variations in frequency are due to their relative speed. Such fading can be characterized as a function of fundamental parameters such as coherence bandwidth, coherence time and Doppler spread.

3.2.3.2 Fundamental Parameters

The fundamental parameters for the characterization of the multipath fading can be derived from its power delay profile. The presence of reflectors in a wireless communication link leads to n_{se} replicas of the transmit signal with different delays. The root mean square delay spread is an indicator of the severity of the fading and can be expressed as

$$\sigma_\tau = \sqrt{\bar{\tau^2} - (\bar{\tau})^2} \tag{3.16}$$

where $\bar{\tau}$ denotes the mean excess delay and is given by

$$\bar{\tau} = \frac{\sum_m a_m^2 \tau_m}{\sum_m a_m^2} \tag{3.17}$$

and

$$\bar{\tau^2} = \frac{\sum_m a_m^2 \tau_m^2}{\sum_m a_m^2} \tag{3.18}$$

In the case of short delay spread, the significant propagation paths typically span a low delay and do not affect the frequencies of wideband signals in a significantly different manner. However, the presence of multipath components with significant delays may cause an unpredictable combination at the receiver side of the phase and amplitude of the frequency components of the wideband signal, due to their different wavelength. The delay spread of the channel can then be related to the concept of *coherence bandwidth* B_C, a statistical measure of the range of frequencies over which the channel maintains the amplitude of the signal components approximately constant and their phase linear. Frequency components located at a shorter spectral distance than the coherence bandwidth are expected to be affected similarly by the channel and are then correlated. Conversely, frequencies whose spectral distance is larger than the coherence bandwidth may be affected differently by the channel. The coherence bandwidth is approximately given by Rappaport (1996):

$$B_C \approx= \frac{1}{5\sigma_\tau} \tag{3.19}$$

This expression holds by assuming that the frequency components of the signal can be considered correlated when the value of their correlation function is below 0.5.

The coherence bandwidth is linked to the delay spread of the channel and represents a measure of its time-dispersive behavior.

The time variations of the channel due to the relative speed between transmitter and receiver can be characterized by the Doppler spread, a measure of the spectral broadening of the signal bandwidth. In the case of a sinusoidal tone at frequency f sent by a transmitter moving at constant speed v, the received signal may have components distributed over the range $f - f_\mathrm{d}$ and $f + f_\mathrm{d}$, where

$$f_\mathrm{d} = \frac{v}{\lambda}. \tag{3.20}$$

λ denotes the wavelength of the tone. It is straightforward to observe that high carrier frequencies are more sensitive to speed and can experience a larger Doppler spread. Given the time–frequency duality, the spectral broadening characterized by the Doppler spread can be related to the *coherence time* T_C definition. With analogy with the coherence bandwidth concept presented above, the coherence time can be defined as a statistical measure of the time duration over which the channel impulse response has no significant variations. Signals transmitted at an interval shorter than the coherence time of the channel then experience approximately the same channel response, which may vary if there is a longer interval between transmissions. The coherence time can be approximated as (Steel 1994):

$$T_\mathrm{C} \approx \frac{9}{16\pi f_\mathrm{d}}. \tag{3.21}$$

This expression holds with the assumption that two signals are considered correlated when their time correlation function is above 0.5.

3.2.3.3 Types of Small-scale Fading

The time–dispersion characteristics of the channel affect its coherence bandwidth, while the Doppler spread has an impact on the coherence time. Different fading types can be described depending on the channel behavior in terms of coherence bandwidth and coherence time.

The signal transmitted by a radio source undergoes a *flat* fading when its bandwidth is lower than the coherence bandwidth of the channel. The radio channel only introduces a constant gain and a linear phase shift on the transmitted signal, preserving its spectral characteristics. Narrowband signals are likely to experience a flat fading.

Current radio access technologies adopt a pilot-based channel estimation where reference sequences scattered across the operational bandwidth are used for estimating the channel frequency response, enabling coherence detection at the receiver. Flat fading can reduce the overhead of such reference sequences due to the predictable behavior of the channel across the operational bandwidth.

A frequency selective fading is instead experienced by a signal whose bandwidth is significantly larger than the coherence bandwidth of the channel. In this case, the channel is introducing different attenuations and phase distortion to the frequencies of a wideband signal. The time dispersion of the channel induces intersymbol interference (ISI) in a sequence of transmitted signals, which can be combated by introducing a guard period between the signals.

Frequency-selective channels may require a large reference sequence overhead for the channel estimation in order to capture the frequency variations and reproduce accurately the channel frequency response at the receiver. Nonetheless, frequency-selective channels may exploit the frequency diversity of the channel; mapping the same information over frequency chunks whose distance is larger than the coherence bandwidth of the channel increases the chances of correctly retrieving such information. This is due to the uncorrelated channel responses over the different frequency chunks.

The coherence time of the channel can lead to slow or fast fading. In a slow fading channel, the impulse response is varying at a slower rate than the transmitted baseband signal, and can then be considered static for the transmission of several signals. In the case of fast fading, the coherence time of the channel is shorter than the transmission time of a certain signal, leading to a variation in the channel impulse response.

Obviously, signals may also experience double-selective channels, with fast and frequency-selective channels. Dealing with such channels is a major challenge for the wireless systems designer.

3.2.3.4 Rayleigh and Rice Distributions

The channel impulse response can be expressed as in Equation (3.15) as a combination of multiple paths with different delays. The amplitude of each individual multiple component is not deterministic and can be characterized with a random distribution, the best known being the Rayleigh distribution. Such distribution corresponds to the envelope of the sum of two quadrature Gaussian noise signals, and refers to a propagation environment characterized by a large number of scatterers without a significant dominant component. In this respect, it is often used for modeling NLOS propagation conditions. The probability density function of the Rayleigh distribution is given by

$$f(r) = \frac{r}{\sigma_\tau^2} e^{-\frac{r^2}{2\sigma_\tau^2}} \tag{3.22}$$

In the case of a dominant propagation component, for example in LOS conditions, the envelope distribution is typically modeled as a Ricean variable, whose distribution is given by

$$f(r) = \frac{r}{\sigma_\tau^2} e^{-\frac{(r^2+A^2)}{2\sigma_\tau^2}} I_0\left(\frac{Ar}{\sigma_\tau^2}\right) \tag{3.23}$$

where A denotes the peak amplitude of the dominant signal and $I_0(\bullet)$ is the modified Bessel function of the first kind and zero-order. The Ricean fading refers then to the case in which random multipath components with different amplitude and phase are added on a stationary dominant signal. It can be shown that, when $A \rightarrow 0$, that is, the dominant component decreases in amplitude, the Ricean distribution degenerates to a Rayleigh distribution.

3.2.4 *Radio Propagation and Mobile Positioning*

Knowledge of radio propagation characteristics is the basis of several mobile positioning methods, ranging from low-accuracy methods based on cell identification to high-accuracy methods combining wireless network information and satellite positioning. Such methods exploit measurements which relate the MS position to the position of fixed reference points (FRPs), for example the positions of BSs, or to the specific behavior of the MS and its surrounding environment. For instance, while measurements of the AOA (Section 4.3.5) give the direction between the MS and the FRP, measurements of the time of arrival (TOA), time difference of arrival (TDOA) and received signal strength (RSS) (Sections 4.3.2–4.3.4) provide information about the distance between the MS and the FRP (Miao 2007).

Formally, position estimation can be obtained by solving a nonlinear system of equations. In general, the solutions are obtained by applying a stochastic gradient algorithm or by numerically approximating the nonlinear least squares (NLLS) problem using Monte Carlo-based techniques. Although the positioning concepts will be thoroughly covered in subsequent chapters, we present here a general overview of the measurements used for positioning in most wireless networks, as well as the main associated positioning algorithms.

3.2.4.1 Measurements

Let $X = [x, y]^{\mathrm{T}}$ denote the two-dimensional position of the mobile device. The ith known BS position is defined by $X_i = [x_i, y_i]^{\mathrm{T}}$. We assume that the BS can move with time; this can be the case for ad hoc and sensor networks. A generic measurement z_i obtained at the ith BS is a function $h_i(X)$ of both the MS position and the ith BS position, and is affected by noise η_i, that is,

$$z_i = h_i(X) + \eta_i. \tag{3.24}$$

As stated later in Section 4.3, typical measurements include RSS, TOA, TDOA and AOA, database correlation (digital map information), and position estimates (Miao 2007).

3.2.4.2 Position Estimation

Given a measurement vector $Z = [z_1, \dots, z_n]^{\mathrm{T}}$ from n different BSs and the associated measurement transform $\mathcal{H}(X) = [h_1(X), \dots, h_n(X)]^{\mathrm{T}}$ and assuming that the measurement noise vector is $\eta = [\eta_1, \dots, \eta_n]^{\mathrm{T}} \sim \mathrm{Norm}(0, R)$, where $\mathrm{Norm}(0, R)$ denotes a normal distribution with zero mean and autocorrelation matrix R, a position estimation can be obtained by various algorithms, as described below. See Chapters 4 and 5 for further details.

3.2.4.2.1 Nonlinear Least-squares Algorithm

When the measurement noise vector η is white, the NLLS position estimator provides an optimal approach equivalent to the maximum likelihood method. The NLLS-based position estimate is obtained from

$$\hat{X}_{\mathrm{nlls}} = \arg\min_{X} \left\{ (Z - \mathcal{H}(X))^{\mathrm{T}} (Z - \mathcal{H}(X)) \right\}. \tag{3.25}$$

3.2.4.2.2 Weighted NLLS

When η is a colored noise vector, the NLLS estimation can be improved by use of the weighted NLLS (WNLLS) estimate, formulated as

$$\hat{X}_{\mathrm{wnlls}} = \arg\min_{X} \left\{ (Z - \mathcal{H}(X))^{\mathrm{T}} R^{-1} (Z - \mathcal{H}(X)) \right\}. \tag{3.26}$$

3.2.4.2.3 ML Estimation

In the more general case where the Gaussian noise vector is dependent on the position, that is, $\eta \sim \text{Norm}(0, R(X))$, the ML estimate is formulated as

$$\hat{X}_{\text{ml}} = \arg\min_{X} \left\{ (Z - \mathcal{H}(X))^{\text{T}} R^{-1} (Z - \mathcal{H}(X)) + \log|R(X)| \right\}, \tag{3.27}$$

where the term $\log|R(X)|$ prevents the selection of positions with large uncertainty (i.e., large $\log|R(X)|$), which could be the case with the WNLLS algorithm.

In general, as there is no closed-form solution for the algorithms above, numerical search methods are natural choices, for example the steepest-gradient method and Newton-type methods (Dennis and Schnabel 1983; Peressini et al. 1988). The success of such local search algorithms depends on the choice of initialization, since a bad initialization may lead to only a local optimum of the criterion function.

3.2.4.3 NLOS Positioning Error Mitigation

All of the algorithms considered above assume the existence of LOS propagation paths between the MS and the BS. In the presence of NLOS propagation, the major positioning errors result from the measurement noise and the NLOS propagation error, with the latter being the dominant factor (Caffery and Stüber 1998a). Compared to an LOS situation, a time estimate in an NLOS situation may have a positive bias μ and probably a larger variance. The probability density function of the NLOS measurement error can be described by a Gaussian distribution (Cong and Zhuang 2005) as

$$\eta_i \sim \text{Norm}\left(\mu, \sigma^2_{\text{nlos}}\right), \tag{3.28}$$

where μ defines the bias introduced by the NLOS propagation and σ^2_{nlos} denotes the variance of the measurement error. The probability density function of the measurement error can be modeled by a mixture of two Gaussian distributions (Cong and Zhuang 2005) as

$$\eta_i \sim \gamma\, \text{Norm}\left(0, \sigma^2_{\text{los}}\right) + (1-\gamma)\, \text{Norm}\left(\mu, \sigma^2_{\text{nlos}}\right), \tag{3.29}$$

where γ stands for the probability of the measurement η_i resulting from LOS paths. To mitigate the effects of NLOS paths, a robust error distribution as in Equation (3.29) can be used to design the positioning algorithm. An alternative option is to include a searching mechanism for excluding the NLOS outliers (Chen 1999; Cong and Zhuang 2001; Xiong 1998). See Chapter 7 for further mitigation algorithms.

3.2.5 RSS-based Positioning

In RSS-based positioning, the target MS measures the received signal strength from the surrounding BSs and the measured data is used to estimate the location of the MS (Soork et al. 2008). Let $X = [x, y]^{\text{T}}$ denote the target MS coordinates and let $\{X_i = [x_i, y_i]^{\text{T}}, i = 1, \ldots, n\}$ denote the coordinates of the n BSs (i.e., $BS_1, \ldots, \text{BS}_n$). The received signal strength from the ith BS (BS_i) is modeled, based on Equation (3.7), as

$$p_i = \alpha_i - 10\beta \log(d_i) + \eta_{i\text{shad}}, \tag{3.30}$$

where d_i is the separation distance between BS_i and the MS, and p_i is the received power strength at BS_i. In order to estimate the position of the target MS using Equation (3.30), a least-squares objective function can be defined as

$$\mathcal{J}(X) = \sum_{i=1}^{n} (\eta_{i\text{shad}})^2 = \sum_{i=1}^{n} [p_i - \alpha_i + 10\beta \log(d_i)]^2, \tag{3.31}$$

or, in matrix form, as

$$\mathcal{J}(X) = [Z - \mathcal{H}(X)]^{\mathrm{T}}[Z - \mathcal{H}(X)], \tag{3.32}$$

where

$$Z = [p_1, \ldots, p_n]^{\mathrm{T}}, \tag{3.33}$$

$$\mathcal{H}(X) = [\alpha_1 - 10\beta \log(d_1), \ldots, \alpha_n - 10\beta \log(d_n)]^{\mathrm{T}}. \tag{3.34}$$

The position estimate will be the optimum point of the nonlinear function $\mathcal{J}(X)$ and is calculated numerically. Numerical algorithms start their operations from an initial guess and iteratively update it to reach the solution. In order to avoid convergence of the numerical algorithm to a local minimum, the starting point must be chosen not far away from the optimum. In order to specify a suitable initial point, one first calculates an estimate of the distance between the MS and all BSs. For further information about least-squares methods and the associated formalism, see Section 5.2.

3.3 Multiple-antenna Techniques

Research on multiple-antenna systems has attracted great attention in the last two decades given the promise of boosting the performance of wireless networks without consuming extra bandwidth or transmitting power. The degrees of freedom offered by multiple-antenna transmission can indeed be used for increasing the data rate or achieving higher robustness to the detrimental effect of the wireless channel, providing better link quality. Some of the significant benefits of multiple-antenna techniques are array gain, interference reduction, diversity gain and/or multiplexing gain (Paulraj et al. 2003).

The downside of multiple-antenna techniques is the additional cost, the larger size and power consumption of the hardware circuitry, as well as the extra complexity of the processing.

Multiple-antenna techniques can be broadly classified into three categories: (1) smart array processing, such as beamforming, aimed at increasing received power and rejecting unwanted interference, (2) spatial diversity, employed to mitigate fading and enhance link reliability, and (3) spatial multiplexing, used to boost the data rate.

3.3.1 Spatial Diversity

Spatial diversity is a technique that exploits the physical separation of the antennas for obtaining multiple independent versions of the radio signal. The antenna set enabling spatial diversity can be located at the receiver (*receive diversity*), at the transmitter (*transmit diversity*) or at both sides.

The level of cross-correlation between the signals transmitted or received by antennas belonging to the same device depends on the distance between the antennas. In order to experience uncorrelated spatial fading, the minimum distance between the antennas must be larger than the coherence distance of the channel (Paulraj et al. 2003). Ideally, an antenna separation of half of the wavelength ($\lambda/2$) should be sufficient (Jakes 1994) for receiving uncorrelated signals. If the antenna separation is sufficient, spatial diversity can be employed to reduce fading. This is due to the statistical independence of the received signals, which makes it unlikely that all of them experience a deep fade at the same time. Spatial diversity is relatively simple to implement and does not require an additional frequency spectrum. When multiple copies of the transmitted signal arrive at the receiver, they have to be precombined in order to obtain a certain diversity gain.

3.3.2 *Spatial Multiplexing*

Differently from diversity techniques, spatial multiplexing aims to increase the data rate of the system by using the antennas for transmitting multiple streams in parallel. Spatial multiplexing techniques require multiple antennas at both the transmitter and the receiver; such a system is often referred to as a MIMO system, and is shown in Figure 3.2. The numbers of transmit and receive antennas are denoted by n_{tx} and n_{rx}, respectively. At the transmitter, n_{tx} independent data streams, $\{s_0^{\text{tx}}(t), s_1^{\text{tx}}(t), \ldots, s_{n_{\text{tx}}-1}^{\text{tx}}(t)\}$, are sent from different antennas at the same time and at the same frequency. For the sake of simplicity, flat fading channels are assumed between pairs of transmit and receive antennas here. At the qth receiving antenna, the independent signals $\{s_0^{\text{tx}}(t), s_1^{\text{tx}}(t), \ldots, s_{n_{\text{tx}}-1}^{\text{tx}}(t)\}$ are linearly combined to produce the received signal $s_q^{\text{rx}}(t)$, which is given by

$$s_q^{\text{rx}}(t) = h_{q,0}s_0^{\text{tx}}(t) + h_{q,1}s_1^{\text{tx}}(t) + \cdots + h_{q,n_{\text{tx}}-1}s_{n_{\text{tx}}-1}^{\text{tx}}(t) + \eta_q^{\text{therm}}(t), \tag{3.35}$$

for $q = 0, 1, \ldots, n_{\text{rx}} - 1$. In Equation (3.35), $h_{q,p}$ denotes the envelope of the flat fading channel between the pth transmit and the qth receive antenna, and $\eta_q^{\text{therm}}(t)$ is the thermal noise at the qth receive antenna. The receiver is expected to estimate $\{s_0^{\text{tx}}(t), s_1^{\text{tx}}(t), \ldots, s_{n_{\text{tx}}-1}^{\text{tx}}(t)\}$ from $\{s_0^{\text{rx}}(t), s_1^{\text{rx}}(t), \ldots, s_{n_{\text{rx}}-1}^{\text{rx}}(t)\}$. Assuming that the channel responses are known, this problem is

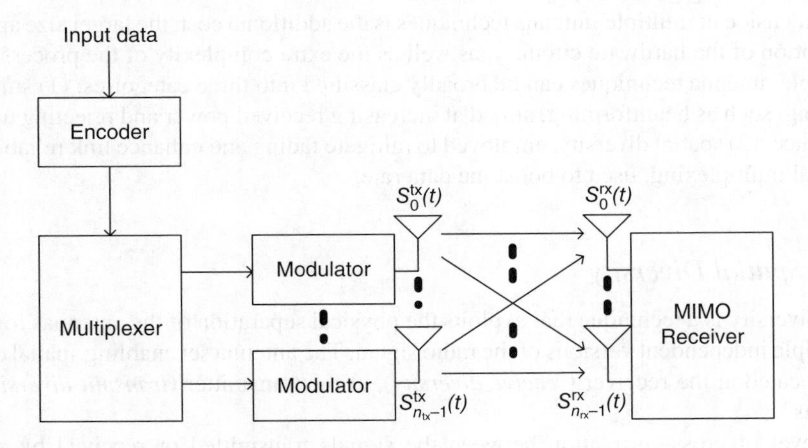

Figure 3.2 A spatial multiplexing system.

equivalent to solving a set of n_{rx} linear equations in n_{tx} unknown variables. In a rich scattering environment, the channels between the pairs of transmit and receive antennas experience different multipath conditions, and thus are statistically independent. In this case, the system is determined and can be solved for $n_{rx} \geq n_{tx}$ (Paulraj et al. 2003).

As a result, spatial multiplexing promises higher throughput compared with single-input single-output (SISO) systems, without the need to increase the system bandwidth or the transmit power. As Foschini and Gans (1998) have shown, under certain circumstances the system capacity increases linearly with the number of antennas used.

3.3.3 Gains Obtained by Exploiting the Spatial Domain

In order to address the possible gain mechanisms of multiple-antenna transceivers, let us first define a nomenclature for the different transceiver architecture configurations. SISO refers to a system with one transmit and one receive antenna, single-input multiple-output (SIMO) refers to one antenna at the transmitter and multiple antennas at the receiver, multiple-input single-output (MISO) refers to the converse and MIMO indicates an architecture with multiple antennas both at the transmitter and at the receiver.

In the following, we shall refer to the numbers of transmit and receive antennas as n_{tx} and n_{rx}, respectively.

3.3.3.1 Array Gain

This term indicates the average increment in signal-to-noise ratio (SNR) derived from the effect of coherent combination of multiple copies of a signal at the transmitter, receiver or both. In the case of SIMO, the signals at the receive array have different amplitudes and phases: when the channel response is known, the receiver can combine the incoming signals coherently, such that the resulting signal power is enhanced. This increase in the average signal power is proportional to n_{rx}. In the case of systems with multiple antennas at the transmitter, knowledge about the channels is again required for exploiting array gain (Paulraj et al. 2003). While estimating the channel response at the receiver is straightforward when a set of reference sequences is used, a feedback message is typically needed for acquiring channel knowledge at the transmitter.

3.3.3.2 Diversity Gain

In a wireless channel, the signal power is subject to fluctuations due to the fading phenomenon; when the signal power decreases by a considerable amount, the channel is said to be in a fade and the signal detection is likely to be compromised. Diversity techniques are intended to counteract the performance degradation due to fading.

In a SIMO channel, receive diversity techniques exploit the fact that different receive antennas see independently faded versions of the same signal. The receiver combines these signals in such a way that the resultant signal has a reduced amplitude variation compared with the signals at each receive antenna (Jakes 1994). The diversity gain, also called the diversity order, is equal to the number of receive antennas in the SIMO case.

In the case of a MISO channel, transmit diversity techniques can be applied. Probably the best-known example of a transmit diversity technique is the space time coding (STC) scheme (e.g., Alamouti 1998; Tarokh et al. 1998), which encodes the data across the spatial domain (with the transmit antennas) and the time domain, and does not make use of any channel knowledge at the transmitter. If the links between all n_{tx} transmit antennas and the receive antenna experience independent fading conditions, the diversity order of the MISO channel is equal to n_{tx}.

In the case of a MIMO channel, one can combine the above-mentioned transmit and receive diversity schemes. When the links between each transmit–receive antenna pair fade independently, then the diversity order becomes $n_{tx}n_{rx}$ (the product of the numbers of transmit and receive antennas) (Paulraj et al. 2003).

3.3.3.3 Multiplexing Gain

Spatial multiplexing can offer a linear increment $\min(n_{tx}, n_{rx})$ in the capacity, with no extra bandwidth or power expense; this gain is called the multiplexing gain, and can only be obtained with multiple antennas at both the transmitter and the receiver sides (Foschini 1996; Paulraj and Kailath 1994; Telatar 1999).

In the case of $n_{tx} = n_{rx} = 2$, that is, a 2×2 MIMO channel, the bit stream to be transmitted is demultiplexed into two half-rate substreams, simultaneously modulated and transmitted from each transmit antenna. Under certain favorable channel conditions, the spatial signatures of these signals induced at the two receive antennas are well separated, and thus a receiver having knowledge about the channel can differentiate between the two co-channel signals and extract both of them. At this point, demodulation yields the original substreams, which can then be combined for retrieving the original bit stream (Paulraj et al. 2003).

While transmit and receive diversity are the means to combat fading, in a MIMO channel fading can in fact be beneficial, leading to an increase in the degrees of freedom available for communication (Foschini 1996; Telatar 1999). Essentially, if the path gains between individual transmit–receive antenna pairs fade independently, multiple parallel spatial channels are created; as a consequence, in a rich scattering environment the independent spatial channels can be exploited to send multiple signals in parallel, at the same time and frequency. This results in higher spectral efficiency, which translates to a higher data rate for the same bandwidth. This effect is called spatial multiplexing (Heath and Paulraj 2000) and is particularly useful in the high-SNR regime, where the system is degrees-of-freedom-limited rather than power-limited. Foschini (1996) has shown that in the high-SNR regime, the capacity of a channel with n_{tx} transmit antennas, n_{rx} receive antennas and i.i.d. Rayleigh-faded gains between each antenna pair is given by

$$C(\rho) = \min(n_{tx}, n_{rx}) \log \rho + O(1), \tag{3.36}$$

where ρ is the SNR and $O(1)$ means that at infinity this component behaves like a constant. The number of degrees of freedom is thus the minimum of n_{tx} and n_{rx}.

3.3.3.4 Interference Reduction

Frequency reuse in wireless networks may give rise to co-channel interference among neighboring cells; a possible way to counteract this effect is to use multiple antennas to distinguish the spatial signature of the desired signal from the signatures of the interfering co-channel signals. Some knowledge about the channel of the desired signal is required to perform interference reduction.

Interference reduction or avoidance techniques can also be implemented at the transmitter; the objective in this case is to maximize the energy sent to the desired user and, at the same time, to minimize the energy sent to the co-channel users. Some positive effects of interference reduction are the possibility of using aggressive reuse factors, leading to an improvement in the network capacity (Paulraj et al. 2003).

3.3.4 MIMO and Mobile Positioning

In recent years various positioning technologies have been proposed using cellular network-based, mobile-based or hybrid approaches. The best-known and most widely used positioning technique is the GPS technique, based on measurements of time of arrival (TOA). The propagation times of signals from satellites at known locations are measured simultaneously, and the distance between a satellite and a user receiver is obtained by multiplying the propagation time by the speed of light, assuming LOS conditions. In most applications, however, the LOS signal is followed by multipath components that arrive at the receiver with a delay, introducing significant errors into the LOS-path TOA and into the gain estimation, especially in urban environments, which are characterized by several reflections from buildings and other objects. MIMO systems can use information from multipath components to improve the accuracy of the estimation. The key issue for positioning in MIMO settings is selecting an appropriate model for the propagation channel. With the proper channel model, the location problem can be solved using sequential Monte Carlo methods, also known as particle filtering (Doucet et al. 2001).

3.4 Duplexing Methods

Mobile radio systems typically involve communication over both directions, that is, from the MS to the BS and vice versa. They can then be classified as simplex, half duplex or full duplex according to the approach adopted for accommodating resources for both transmission directions.

3.4.1 Simplex Systems

In simplex systems, communication is only possible in one direction. Examples of simplex systems are TV and radio broadcasting, where the information flows from one source to many destination devices. Paging systems where the messages are received but not acknowledged are also examples of simplex systems.

3.4.2 Half-duplex

In half-duplex radio systems, two-way communication is possible, allowing for information exchange over both directions. Such systems can be frequency division duplex (FDD) or time division duplex (TDD).

3.4.2.1 FDD

In FDD systems, two-way communication can happen simultaneously but over different carrier frequencies. Separate antennas and radio-frequency circuits can be used for transmitting and receiving, respectively. When a single antenna is adopted, a duplexer is needed to separate from the receiver chain the signals coming from the transmitter circuits. The duplexer can be designed in a cost-effective manner when there is sufficient frequency separation between the transmit and receive carriers (typically, at least 5% of the largest carrier frequency). Filters are needed for separating the uplink and downlink channels. FDD systems are efficient where there is symmetric traffic. FDD is widely used in large cells where the high number of users leads to a large traffic flow aggregation in the uplink, compensating for the traditional disparity between uplink and downlink loads.

However, FDD systems require paired bands for accommodating a bidirectional communication link, and are then inherently frequency inefficient.

3.4.2.2 TDD

In TDD systems, two-way communication occurs over the same frequencies but at different time instants. TDD has inherently lower cost than FDD since no duplexer is needed to isolate the transmitter and receiver chains. TDD can accommodate asymmetric traffic efficiently since the amount of resources to be allocated to the uplink and the downlink can be dynamically adjusted. In this respect, TDD is particularly useful for small cells serving a low number of users, with disparity between uplink and downlink traffic demands. Since the same bandwidth is used for both communication directions, unpaired frequency bands can be exploited, leading to a flexible usage of the available spectrum. Another significant advantage of the TDD mode is the possibility of exploiting the reciprocity of the radio channel, for instance knowledge of the channel response in the uplink direction can be exploited by channel-aware transmission schemes also in the downlink.

However, TDD also has significant drawbacks. It requires tight time synchronization in order to work efficiently, and it may significantly increase the transmission delay in one direction due to the necessity of alternating time uplink and downlink transmissions. Moreover, TDD requires a guard period (GP) for accommodating the switching between the transmission directions. The GP is meant to cope with the turnaround time of the radio circuitry, as well as with the propagation delay. In this respect, the overhead of the required GP may significantly increase for large cells serving users located at a greater distance from the BS.

3.4.3 Full Duplex

In full-duplex systems, transmission and reception occur simultaneously over the same band. Full-duplex transmission was historically considered impractical due to the overwhelming loopback interference from the transmitter chain to the receiver chain of the device. Nonetheless, recent advances in self-interference cancellation (SIC) over both analog and digital domains (Hong et al. 2014) have shown the possibility of suppressing such self-interference power down to tolerable levels, making full-duplex communication feasible with affordable costs. Full duplex is indeed considered a potential breakthrough for upcoming fifth-generation (5G) wireless communication systems.

Ideally, full duplex can double the capacity of a bidirectional radio link. Nonetheless, there are factors that may jeopardize its expected performance. First, SIC is non-ideal and can lead to an non-negligible decrease in the SNR. Second, full-duplex transmission doubles the interference level in a certain band due to the simultaneous transmission and reception. This may severely compromise the performance of interference-limited scenarios such as densely-deployed small cells. Note that full duplex can only be exploited when there is data available simultaneously for both uplink and downlink, which may not be the case for realistic networks serving users with finite buffer queues. Addressing the effective potential of full-duplex transmission in realistic scenarios is currently a hot topic in 5G research.

3.5 Modulation and Multiple-access Techniques

3.5.1 Modulation Techniques

3.5.1.1 Multi-carrier Modulation

In a traditional single-carrier modulation, the information bits are mapped over a predefined symbol constellation, for example quadrature amplitude modulation (QAM) and 16QAM, and such symbols are then transmitted serially over time by modulating a radio waveform at a certain carrier frequency. The transmit signals occupy a certain bandwidth whose dimension depends on the symbol rate. In multi-carrier modulation, the bandwidth is instead divided into a number of small chunks and the constellation symbols are mapped over them. The transmit signal is then a linear combination of symbols transmitted over different frequencies. This allows the possibility of achieving a number of gain mechanisms in terms of multipath mitigation, as will be described next. Figure 3.3 highlights the difference between single- and multi-carrier modulation.

3.5.1.1.1 Orthogonal Frequency Division Multiplexing

Orthogonal frequency division multiplexing (OFDM) is the best-known multi-carrier transmission technique and has been adopted by several current radio access technologies. The multiple frequency chunks where the information symbols are mapped, called *subcarriers*, are placed at the minimum frequency distance, which still guarantees their mutual orthogonality,

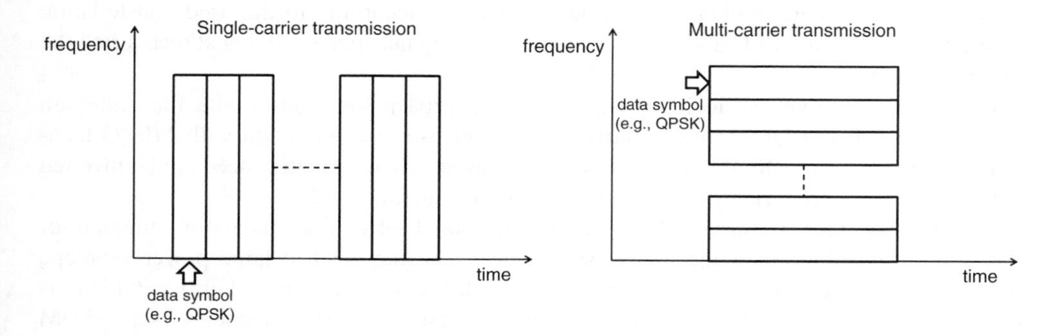

Figure 3.3 Single-carrier vs. multi-carrier modulation.

with the aim of enhancing the spectrum utilization (Prasad et al. 2009). OFDM promises high-user-data-rate transmission capabilities with reasonable complexity and precision.

At high data rates, the data distortion caused by the multipath channel may be significant. In traditional single-carrier systems a very complex receiver structure, which makes use of computationally expensive equalizers and channel estimation algorithms, is needed to correctly retrieve the originally transmitted data. OFDM can drastically simplify the equalization problem by turning a frequency-selective channel into multiple flat channels. In other words, each data symbol is independently affected by the channel and a simple one-tap equalizer is needed to recover the data. Such one-tap equalization is enabled by the insertion of a cyclic prefix, obtained as a copy of the last part of the OFDM symbol, at the beginning. The cyclic prefix translates the linear convolution with the channel to a cyclic convolution, which in turn converts the selective channel in multiple flat channels in the frequency domain (Bolcskei et al. 2002).

Figure 3.4 shows the transceiver chain for an OFDM system, though several of its building blocks can also be used for a single-carrier system. In the transmitter, forward error correction (FEC) coding and interleaving are added to obtain robustness against burst errors. An OFDM system with this addition is referred to as coded OFDM (COFDM). In the digital domain, the binary input data are collected and FEC coded with schemes such as convolutional codes or turbo codes. The coded bit stream is interleaved to obtain diversity gain. After this, a group of channel-coded bits is gathered (one bit per symbol for binary phase shift keying (BPSK) modulation, two for quadrature phase shift keying (QPSK), four for 16QAM, etc.) and mapped to the corresponding constellation points. Known reference symbols mapped with known modulation schemes are then typically inserted to enable channel estimation at the receiver. A serial-to-parallel converter is applied and an inverse fast Fourier transform (IFFT) operation is performed on the parallel complex data. The cyclic prefix is inserted into every block of data according to the system specification and the data is demultiplexed in a serial fashion. At this point, the data is OFDM-modulated and ready to be transmitted. A digital-to-analog converter (DAC) is used to transform the time-domain digital data to analog data. Radio frequency (RF) modulation is then performed, and the signal is up-converted to the transmission frequency.

The receiver down-converts the signal to an intermediate frequency and then converts it to the digital domain by using an analog-to-digital converter (ADC). At the time of down-conversion of the received signal, carrier frequency synchronization is performed, and after ADC conversion symbol-timing synchronization is carried out. A fast Fourier transform (FFT) block is used to demodulate the OFDM signal. After that, channel estimation is performed using the demodulated reference symbols. Using the resulting estimates, the complex received data is obtained and then demapped according to the used constellation diagram. At this point, FEC de-interleaving and decoding take place, aiming at recovering the original bit stream.

OFDM is attractive due to its low-complexity approach for dealing with the multipath channel. This makes OFDM particularly suitable for use in conjunction with MIMO techniques, since for each OFDM subcarrier only a constant channel matrix needs to be inverted at the receiver for retrieving the mapped symbols (see Section 3.3.2).

Nonetheless, OFDM suffers from significant drawbacks. The linear combination of the signal transmitted over the narrowband subcarriers may lead to large power envelope fluctuations, which may reduce the efficiency of the power amplifiers. Further, OFDM is rather sensitive to hardware impairments such as phase noise or frequency offset. OFDM also features high out-of-band (OOB) emission, which may generate significant interference levels over neighbor bands.

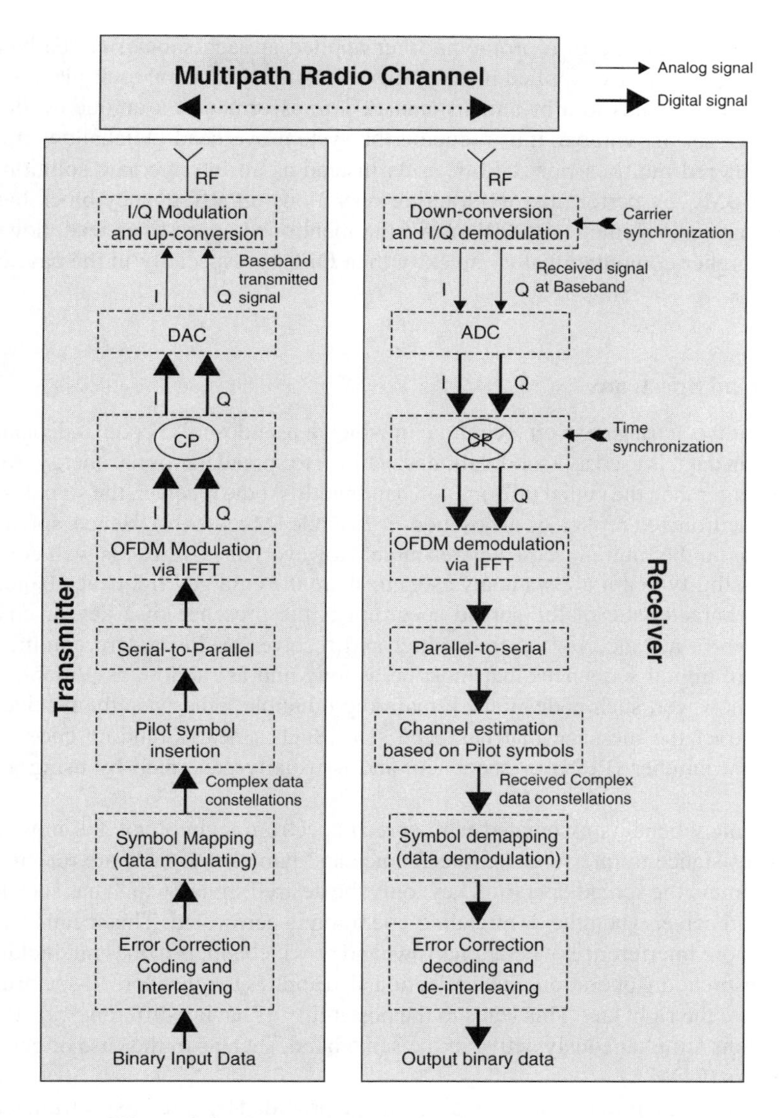

Figure 3.4 OFDM transceiver chain (Prasad et al. 2009).

3.5.1.1.2 Other Multi-carrier Techniques

Radio spectrum is a scarce and expensive resource, and future wireless technologies are expected to improve the efficiency of its utilization, for instance by opportunistically utilizing the spectrum holes in the time instants where transmissions may not occur. The large OOB emissions of OFDM are a significant drawback in that respect due to the power leakage generated over the neighboring bands. In the last few years, a number of novel multi-carrier techniques have been proposed in the literature with the aim of overcoming these OFDM disadvantages. Filter bank multi-carrier (FBMC) waveforms feature an excellent spectral containment due to the use of a prototype filter applied at each subcarrier (Farhang 2011). A pulse-shaping filter is also applied in generalized frequency division multiplexing (GFDM) modulation, where thanks to a bi-dimensional FFT transformation a unique cyclic prefix is used for a large set of symbols, thus reducing the system overhead (Michailow et al. 2012). A universal filtered multi-carrier (UFMC) acts instead as an intermediate solution between OFDM and FBMC by performing the filtering operations on a frequency block basis rather than per subcarrier (Vakilian et al. 2013). All the mentioned approaches lead, however, to a significantly higher computational complexity than OFDM, especially in the case of MIMO transmission.

3.5.1.2 Spread Spectrum

Spread spectrum is a transmission technique in which a pseudo-noise code, independent from the information data, is used as a modulation waveform to spread the signal energy over a bandwidth much larger than the signal information bandwidth. At the receiver, the signal is despread by using a synchronized replica of the pseudo-noise code (Meel 1999). Hence, spreading does not economize on the limited frequency resource. However, this overuse is well compensated for by the possibility that it allows many users to share the enlarged frequency band.

The main characteristic of the spread spectrum is the presence of a key or code, which must be known in advance by the transmitter and receiver(s). In modern communications, these codes are digital sequences that must be as long and as random as possible to appear "noise-like"; however, such codes must remain reproducible, otherwise the receiver will not be able to extract the message that has been sent. Such a nearly random code is called a pseudo-random number (PRN) or sequence, and is usually generated by using a feedback shift register.

There are many benefits associated with spread-spectrum technology, the most important of which is resistance to interference. Interference and jamming signals are rejected because they do not contain the spread-spectrum key: only the desired signal, which has such a key, will be seen at the receiver when the despreading operation is performed. This means that one can practically ignore interference, whether narrowband or wideband, if it does not include the key used in the despreading operation. Such rejection also applies to other spread-spectrum signals that do not have the right key. This leads to the possibility of having different spread-spectrum communications simultaneously active in the same band, such as in the case of code division multiple access (CDMA).

Resistance to interception is the second advantage provided by spread-spectrum techniques; since unauthorized listeners do not have the key used to spread the original signal, they cannot decode the signal. Without the right key, the spread-spectrum signal appears as noise or interference.

The spread spectrum also provides resistance to multipath fading. Since the despreading process is synchronized to the direct-path signal, the reflected-path signal is rejected even if it contains the same key.

Different spread-spectrum techniques are distinguished according to the point in the system at which a PRN is inserted into the communication channel. If the PRN is inserted at the data level, the technique is called direct sequence spread spectrum (DSSS). If the PRN acts at the carrier-frequency level, the technique is called frequency hopping spread spectrum (FHSS). Finally, if the PRN is applied at the local-oscillator stage, FHSS PRN codes force the carrier to change (hop) according to the pseudo-random sequence. Analogously, if the PRN acts as an on/off gate to the transmitted signal, the technique is referred to as time hopping spread spectrum (THSS). One can also mix the above techniques, forming a hybrid spread-spectrum technique. DSSS and FHSS are the spread-spectrum techniques that are mostly used in third-generation (3G) radio systems (Maxim 2003).

3.5.2 Multiple-access Techniques

3.5.2.1 TDMA

In time division multiple access (TDMA), users transmit over the same bandwidth but different time slots (hence the words "time division") (Celusion 2009); this multiple-access technique is used, for instance, in GSM radio access technology.

3.5.2.2 FDMA/OFDMA

In frequency division multiple access (FDMA), the available bandwidth is divided into channels, with each allocated to a different user for transmission (Celusion 2009). As an example of FDMA, we shall mention orthogonal FDMA (OFDMA), which is used in the worldwide interoperability for microwave access (WiMAX) system and in long-term evolution (LTE) downlink.

In OFDMA, the available frequency subcarriers are distributed among the users for transmission. The subcarrier assignment is made for the lifetime of the user connection or at least for a considerable time frame. This scheme was first proposed for cable television systems (Sari and Karam 1996) and later adopted for wireless communication. OFDMA can support a number of identical or different user data rates, for example by assigning a different number of subcarriers to each user. Based on the channel conditions experienced on the group of allocated subcarriers, different modulation schemes can be used for the individual subcarriers, for example QPSK, 16QAM or 64QAM. This is referred to as adaptive subcarrier/subchannel, bit and power allocation or QoS-based allocation (Kivanc and Liu 2000; Rhee and Cioffi 2000; Wong et al. 1999a,b). Frequency hopping can be employed to provide security and resilience to the intercell interference.

In OFDMA, the granularity of the resource allocation is higher than in OFDM-TDMA, where flexibility can be accomplished by suitably choosing the subcarriers associated with each user at different times. Here, the fact that each user experiences a different radio channel can be exploited by allocating only the set of subcarriers experiencing a high SNR to each user. In addition to this, the number of subcarriers for a specific user can be varied according to the required data rate. Thus, a multirate transmission can be achieved without significantly increasing the complexity of the system. Note that subcarriers can also be assigned arbitrarily among the users.

3.5.2.3 CDMA

In a code division multiple access (CDMA) system, as utilized by 3G systems, all users occupy the same frequency and are separated from each other by means of a code. Each user is assigned a code that is used to transform the user's signal into a spread-spectrum-coded version of the user's data stream. The receiver uses the same spreading code in order to transform the spread-spectrum signal back into the original user's data stream (Celusion 2009).

3.5.2.4 SDMA

Space division multiple access (SDMA) is a multiple-antenna-based scheme which enables access by identifying the user location. This enhances the efficiency of network bandwidth utilization, since users can be distinguished depending on their location. SDMA can be applied on top of most of the well-known mobile access schemes such as CDMA, TDMA and FDMA.

SDMA enables the channeling of radio signals based on the mobile devices' locations. In this way, SDMA not only protects the quality of radio signals, avoiding interference from adjacent cells, but also saves unnecessary transmissions in areas where mobile devices are not currently active or available.

SDMA works through simultaneous transmission of streams at the same frequency from a transmitter that is equipped with multiple antennas (potentially, SDMA can be enhanced if the receiver is also equipped with multiple antennas) (SDMA Toolbox for IT 2009).

3.5.2.5 CSMA/CA

Carrier-sense multiple access (CSMA) protocols are well known in the industry and have been adopted in several standards. The popular Ethernet protocol, for example, is a CSMA/collision detection (CD) protocol. A CSMA protocol works as follows: a station desiring to transmit senses the medium; if the medium is not busy (i.e., if some other station is not transmitting), the station is allowed to transmit. This multiple-access technique is utilized by WiFi.

Such protocols are very effective when the medium is not heavily loaded, since they allow transmission with minimum delay. However, there is always a chance of collisions, which occur when other stations sense the medium is free and decide to transmit at the same time. In the case of Ethernet, these collisions are recognized by the transmitting stations, which perform a retransmission phase based on an exponential random backoff algorithm, for example retransmissions are done after waiting after a failed transmission for a time that is chosen randomly from a set of possible waiting times, and the cardinality of this set increases exponentially with time. The usage of random retransmission times reduces the risk of persistent collisions. In a wired LAN, CD mechanisms are based on their own frame's echo. Nonetheless, they cannot be used in a WLAN environment for the following reasons:

- Implementing a CD mechanism would require a full-duplex radio, which has significant costs.
- The basic assumption of a CD scheme is that all the stations can hear each other. This may not hold in a wireless network, since a station willing to transmit may not be able to sense a transmission that is happening towards its intended receiver.

In order to overcome such drawbacks, the 802.11 standard uses a CSMA/collision avoidance (CA) mechanism together with a positive acknowledge scheme. A station willing to transmit senses the medium; if the medium is busy, then the station defers its transmission.

If the medium is free for a specified time, then the station is allowed to transmit; the receiving station will check the cyclic redundancy check (CRC) code of the received packet and send an acknowledgment (ACK) packet. The receipt of the acknowledgment will indicate to the transmitter that no collision has occurred. If the sender does not receive an acknowledgment, then it will retransmit the fragment until it is acknowledged, or discard the packet after a certain number of retransmissions.

In order to reduce the probability of two stations colliding due to the fact that they cannot hear each other, a *virtual carrier sense* mechanism has been defined: a station willing to send a packet first transmits a short control packet called the request-to-send (RTS). If the medium is free, the destination station responds with a control packet called the clear-to-send (CTS). All stations receiving either the RTS or the CTS set their virtual-carrier-sense indicator – called the network allocation vector (NAV) – for a given duration, and use this information together with the physical carrier sense when sensing the medium. This mechanism reduces the probability of a collision in the receiver area caused by a station that is "hidden" from the transmitter for the duration of the RTS transmission, as that station will hear the CTS and "reserve" the medium until the end of the transaction. It should also be noted that since the RTS and CTS are short frames, such a mechanism reduces the overhead of collisions, as the latter are recognized faster than they would be if the whole packet were to be transmitted (Brenner 1997).

3.5.3 OFDMA and Mobile Positioning

As an example of the relation of multiple-access techniques to mobile positioning, we will consider an OFDMA-based system, namely WiMAX. In Jiao et al. (2008), a mobile WiMAX positioning scheme that adopts a layered infrastructure was proposed. There are three function modules, located in the physical layer (PHY), the medium access control (MAC) layer and the application layer, which communicate with each other through a cross-layer mechanism. The PHY module is the key part of the structure, which performs preamble detection and measurement in the MS and selects several approaches for positioning, for example TDOA. The MAC module performs signaling and provides opportunities for channel measurements according to positioning requests from the application layer. Position and optional velocity computations are performed in the application layer based on the measurements. Positioning algorithms are embedded either in the location computation server in the cognitive-service network or in the MS. The communication among the components in the application layer (including the communication between the MS or the cognitive-service network and the location computation server) follows the mobile positioning protocol, which is an Internet-based protocol used by location-dependent applications to interface with a mobile positioning server. Through this protocol, it is possible to request the position of mobile terminals (Rios 2009).

3.6 Radio Resource Management and Mobile Positioning

3.6.1 Handoff, Channel Reuse and Interference Adaptation

A possible solution for increasing the capacity of a cellular system is to adapt to the dynamic changes of the traffic loads or to the interference experienced in the various cells. The aim of such adaptation is to enhance cell reuse without producing a detrimental level of co-channel interference (Chiu and Bassiouni 2000). The average signal strength p (in linear units, i.e., watts) at a distance d from the transmitter is usually modeled as (Zander and Erikkson 1993)

$$p = \frac{10^{\alpha/10}}{d^{\beta}}, \tag{3.37}$$

where α and the path loss exponent β are taken from Equation (3.7). By assuming no noise, this relation can be used to compute the carrier-to-interference ratio (CIR), written as C/I, for the downlink channel. Let BS_0 represent the BS that currently communicates with the MS and let BS_i represent an interfering BS, that is, a neighboring BS which shares the same downlink channel with BS_0. Then, the downlink C/I (excluding the effect of shadowing) is given by

$$\frac{C}{I} = \frac{d_0^{-\beta}}{\sum_{i \neq 0} d_i^{-\beta}} \geq \xi, \tag{3.38}$$

where d_i is the distance between BS_i and the MS, and ξ is a threshold. The above constraint implies that to achieve a good downlink connection, BS_0 should provide the highest average signal level for the MS. Moreover, the signal from BS_0 should be sufficiently strong with respect to the sum of the interference signals from the other BSs. A similar constraint applies to the uplink case.

3.6.1.1 Handoff Prioritization

The mechanism of handoff in cellular networks transfers an ongoing call from the current cell to a neighbor cell as the MS moves through the coverage area of the cellular system. A successful handoff provides continuation of the call, which is crucial to keeping a satisfactory perceived quality of service (QoS) (Chiu and Bassiouni 2000).

Handoff mechanisms are based on the reservation of radio resources. A well-known option is bandwidth reservation using static guard channels (Hong and Rappaport 1986; Tripathi et al. 1998). The shadow cluster concept was proposed by Levine et al. (1997) to estimate future resource requirements and perform admission control, with the aim of limiting the handoff-dropping probability. The potential influence of the active MSs on the channel resources of each cell in the cluster is determined probabilistically by the BS of the current cell based on previous knowledge of the mobility pattern of the active MSs. Lu and Bharghavan (1996) explored mobility estimation for an indoor wireless system based on both mobile-specific and cell-specific observation histories. Talukdar et al. (1997) proposed a bandwidth reservation scheme that assumes that the mobility of a MS can be characterized by the set of cells that the MS is expected to visit during its connection's lifetime, reserving bandwidth in each of these cells, which is often excessive. Choi and Shin (1998) introduced a predictive and adaptive scheme which also performs bandwidth reservation for handoffs. The aggregate history of mobility observed at the cell level is used to predict probabilistically the direction of the MS and the time of the expected handoff.

One problem with all history-based schemes is the overhead related to the creation and update of traffic histories for the multiple cells in the network. Such histories are never fully reliable, since they continuously experience short-term changes (e.g., diversion of traffic due to accidents), medium-term changes (e.g., traffic rerouting during road construction) and long-term changes (e.g., after a shopping center has opened or closed). The predictive channel reservation (PCR) approach (Chiu and Bassiouni 2000) uses real-time position measurements to predict the future path of a MS. The utilization of real-time measurements introduces considerations not addressed in history-based schemes (e.g., the monitoring of changes in motion).

3.6.1.2 Channel Reuse and Interference Adaptation

Based on the C/I constraint, we can view the cell served by a BS as the ensemble of points where the downlink and uplink C/I constraints are satisfied, that is, the area where a good connection can be established. Traditionally, the interference factor has been modeled by a parameter called the *co-channel reuse distance*, which represents the minimum distance between two BSs sharing the use of a common channel. Fixed systems that do not employ adaptive designs usually adopt safe worst-case scenario margins of capacity and reuse distance, with the aim of avoiding severe performance degradation (Chiu and Bassiouni 2000).

3.6.1.3 Predictive Channel Reservation

In the PCR scheme, channel allocation decisions are based on prediction (extrapolation) of the motion of the MSs. Each MS periodically measures its position and orientation. The latter can be easily obtained from the vector of two consecutive position measurements. This information may be sent to the BS in uplink or may be readily available if the positioning is carried out by the BS itself. The BS uses the position/orientation information to predict the projected future path of the MS (Chiu and Bassiouni 2000). When a more advanced positioning technology (e.g., one that provides accurate estimation of velocity and acceleration) is adopted, complex extrapolation methods can be used, similar to the methods developed for predicting the movement of simulated vehicles in networked virtual-reality training exercises (Bassiouni et al. 1997). Based on the projected path, the next cell that the mobile is moving towards is determined. When the MS is within a certain distance from that cell, the current BS sends a reservation request to the new BS in order to preallocate a channel and to accommodate the expected handoff event. A reservation may be deemed invalid (a false reservation) at a later time. In this case, a cancellation of the reservation must be sent to deallocate the reserved channel (Chiu and Bassiouni 2000).

3.6.2 Power Control

The signal-to-interference-plus-noise ratio (SINR) is a limiting performance factor due to its direct relation to the variance of the TOA estimate. The SINR can be increased using two techniques (Prasad et al. 2000): (1) a technique based on the power-up phase in the uplink local positioning (LP) scenario and (2) another based on an idle slot in the downlink LP scenario. However, these techniques have the disadvantage of reducing the network capacity. To avoid this phenomenon, estimation methods resistant to the near–far problem could theoretically be applied directly to the received data to avoid the near–far problem. Unfortunately, in practice these methods are very expensive to implement (Muhammad et al. 2002).

The near–far problem occurs when two transmitters, say A and B, one close to the receiver (e.g., A) and the other far away (e.g., B), transmit at the same time. As the useful transmitted signal of A is seen as noise by B, and vice versa, one can see that the SNR for B is much lower than that for A, assuming that the transmit powers of the two transmitters are comparable. This makes it more difficult, if not impossible, for transmitter B to be understood by the common receiver.

3.7 Synchronization

In current radio access technologies based on scheduled access to radio resources, the MSs are required to be synchronized to the BSs in order to perform coherent data detection. The MSs are typically retrieving the transmission timing from the BSs by a set of reference sequences. Further, network synchronization in a set of neighboring BSs is recognized as an important enabler of interference coordination and mitigation techniques, since the BSs may coordinate their transmission with the aim of limiting their mutual harm.

3.7.1 Centralized Synchronization

The most straightforward approach for synchronizing multiple nodes in a network consists of locking them to a common high-precision reference clock. Nowadays communication networks offer multiple options for centralized synchronization. For instance, in packed switched networks with variable latency, the network time protocol (NTP) exploits a hierarchical system of clock sources, whose first stratum consists of atomic clocks with estimated error of 1 ns per day (Rybaczyk 2005). Synchronization at the lower layers is obtained through the exchange of synchronization packets containing a timestamp; an NTP client estimates the round trip time with the source and adds a correspondent offset to the received timestamp. The accuracy of NTP improves if the client and server have symmetrical nominal delays, which may not often be the case for current packet networks given the multiple routing options. In practice, the accuracy of NTP is in the order of 10 ms.

Similarly to NTP, the IEEE precision time protocol (PTP) uses a hierarchical master-slave architecture for clock distribution through timestamp messages, but it assumes that each intermediate node corrects the local timestamp with the cumulated error at each hop. IEEE PTP performance is unfortunately dependent on the network load; an accuracy of around 2.5 μs can be achieved for a lightly loaded network. The accuracy can considerably improve if PTP is supported by the hardware switches; errors in the sub-μ range are expected in this case, but costs may considerably increase (Ohly et al. 2009).

In current cellular systems, synchronization across multiple BSs can be obtained instead by locking their local clocks to the timing signals transmitted by GPS satellites. Since a cellular base station is typically a stationary receiver, a single GPS satellite needs to be tracked for a correct time/frequency estimation. The synchronization accuracy of GPS is nowadays limited to 340 ns due to a deliberate distortion of the satellite signals for military accuracy. GPS signals are by design extremely weak (–130 dBm in open sky) and therefore may be correctly retrieved indoors only in LOS conditions. In principle, macrocell BSs can relay the synchronization signals received from GPS satellites.

The assisted GPS (AGPS) concept has been proposed with the aim of improving the indoor performance of GPS. Basically, AGPS allows the satellite signals to operate not in a self-contained way, since the amount of data they carry (e.g., orbital coefficient, time offset correction with respect to Coordinated Universal Time) is supplied by a parallel backhaul network. This removes the requirement of demodulating the unknown data when processing the satellite signals, enabling the GPS to operate at signal levels below –145 dBm. However, this may not be sufficient for general indoor deployment. Recently, a hybrid AGPS-PTP concept has been proposed for femtocell networks (Small Cell Forum 2012). Each node recovers time and frequency from the AGPS; however, in case of poor signal quality, it may switch to PTP. All femtocells have a neighbor list that contains information on how to make a cluster, that is, a subnetwork with one or more PTP masters synchronized with GPS and multiple PTP slaves. The network distances between nodes belonging to the same cluster should be rather short in order to maintain good timing. AGPS-PTP can then obtain GPS accuracy but in the worst case it is subject to PTP limitations.

3.7.2 Distributed Synchronization

When a distributed synchronization approach is used, the nodes in a network need to autonomously agree on a common timeline by periodically exchanging messages. Distributed synchronization has been widely studied in the literature in the framework of self-organized networks and is meant to cope with situations in which a high-precision reference clock is not available, for example in indoor situations due to the penetration losses of the GPS signals. A common distinction persists between initial synchronization and runtime synchronization problems. While initial synchronization aims to align the timing functions of multiple nodes, the goal of runtime synchronization is to maintain the time alignment in the long term by compensating the time drifts due to the non-idealities of the hardware clocks. The initial synchronization problem can be considered straightforward for sequential activation of the BSs, since each new node joining the network can simply align its transmission to the already active nodes. However, robust initial synchronization techniques are required for recovering from critical situations such as simultaneous re-activation of the nodes after a power break. The best-known initial synchronization solution is probably the firefly algorithm (Mirollo et al. 1990), which takes inspiration from the observation of the blinking behavior of fireflies in South-East Asia. Each node periodically sends a beacon message according to an inner state function, and during the waiting time it listens to the eventual transmissions from the neighboring nodes. If a certain beacon is received, the node will increase its internal timing by a certain time step ϵ. It has been proved that all the network nodes end up transmitting their beacon at the same instant within a limited time. In its original formulation, firefly synchronization assumes full-duplex transmission and neglects non-idealities such as propagation and processing delays. Further improvements have been proposed to cope with more realistic scenarios and removing the half-duplex constraints (e.g., Tyrrell et al. 2006; Nagpal 2005). However, the convergence of firefly synchronization is ensured when inner functions and step parameters are set according to the network topology: this may jeopardize the definition of self-organized networks. Recently, a different approach based on the circular averaging of the network timings has been proposed (Hu et al. 2008). Basically, each node updates its internal timing with the aim of minimizing the average time difference to its neighbors. This solution does not require any network-level parameter to be set, and has been also proved to be robust to additive node disturbances.

Even though initial synchronization has attracted more attention in academia, runtime synchronization is the key problem hindering the achievement of high-precision timing targets. For example, current commercial voltage-controlled oscillators (VCOs) feature a nominal accuracy of 10 parts per million (ppm); this means that, even when an initial synchronization is achieved, it may take only 5 s before the timing functions of different nodes diverge for an error equal to 10 μs. Several solutions for correcting the clock drift in a master–slave scenario have been proposed. Solutions dealing with network-wide clock corrections are usually based on the local collection of timing tables of multiple nodes (Choi et al. 2010).

In IEEE 802.11, which represents the current de facto local area standard, both initial and runtime synchronization problems are solved with the timing synchronization function (TSF) algorithm (IEEEstd 2002a). The TSF assumes that all the nodes compete in sending beacons within a certain time window; each node computes a random delay before sending its beacon and cancels its transmission when another beacon is received. Moreover, it updates its internal timing with that of the received beacon in case its timestamp is higher than the local timing. The algorithm clearly tracks the fastest node in the network. However, its accuracy is obviously dependent on the network size, since a large number of nodes may increase the number of collisions and relax the pace of the clock update.

The clock sampling–mutual network synchronization (CS-MNS) algorithm proposed in Rentel et al. (2008) provides an attractive solution for the clock update, since it assumes that each node is able to update its clock timing at each frame on reception of a single beacon, rather than keeping track of all its neighbors before each update. The clock correction procedure is based on automatic control principles. CS-MNS convergence has been proven analytically for a two-node network, and with a larger network with Monte Carlo simulations. The shorter convergence time with respect to IEEE 802.11 TSF has also been proven.

3.8 Cooperative Communications

Cooperative wireless networks aim to improve the delivered QoS by enabling cooperation among multiple wireless agents. Each network node, for example a BS or an MS, is not only expected to transmit or receive dedicated data, but also acts as a cooperative agent for other entities in the network (Figure 3.5).

Cooperation leads inevitably to a reduction in the power allocated to the user data since part of the radio resources are spent on exploiting the cooperative tasks. However, the presence of multiple agents conveying the same information boosts the diversity gain at the end receiver, with significant SNR benefits. Similarly, while each user needs to reduce part of its own data rate for conveying other user messages, the final data rates can be improved due to the cooperation diversity, which allows the channel code rates to be increased (Nosratinia et al. 2004).

The best-known cooperation techniques are decode and forward, amplify and forward, and coded.

Decode and forward refers to a scheme where the cooperative agent detects the message of the cooperative user, encodes it again and retransmits it. This approach exploits the error correction capabilities of the cooperative agent for re-transmitting an error-free message and therefore improves the chances of correct detection at the final communication node. The obvious drawbacks are latency due to the necessity of decoding and re-encoding the message of the user, as well as the risk of error propagation in the case of incorrect detection (Laneman 2001).

In the amplify and forward methods, the cooperative agent transmits an amplified version of the retrieved noisy version of the signal transmitted by its partner. Such noise enhancement

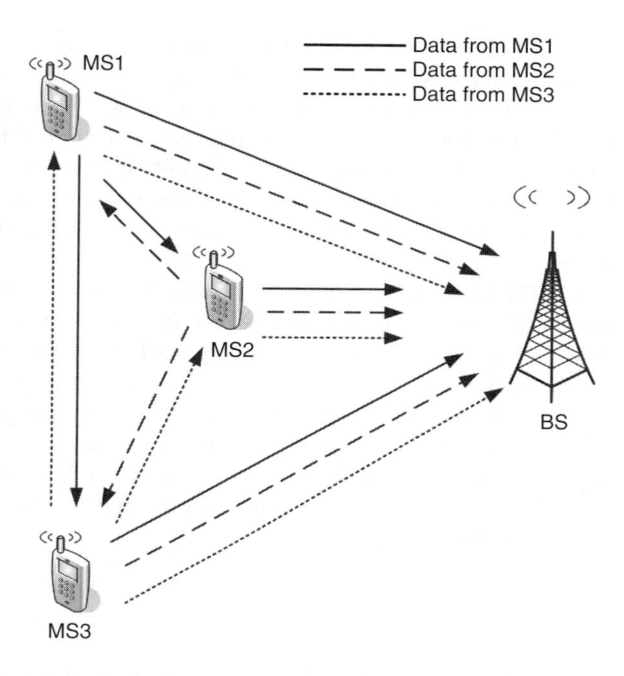

Figure 3.5 An example of a cooperative communication scenario.

is counteracted by the diversity gain due to the multiple receive copies of the same signal. Further, the advantage of this approach is the low latency due to the minimal processing run by the cooperative agent.

Coded cooperation refers to more complex solutions where the cooperative agents transmit part of the codeword of the partner, with the aim of increasing the coding gain at the receiver (Hunter et al. 2006).

3.8.1 Cooperative MIMO

Historically, cooperative communication has been conceived as a solution for harvesting the spatial diversity benefits of MIMO transmission in a network of devices equipped with a single antenna. The evolution of the technology components and the transition towards higher carrier frequencies makes the placement of multiple antennas in the same hand-held device possible, paving the way for cooperative MIMO applications (Nguyen et al. 2013).

Cooperative MIMO offers several gain mechanisms with respect to individual MIMO links. First, cooperation allows the multiplexing gain to be boosted, which in an individual MIMO link is upper-bounded by the number of antennas per node. In cooperative MIMO, the number of antennas which can serve a certain user is not fixed and can be dynamically chosen. Ideally, it can equal the total number of antennas in the network. Nonetheless, such large multiplexing gain comes at the expense of a significant increase in the inter-node signaling for coordinating the operations of the nodes.

Cooperative MIMO also allows the diversity gain to be boosted by transmitting multiple correlated versions of the same signal by a number of cooperative nodes. Distributed space-time block codes (STBCs) are a variant of single-link STBCs where the codewords are stored at different nodes, which jointly encode the signal before forwarding it to the next node (Laneman 2003). Similarly to the multiplexing gain, exploiting diversity gain also requires significant signaling overhead. A further benefit of the cooperative MIMO diversity gain is the possibility of extending the transmission range, enhancing the network connectivity or improving the coverage.

Finally, cooperative MIMO has the potential of combating the increased interference levels which affect the capacity of, for example, dense networks. Information theorists have pointed out that in dense networks the capacity of each link decreases in proportion to the root square of the node density. Interference alignment and interference cancellation techniques are recognized as effective approaches for managing interference. Interference alignment techniques aim to coordinate the transmissions of the multiple nodes such that an unintended receiver perceives the interference from the multiple streams as an unique interfering stream. This allows the MIMO degrees of freedom, which can be used for suppressing such interference, to be relinquished at the unintended receivers.

3.8.2 Clustering

In communication networks, the concept of clustering refers to a logical grouping of nodes sharing common tasks and activities. In this respect, clustering is of fundamental importance in wireless sensor networks, characterized by a large set of inexpensive low-power nodes deployed in a certain geographical area. Sensors collect data on the environment and need to establish dedicated networks to transmit their data to a BS. The nodes collaborate to get data and extend the operating lifetime of the entire system. Clustering is meant to provide network scalability and energy-efficient communication by reducing transmission overheads and enhancing reliability. Moreover, clustering can stabilize the network topology at the level of sensor nodes, thus saving topology maintenance overheads. A *cluster head* is responsible for the formation of the clusters and maintenance of the topology of the network, as well as for the resource allocation for all the nodes in the cluster. Due to the dynamic nature of the mobile nodes, their association and dissociation to and from clusters may perturb the stability of the network. In this respect, a good clustering scheme should preserve its structure over time, avoiding frequent reconfigurations.

We can distinguish between centralized and self-configuring cluster formation. In centralized cluster formation, the BS is in charge of electing one or more cluster heads. Centralized protocols often need sensor nodes to be equipped with high-sensitivity global positioning receivers to gather position information on sensor nodes. In self-configuring cluster formation, the cluster head is elected by the sensor nodes using a distributed algorithm. The advantage of the latter approach is that long-distance communication with the BS is not required, and the cluster formation can be done without the exact location information of the sensor nodes in the network. Further, no global communication is needed to set up the clusters. Several algorithms for self-configuring clustering have been proposed in literature, the best-known being the low-energy adaptive clustering hierarchy (LEACH), which is shown to achieve good performance in terms of system lifetime, latency and application perceived quality (Heinzelman et al. 2002).

3.8.3 Cooperative Routing

In cooperative ad hoc networks, mobile nodes are arbitrarily located and communicate with each other through one or more cooperative agents. A routing protocol is needed to discover routes between a pair of nodes such that the information can be efficiently delivered. This route construction should be done with minimum overhead and bandwidth consumption. An efficient routing is expected to have a significant impact on the transmission power since the signals can only travel through shorter distances and/or high-quality paths. Energy saving can further improve with larger network/clusters, given the higher number of routing options. In general, finding the minimum energy route between two nodes is equivalent to finding the shortest path in a graph in which the cost associated with a link between two nodes is proportional to the distance between those nodes raised to the path-loss exponent. In wireless networks, path breakage can occur more often than in wireline network due to time-variant fading, shadowing and interference, as well as node mobility and power failure. In this respect, rerouting or alternative routing may be necessary and should be performed promptly in order to reduce the communication latency towards the final end point (Khandani et al. 2007).

3.8.4 RSS-based Cooperative Positioning

Cooperative positioning relies on received signal strength (RSS) data for both the BS–MS and the MS–MS links. Suppose MS_1 is the target MS, $\{X^{(i)} = [x^{(i)}, y^{(i)}]^T, i = 1, \dots, n_{ms}\}$ are the coordinates of the MS and $\{X^{[j]} = [x^{[j]}, y^{[j]}]^T, j = 1, \dots, n_{bs}\}$ are the coordinates of the BSs. Similarly to the case of noncooperative positioning described in Section 3.2.5, Soork et al. (2008) have proposed a least-squares objective function as follows:

$$\mathcal{J}(X) = \sum_{i=1}^{n_{ms}} \sum_{j=1}^{n_{bs}} [p^{(i)[j]} - \alpha^{(i)[j]} + 10\beta \log(d^{(i)[j]})]^2 + \sum_{i=1}^{n_{ms}} \sum_{m=i+1}^{n_{ms}} [p^{(i)(m)} - \alpha^{(i)(m)} + 10\beta \log(d^{(i)(m)})]^2, \quad (3.39)$$

where $p^{(i)[j]}$, $\alpha^{(i)[j]}$ and $d^{(i)[j]}$ are, respectively, the received power, the α term and the distance for the link MS_i–BS_j. Similarly, $p^{(i)(m)}$, $\alpha^{(i)(m)}$ and $d^{(i)(m)}$ are, respectively, the received power, the α term and the distance for the cooperative link MS_i–MS_m.

The optimum point of $\mathcal{J}(X)$ in Equation (3.39), which is again obtained numerically, gives a position estimate for all MSs. As before, for further information about least-squares methods and the associated formalism, see Section 5.2.

3.9 Cognitive Radio and Mobile Positioning

Conventional static spectrum allocation results in a low spectrum efficiency when considering the continuously increasing demand for bandwidth. This has encouraged research in the field of cognitive radio (CR), which is seen as a promising technology for maximizing spectrum utilization. Fundamentally, CR tries to utilize radio resources by respecting the following rules: (1) sensing the spectrum utilization in the surrounding environment and (2) making intelligent decisions on its usage. Mitola and Maguire (1999) introduced the concept of the cognitive

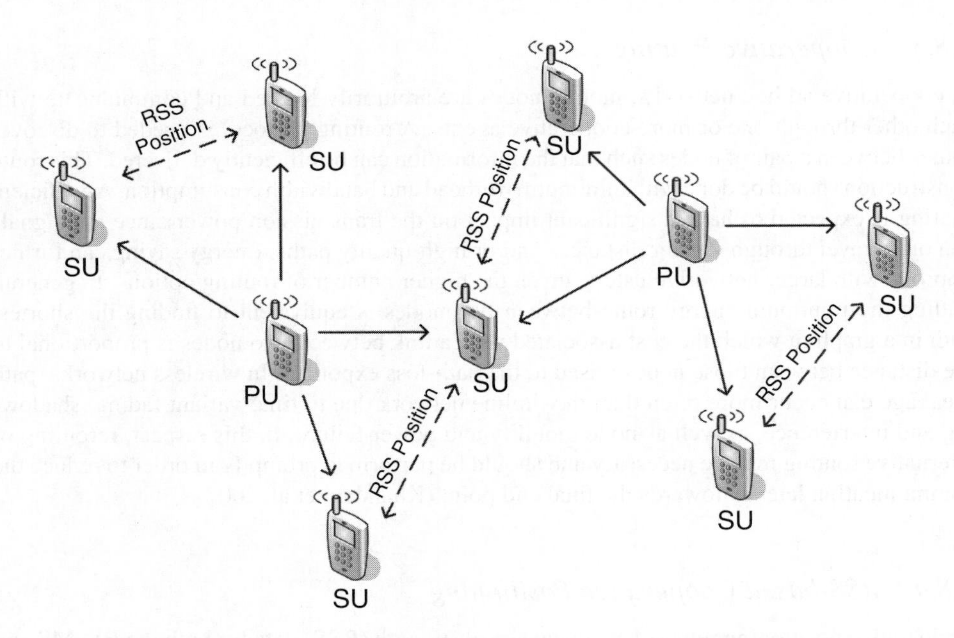

Figure 3.6 Cognitive radio and mobile positioning. PU, primary user; SU, secondary user.

cycle, which consists of *radio scene sensing and analysis*, including *channel state estimation and prediction*, and *action* to transmit a signal. All actions in CR are based on the result of radio scene sensing and analysis, which analyzes RF stimuli from the radio environment, and then identifies the white spaces (available unused portions of the spectrum) and the position and transmission parameters of the various systems in use.

In the CR framework, two types of users are typically considered: primary users, who have priority to access the spectrum, and secondary users, who have no priority or a lower priority than primary users. Hence, one of the challenging issues in CR is how to identify and analyze a primary user and the systems that coexist with a CR system (Kim et al. 2008).

Kim et al. (2008) proposed a scheme to be used by secondary users to estimate the unknown primary users' positions and transmission power, under the following assumptions (Figure 3.6):

1. The secondary users' positions are known.
2. The secondary users measure RSS values from primary users.
3. At least four secondary users receive a signal from each primary user (this assumption will be justified later, when the equations of the model are presented).
4. The shadowing effect is independent for each secondary user.

Let the real position of the primary user be $X = [x, y]^{\mathrm{T}}$ and the coordinates of the ith secondary user be $X_i = [x_i, y_i]^{\mathrm{T}}$, $i = 1, \ldots, n$, where n is the total number of secondary users receiving the primary user's signals. Let d_i denote the distance between the primary user and the ith secondary user:

$$d_i = \sqrt{(X - X_i)^{\mathrm{T}}(X - X_i)}, \quad i = 1, \ldots, n. \tag{3.40}$$

The ideal RSS or received power at the ith secondary user, denoted by p_i, is expressed as

$$p_i = RSS^{(i)} = \phi_i \frac{p_i^{\text{tx}}}{d_i^{\beta}}, \quad i = 1, \dots, n, \tag{3.41}$$

where p_i^{tx} is the transmission power at the transmitter, ϕ_i represents all other factors that affect the received signal power, including antenna gain and antenna height, and β is the path loss exponent. We take the shadowing effect into account by using a log-normal path loss model:

$$p_i = RSS^{(i)} = \phi_i \frac{p_i^{\text{tx}}}{d_i^{\beta} \eta_i^{\text{shad}}}, \quad i = 1, \dots, n, \tag{3.42}$$

where $\eta_i^{\text{shad}} = 10^{0.1\eta_i}$ is a log-normal random variable and η_i is a Gaussian random variable with zero mean and variance σ^2 (Kim et al. 2008).

The RSS values at each user have severe disturbances caused by the shadowing effect, so after measuring the raw RSS values we have to refine the sample data by using the sample mean:

$$\overline{p_i}[\text{dBm}] = \frac{\sum_{k=1}^{m} \text{RSS}_k^{(i)}[\text{dBm}]}{m}, \tag{3.43}$$

where $RSS_k^{(i)}$ is the kth sample RSS value at the ith secondary user in dBm, and m is a predefined number of samples. Then, we can use a measurement model such as

$$r_i = \frac{\phi_i}{\overline{p_i}}, \tag{3.44}$$

where $\overline{p_i}$ is the sample mean of the RSS values at the ith node and ϕ_i is the propagation factor in Equations (3.41) and (3.42). The measurement model is based on the sample mean of the RSS values at each node. The above preprocesses reduce the shadowing disturbances in the RSS values in Equation (3.42).

When the number of samples for the measurement is m, the relationship between the measurement and the propagation model of Equation (3.42) can be expressed as

$$r_i = \frac{\phi_i}{\overline{p_i}} \approx \frac{\phi_i}{\text{RSS}^{(i)}} = \frac{d_i^{\beta}}{p_i^{\text{tx}}} \eta_{\text{shad}i}, \quad i = 1, \dots, n, \tag{3.45}$$

where $\eta_{\text{shad}i}$ has the same meaning as in Equation (3.42) (Kim et al. 2008).

If there is no disturbance in the RSS measurement (no shadowing effect), Equation (3.42) can be expressed as

$$\frac{\phi_i}{RSS^{(i)}} = \frac{d_i^{\beta}}{p_i^{\text{tx}}} = \frac{[(x - x_i)^2 + (y - y_i)^2]^{\beta/2}}{p_i^{\text{tx}}}, \quad i = 1, \dots, n. \tag{3.46}$$

By raising both sides of Equation (3.46) to the power of $2/\beta$, after some manipulations we can obtain

$$x_i^2 + y_i^2 = 2xx_i + 2yy_i + \rho \left(\frac{\phi_i}{RSS^{(i)}} \right)^{2/\beta} - r, \quad i = 1, \dots, n, \tag{3.47}$$

where $r = x^2 + y^2$ and $\rho = (p_i^{\text{tx}})^{2/\beta}$. Equation (3.47) holds for each secondary user that is receiving the primary user's signal. We can then express Equation (3.47) in matrix form,

$$HX = B, \tag{3.48}$$

where

$$H = \begin{bmatrix} 2x_1 & 2y_1 & \left(\frac{\phi_1}{\text{RSS}^{(1)}}\right)^{2/\beta} & -1 \\ 2x_2 & 2y_2 & \left(\frac{\phi_2}{\text{RSS}^{(2)}}\right)^{2/\beta} & -1 \\ 2x_3 & 2y_3 & \left(\frac{\phi_3}{\text{RSS}^{(3)}}\right)^{2/\beta} & -1 \\ 2x_4 & 2y_4 & \left(\frac{\phi_4}{\text{RSS}^{(4)}}\right)^{2/\beta} & -1 \end{bmatrix}, \tag{3.49}$$

$$X = \begin{bmatrix} X \\ \rho \\ r \end{bmatrix} = \begin{bmatrix} x \\ y \\ \rho \\ r \end{bmatrix} \tag{3.50}$$

and

$$B = \begin{bmatrix} x_1^2 + y_1^2 \\ x_2^2 + y_2^2 \\ x_3^2 + y_3^2 \\ x_4^2 + y_4^2 \end{bmatrix}. \tag{3.51}$$

Since Equation (3.50) expresses that four parameters need to be estimated, we can justify the earlier assumption that at least four secondary users receive the signal from each primary user. The above estimation problem can be solved by using the LS method, with the aim of minimizing the disturbance caused by shadowing to the measurements. Let $\hat{X}$ be the estimated position; then the solution can be computed as

$$\hat{X} = \arg\min_{X} \left\{(HX - B)^{\text{T}}(HX - B)\right\} \tag{3.52}$$

$$= (H^{\text{T}}H)^{-1}H^{\text{T}}B. \tag{3.53}$$

In Kim et al. (2008), a constrained, weighted optimization method to solve Equation (3.48) taking into account disturbances in measurements was also proposed.

3.10 Conclusions

In this chapter, the basics of wireless communications and their relation to mobile positioning have been discussed. The radio propagation scenario has been introduced and the various physical phenomena that underlie wireless communications have been classified. We have introduced the most known multiple-antenna techniques, highlighting the possibilities that they offer in terms of mobile positioning. The duplexing methods for two-way communication have been presented. Modulation and multiple-access techniques considered relevant for future wireless positioning systems have been reviewed. The principles of network synchronizations have been discussed for both centralized and distributed approaches. Some radio resource management techniques have been presented, and we have outlined how localization can benefit from them, and vice versa. Finally, two of the emerging areas in wireless communications, cooperative communications and cognitive radio technology, have been discussed, and their relation to mobile positioning has been highlighted.

4

Fundamentals of Positioning

João Figueiras
Aalborg University, Aalborg, Denmark

4.1 Introduction

Whereas previous chapters have presented position-related aspects of wireless services and wireless communications, the current chapter will focus on positioning methods. Positioning in wireless networks is possible by exploiting the characteristics of the propagation signals. Certain characteristics of the signal must be measured, such as the received power, in order to permit the estimation of position. The strict requirement is that the measured information has to have a physical relation to the position of the device; otherwise, the information is irrelevant for positioning purposes. The relation between the measurements and the position is always corrupted by a noise component, adding unavoidable complexity to a positioning system. For this reason, it is necessary to have mathematical mechanisms that are able to manipulate the measurements in order to extract position coordinates. The more advanced methods include a statistical treatment of the errors in the measurements.

This chapter starts by describing the types of measurements that are commonly possible in typical wireless communication systems, and it then introduces some techniques for estimating position based on those wireless measurements. The chapter continues by describing some of the errors that exist in the measurements and presenting some metrics of accuracy commonly used within the context of positioning.

4.2 Classification of Positioning Infrastructures

Positioning systems can be classified according to several different key characteristics. The typical classification concerns the topology of the system, in particular the methodology used to obtain measurements and subsequently process them in order to determine location information. Some alternative classifications are, for instance, related to the properties of the communication link or to the *type of integration into the communication technology*.

Mobile Positioning and Tracking: From Conventional to Cooperative Techniques, Second Edition.
Simone Frattasi and Francescantonio Della Rosa.

In the classification based on the communication link, systems are grouped according to their similarities with respect to the propagation conditions. The classification according to the type of integration groups systems depending on the approach used to integrate positioning systems and solutions into the communication technology.

4.2.1 Positioning-system Topology

As defined by Drane et al. (1998) and Vossiek et al. (2003), the classification of systems with respect to system topology groups systems according to where measurements are made and where those measurements are processed. Figure 4.1 shows this classification. In general, the entity responsible for the measurements and the entity responsible for its subsequent processing are not necessarily the same. When the measurements are made and processed in the mobile station (MS), the system is classified as a self-positioning or mobile-based positioning system (Figure 4.1(a)). In order to allow the MS to self-position itself, it is necessary to inform the MS about the location of the base station (BS). This location can either be widely known in the network or be sent by each BS during the MS–BS communication. The major advantage of this solution is that it is scalable and simple, but it does not allow tracking of the MS from the point of view of the network. When measurements are both obtained and subsequently processed on the network side, the solution is referred to as remote positioning or network-based

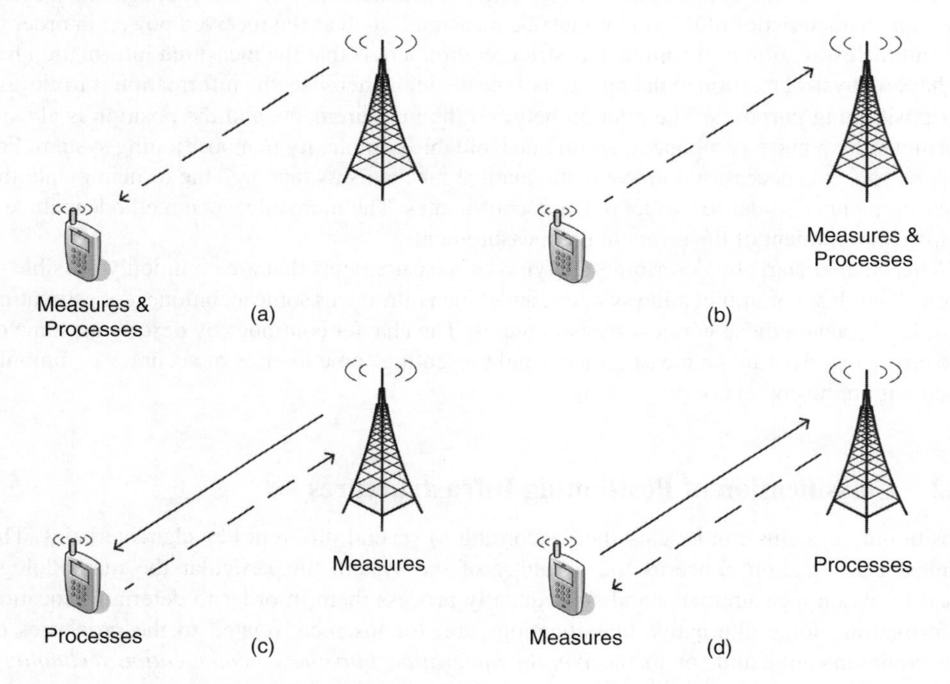

Figure 4.1 Classification according to system topology: (a) self-positioning; (b) remote positioning; (c) indirect self-positioning; (d) indirect remote positioning. Dashed lines represent a communication signal used for measuring the wireless channel, and solid lines represent an actual transfer of measured data.

Table 4.1 Classification of positioning solutions based on system topology.

Concept	Description
Self-positioning	Measurement and processing done by the MS
Remote positioning	Measurement and processing done by the network
Indirect self-positioning	Measurement done by the network and processing done by the MS
Indirect remote positioning	Measurement done by the MS and processing done by the network

positioning (Figure 4.1(b)). The system works in such a way that signals emitted by the MS or reflected from it are received by the network BSs. In this case the position solution suffers from scalability problems because the network may be overloaded with signaling and processing when the number of MSs increase. When measurements are obtained in one entity but processed in another, the systems are said to be "indirect", either an indirect self-positioning system (Figure 4.1(c)) if the network measures the signals and the MS processes them, or an indirect remote positioning system (Figure 4.1(d)) in the other case. These solutions are also referred to as mobile-assisted positioning. Owing to the differentiation between the entities that measure and the entities that process information, it is necessary to establish a temporary link between the MS and the BS so that information can be exchanged.

Table 4.1 shows a summarized description of this classification. Despite the class of the system, position information may be needed in any element of the network, meaning that it may be necessary to establish additional links after the estimation of position so that the information is exchanged.

4.2.2 Physical Coverage Range

Another type of classification concerns the physical coverage range of the communication technology. This classification groups systems according to the properties of the propagation conditions of the signals. Thus, the classification is generally into satellite, cellular and short-range positioning. Table 4.2 shows an overview of each solution. Satellite solutions are based on space–earth communication. For this reason, problems such as propagation in the ionosphere have to be considered. Regarding cellular solutions, the most important problems concern non-line-of-sight (NLOS), scattering, shadowing and multipath effects. For short-range positioning, the propagation statistics are considerably dependent on the scenario.

Satellite positioning systems, such as the Global Positioning System (GPS), are aimed at providing positioning information on a global scale. The satellites, orbiting around the Earth, have a communication range of *virtually everywhere*. Of course, it is necessary to take into account the limitations of such systems, such as their inability to position users when they are indoors. Cellular positioning concerns methods for positioning MSs within the range of technologies such as GSM and Universal Mobile Telecommunications System (UMTS) (see Chapter 8). For these systems, the communication is limited by the range of the cells. The propagation conditions are commonly urban scenarios. The short-range positioning solutions

Table 4.2 Classification of positioning solutions based on physical coverage range.

Concept	Description
Satellite	Positioning solution requires satellite communication
Cellular	Cellular networks are used for obtaining position information
Short-range	Positioning in WLAN/WPAN networks

run over wireless local area network/wireless personal area network (WLAN/WPAN) technologies such as IEEE 802.11 or Bluetooth. For short-range communications, the typical scenario is indoors.

In this classification, the radio frequency identification (RFID) technology can be considered as an extra case, since propagation conditions are rarely treated in the context of positioning information (Section 8.4.2.1).

4.2.3 Integration of Positioning Solutions

As suggested by Figueiras (2008), positioning solutions can also be classified according to the way that the positioning system is integrated into the communication technology. Systems can be classified, according to Table 4.3, into integrated, opportunistic and hybrid systems. The integrated solutions relate to stand-alone positioning systems that are adapted to interact with communication solutions. The current integration of GPS receivers into smartphones is a typical case of this class of solutions. The second group, the opportunistic solutions, are generally implemented in software. These solutions exploit the properties of the communication technology in order to estimate the user's position. In communication networks making use of power control techniques, these solutions can be easily applied because the technology itself has already implemented a mechanism for measuring signal strength. The price to pay is that the opportunistic solutions are commonly less accurate than integrated solutions. Finally, the hybrid solutions make use of both integrated and opportunistic solutions in the same system. These hybrid solutions are appropriate for combining results or for adapting the positioning scheme according to the most accurate solution given the conditions.

Table 4.3 Classification of positioning solutions based on the type of integration.

Concept	Description
Integrated	The positioning system is integrated with the wireless communication technology
Opportunistic	The available communication links are used opportunistically for estimating location information
Hybrid	Both of the above types are used in a cooperative fashion or alternately, depending on the scenario/conditions

4.3 Types of Measurements and Methods for their Estimation

Despite the large number of different positioning systems and solutions for wireless networks, the types of measurements used to determine user positions are rather few. The measurements can be BS-dependent identity labels and/or link-dependent properties such as received signal strength (RSS). Whatever their type, these measurements are the output of the measurement layer defined by the location stack in Section 2.2.

4.3.1 Cell ID

Cell identification, also called cell ID (or CID), is a technique for localization which estimates the location of the MSs as the position of the fixed reference points (FRPs) which the MSs are connected to. In practice, "cell ID" means an actual measurement obtained in the positioning system. This cell ID is then indexed in a database in order to determine the position and the accuracy range of the FRP. The accuracy of this technique is therefore limited by the physical communication range of the FRPs. While in short-range networks the physical range is some tens of meters, in cellular networks the range can be some few thousands of meters. Thus, if no additional information is available, the accuracy of the system is given by the communication range of the wireless technology. Also, in order to perform unambiguous positioning, the FRPs need to be uniquely identified in the whole network. For instance, in a WLAN or Bluetooth network, the medium access control (MAC) address is a commonly used identifier. In cellular networks, FRPs can be identified by their mobile country code (MCC), mobile network code (MNC), local area code (LAC) and cell ID. In theory, for WLAN networks the MAC address is virtually unique (IEEE 2002b), but in practice the address can be programmed and changed by software (Kingpin 1998). Regarding the identifier in cellular networks, it is common that not all of the required fields are available. This can introduce some indecision into the identification of the FRP. Additionally, it is common that FRPs in cellular systems have several (commonly three) sections, which increases the complexity of the system.

Cell ID is often used in combination with other types of information, such as range and/or angle measurements, coverage map information or cartographic mapping (Gustafsson and Gunnarsson 2005). One of the major problems with such information is that the positions of the FRPs are typically treated as confidential by the network operators. This implies that such kinds of solutions must often be used in an opportunistic fashion, where the localization service is provided by third-party entities.

4.3.2 Signal Strength

The RSS is a measure of the magnitude of the power of the signal that the receiver sees at its terminals. Owing to the destructive propagation effects presented in Section 3.2, this measure is typically subject to a noise component with variations of tens of dB for the typical scenarios of interest, that is, indoor short-range localization and outdoor long-range localization.

The RSS can be modeled by a constant component and a variable component. Generally, the constant component is modeled by a path loss propagation model and the variable component is modeled by a range of complex propagation effects (see Section 3.2). These propagation effects can be, for instance, signal attenuation, shadowing, multipath effects, scattering and diffraction. For an overview of these propagation effects see Section 3.2, and for a deeper insight see, for example, Parsons (2000) and Goldsmith (2005).

To model the path loss effect, the following equation is considered:

$$p = \alpha - 10\beta \log(d), \tag{4.1}$$

where p is the received signal strength, α is a parameter dependent on the transmission power, additional losses and other system-component-dependent constants, β is the path loss exponent, and d is the physical length of the communication link, that is, the distance between transmitter and receiver.

Regarding the variable component, it is complex to obtain a model that takes into account the stochastic behavior of the propagation effects involved. The reason is that these effects strongly depend on the scenario. Thus, the most generic approach is to consider all these effects as a single additive Gaussian-distributed noise component with a standard deviation of σ_{shad}:

$$p = \alpha - 10\beta \log(d) + \eta_{\text{shad}}, \quad \eta_{\text{shad}} \sim \text{Norm}(0, \sigma_{\text{shad}}). \tag{4.2}$$

Although this noise component is generally considered as Gaussian throughout the present section, more detailed models are considered in Section 4.5.1.

4.3.3 *Time of Arrival*

The time of arrival (TOA) is the timestamp that a receiver sees on its internal clock when a signal is received at its terminals. With this type of measurement, one of the factors that most influences the error that corrupts TOA observations is the granularity of the internal clock. Given that the clock increments only in the instants when a clock tick occurs, any signal trigger between two consecutive clock ticks will result in the same timestamp reading. Thus, for every instant within this period, the estimated link length will be the same. In terms of distance, the maximum error due to clock granularity can be easily calculated as the quotient of the speed of light c and the clock frequency f:

$$\varepsilon_{\text{clock}} = \frac{c}{f}. \tag{4.3}$$

When this error has the same magnitude as the typical length of the communication channel, it becomes unfeasible to estimate the link length based on TOA measurements. Although the TOAis strongly dependent on the technology, it is common to consider it, in general, as modeled by the sum of a constant component and a variable noise component.

The constant component t of the TOA model is equivalent to the sum of the transmission time t_{tx} (the timestamp in the transmitter clock when the signal is transmitted) and the propagation delay (the time that the signal needs to travel from the transmitter to the receiver):

$$t = t_{\text{tx}} + \frac{d}{c} + \varepsilon_{\text{sync}}. \tag{4.4}$$

In Equation (4.4), c is the propagation velocity in the medium, commonly the speed of light, and d is the physical length of the communication link. The component $\varepsilon_{\text{sync}}$ is a deterministic quantity that compensates any mis-synchronization between the transmitter and the receiver. Note that Equation (4.4) requires the transmitter and the receiver to be either clock synchronized (where $\varepsilon_{\text{sync}} = 0$) or nonsynchronized but with a known difference (where $\varepsilon_{\text{sync}} \neq 0$), otherwise the transmitted timestamp would not have the same reference as the received timestamp. A major complexity related to TOA is knowing $\varepsilon_{\text{sync}}$, which often is not possible to obtain. This can represent a strong disadvantage of TOA observations because clock-synchronizing a whole network is commonly complex to implement and expensive to maintain.

Regarding the variable component, it is possible to divide the problem into clock granularity and propagation effects. While the clock granularity introduces a uniform error between 0 and the value given by Equation (4.3), the propagation effects require a complex treatment, owing to the unpredictability of the impact that such effects have on communications. In other words, the propagation effects depend strongly on the scenario considered. For this reason, the error due to these effects can be generalized and simplified into a Gaussian distribution:

$$t = t_{\text{tx}} + \frac{d}{c} + \varepsilon_{\text{sync}} + \eta_{\text{clock}} + \eta_{\text{prop}}, \qquad \begin{cases} \eta_{\text{clock}} \sim \text{Unif}(0, c/f), \\ \eta_{\text{prop}} \sim \text{Norm}(0, \sigma_{\text{prop}}), \end{cases} \tag{4.5}$$

where σ_{prop} represents the standard deviation of the errors introduced by the propagation effects. The value of σ_{prop} is commonly taken from the literature or obtained by a precalibration phase. Note that in Equation (4.5), assuming that $\varepsilon_{\text{sync}}$ is known is mathematically equivalent to assuming that the clocks are synchronized between transmitter and receiver.

4.3.4 *Time Difference of Arrival*

The time difference of arrival (TDOA) is the difference between two TOA measurements obtained from two equivalent signals emitted at exactly the same time. This technique was proposed as a solution that allows one to drop the requirement that TOA measurements impose clock synchronization between transmitter and receiver. Since the TDOA is the difference between two equivalent signals sent at the same time, the dependency shown in Equation (4.4) on the time of transmission of the signal is lost. Thus, assuming that at time t_{tx} two equivalent signals are sent from BS_1 and BS_2, and given that the MS is d_1 meters from BS_1 and d_2 meters from BS_2, the TDOA t seen at the MS is given by:

$$t = t_{\text{tx}} + \frac{d_1}{c} + \varepsilon_{\text{sync}} - \left(t_{\text{tx}} + \frac{d_2}{c} + \varepsilon_{\text{sync}} \right) = \frac{d_1 - d_2}{c}. \tag{4.6}$$

This technique assumes two possible modes, uplink and downlink. In the uplink mode, the MS produces a signal which is received at two different BSs. This mode may have scalability problems, since all the operations are performed in the network. In contrast, in the downlink mode, two BSs transmit, at the same time, two equivalent signals, which are received by the MS. This mode is more scalable than the uplink mode since calculations are performed in the MS. Whatever the mode of operation is, the BSs need to be clock synchronized, though BSs and MSs do not need to be reciprocally synchronized.

Introducing a variable component into the TDOA is straightforward if we use Equation (4.5):

$$t = \frac{d_1 - d_2}{c} + \eta_{\text{clock}} + \eta_{\text{prop}}, \qquad \begin{cases} \eta_{\text{clock}} \sim \text{Tri}(-c/f, c/f), \\ \eta_{\text{prop}} \sim \text{Norm}(0, 2\sigma_{\text{prop}}). \end{cases} \tag{4.7}$$

In Equation (4.7), the subtraction of two uniform distributions results in a triangular distribution between $-c/f$and c/f centered at 0. Similarly, the subtraction of two equal normal distributions results in a normal distribution with variance $2\sigma_{\text{prop}}$ and centered at 0. In Equation (4.7), it is assumed that the noise components are equal in the communication channels between the MS and BS_1 and between the MS and BS_2. The extension to unequal distributions is straightforward.

4.3.5 *Angle of Arrival*

The angle of arrival (AOA) is a measure of the angle at which the signal arrives at the receiver. The technical procedure for obtaining AOA measurements requires directional antennas, and for this reason its implementation may represent additional cost and system complexity. In general, the AOA is related to an internal reference, which is commonly chosen based on the topology of the network. For instance, in multi-BS scenarios, the angles are commonly referenced to a virtual line that connects two BSs. A major problem with this type of measurement is the shadowing effect (see Section 3.2.2) present in the propagation channel. Since the shadowing effect biases the trajectory described by the main signal component, the signal received will arrive from an angle different from that described by the physical positions of the transmitter and the receiver. Figure 4.2 shows this problem. Owing to the complexity or even impossibility of detecting the influence of this effect, it is acceptable to use an estimate of the angle as a rough indication of position in terrestrial systems.

A simple, general model of the AOA is to assume an internal reference which all angles and coordinates are related to, that is, all angles and the entire coordinate system are relative to that reference. Then, assuming that the receiver is placed at position (x, y) and the transmitter at position $(x_{\text{tx}}, y_{\text{tx}})$, the AOA θ can be modeled by:

$$\theta = \operatorname{atan}\left(\frac{y_{\text{tx}} - y}{x_{\text{tx}} - x}\right). \tag{4.8}$$

Note that the model of Equation (4.8) does not give a unique solution for (x, y). For this reason, it is necessary to implement angulation techniques for estimating a single position of the MS; these are surveyed in Section 4.4.2.3.

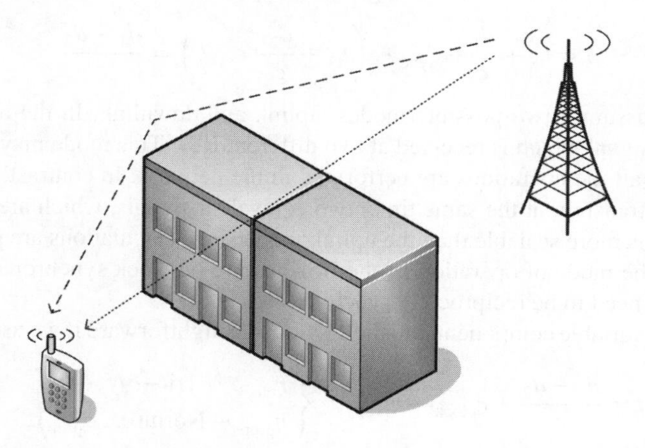

Figure 4.2 Effect of shadowing on AOA measurements. The dotted line represents the direct angle between transmitter and receiver, and the dashed line represents the propagation path influenced by the shadowing effect.

Regarding the noise component of AOA measurements, Goldsmith (2005) defines the noise as Gaussian. Thus, by assuming that this component has a zero-mean Gaussian distribution with standard deviation η_{aoa}, the model of the AOA can be written as:

$$\theta = \text{atan}\left(\frac{y_{\text{tx}} - y}{x_{\text{tx}} - x}\right) + \eta_{\text{aoa}}, \quad \eta_{\text{aoa}} \sim \text{Norm}(0, \sigma_{\text{aoa}}). \tag{4.9}$$

4.3.6 Personal-information Identification

Personal-information identification is based on another type of data, not directly related to wireless positioning, but is of wide use, especially in positioning for forensic investigation (Mohay et al. 2003). The type of data used can be data from airplane tickets, government census data, car identification information, credit card data, data from security cameras, computer-based information (such as wireless MAC addresses, login user names or Internet Protocol (IP) addresses), person-dependent characteristics (such as fingerprints, eye data or DNA data) or words of other people, among many other sources. This type of positioning typically relies on robust data-mining techniques capable of extracting meaningful relationships that can be used to estimate or predict movement trends and patterns both in the past and in the future. This topic will not be further discussed, since it is not the main focus of this book.

4.4 Positioning Techniques

Based on the various measurements obtained in a wireless communication, as reviewed in the previous section, the current section introduces the algorithms and methods used in order to estimate the position of an MS. The methods presented in this section are the basic methods used in positioning applications. More complex methods are introduced in later chapters.

4.4.1 Proximity Sensing

Proximity-sensing techniques are techniques that do not use the range or angles of communication links. Instead, they identify the mobile user's position once the user is within the physical communication range of a device.

4.4.1.1 Physical Contact

This is the simplest technique, which localizes the MS when it establishes physical contact with the BS. Here, the use of the terms "MS" and "BS" may be somewhat inaccurate, since this techniques relies basically on sensors of pressure, contact or distance. If the positions of the sensors are known, it is possible to localize the user once a contact is established.

4.4.1.2 Identity Methods

Identity methods are methods that rely on the position of a reference point in order to estimate the position of the MS. This technique is common in RFID, but is also widely used in

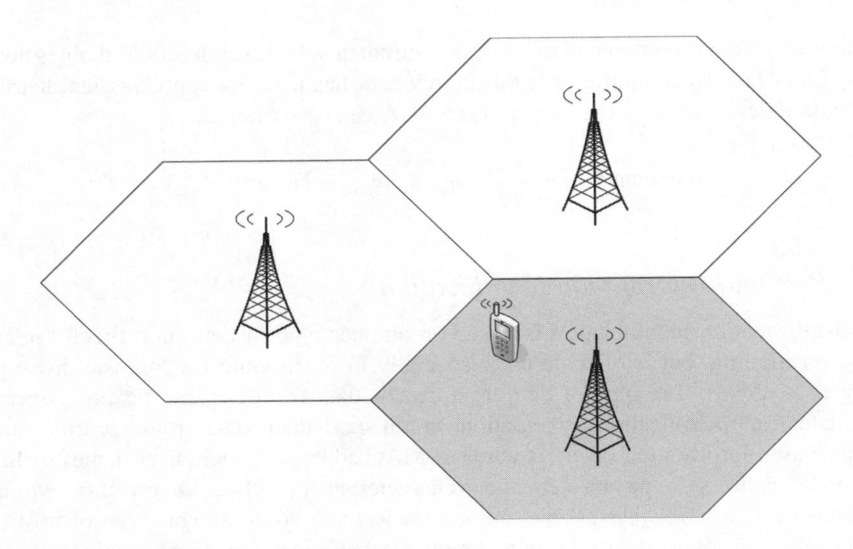

Figure 4.3 Cell ID technique.

all types of cellular system. Figure 4.3 shows a representation of this technique. Once the MS enters the range of a specific BS, the cell ID of the BS is read and is subsequently used to index a database that returns the position of the BS and possibly its communication range. The position of the BS is then assumed to be the position of the MS given that no other information is available. The communication range of the BS is assumed to be the best estimator for the accuracy of the position estimator. Section 8.3.2.1 introduces some more advanced techniques.

Identity methods are the simplest positioning techniques, but, at the same time, they are the least accurate methods for wireless positioning. A typical use of this technique is to support other positioning methods in hybrid approaches or establish geometrical constraints on the estimators.

A simple algorithm capable of estimating the position of MSs based on the locations of all the BSs that each MS can communicate with is the centroid method (Cheng et al. 2005). This algorithm calculates an arithmetic mean of the coordinates of all the BSs detected by the MS. Thus, assuming that the MS detects n BSs located at positions $\{X_i = [x_i, y_i]^T, i = 1, \ldots, n\}$, the estimated position $X = [x, y]^T$ of the MS is:

$$X = \frac{1}{n}\sum_{i=1}^{n} X_i. \tag{4.10}$$

As an alternative to the case when more information is available, the centroid algorithm can be weighted either by a measurement of range or by the actual communication range of each BS:

$$X = \frac{\sum_{i=1}^{n} (d_i^{\max})^{-1} X_i}{\sum_{i=1}^{n} (d_i^{\max})^{-1}}. \tag{4.11}$$

4.4.1.3 Macropositioning

Macropositioning concerns data-mining techniques commonly used together with personal-information identification. These methods are not directly related to wireless positioning, but represent an important group of methods in the wide field of positioning.

4.4.2 Triangulation

In contrast to the identification methods mentioned in Section 4.4.1.2, triangulation methods use range or angle measurements in order to estimate the position of the MS. These methods use trigonometry in order to combine data from several sources. Generally, these type of methods are supported by identity methods, since range and angle measurements require the position of the BSs to be known.

4.4.2.1 Lateration

Lateration is a technique that performs localization by using the distances between BSs and MSs. The calculation of the predicted location can be seen as the problem of determining the third vertex of a triangle given the lengths of the three edges and the location of two vertices. As Figure 4.4 shows, the two known vertices of the triangle are given by the positions of two BSs with known locations, one of the edges is given by the distance between the two BSs and the remaining two edges are given by the estimated length of the MS–BS communication channel. The estimated length can be calculated based on range observations such as RSS (see Section 4.3.2) or TOA (see Section 4.3.3) measurements. Although two BSs define the concept of triangulation, there is still a "flipping" problem

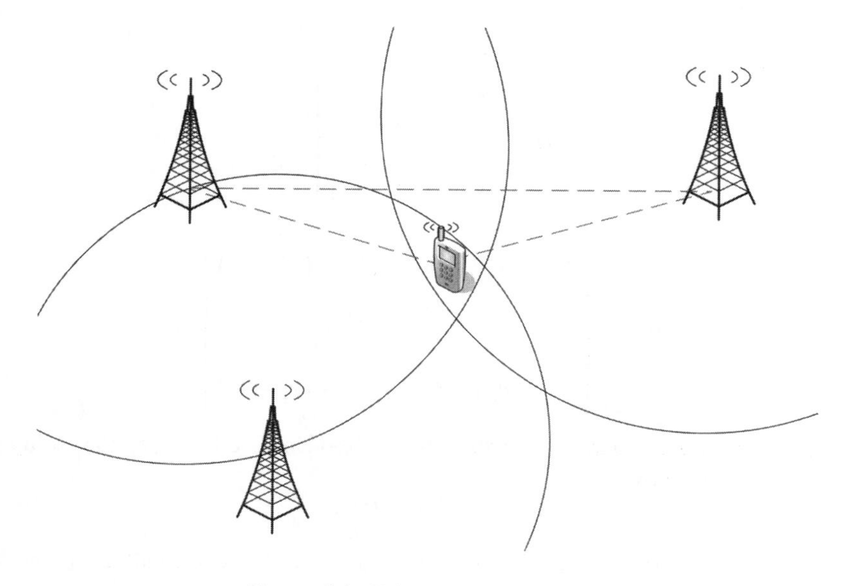

Figure 4.4 Triangulation technique.

around the axis connecting the two BSs. For this reason, with only two BSs, the predicted location may result in two points for a two-dimensional (2D) space. Thus, the minimum number of BSs necessary is three (placed at noncollinear points), where the third BS helps in determining which of the two positions is more probable. A similar idea can be used to argue that four noncoplanar points are the minimum number of BSs for three-dimensional (3D) localization.

As Figure 4.4 shows, the triangulation method requires in a 2D scenario three BSs with known locations. However, owing to the noisy behavior of the measurements, it is not possible to determine a single estimate of the MS position using trigonometric manipulation. The problem is that the link length estimators are corrupted by noise (see Section 4.3) and consequently they do not intersect in a single location. Thus, in order to estimate the position of the MS, it is necessary to combine data from the three BSs. The straightforward solution, although not the most elegant, would be to use two BSs to determine the intersection points and then use the third BS to choose one of these two points. This approach has two problems: firstly, it does not use data from the third BS, except for using it as a decision criterion; and, secondly, it does not handle the case when the circles do not intersect. A more elegant solution can be obtained as follows. Let us assume that the distance between the MS, located at position $X = [x, y]^{\mathrm{T}}$, and BS_i, located at $\{X_i = [x_i, y_i]^{\mathrm{T}}\}$, is given by d_i, where:

$$d_i^2 = (x_i - x)^2 + (y_i - y)^2. \tag{4.12}$$

Then, by subtracting Equation (4.12) for BS_i from the equivalent expression for BS_1, it is possible to obtain:

$$d_i^2 - d_1^2 = x_i^2 + y_i^2 - x_1^2 - y_1^2 - 2x(x_i - x_1) - 2y(y_i - y_1). \tag{4.13}$$

We can write this in matrix notation as:

$$HX = B, \tag{4.14}$$

where

$$H = \begin{bmatrix} x_2 - x_1 & y_2 - y_1 \\ x_3 - x_1 & y_3 - y_1 \\ \vdots & \vdots \\ x_n - x_1 & y_n - y_1 \end{bmatrix}, \tag{4.15}$$

and

$$B = \frac{1}{2} \begin{bmatrix} (d_1^2 - d_2^2) + (x_2^2 + y_2^2) - (x_1^2 + y_1^2) \\ (d_1^2 - d_3^2) + (x_3^2 + y_3^2) - (x_1^2 + y_1^2) \\ \vdots \\ (d_1^2 - d_n^2) + (x_n^2 + y_n^2) - (x_1^2 + y_1^2) \end{bmatrix}. \tag{4.16}$$

According to the least-squares algorithm formalized in Section 5.2, the solution for Equation (4.14) is given by:

$$X = (H^{\mathrm{T}}H)^{-1}H^{\mathrm{T}}B. \tag{4.17}$$

For more advanced algorithms, see Chapter 5 or Kailath et al. (2000) and Sayed and Yousef (2003).

4.4.2.2 Hyperbolic Localization

Hyperbolic localization is an alternative to the triangulation technique when link lengths are estimated using TDOA measurements. Since these types of measurements do not depend on the link length but on the difference between two measures of link lengths, the TDOA measurements describe a hyperbola as the possible location of the user position (see Figure 4.5). For a 2D space, it is possible to estimate the (x, y) position of the MS from a system of equations with three different TDOA readings. For the same reason for which we need a third BS in order to obtain a unique solution using triangulation, hyperbolic localization requires a fourth BS.

Similarly to the lateration technique, hyperbolic localization is subject to the problem of non-intersecting curves. The same approaches as those used to solve the problem in the case of triangulation can be used in this case. Thus, assuming that TDOA measurements t_i result in a difference of distances d_i between BS_i and BS_1 [1], the problem can be solved as follows:

$$d_i = d_i - d_1, \tag{4.18}$$

$$(d_i + d_1)^2 = d_i^2, \tag{4.19}$$

$$d_i^2 + 2d_i d_1 = x_i^2 + y_i^2 - x_1^2 - y_1^2 - 2x(x_i - x_1) - 2y(y_i - y_1). \tag{4.20}$$

We can write this in matrix notation as:

$$H\mathsf{X} = B, \tag{4.21}$$

where

$$\mathsf{X} = [X^{\mathrm{T}}, d_1]^{\mathrm{T}}, \tag{4.22}$$

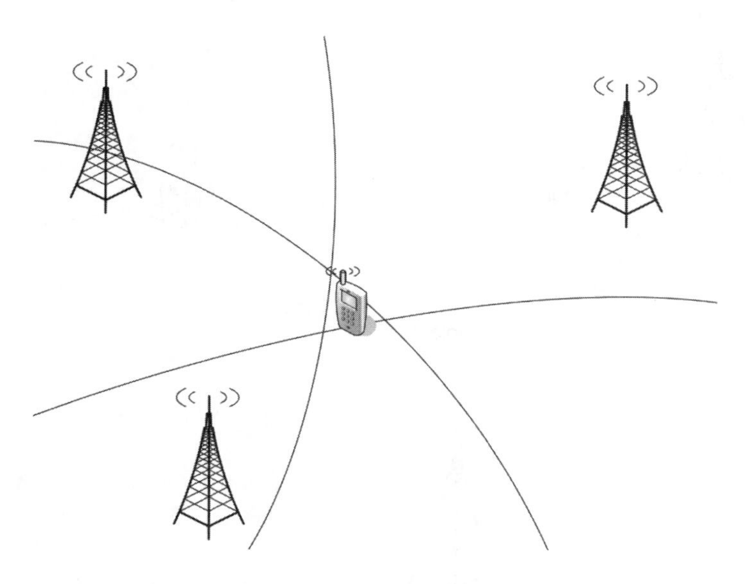

Figure 4.5 Hyperbolic localization technique.

[1] Referenced to BS_1.

$$H = \begin{bmatrix} x_2 - x_1 & y_2 - y_1 & \mathsf{d}_2 \\ x_3 - x_1 & y_3 - y_1 & \mathsf{d}_3 \\ \vdots & \vdots & \vdots \\ x_n - x_1 & y_n - y_1 & \mathsf{d}_n \end{bmatrix}, \tag{4.23}$$

$$B = \frac{1}{2} \begin{bmatrix} (x_2^2 + y_2^2) - (x_1^2 + y_1^2) - \mathsf{d}_2^2 \\ (x_3^2 + y_3^2) - (x_1^2 + y_1^2) - \mathsf{d}_3^2 \\ \vdots \\ (x_n^2 + y_n^2) - (x_1^2 + y_1^2) - \mathsf{d}_n^2 \end{bmatrix}. \tag{4.24}$$

The solution is then given by:

$$\mathsf{X} = (H^{\mathrm{T}} H)^{-1} H^{\mathrm{T}} B. \tag{4.25}$$

Note that in Equation (4.21), the distance between the MS and BS_1, d_1, is included in the vector X on the left because it is itself also unknown.

4.4.2.3 Angulation

Angulation is a technique that makes use of measurements of AOA. The algorithm relies in a 2D scenario on two BSs at known locations capable of measuring AOA (see Figure 4.6). In a 3D space, it is possible to use this technique if another measurement of azimuth is available. In this technique, a possible convention is to predefine the direction of one of the vectors that

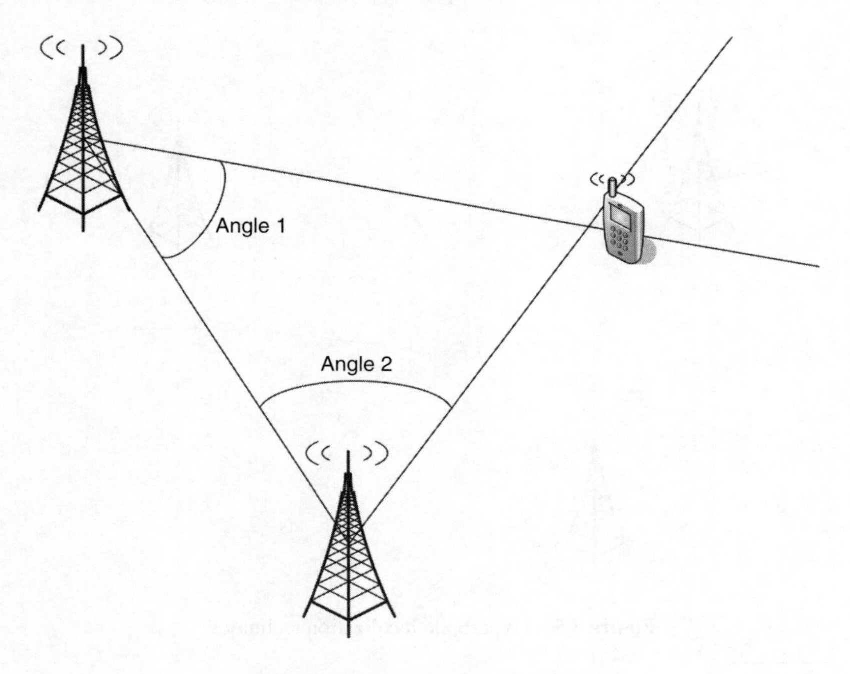

Figure 4.6 Angulation technique.

connects two BSs as the magnetic north. This vector is defined as the reference at 0°. Consequently, the magnetic south is defined as 180°. Each BS must be able to identify the angle that the transmitted signal originated from, referenced to the predefined magnetic north. Several methods can be used to measure this angle, such as the use of directive antennas (Vanderveen et al. 1997) or methods making use of the phase properties of the signal (Departments of Defense and of Transportation 2001). After measuring the angles and knowing the location of the BSs, it is possible to predict the location of the MS.

Due to the fact that angulation uses AOA, the problem of systematic errors introduced by shadowing is present here. Since this problem may be not noticeable by the BSs, the errors may be prohibitive for pinpointing a position accurately. However, for a rough indication of the area where the MS is placed, angulation can be a good candidate solution for localization.

Assuming that each BS i is capable of measuring the AOA θ_i of the signal transmitted from the MS, based on Equation (4.8), it is possible to obtain:

$$(x_i - x)\sin(\theta_i) = (y_i - y)\cos(\theta_i), \tag{4.26}$$

and in matrix form:

$$HX = B, \tag{4.27}$$

where

$$H = \begin{bmatrix} -\sin(\theta_1) & \cos(\theta_1) \\ -\sin(\theta_2) & \cos(\theta_2) \\ \vdots & \vdots \\ -\sin(\theta_n) & \cos(\theta_n) \end{bmatrix}, \tag{4.28}$$

and

$$B = \begin{bmatrix} y_1\cos(\theta_1) - x_1\sin(\theta_1) \\ y_2\cos(\theta_2) - x_2\sin(\theta_2) \\ \vdots \\ y_n\cos(\theta_n) - x_n\sin(\theta_n) \end{bmatrix}. \tag{4.29}$$

The solution is given by:

$$X = (H^{\mathrm{T}}H)^{-1}H^{\mathrm{T}}B. \tag{4.30}$$

4.4.3 Fingerprinting

This group of techniques performs localization of MSs based on information obtained from the surroundings of the MSs. For these techniques, it is important to have good knowledge about the surroundings in order to correlate the information obtained during the location process. This requires effort in the calibration of the system, which may not be desirable or not even implementable for some specific applications.

In this technique, the accuracy localizing the MS depends strongly on the information contained in the database. This database must be as complete as possible, with as much variety of measurements as possible, such as received power, time of flight and angle of arrival. As we can see in Figure 4.7, it is also possible to consider measurements from several systems such as GSM, UMTS systems or any other technology enabled into the MS.

The greater the amount of information in the database, the more complex are the algorithms for localizing MSs and also the greater is the complexity of managing the database.

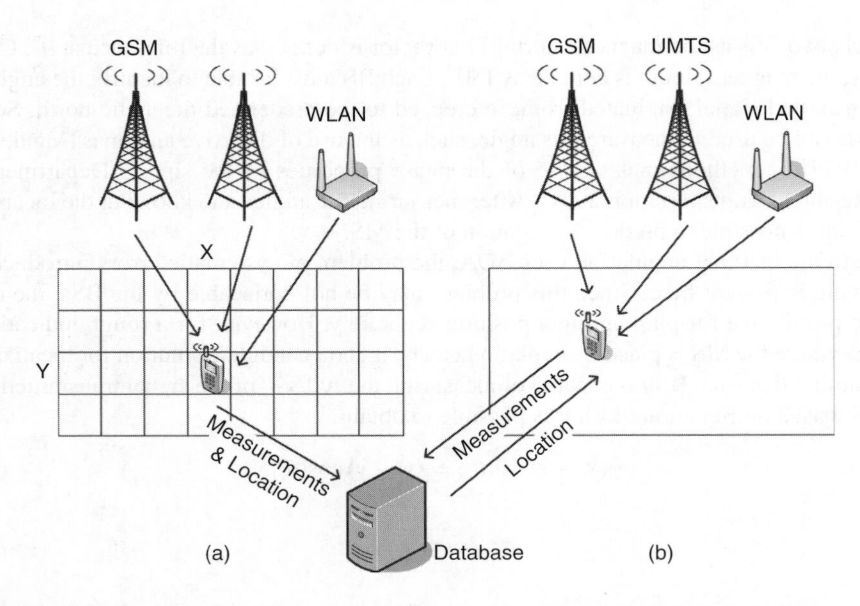

Figure 4.7 Database correlation technique: (a) calibration procedure; (b) localization procedure.

Fingerprinting techniques are very dependent on the scenario. If something that may influence the propagation conditions changes, for example a door opening in an indoor scenario, the database must be recalibrated all over again.

As we can see in Figure 4.7, during the calibration phase the measurements must be performed over the possible area with a certain step size. The locations and measurements are then passed into the database for further processing. During the localization phase, the MS must be able to measure some of the possible signals and relay them to a central server. This server, after correlating these measurements with the information included in the database, will then determine the predicted location. Section 5.5.1 presents an example of this technique.

4.4.3.1 Calibration Phase for Database Creation

In general, fingerprinting techniques are of high accuracy, but it is necessary to have a database with as much information as possible. Obtaining this information can be a major problem and virtually impossible for global positioning. In small areas, the creation of the database can be done by dividing the entire area into small clusters. Then, in each cluster, measurement samples are obtained and stored in the database. The smaller the clusters, the higher the accuracy. This precalibration phase is commonly cumbersome and time-consuming, and becomes unfeasible for some applications such as positioning in large areas. In the extreme case of global positioning, it is not possible to cover the whole world. In this case, the approach of creating the database is different; the database is often composed of information which identifies the FRPs. In this case, the methods used are commonly not ambitious enough to aim at characterizing subcell-scale areas.

Besides the difficult phase of creating the database, there is a problem related to the maintenance of the database itself. If the propagation conditions change during the operation of the system, a recalibration phase needs to be run. This problem makes it considerably harder to use such fingerprinting positioning systems. In the particular case of global positioning, possible solutions to this problem of database maintenance may rely on large-scale collaborative schemes.

4.4.3.2 Image/Video Approaches

These techniques use video or image techniques in order to locate devices. In order to localize an MS, the surroundings of it are filmed or photographed, and analyzed. Based on a priori knowledge of the environment or robust artificial intelligence techniques, it is possible to infer the position of the MS. For instance, as we can see in Figure 4.8, an extrapolation of the shapes of the silhouettes of landmarks can give some information about the location of the MS.

The video analysis technique, although it is conceptually simple to understand as it is the technique closest to the sense of orientation of a human being, may be extremely complex to implement due to the huge amount of computation and the complex algorithms that are involved in the procedure. However, this technique has the strong advantage of still being operative even if the MSs are in idle mode because no emission of signals is required. For this reason, this technique is widely used in the context of robotics.

This technique encompasses two approaches: the use of a static scene, where only one image is analyzed at a time, and the use of a differential scene, where the analysis is done on the differences observed between sequential images. If the static scene is used, it is necessary to have a predefined database that maps the observed features in the image to the positions of the MSs. On the other hand, if the differential scene is used, the differences in the scene are related to the movement of an MS, allowing the MS to compute its location if the positions of the features in the image are known.

Figure 4.8 Video analysis technique.

4.4.3.3 Collaborative Approach for Database Maintenance

Maintaining the database with information about wireless signals is not a simple task. One of the few automatic solutions relies on a large-scale collaborative approach. In a collaborative fashion, MSs which know their own locations report from time to time their locations and measurements obtained from the communication channel. This information, after processing, is then saved in the database as a fingerprint. Then, when a new user passes through the same location and requests his own location, this user is subsequently asked to report measurements obtained in its location. The system, by matching these measurements with known fingerprints, is able to return to the user the position stored in the database, if this position has previously been reported by another user. This technique is, for example, the approach used by My Location, which is a service provided by Google Maps for Mobile (Google 2007).

4.4.4 Dead Reckoning

Dead reckoning (DR) is a technique that was widely used in the past in naval navigation (Bowditch 2002). It is currently used in advanced inertial systems (Farrell and Barth 1999) and, generally, is used in a range of different fields (Pantel and Wolf 2002). This technique is aimed at estimating the position of an MS or of any other type of target by predicting the position based on previously estimated positions or on FRPs. DR relies on pre-estimated or pre-acquired information regarding direction, speed, elapsed time and external factors such as frictional forces.

The DR concept is still widely used, integrated into more complex algorithms such as the Kalman filter described in Section 5.3.1. The prediction steps in movement-tracking algorithms often use a concept that resembles the DR method. Moreover, since terminal suppliers started to integrate inertial sensors into their smartphones, the range of applications of DR has increased. The combination of inertial sensors and wireless positioning solutions results in more complex systems, which certainly require some form of DR.

A serious problem of the technique when used without external information from FRPs is integration drift. This problem is related to a divergence of the predicted movement as a consequence of the accumulated errors that the DR returns in the various iterations of the algorithm. Thus, if no external information is available, the open-loop system that the DR represents is complex to handle.

In a hybrid scenario, this technique could be used, for instance, in combination with cell ID information. The DR algorithm would be responsible for predicting the movement of the MS while the cell ID information would function as FRPs in order to assist movement estimation.

4.4.5 Hybrid Approaches

Hybrid approaches generally mean any combination of two or more techniques, not necessarily wireless positioning techniques. For instance, combining RSS-based triangulation methods with video camera motion tracking is considered a hybrid positioning technique. The major challenge is often the combination of data obtained from different techniques.

4.4.5.1 Hybrid Angulation and Lateration

The hybrid angulation and lateration (Venkatraman and Caffery 2004) technique is a straightforward technique that combines the two types of measurements by means of simply merging the formulas of each method.

Let us assume that n BSs, placed at $\{X_i = [x_i, y_i]^T, i = 1, \ldots, n\}$, obtain TOA and AOA measurements from a communication channel with an MS and that each TOA measurement between BS_i and the MS is denoted by d_i and each AOA measurement between BS_i and the MS is denoted by θ_i. The hybrid angulation and lateration method can then be defined as:

$$X = (H^T H)^{-1} H^T B, \tag{4.31}$$

where

$$H = \begin{bmatrix} x_2 - x_1 & y_2 - y_1 \\ x_3 - x_1 & y_3 - y_1 \\ \vdots & \vdots \\ x_n - x_1 & y_n - y_1 \\ -\sin(\theta_1) & \cos(\theta_1) \\ -\sin(\theta_2) & \cos(\theta_2) \\ \vdots & \vdots \\ -\sin(\theta_n) & \cos(\theta_n) \end{bmatrix}, \tag{4.32}$$

and

$$B = \frac{1}{2} \begin{bmatrix} (d_1^2 - d_2^2) + (x_2^2 + y_2^2) - (x_1^2 + y_1^2) \\ (d_1^2 - d_3^2) + (x_3^2 + y_3^2) - (x_1^2 + y_1^2) \\ \vdots \\ (d_1^2 - d_n^2) + (x_n^2 + y_n^2) - (x_1^2 + y_1^2) \\ y_1 \cos(\theta_1) - x_1 \sin(\theta_1) \\ y_2 \cos(\theta_2) - x_2 \sin(\theta_2) \\ \vdots \\ y_n \cos(\theta_n) - x_n \sin(\theta_n) \end{bmatrix}. \tag{4.33}$$

4.4.5.2 Hybrid Angulation and Hyperbolic Localization

Another example of hybrid localization is the combination of hyperbolic localization and angulation (Cong and Zhuang 2002). Let us assume that n BSs, placed at $\{X_i = [x_i, y_i]^T, i = 1, \ldots, n\}$, obtain TDOA and AOA measurements from a communication channel with an MS and that each TDOA measurement between BS_i and BS_1 is denoted by t_i and can be related to a difference of distances d_i, and each AOA measurement between BS_i and the MS is denoted by θ_i. The hybrid angulation and hyperbolic-localization method can be defined as:

$$\mathsf{X} = [x, y, d_1]^T = (H^T H)^{-1} H^T B, \tag{4.34}$$

where

$$H = \begin{bmatrix} x_2 - x_1 & y_2 - y_1 & \mathsf{d}_2 \\ x_3 - x_1 & y_3 - y_1 & \mathsf{d}_3 \\ \vdots & \vdots & \vdots \\ x_n - x_1 & y_n - y_1 & \mathsf{d}_n \\ -\sin(\theta_1) & \cos(\theta_1) & 0 \\ -\sin(\theta_2) & \cos(\theta_2) & 0 \\ \vdots & \vdots & \vdots \\ -\sin(\theta_n) & \cos(\theta_n) & 0 \end{bmatrix}, \tag{4.35}$$

and

$$B = \frac{1}{2} \begin{bmatrix} (x_2^2 + y_2^2) - (x_1^2 + y_1^2) - \mathsf{d}_2^2 \\ (x_3^2 + y_3^2) - (x_1^2 + y_1^2) - \mathsf{d}_3^2 \\ \vdots \\ (x_n^2 + y_n^2) - (x_1^2 + y_1^2) - \mathsf{d}_n^2 \\ y_1 \cos(\theta_1) - x_1 \sin(\theta_1) \\ y_2 \cos(\theta_2) - x_2 \sin(\theta_2) \\ \vdots \\ y_n \cos(\theta_n) - x_n \sin(\theta_n) \end{bmatrix}. \tag{4.36}$$

4.5 Error Sources in Positioning

The aim of this section is to briefly describe some of the most common types of errors that exist in the measurements of the communication channel and models for these errors. This section is based on the concepts introduced in Chapter 3.

4.5.1 *Propagation*

In wireless positioning, the most common problems and also the most complex problems to deal with are related to propagation effects, which often have a destructive effect on the measurements of the communication channel (Caffery and Stüber 1998b). This section highlights some of these problems and describes some of their statistical properties.

4.5.1.1 Non-line-of-sight

The problem of NLOS paths is severe in range-based systems. The problem is due to obstructions between the transmitter and the receiver. These obstructions cause the signals to behave differently, as a consequence of the presence of several propagation media. The reason for this problem is related to differences in the properties characterizing the media that the signal propagates in. These properties include the velocity of propagation and the indexes of attenuation.

For instance, in RSS models such as those represented by Equation (4.2), the NLOS effect would be considered by including an additional attenuation l_{obstr}:

$$p = \alpha - 10\beta \log(d) + l_{\text{obstr}} + \eta, \quad \eta \sim \text{Norm}(0, \sigma). \tag{4.37}$$

This attenuation parameter l_{obstr} is strongly dependent on the propagation characteristics of the media that obstructs the line-of-sight communication. In more detail, this attenuation parameter can be defined as a loss given by a law similar to $l_{obstr} = \alpha_{obstr} - 10\beta_{obstr}\log(d_{obstr})$. Although the distance d_{obstr} of propagation through the obstruction maybe only of the order of few centimeters, for instance in the case of a wall section, the propagation parameter in the wall material may have such a large propagation exponential β_{obstr} that the attenuation can be of the order of some tens of dB.

The NLOS effect is a difficult problem to solve in scenarios with several obstructions. A typical case is indoor scenarios, owing to the multiple obstructions among the rooms of a building, which cause effects that are difficult or even impossible to estimate in NLOS models. For this reason, NLOS effects strongly influence the positioning accuracy of range-based systems in indoor scenarios. Techniques for mitigating NLOS conditions in such cases are presented in Chapter 7.

4.5.1.2 Multipath Fading

Multipath fading is an effect that corrupts signals at the receiver due to multiple reflections of copies of the same transmitted signal. These copies, because of their different propagation trajectories, arrive at the receiver with different attenuations, delays and phases compared with the direct signal. The copies of the signal are superimposed at the receiver, causing the properties of the signal to change.

Under NLOS conditions, it is common to model the envelope of the received signal by a Rayleigh distribution, that is, the envelope tends with time to follow a Rayleigh distribution:

$$f_{\text{Rayleigh}}(r) = \frac{r}{\sigma^2}\exp\left(-\frac{r^2}{2\sigma^2}\right). \tag{4.38}$$

Given that the relation between the signal and the received power is quadratic (Goldsmith 2005), the Rayleigh distribution of the envelope results in an exponential distribution of the received power:

$$f_{\text{Rayleigh}}(\gamma) = \frac{1}{\bar{\gamma}}\exp\left(-\frac{\gamma}{\bar{\gamma}}\right). \tag{4.39}$$

According to this distribution of the received power, the model for the RSS in Equation (4.2) can be written as:

$$p = \alpha - 10\beta\log(d) + \eta, \quad \eta \sim Exp(\bar{\gamma}). \tag{4.40}$$

Under line-of-sight (LOS) conditions, the envelope of the received signal follows a Ricean distribution:

$$f_{\text{Rice}}(r) = \frac{r}{\sigma^2}\exp\left(-\frac{r^2+a^2}{2\sigma^2}\right)I_0\left(\frac{ar}{\sigma^2}\right), \tag{4.41}$$

$$\kappa = \frac{a^2}{2\sigma^2}, \tag{4.42}$$

where $I_0()$ is the modified Bessel function of the first kind of order 0, and the parameter a depends on the magnitude of the LOS ray. When $a = 0$, the distribution reduces to a Rayleigh distribution. In this case the Ricean distribution of the envelope results in a complex distribution of the received power, which can be defined by:

$$f_{\text{Rice}}(\gamma) = \frac{2(1+\kappa)}{\bar{\gamma}} \exp(\kappa) \exp\left(-\frac{(1+\kappa)\gamma}{\bar{\gamma}}\right) I_0\left(\sqrt{\frac{4\kappa(1+\kappa)\gamma}{\bar{\gamma}}}\right). \tag{4.43}$$

4.5.1.3 Shadowing

Shadowing is an effect that is present when LOS conditions do not exist, but a nonobstructed ray is still possible by an indirect path. The signal needs to be reflected or refracted in order to travel along a nonobstructed path. This results in a longer distance traveled by the signal and a change in the angle from which the receiver listens to the signal (see Section 4.3.5). The measurements mostly affected by this effect are AOA measurements, although range-based measurements are also affected. Assuming that the target is static, an additional bias due to the diffraction of the signal is added to the AOA model represented by Equation (4.8):

$$\theta = \operatorname{atan}\left(\frac{y_{\text{tx}} - y}{x_{\text{tx}} - x}\right) + \theta_{\text{bias}}, \tag{4.44}$$

where θ_{bias} is the bias introduced by the shadowing factor. When the device moves, the shadowing effect shows a probabilistic behavior. This behavior, however, is only visible for long movement distances, that is, distances long enough that the major obstructions in the scenario (e.g., buildings, walls, vehicles) change. In this situation, the shadowing effect shows a log-normal distribution (Liberti and Rappaport 1992):

$$f_{\text{Shadowing}}(\gamma) = \frac{1}{\gamma\sigma\sqrt{2\pi}} \exp\left(-\frac{(\ln(\gamma) - \mu)^2}{2\sigma^2}\right), \tag{4.45}$$

where μ and σ are, respectively, the mean and standard deviation of the logarithm of the variable γ. Since Equation (4.45) assumes γ to be in linear units, the corresponding equation in units of dB is given by a normal distribution:

$$f_{\text{Shadowing}}(\gamma) = \frac{1}{\sigma\sqrt{2\pi}} \exp\left(-\frac{(\gamma - \mu)^2}{2\sigma^2}\right). \tag{4.46}$$

4.5.1.4 Body Shadowing

Body shadowing is also a severe problem for wireless communication technologies, in particular for wireless positioning (Obayashi and Zander 1998). This effect is due to the presence of the human body in close proximity to the MS. This presence introduces additional attenuations, commonly complex to model and characterized by large variability with time. Owing to this complexity, the destructive effect of body shadowing is commonly counteracted by increasing the transmission power. In fact, the impact of this effect in wireless positioning is currently a field of research under exploration.

4.5.1.5 Interference

Interference is another effect in wireless communications which has a complex statistical treatment, mainly because it strongly depends on the systems active in the vicinity of the MSs and the BSs. Also, the same technology is often a source of interference in the same system. Several aspects of the communication technology in use influence the communication channel, such as the transmitted power, MAC layer protocols and traffic, among other aspects. Besides the technology itself, the positions of all the transmitters and receivers also have an influence the interference. Owing to the complexity of the problem, interference is commonly studied in very specific scenarios, with strong assumptions, and it is typically treated in terms of average values.

4.5.1.6 The Ionosphere

Ionospheric and tropospheric effects are strongly present in satellite positioning systems (Grewal et al. 2007). These effects are due to gases ionized by solar radiation. This ionization, besides producing ions, also produces clouds of free electrons, which impact on the propagation of the signals. Due to the periodic variation of the solar radiation intensity with the daily cycles of the Earth, the effects on the propagation also vary on a periodic basis, assuming that additional variations in solar activity are ignored. One aspect worth mentioning is that when a satellite is on the horizon, the signal travels a longer distance than in the case when the satellite is at the zenith. For this reason, robust ionospheric models must take into account the positions of the satellites.

The Klobuchar model, introduced by Klobuchar (1976) and analyzed in detail by Feess and Stephens (1987), is given by:

$$t_g = t_{\mathrm{off}} + \mathsf{a}\left[1 - \frac{\omega^2}{2} + \frac{\omega^4}{24}\right], \quad \text{for } |\omega| \leq \frac{\pi}{2}, \tag{4.47}$$

where

$$\omega = \frac{2\pi\left(t_{\mathrm{earth}} - t_{\mathrm{phase}}\right)}{\tau}, \tag{4.48}$$

with ω in radians. In Equation (4.47), the term t_{off} is a constant offset, generally assumed to be 5 ns, t_{phase} is a phase equal to 50 400 s (14 h), a is the amplitude and τ is the period of the ionospheric time delay, and t_{earth} the local time of the point on the Earth beneath the intersection of the signal with a mean ionospheric height of 350 km. The terms a and τ can be modeled by:

$$\mathsf{a} = \sum_{k=0}^{3} a_k \phi_m^k, \quad \tau = \sum_{k=0}^{3} b_k \phi_m^k, \tag{4.49}$$

where a_k and b_k are a set of coefficients (out of 370 sets) sent by the satellite to the user. The terms ϕ_m are the geomagnetic latitude of the ionospheric subpoint and can be modeled by:

$$\phi_m = \phi_I + 11.6^\circ \cos(\lambda_I - 291^\circ), \tag{4.50}$$

where

$$\phi_I = \phi_{\text{user}} + \theta_{\text{ea}} \cos(\theta_{\text{az}}),$$
$$\lambda_I = \lambda_{\text{user}} + \theta_{\text{ea}} \frac{\cos(\theta_{\text{az}})}{\cos(\phi_I)}, \tag{4.51}$$

and

$$\theta_{\text{ea}} = \frac{445}{l + 20^{-4}},$$

where $(\phi_{\text{user}}, \lambda_{\text{user}})$ are the geodesic latitude and longitude of the user, and (ϕ_I, λ_I) are the equivalent measures of the ionospheric subpoint. The terms θ_{ea}, θ_{az} and l are, respectively, the Earth angle, the azimuth and the elevation of the satellite with respect to the user.

4.5.2 Geometry

Another type of error widely known in the literature concerns the geometric distribution of the static FRPs. The problem is related to the relative positions between these FRPs and the target itself. This problem has been analyzed in the context of satellite positioning (Grewal et al. 2007). With the advent of positioning in cellular networks, the problem now also exists in this context.

Let us assume the position of the target to be X and the position of the n FRPs to be X_i, with $i = 1, \ldots, n$:

$$X = [x \quad y \quad z]^{\text{T}}, \quad X_i = [x_i \quad y_i \quad z_i]^{\text{T}} . \tag{4.52}$$

Since the clocks of the satellites and the receiver commonly have a constant offset, this offset c_{off} also needs to be calculated. Thus, the unknown position X of the receiver is extended with the clock offset c_{off}:

$$Y = [X^{\text{T}} \quad c_{\text{off}}]^{\text{T}} . \tag{4.53}$$

It is possible to define the Euclidean distance between each FRP and the target as:

$$d_i(Y) \equiv d_i(X, c) = \sqrt{(X - X_i)^{\text{T}}(X - X_i)} + c_{\mathit{off}}, \tag{4.54}$$

$$D(Y) = [d_1(Y) \quad \cdots \quad d_n(Y)]^{\text{T}} . \tag{4.55}$$

We take into account the fact that measurements are never noiseless, that is, the measured ranges $\hat{D}(Y)$ are corrupted by a noise component η:

$$\hat{D}(Y) = D(Y) + \eta, \quad \eta = [\eta_1 \quad \cdots \quad \eta_n]^{\text{T}} . \tag{4.56}$$

Expanding the term $D(Y)$ of Equation (4.56) as a Taylor series, we obtain:

$$\hat{D}(Y) = D(\hat{Y}) + \left.\frac{\partial D}{\partial Y}\right|_{\hat{Y}} (Y - \hat{Y}) + \eta, \quad \left.\frac{\partial D}{\partial Y}\right|_{\hat{Y}} = \left[\left.\frac{\partial D}{\partial X}\right|_{\hat{X}} \mathbf{1} \right], \tag{4.57}$$

where $\partial D / \partial Y|_{\hat{Y}}$ is the Jacobian matrix of $D(Y)$ for the values $\hat{Y}$, and $\mathbf{1}$ is the $n \times 1$ column vector of ones. By defining

$$\tilde{D} = \hat{D}(Y) - D(\hat{Y}), \quad \tilde{Y} = Y - \hat{Y}, \tag{4.58}$$

we have:

$$\tilde{D} = \frac{\partial D}{\partial Y}\Big|_{\hat{Y}} \tilde{Y} + \eta. \quad (4.59)$$

From the least-squares algorithm (presented in detail in Section 5.2), the value for $\tilde{Y}$ that minimizes the error η is given by:

$$\tilde{Y} = (H^{\mathrm{T}}H)^{-1}H^{\mathrm{T}}\tilde{D}, \quad H \equiv \frac{\partial D}{\partial Y}\Big|_{\hat{Y}}. \quad (4.60)$$

In order to understand the problem related to the geometry of the FRPs, let us assume that all the FRPs are concentrated in the same area with distances between them at least one order of magnitude smaller than the distance to the target. In this case, the entries in the Jacobian matrix H are of similar magnitude and, consequently, are close to singular. As a consequence, $\tilde{Y}$ increases. On the other hand, if the target is in the middle of the FRPs, the entries in the Jacobian H will have different magnitudes, and the result is that H is far from singularity; the result is then a lower value for $\tilde{Y}$.

4.5.3 Equipment and Technology

Another type of problem that influences wireless positioning accuracy concerns the equipment itself and the technologies with which positioning is performed. The impact of these errors can be seen in several different ways, such as in variability in the amplitude of errors, biases, non-stationary measurement processes and variations in the distribution of errors. The most common type of error that influences the communication is thermal noise, which causes fluctuations in the measurements used later in position estimation.

The technology itself can also introduce errors in the measurements. For instance, MAC protocols with randomized processes, such as collision avoidance or listen-before-talk techniques, commonly used to circumvent interference problems, can also introduce additional delays in the measurements used for wireless positioning. The IEEE 802.11 technology is one case. These delays are related to the timestamps of the measurements rather than to the actual time measurements used for positioning. Furthermore, the resolution of the measurements permitted in the technology may also be a source of errors. See, for instance, the Bluetooth case (see Section 8.4.1.2), where by specification it is only possible to get measurements with a granularity of 1 dB. Concerning time measurements, clock synchronization errors are a major problem, introducing reading errors into TOA and TDOA measurements. Cellular communication systems such as GSM and UMTS are subject to clock errors.

Apart from the communication technology, even the network management strategy can introduce errors or misinformation into positioning applications. A typical example occurs in cellular networks where the operator often treats the locations of the BSs as confidential, meaning that the estimation of those locations commonly has to be performed using opportunistic positioning solutions. Furthermore, the BS cell IDs provided by the operators to the MSs are often incomplete, which may result in confusion between measurements obtained from different BSs *that share the same information about the known part of the incomplete information*.

The range of sources of errors is indeed large and sometimes even unexpected by positioning-system designers. Besides the examples given, there are many others, such as governmental requirements to degrade the accuracy of GPS data and design decisions of MS providers that do not provide application programming interfaces (APIs) for obtaining RSS measurements, or restrict those APIs.

4.6 Metrics of Location Accuracy

This section describes some of the metrics used to estimate accuracy levels for positioning applications. There are several ways of characterizing accuracy in the literature, but in this section we state some standard metrics used in positioning applications.

4.6.1 *Circular Error Probability*

The circular error probability (CEP) is a metric that was first defined in the context of military ballistics. It is a simple measure of precision that defines a circle within which the probability of there being a correct estimate of the position is 50%:

$$d\colon \mathrm{P}(\| X, \hat{X} \| < d) = 0.5. \tag{4.61}$$

4.6.2 *Dilution of Precision*

Dilution of precision (DOP) is a concept that is tightly related to GPS positioning, but extendable to any range-based positioning solution. DOP gives information about the impact that the configuration of the satellites or FRPs has on positioning accuracy. In order to understand the concept of DOP (Grewal et al. 2007), it is necessary to start by calculating the error covariance of $\tilde{Y}$ in Equation (4.60):

$$\mathrm{E}[\tilde{Y}\tilde{Y}^{\mathrm{T}}] = \mathrm{E}[(H^{\mathrm{T}}H)^{-1}H^{\mathrm{T}}\tilde{D}[(H^{\mathrm{T}}H)^{-1}H^{\mathrm{T}}\tilde{D}]^{\mathrm{T}}], \tag{4.62}$$

$$\mathrm{E}[\tilde{Y}\tilde{Y}^{\mathrm{T}}] = (H^{\mathrm{T}}H)^{-1}H^{\mathrm{T}}\mathrm{E}[\tilde{D}\tilde{D}^{\mathrm{T}}]H(H^{\mathrm{T}}H)^{-1}. \tag{4.63}$$

Let us assume that there is no correlation between the measurements obtained from each of the n FRPs and that such measurements are corrupted by a noise component with variance σ^2:

$$\mathrm{E}[\tilde{Y}\tilde{Y}^{\mathrm{T}}] = \sigma^2(H^{\mathrm{T}}H)^{-1}\underbrace{H^{\mathrm{T}}H(H^{\mathrm{T}}H)^{-1}}_{I}, \tag{4.64}$$

$$\mathrm{E}[\tilde{Y}\tilde{Y}^{\mathrm{T}}] = \sigma^2(H^{\mathrm{T}}H)^{-1}. \tag{4.65}$$

As Equation (4.65) shows, the term $(H^{\mathrm{T}}H)^{-1}$ establishes a direct relationship between the error covariance in the measurements $d_i(Y)$ and the error covariance in the position coordinates Y. For this reason, the term $(H^{\mathrm{T}}H)^{-1}$ is responsible for the dilution of precision in the positioning of the receiver. Depending on the entries in the term $(H^{\mathrm{T}}H)^{-1}$, it is possible to determine several different types of DOP:

$$\epsilon_{\mathrm{GDOP}} = \sqrt{a_{11} + a_{22} + a_{33} + a_{44}}, \tag{4.66}$$

$$\epsilon_{\mathrm{PDOP}} = \sqrt{a_{11} + a_{22} + a_{33}}, \tag{4.67}$$

$$\epsilon_{\text{HDOP}} = \sqrt{a_{11} + a_{22}}, \tag{4.68}$$

$$\epsilon_{\text{VDOP}} = \sqrt{a_{33}}, \tag{4.69}$$

and

$$\epsilon_{\text{TDOP}} = \sqrt{a_{44}}, \tag{4.70}$$

where the scalars a are the entries in the term $(H^{\text{T}}H)^{-1}$. From Equations (4.66)–(4.70), ϵ_{GDOP}, ϵ_{PDOP}, ϵ_{HDOP}, ϵ_{VDOP} and ϵ_{TDOP} are the geometric, position, horizontal, vertical and time DOP metrics, respectively. While, for instance, ϵ_{PDOP} accounts for 3D position errors, ϵ_{TDOP} accounts only for time errors.

4.6.3 Cramér–Rao Lower Bound

The Cramér–Rao lower bound (CRLB) establishes a lower limit for the variance that an unbiased estimator of a parameter of a distribution can attain (Deffenbaugh et al. 1996). Let θ be an unbiased, unknown column vector of parameters of the distribution $p(r, \theta)$ of a random variable r. If an estimator of θ is denoted by $\hat{\theta}$, the CRLB is given by:

$$\text{var}(\hat{\theta}) \geq I(\theta)^{-1}, \tag{4.71}$$

where the inequality $A \geq B$ is equivalent to saying that the matrix $A - B$ is positive semi-definite. The term $I(\theta)$ represents the Fisher information matrix (FIM), which is given by:

$$I(\theta) = -E\left[\frac{\partial^2 \log p(r, \theta)}{\partial \theta^2}\right]. \tag{4.72}$$

Since by definition the mean squared error (MSE) is given by the variance of an unbiased estimator, Equation (4.71) corresponds also to establishing a limit on the MSE. In a more general case where instead of estimating θ one is interested in estimating a function $g(\theta)$, assuming that this estimator has a bias $b(\theta)$, the MSE is given by:

$$\text{MSE} \leq \frac{\partial(g+b)(\theta)}{\partial \theta} I(\theta)^{-1} \left(\frac{\partial(g+b)(\theta)}{\partial \theta}\right)^{\text{T}} + b^2(\theta). \tag{4.73}$$

4.7 Conclusions

This chapter reviewed the fundamentals of wireless positioning. First, positioning techniques were classified according to various criteria: the topology of the positioning system, the physical coverage range and the level of integration in the communication technology. The types of measurements are typically RSS, TOA, TDOA and AOA, while the typical positioning techniques rely on cell identification, triangulation or fingerprinting strategies. Several error sources and models were illustrated mainly for propagation-related effects. The chapter closed by defining some of the most common accuracy metrics. Practically, the aim of this chapter was to give the basics needed to prepare the reader for the forthcoming chapters, which deal with more advanced techniques for wireless positioning.

5

Data Fusion and Filtering Techniques

João Figueiras
Aalborg University, Aalborg, Denmark

5.1 Introduction

As has been widely presented in previous chapters, wireless channel measurements for position estimation are unavoidably corrupted by noise. As described in Chapter 3, the various types of noise sources span a wide range of effects, commonly causing the channel measurements to present statistical behaviors that are difficult to model and complex to circumvent. Such complexity results in observable impairments in the positioning mechanisms, which impact on the accuracy of the system. There are several approaches to wireless positioning that are aimed at handling the statistical behavior of the measurement noise. In this chapter, two major approaches are presented: the least squares (LS) approach and the Bayesian framework.

The LS approach is an algorithm that estimates the position of a mobile station (MS) by minimizing the squared error between the actual measurements observed in the wireless channel and the expected measurements resulting from the estimated position. In order to execute this algorithm, it is necessary to model the relation between the position and the actual measurements. We present both the linear and the nonlinear case of the LS approach. As a particular case of the linear LS, the recursive version is presented.

In the Bayesian framework, the position is determined as an estimator that minimizes the mean square error between the actual measurements and the expected measurements. Three major implementations of the Bayesian framework are presented: the Kalman filter (KF), the particle filter (PF) and grid-based methods. The KF is a group of methods that assumes the measurements to be corrupted by white Gaussian noise, the PF is a Monte Carlo type of algorithm that does not constrain the noise component of the Gaussian distribution, and the grid-based methods assume that the state space is discrete and finite.

Mobile Positioning and Tracking: From Conventional to Cooperative Techniques, Second Edition.
Simone Frattasi and Francescantonio Della Rosa.

In addition to these two major approaches to data fusion, alternative solutions are also analyzed. The first alternative solution relies on fingerprinting techniques. The fingerprint technique follows a prephase of calibration where measurements are obtained at several strategic positions. These measurements are then used in the operational mode as reference values to be correlated with future measurements. The second alternative relies on time series theory, which by averaging pre-determined position estimators over time tries to minimize the effect of errors.

5.2 Least-squares Methods

LS methods are used for calculating the solutions of overdetermined systems. In statistical contexts, LS methods can be used to approximate a given set of data points by estimating the parameters of a predefined model. In the context of wireless positioning, these methods can be used to approximate a set of measurements from the wireless channel by estimating the position of a certain MS. The estimated position is obtained by minimizing the sum of the squared errors between a model that relates the position to wireless measurements and the actual data measured from the wireless channel.

Let us assume that the position coordinates $X = [x, y]^{\mathrm{T}}$ in a two-dimensional space correspond to the channel measurements $Z = [z_1, \dots, z_n]^{\mathrm{T}}$ at the ith BS, where n is the number of base stations (BSs). Let $h_i(X)$ be a model that relates the two variables such that

$$z_i = h_i(X) + \eta_i. \tag{5.1}$$

The parameter η_i represents a noise component added to the model. The best fit according to the LS approach is when the cost function $\mathcal{J}(X)$ is minimum:

$$\mathcal{J}(X) = \sum_{i=1}^{n} \eta_i^2 = \sum_{i=1}^{n} [z_i - h_i(X)]^2 = [Z - \mathcal{H}(X)]^{\mathrm{T}}[Z - \mathcal{H}(X)], \tag{5.2}$$

where $\mathcal{H}(X) = [h_1(X), \dots, h_n(X)]^{\mathrm{T}}$. The estimated position $\hat{X}$ is determined as the argument that minimizes Equation (5.2), that is,

$$\hat{X} = \arg\min_{X} \{\mathcal{J}(X)\}. \tag{5.3}$$

The solution to Equation (5.3) can be obtained by setting the gradient of $\mathcal{J}(X)$ to zero, where the partial derivatives of $\mathcal{J}(X)$ are taken with respect to X:

$$\nabla\mathcal{J}(X) = -2[Z - \mathcal{H}(X)]^{\mathrm{T}}\nabla\mathcal{H}(X) = 0. \tag{5.4}$$

Equation (5.4) defines the framework to be used in an LS method. Once the model $h_i(X)$ is known, it may be possible to manipulate Equation (5.4) further into a simpler solution; this is the case when the model is linear. In other cases, numerical optimization may be necessary.

A more general approach to the LS method assumes that the various sources of data produce measurements corrupted by error components with different statistical properties with respect to the other sources. In this general case, the method is given the name of weighted least squares (WLS). Thus, assuming that the measurement z_i is corrupted by white Gaussian noise with standard deviation σ_i, that is, $\eta_i \sim \mathrm{Norm}(0, \sigma_i)$ with $\eta = [\eta_1, \dots, \eta_n]^{\mathrm{T}}$ and $[\eta\eta^{\mathrm{T}}] = R$, the best fit in the LS sense is when the cost function $\mathcal{J}(X)$ is a minimum:

$$\mathcal{J}(X) = \sum_{i=1}^{n} R_{i,i}^{-1}[z_i - h_i(X)]^2 = [Z - \mathcal{H}(X)]^{\mathrm{T}} R^{-1} [Z - \mathcal{H}(X)]. \tag{5.5}$$

The minimum is then obtained from:

$$\nabla \mathcal{J}(X) = -2[Z - \mathcal{H}(X)]^{\mathrm{T}} R^{-1} \nabla \mathcal{H}(X) = 0. \tag{5.6}$$

In Equation (5.6), it is assumed that R is a symmetric matrix. Commonly, R is in fact defined as a diagonal matrix, that is, the measurements are assumed to be independent from each other.

5.2.1 Linear Least Squares

A special case of the LS methods is the linear least squares (LLS) method. The peculiarity of this method is that the model $\mathcal{H}(X)$ in Equation (5.4) is assumed to be linear, that is,

$$\mathcal{H}(X) = HX. \tag{5.7}$$

This linearity of the model simplifies the mathematical formulation of the method, in the sense that it is possible to get closed-form solutions for the parameters that minimize the cost function of Equation (5.2), that is, it is possible to obtain a closed-form solution for Equation (5.4). Thus, by using Equation (5.7) in Equation (5.4), we have:

$$\nabla \mathcal{J}(X) = -2[Z^{\mathrm{T}} - X^{\mathrm{T}} H^{\mathrm{T}}] \nabla HX = 0. \tag{5.8}$$

Since the gradient of $\mathcal{H}(X)$ with respect to the parameters X equals H, Equation (5.8) results in:

$$\nabla \mathcal{J}(X) = -2[Z^{\mathrm{T}} - X^{\mathrm{T}} H^{\mathrm{T}}] H = 0. \tag{5.9}$$

Then, after further manipulation:

$$X^{\mathrm{T}} H^{\mathrm{T}} H = Z^{\mathrm{T}} H, \tag{5.10}$$

and, after transposing both sides of the equation, we obtain:

$$H^{\mathrm{T}} H X = H^{\mathrm{T}} Z. \tag{5.11}$$

It is now possible to get a closed-form solution for the LLS method:

$$X = (H^{\mathrm{T}} H)^{-1} H^{\mathrm{T}} Z. \tag{5.12}$$

As Equation (5.12) shows, the LLS method can be used easily as a minimization tool. In particular, for positioning applications, the LLS algorithm can be used to determine the position of an MS when there are measurements available from several BSs. This is the example illustrated in Sections 3.8 and 3.9.

For the weighted formulation of the LLS method, similar manipulations can be done. Owing to its straightforwardness, the calculations will not be presented here. The weighted LLS solution is given by:

$$X = (H^{\mathrm{T}} R^{-1} H)^{-1} H^{\mathrm{T}} R^{-1} Z. \tag{5.13}$$

An important property of the LLS method is the covariance matrix of the LS estimator. This covariance matrix represents a measurement of the certainty in the estimation of the position:

$$P = [\{\hat{X} - [\hat{X}]\}\{\hat{X} - [\hat{X}]\}^{\mathrm{T}}] \tag{5.14}$$

$$= [\{\hat{X} - X\}\{\hat{X} - X\}^{\mathrm{T}}] \tag{5.15}$$

$$= (H^{\mathrm{T}}R^{-1}H)^{-1}H^{\mathrm{T}}R^{-1}[\eta\eta^{\mathrm{T}}]R^{-1}H(H^{\mathrm{T}}R^{-1}H)^{-1} \tag{5.16}$$

$$= (H^{\mathrm{T}}R^{-1}H)^{-1}H^{\mathrm{T}}R^{-1}RR^{-1}H(H^{\mathrm{T}}R^{-1}H)^{-1}. \tag{5.17}$$

Equation (5.17) results in:

$$P = (H^{\mathrm{T}}R^{-1}H)^{-1}. \tag{5.18}$$

5.2.2 *Recursive Least Squares*

The recursive least squares (RLS) algorithm is a recursive method for iteratively determining the LLS estimator as a function of time. The RLS algorithm typically updates the current estimator by running each iteration at every discrete time k when a newly obtained measurement is available. The following matrices are used to ease the manipulation:

$$Z_{0:k+1} = \begin{bmatrix} Z_{0:k} \\ Z_{k+1} \end{bmatrix}, \tag{5.19}$$

$$H_{0:k+1} = \begin{bmatrix} H_{0:k} \\ H_{k+1} \end{bmatrix} \tag{5.20}$$

and

$$R_{0:k+1} = \begin{bmatrix} R_{0:k} & 0 \\ 0 & R_{k+1} \end{bmatrix}. \tag{5.21}$$

Taking the covariance matrix as a starting point, it is possible to write:

$$P_{k+1}^{-1} = H_{0:k+1}^{\mathrm{T}}(R_{0:k+1})^{-1}H_{0:k+1} \tag{5.22}$$

$$= \begin{bmatrix} H_{0:k}^{\mathrm{T}} & H_{k+1}^{\mathrm{T}} \end{bmatrix} \begin{bmatrix} R_{0:k} & 0 \\ 0 & R_{k+1} \end{bmatrix}^{-1} \begin{bmatrix} H_{0:k} \\ H_{k+1} \end{bmatrix} \tag{5.23}$$

$$= H_{0:k}^{\mathrm{T}}R_{0:k}^{-1}H_{0:k} + H_{k+1}^{\mathrm{T}}R_{k+1}^{-1}H_{k+1} \tag{5.24}$$

$$= P_k^{-1} + H_{k+1}^{\mathrm{T}}R_{k+1}^{-1}H_{k+1}, \tag{5.25}$$

where the errors in the measurements are not correlated in time. Using the matrix inversion lemma, Equation (5.25) can be rewritten as:

$$P_{k+1} = [P_k^{-1} + H_{k+1}^{\mathrm{T}}R_{k+1}^{-1}H_{k+1}]^{-1} \tag{5.26}$$

$$= P_k - P_kH_{k+1}^{\mathrm{T}}[H_{k+1}P_kH_{k+1}^{\mathrm{T}} + R_{k+1}]^{-1}H_{k+1}P_k. \tag{5.27}$$

This results in the following equation for P_{k+1}:

$$P_{k+1} = [I - K_{k+1}H_{k+1}]P_k, \tag{5.28}$$

where K_{k+1} is the update gain, which is given by:

$$K_{k+1} = P_k H_{k+1}^{\mathrm{T}} [H_{k+1} P_k H_{k+1}^{\mathrm{T}} + R_{k+1}]^{-1}. \tag{5.29}$$

In Equation (5.29), the term $H_{k+1} P_k H_{k+1}^{\mathrm{T}} + R_{k+1}$ is the covariance of the residuals. As demonstrated in Bar-Shalom et al. (2001), the update gain K_{k+1} can also be written as $K_{k+1} = P_{k+1} H_{k+1}^{\mathrm{T}} R_{k+1}^{-1}$, which formulation is of clear interest in the manipulation of the update formulas for the position estimator X shown below.

Using Equations (5.18) and (5.19)–(5.21), the batch LS estimator of Equation (5.13) can be rewritten as:

$$\begin{aligned}
\hat{X}_{k+1} &= P_{k+1} H_{0:k+1}^{\mathrm{T}} (R_{0:k+1})^{-1} Z_{0:k+1} &(5.30)\\
&= P_{k+1} \begin{bmatrix} H_{0:k}^{\mathrm{T}} & H_{k+1}^{\mathrm{T}} \end{bmatrix} \begin{bmatrix} R_{0:k} & 0 \\ 0 & R_{k+1} \end{bmatrix}^{-1} \begin{bmatrix} Z_{0:k} \\ Z_{k+1} \end{bmatrix} &(5.31)\\
&= P_{k+1} H_{0:k}^{\mathrm{T}} (R_{0:k})^{-1} Z_{0:k} + P_{k+1} H_{k+1}^{\mathrm{T}} R_{k+1}^{-1} Z_{k+1} &(5.32)\\
&= [I - K_{k+1} H_{k+1}] P_k H_{0:k}^{\mathrm{T}} (R_{0:k})^{-1} Z_{0:k} + K_{k+1} Z_{k+1} &(5.33)\\
&= [I - K_{k+1} H_{k+1}] \hat{X}_k + K_{k+1} Z_{k+1}, &(5.34)
\end{aligned}$$

and, consequently,

$$\hat{X}_{k+1} = \hat{X}_k + K_{k+1} [Z_{k+1} - H_{k+1} \hat{X}_k]. \tag{5.35}$$

The RLS algorithm is defined by Equations (5.35), (5.28) and (5.29).

5.2.3 Weighted Nonlinear Least Squares

The previous sections have shown examples of LS methods for the particular case where the model in Equation (5.4) is linear. While for linear models the LS method can be mathematically manipulated in order to obtain a closed-form solution, in the generic case of nonlinear models, the mathematical formulations are more complex. The typical solution for the nonlinear case is to solve the LS problem using a numerical method for optimization. The problem can be seen as a numerical optimization of:

$$\hat{X} = \arg\min_X \{\mathcal{J}(X)\} = \arg\min_X \left\{ \sum_{i=1}^{n} \gamma_i [z_i - h_i(X)]^2 \right\}, \tag{5.36}$$

where γ_i is a weight, commonly defined as the inverse of the variance.

For instance, the cost function to be minimized when time of arrival (TOA) and angle of arrival (AOA) measurements are available is:[1]

$$\mathcal{J}(X) = \sum_{i=1}^{n} \gamma_t^{[i]} (\eta_t^{[i]})^2 + \gamma_\theta^{[i]} (\eta_\theta^{[i]})^2, \tag{5.37}$$

[1] The unconstrained minimization procedure employed to search for the minimum of the objective function is the trust region method defined in Coleman and Li (1996).

where $\eta_t^{[i]}$ and $\eta_\theta^{[i]}$ represent the error components of the measurements of TOA and AOA, respectively, and $\gamma_t^{[i]}$ and $\gamma_\theta^{[i]}$ are weights appropriately selected to represent the reliability of the TOA and AOA measurements, respectively. When time difference of arrival (TDOA) and AOA measurements are available, the cost function is:

$$\mathcal{J}(X) = \sum_{i=1}^{n} \gamma_{\mathrm{t}}^{[i]}(\eta_{\mathrm{t}}^{[i]})^2 + \gamma_\theta^{[i]}(\eta_\theta^{[i]})^2, \tag{5.38}$$

where $\eta_{\mathrm{t}}^{[i]}$ is the error component for a measurement of TDOA and $\gamma_{\mathrm{t}}^{[i]}$ is the corresponding reliability weight.

In Equations (5.37) and (5.38), the error functions $\eta_t^{[i]}$, $\eta_{\mathrm{t}}^{[i]}$ and $\eta_\theta^{[i]}$ are respectively defined as:

$$\eta_t^{[i]} = t^{[i]} - \sqrt{[X - X^{[i]}]^{\mathrm{T}}[X - X^{[i]}]}/c, \tag{5.39}$$

$$\eta_{\mathrm{t}}^{[i]} = \mathrm{t}^{[i]} - \left(\sqrt{[X - X^{[i]}]^{\mathrm{T}}[X - X^{[i]}]} - \sqrt{[X - X^{[1]}]^{\mathrm{T}}[X - X^{[1]}]}\right)/c \tag{5.40}$$

and

$$\eta_\theta^{[i]} = \theta^{[i]} - \arctan(X - X^{[i]}), \tag{5.41}$$

where c is the speed of light, $X^{[i]}$ is the location coordinates of the ith BS, and $t^{[i]}$, $\mathrm{t}^{[i]}$ and $\theta^{[i]}$ are the TOA, TDOA and AOA measurements respectively, at the ith BS.[2] The weights $\gamma_t^{[i]}$, $\gamma_{\mathrm{t}}^{[i]}$ and $\gamma_\theta^{[i]}$ in Equations (5.37) and (5.38) are appropriately selected in order to reflect the reliability of the TOA, TDOA and AOA measurements:

$$\gamma_t^{[i]} = (\sigma_t^{[i]})^{-2}, \tag{5.42}$$

$$\gamma_{\mathrm{t}}^{[i]} = (\sigma_{\mathrm{t}}^{[i]})^{-2} \tag{5.43}$$

and

$$\gamma_\theta^{[i]} = (\sigma_\theta^{[i]})^{-2}, \tag{5.44}$$

where $\sigma_t^{[i]}$, $\sigma_{\mathrm{t}}^{[i]}$ and $\sigma_\theta^{[i]}$ are directly derived from the available measurements and represent the standard deviation of the error in the measurements.

5.2.3.1 Example of Application

Let us consider an MS in range of a geographically distributed number of cellular BSs ($\mathrm{BS}_1, \ldots, \mathrm{BS}_n$). We assume that for each available BS the MS performs m observations of either TOA and TDOA, and that the home BS, BS_1, performs m observations of AOA on the communications link BS_1–MS. After the estimation process, at the location server we have the following set of available data: (i) (**TOA**, **AOA**) or (ii) (**TDOA**, **AOA**), with entries defined by:

$$\mathbf{TOA} = \begin{bmatrix} \mathbf{TOA}_1 \\ \vdots \\ \mathbf{TOA}_n \end{bmatrix} = \begin{bmatrix} TOA_{1,1} \cdots TOA_{1,m} \\ \vdots \\ TOA_{n,1} \cdots TOA_{n,m} \end{bmatrix}, \tag{5.45}$$

[2] If we have available more than a single TOA, TDOA or AOA measurement, $t^{[i]}$, $\mathrm{t}^{[i]}$ and $\theta^{[i]}$ are the median values calculated over each set of measurements.

$$\mathbf{TDOA} = \begin{bmatrix} \mathbf{TDOA}_1 \\ \vdots \\ \mathbf{TDOA}_{n-1} \end{bmatrix} = \begin{bmatrix} TDOA_{2,1} \ldots TDOA_{2,m} \\ \vdots \\ TDOA_{n,1} \ldots TDOA_{n,m} \end{bmatrix} \tag{5.46}$$

and

$$\mathbf{AOA} = [\mathbf{AOA}_1] = [AOA_{1,1} \ldots AOA_{1,m}]. \tag{5.47}$$

Note that since TDOA measurements need two BSs and those measurements are taken with respect to BS_1, the number of actual observations is $n - 1$. For a better understanding, see Section 4.4.2.2. Let us write this problem in a weighted nonlinear least squares (WNLLS) form.

STEP 0: In order to efficiently describe our problem, we need to develop a mathematical formalism. Therefore, we start by introducing the following definitions (Frattasi and Monti 2007b):[3]

$$X^{[j]} = \begin{bmatrix} x^{[j]} \\ y^{[j]} \end{bmatrix}, \quad j = 1, \ldots, n, \tag{5.48}$$

$$\hat{X} = \begin{bmatrix} \hat{x} \\ \hat{y} \end{bmatrix}, \tag{5.49}$$

where $X^{[j]}$ represents the location coordinates of BS_j and $\hat{X}$ represents the estimated location coordinates of the MS, both in a Cartesian coordinate system:

$$\hat{d}^{[j]} = (t^{[j]} - t_{\text{tx}})c, \quad j = 1, \ldots, n, \tag{5.50}$$

$$\hat{\mathsf{d}}^{[j]} = (t^{[j]} - t_{\text{tx}})c - (t^{[1]} - t_{\text{tx}})c = (t^{[j]} - t^{[1]})c = \mathsf{t}^{[j]}c, \quad j \neq 1, \tag{5.51}$$

where $\hat{d}^{[j]}$ is the estimated distance between the MS and BS_j derived from $t^{[j]}$, the TOA of BS_j's signal at the MS, $\hat{\mathsf{d}}^{[j]}$ is the estimated distance difference between the MS and the pair (BS_1, BS_j) derived from $\mathsf{t}^{[j]}$, the TDOA of BS_1 and BS_j's signals at the MS, $c = 3 \times 10^8$ m/s is the speed of light, and t_{tx} is the time instant at which BS_1 and BS_j begin to transmit. In addition,

$$\hat{d}_k^{[j]} = \sqrt{\{\hat{X}_{k-1} - X^{[j]}\}^{\mathrm{T}}\{\hat{X}_{k-1} - X^{[j]}\}}, \quad j = 1, \ldots, n, \tag{5.52}$$

$$\hat{\mathsf{d}}_k^{[j]} = \sqrt{\{\hat{X}_{k-1} - X^{[j]}\}^{\mathrm{T}}\{\hat{X}_{k-1} - X^{[j]}\}} - \sqrt{\{\hat{X}_{k-1} - X^{[1]}\}^{\mathrm{T}}\{\hat{X}_{k-1} - X^{[1]}\}}, \quad j \neq 1, \tag{5.53}$$

$$\hat{\theta}_k^{[1]} = \arctan\left(\frac{\hat{y}_{k-1} - y^{[1]}}{\hat{x}_{k-1} - x^{[1]}}\right), \tag{5.54}$$

[3] For simplicity of presentation, only the x and y coordinates are considered in the derivations given here (this corresponds to the case where the MS and BS_j are located on a relatively flat plane). Although the z coordinate is ignored, the present example can easily be extended to the three-dimensional case.

where $\hat{d}_k^{[j]}$ and $\hat{\mathrm{d}}_k^{[j]}$ have meanings similar to those in Equations (5.50) and (5.51), but with the difference that here, instead of being calculated directly from the TOA or TDOA measurements, they are derived from $\hat{X}_{k-1} = [\hat{x}_{k-1}, \hat{y}_{k-1}]^{\mathrm{T}}$, the estimated location coordinates of the MS at iteration $k-1$ of the minimization routine. The same applies to $\hat{\theta}_k^{[1]}$, the estimated angle between the MS and BS_1.

STEP 1: The WNLLS method requires an initial guess $\hat{X}_0 = [\hat{x}_0, \hat{y}_0]^{\mathrm{T}}$ at the beginning of its minimization routine.

STEP 2: Unlike the KF method (see Section 5.3.1), which picks up in each iteration a new set of measurements (e.g., $(\mathbf{TOA}_{1,1}, \mathbf{AOA}_{1,1})$, $(\mathbf{TOA}_{1,2}, \mathbf{AOA}_{1,2})$, etc.), the WNLLS method uses only one set of measurements at the beginning of its minimization routine. As a consequence, in order for the WNLLS method to be able to exploit all of the available data wisely, we apply the median operator, that is, $\widetilde{(\bullet)}$, to each of the rows in Equations (5.45)–(5.47) and, by using the formalism introduced previously, we obtain:[4]

$$\mathbf{TOA} = \begin{bmatrix} \widetilde{\mathbf{TOA}_1} \\ \vdots \\ \widetilde{\mathbf{TOA}_n} \end{bmatrix} = \begin{bmatrix} t^{[1]} \\ \vdots \\ t^{[n]} \end{bmatrix}, \tag{5.55}$$

$$\mathbf{TDOA} = \begin{bmatrix} \widetilde{\mathbf{TDOA}_1} \\ \vdots \\ \widetilde{\mathbf{TDOA}_{n-1}} \end{bmatrix} = \begin{bmatrix} \mathrm{t}^{[2]} \\ \vdots \\ \mathrm{t}^{[n]} \end{bmatrix} \tag{5.56}$$

and

$$\mathbf{AOA} = [\widetilde{\mathbf{AOA}_1}] = [\theta^{[1]}]. \tag{5.57}$$

STEP 3: Finally, we can write the objective function $\mathcal{J}(X)$ in Equation (5.36) for each of the two envisioned cases, that is, **(TOA, AOA)** and **(TDOA, AOA)**, as (Frattasi 2007):

$$\mathcal{J}(X) = \sum_{j=1}^{n} \gamma_t^{[j]} \{f_t(\hat{X})\}^2 + \gamma_\theta^{[1]} \{f_\theta(\hat{X})\}^2 \tag{5.58}$$

and

$$\mathcal{J}(X) = \sum_{j=1}^{n} \gamma_{\mathrm{t}}^{[j]} \{f_{\mathrm{t}}(\hat{X})\}^2 + \gamma_\theta^{[1]} \{f_\theta(\hat{X})\}^2, \tag{5.59}$$

where the functions $f(\bullet)$ are defined as:

$$f_t(\hat{X}) = \hat{d}^{[j]} - \hat{d}_k^{[j]}, \quad j = 1, \ldots, n, \tag{5.60}$$

$$f_{\mathrm{t}}(\hat{X}) = (\hat{d}^{[j]} - \hat{d}^{[1]}) - (\hat{d}_k^{[j]} - \hat{d}_k^{[1]}) = \hat{\mathrm{d}}^{[j]} - \hat{\mathrm{d}}_k^{[j]}, \quad j = 1, \ldots, n, \tag{5.61}$$

and

$$f_\theta(\hat{X}) = \theta^{[1]} - \hat{\theta}_k^{[1]}. \tag{5.62}$$

The weights γ_t, γ_{t} and γ_θ should be appropriately selected in order to reflect the reliability of the available TOA, TDOA and AOA measurements. In particular, they can be chosen as (Frattasi 2007):

[4] The *median* operator is preferred to the *mean* operator, owing to its higher robustness against biased measurements.

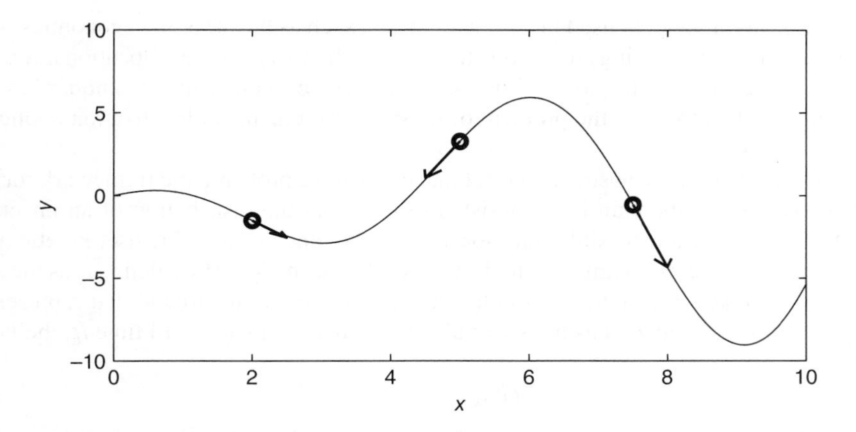

Figure 5.1 Local-minimum problem (the function has been chosen for illustrative purposes only).

$$\gamma_t^{[j]} = \frac{1}{\sigma^2_{\mathbf{TOA}_j}}, \quad j = 1, \ldots, n, \tag{5.63}$$

$$\gamma_{\mathrm{t}}^{[j]} = \frac{1}{\sigma^2_{\mathbf{TDOA}_j}}, \quad j = 1, \ldots, n, \tag{5.64}$$

and

$$\gamma_\theta^{[1]} = \frac{1}{\sigma^2_{\mathbf{AOA}_1}}, \tag{5.65}$$

where $\sigma^2_{\mathbf{TOA}_j}$, $\sigma^2_{\mathbf{TDOA}_j}$ and $\sigma^2_{\mathbf{AOA}_1}$ are variances directly derived from the available sets of data.

5.2.4 The Absolute/Local-minimum Problem

Optimization algorithms typically operate in a given interval and iteratively evaluate the values of the subject function in consecutive smaller subintervals until a certain criteria of decision is met. The requirement is that the subject function shall assure that inside that interval the problem is convex, otherwise the optimization algorithm does not assure that the absolute minimum is determined. For instance, given the interval in the example depicted in Figure 5.1, the optimization algorithm could evolve in the direction indicated by the arrows. As convexity is not assured within that interval, the local minimum around $x = 3$ could be returned by the optimization algorithm instead of the absolute minimum around $x = 9$. As it is complex or even impossible to circumvent this problem and assure convergence to an absolute minimum, the system designed shall take this fact into consideration.

5.3 Bayesian Filtering

Bayesian filtering techniques are techniques that estimate the state of a dynamic system based on measurements from noisy sensors (Arulampalam et al. 2002; Grewal and Andrews 2001).

These techniques are widely used in several contexts, such as business and economics, signals and systems, image processing and mobile tracking. In the specific case of location information estimation in scenarios with static or moving devices, Bayesian filtering techniques have been successfully applied to solve the problem of noisy measurements in localization applications (Fox et al. 2003).

The Bayesian filter is an abstract concept that describes a probabilistic framework for recursive state estimation. Its main idea consists of estimating the distribution of an uncertainty, called the *belief*, over the possible state space (e.g., the range of possible user kinetics) given the sensor measurements obtained up to the time of the estimation. If we define t_k as the instant in time when the kth measurement is obtained, x_k as the random variable that represents the state space at time t_k and $Z_{1:k}$ as the set of all sensor measurements until time t_k, the *belief* is defined as:

$$Bel(x_k) = \mathrm{f}(x_k|Z_{1:k}), \tag{5.66}$$

where $\mathrm{f}(x_k|Z_{1:k})$ is the probability density function of x_k given $Z_{1:k}$. Note that in positioning applications, x_k often represents the user kinetics. However, in this section, x_k is considered as an arbitrary state space.

As Equation (5.66) shows, the belief has a dependency on all measurements obtained from time t_0 until the present time t_k, meaning that the complexity increases with time. To solve the problem presented by this dependency, it is assumed that the Bayesian filter has Markovian properties of the first order, that is, all the information necessary to compute an estimation of position at time t_k is given by the information available at time t_{k-1}.

The Bayesian filter has two main steps:

- **Prediction.** At each time, the current state x_k is updated based on the previous state x_{k-1}:

$$\mathrm{Bel}^-(x_k) = \int \mathrm{f}(x_k|x_{k-1})\mathrm{Bel}(x_{k-1})\ \mathrm{d}x_{k-1}. \tag{5.67}$$

  In Equation (5.67), the term $\mathrm{f}(x_k|x_{k-1})$ describes the *system dynamics*, that is, how the system evolves from the state x_{k-1} to the current state x_k. The belief at time t_0 is usually defined as a reasonable density function within the context of the system in use.
- **Correction.** The predicted belief is corrected every time a new sensor measurement is obtained:

$$\mathrm{Bel}(x_k) = \varrho_k \mathrm{f}(z_k|x_k)\mathrm{Bel}^-(x_k). \tag{5.68}$$

  In Equation (5.68), the term ϱ_k represents the normalization constant that is used in order to ensure that $\int \mathrm{Bel}(x_k)\ \mathrm{d}x_k = 1$. The term $\mathrm{f}(z_k|x_k)$ represents the *observation model*, that is, the probability of having a sensor measurement z_k given that the system is in state x_k.

For localization purposes, the term $\mathrm{f}(x_k|x_{k-1})$ in Equation (5.67) is determined by the motion model, while the term $\mathrm{f}(z_k|x_k)$ in Equation (5.68) is dependent on the technology of the sensor and the location method. For instance, the latter term could be the path loss expression of Section 3.2.1 if we are considering an RSS-based positioning system.

5.3.1 The Kalman Filter

The Kalman filter (Grewal and Andrews 2001; Kailath et al. 2000; Kalman 1960) is one of the most widely used implementations of the Bayesian filter. The basic assumptions of this filter are that the models $f(x_k|x_{k-1})$ in Equation (5.67) and $f(z_k|x_k)$ in Equation (5.68) are linear, and that the noise statistics have a Gaussian distribution, that is, the belief is Gaussian. The filter defines five equations (the set of expressions given by Equations (5.73)–(5.77)), which, in a recursive manner, can estimate the state of an unknown noisy process. It is a very powerful filter because the amount of processing is considerably reduced when compared with other Bayesian filters (such as the particle filter described in Section 5.3.2) and because it estimates past, present and even future states using approximations to the true system model. Since this filter was proposed, it has been successfully applied in several areas of research. An example of such an application is in the tracking of moving devices.

The KF equations are able to predict the state vector X of a discrete-time process that is governed by the following linear system in state space form:

$$X_k = AX_{k-1} + BU_{k-1} + \mathbf{w}_{k-1}, \quad \zeta \sim \mathrm{Norm}(0, Q), \tag{5.69}$$

$$Z_k = HX_k + \mathbf{v}_k, \quad \eta \sim \mathrm{Norm}(0, R). \tag{5.70}$$

In Equation (5.69), the matrix A defines the relation between the previous state and the current one, the matrix B relates the previous external input vector U_k to the current state, and ζ is a random variable representing the process noise. The control input U can be either known or unknown. In Equation (5.70), the matrix H represents the relation between the present state X_k and the measurement Z_k expected for that state X_k. The random variable η represents the measurement noise. In Equations (5.69) and (5.70), the random variables ζ and η are assumed to be independent of each other, to be white and to have a normal distribution with zero mean and covariance matrices Q and R, respectively. Having defined the matrices in Equations (5.69) and (5.70) and the related parameters, we still need to define the starting values for the state $X_{0|0}$ and the error covariance estimator $P_{0|0}$. These parameters are system-dependent. For the state $X_{0|0}$, this value is the expected value of the state space at time t_0. The estimator $P_{0|0}$ is commonly initialized with the covariance matrix Q. Based on the system of Equations (5.69) and (5.70), the KF can be defined as:

Initialization:

$$\hat{X}_{0|0} = [X_0], \tag{5.71}$$

$$P_{0|0} = [(\hat{X}_{0|0} - X_0)(\hat{X}_{0|0} - X_0)^{\mathrm{T}}] \simeq Q. \tag{5.72}$$

Prediction:

$$\hat{X}_{k|k-1} = A_k \hat{X}_{k-1|k-1} + B_k U_{k-1|k-1}, \tag{5.73}$$

$$P_{k|k-1} = A_k P_{k-1|k-1} A_k^{\mathrm{T}} + Q_k. \tag{5.74}$$

Correction:

$$K_k = P_{k|k-1} H_k^{\mathrm{T}} (H_k P_{k|k-1} H_k^{\mathrm{T}} + R_k)^{-1}, \tag{5.75}$$

$$\hat{X}_{k|k} = \hat{X}_{k|k-1} + K_k (Z_k - H_k \hat{X}_{k|k-1}), \tag{5.76}$$

$$P_{k|k} = (I - K_k H_k) P_{k|k-1}. \tag{5.77}$$

From Equations (5.75) and (5.76), the "innovation process" operating in this context can be identified as the difference between the obtained measurements Z_k and the expected measurements $H_k\hat{X}_{k|k-1}$. This process, by definition Gaussian-distributed, defines how much each individual measurement Z_k contributes to the overall estimation of the state space $\hat{X}_k$. The innovation process and its associated covariance are defined as:

$$\tilde{Z}_k = Z_k - H_k\hat{X}_{k|k-1}, \tag{5.78}$$

$$S_k = H_k P_{k|k-1} H_k^{\mathrm{T}} + R_k. \tag{5.79}$$

In Section 6.4.3.1, this innovation process is used as a solution for estimating abrupt changes in the dynamics of the MS, often referred to as "maneuvers".

As an alternative form of the KF for scenarios where the positioning system is not run in real time, we shall now present the fixed-lag Kalman smoother (Cohn et al. 1994; Todling et al. 1997). This smoother runs a KF similar to the one presented above in this section and then a set of additional equations called (Todling et al. 1997) *retrospective-analysis* equations. A smoothed estimation $\hat{X}_{k-\ell|k}$ at a past time $t_{k-\ell}$, $\ell > 0$, can be obtained by running the expressions of the ordinary KF until time $t_{k-\ell}$ and then, beyond that time until the current time t_k, running the *retrospective-analysis* equations. The smoothed estimator $\hat{X}_{k-\ell|k}$ of the hidden process beyond time $t_{k-\ell}$, given observations until time t_k, is calculated from:

Smoothing:

$$\hat{X}_{k-\ell|k} = \hat{X}_{k-\ell|k-1} + K_{k-\ell|k}\tilde{Z}_k, \tag{5.80}$$

$$K_{k-\ell|k} = (P^{\mathrm{fa}}_{k,k-\ell|k-1})^{\mathrm{T}} H_k^{\mathrm{T}} S_k^{-1}, \tag{5.81}$$

$$P^{\mathrm{fa}}_{k,k-\ell|k-1} = A_k P^{\mathrm{aa}}_{k-1,k-\ell|k-1}, \tag{5.82}$$

$$P^{\mathrm{aa}}_{k,k-\ell|k} = P^{\mathrm{fa}}_{k,k-\ell|k-1} - K_k H_k P^{\mathrm{fa}}_{k,k-\ell|k-1}, \tag{5.83}$$

where the quantities $\tilde{Z}_k$ and S_k are given by Equations (5.78) and (5.79), respectively. The matrix $P^{\mathrm{fa}}_{k,k-\ell|k-1}$ is called the forecast retrospective-analysis error cross-covariance matrix and $P^{\mathrm{aa}}_{k,k-\ell|k}$ is the filter analysis retrospective-analysis error cross-covariance matrix. Basically, the matrices $P^{\mathrm{fa}}_{k,k-\ell|k-1}$ and $P^{aa}_{k,k-\ell|k}$ propagate along time the uncertainty in the estimation at time $t_{k-\ell}$, given the measurements until time t_k. The covariance error $P_{k-\ell|k}$ at time $t_{k-\ell}$ is given by:

$$P_{k-\ell|k} = P_{k-\ell|k-1} - K_{k-\ell|k} H_k P^{\mathrm{fa}}_{k,k-\ell|k-1}. \tag{5.84}$$

The set of Equations (5.80)–(5.84), together with the KF Equations (5.73)–(5.77), define the Kalman smoother.

5.3.1.1 Extended KF

As we have seen in the previous section, the KF is an algorithm that requires the models to be linear. The main problem with this approach is that in typical cases where such an algorithm could be useful, the models of the system are nonlinear, that is,

$$X_k = \mathcal{F}(X_{k-1}, U_{k-1}, \zeta_{k-1}), \quad \zeta \sim \mathrm{Norm}(0, Q), \tag{5.85}$$

$$Z_k = \mathcal{H}(X_k, \eta_k), \quad \eta \sim \mathrm{Norm}(0, R). \tag{5.86}$$

Researchers have proposed several extensions of the KF for nonlinear cases. The most used solutions are the extended Kalman filter (EKF), described in detail, for example, by Grewal and Andrews (2001), and the unscented Kalman filter (UKF), proposed by Julier and Uhlmann (1997). These solutions deal with the nonlinearity by approximating it with a Taylor expansion. While the EKF approximates the models with their first derivative, the UKF approximates them with their second derivative.

The EKF is a commonly used solution when the nonlinearity is not "very" strong, meaning that the error in this approximation is several orders of magnitude lower than the typical errors introduced by the noise in the system. The solution uses a Taylor expansion, truncated to the linear terms, resulting in expressions for the models equivalent to Equations (5.69) and (5.70). The difference is that the terms A and H are now Jacobian matrices instead of precise relationships. These Jacobian matrices can be defined as:

$$A = \frac{\partial \mathcal{F}}{\partial X}(X_{k-1}, U_{k-1}, 0), \tag{5.87}$$

$$H = \frac{\partial \mathcal{H}}{\partial X}(X_k, 0). \tag{5.88}$$

Based on the nonlinear Equations (5.85) and (5.86), and the parameters in Equations (5.87) and (5.88), the EKF equations can be defined as:

Initialization:

$$\hat{X}_{0|0} = [X_0], \tag{5.89}$$

$$P_{0|0} = [(\hat{X}_{0|0} - X_0)(\hat{X}_{0|0} - X_0)^{\mathrm{T}}] \simeq Q. \tag{5.90}$$

Prediction:

$$\hat{X}_{k|k-1} = \mathcal{F}(\hat{X}_{k-1|k-1}, U_{k-1|k-1}, 0), \tag{5.91}$$

$$P_{k|k-1} = A_k P_{k-1|k-1} A_k^{\mathrm{T}} + Q_k. \tag{5.92}$$

Correction:

$$K_k = P_{k|k-1} H_k^{\mathrm{T}} (H_k P_{k|k-1} H_k^{\mathrm{T}} + R_k)^{-1}, \tag{5.93}$$

$$\hat{X}_{k|k} = \hat{X}_{k|k-1} + K_k(Z_k - \mathcal{H}(\hat{X}_{k|k-1}, 0)), \tag{5.94}$$

$$P_{k|k} = (I - K_k H_k) P_{k|k-1}. \tag{5.95}$$

One of the drawbacks of the EKF approach for stochastic filtering is, as previously stated, the linearization of the models. This linearization, depending on the application, can represent a high impact on the system concerning the errors in the estimated mean and covariance. The problem becomes more and more evident as higher-order terms become more important in the nonlinear model. To provide a more accurate approximation of the models, the UKF was proposed.

5.3.1.2 Unscented KF

Similarly to the EKF, the UKF assumes the noise in the models follows a Gaussian random distribution. The difference is that the distributions are represented by a minimal set of sample points chosen in such a way that the true mean and covariance of the errors are captured. As was shown by Julier and Uhlmann (1997), these samples, when propagated through the nonlinearity, accurately capture the posterior mean and covariance up to the third-order Taylor expansion of the nonlinearity. This procedure is known as the unscented transformation. The algorithm augments the hidden space X_k with noise components such that the hidden space becomes $X^a = [X^{\mathrm{T}} \quad \zeta^{\mathrm{T}} \quad \eta^{\mathrm{T}}]^{\mathrm{T}}$. Then the algorithm can be described as:

Initialization:

$$\hat{X}_{0|0} = [X_0], \tag{5.96}$$

$$P_{0|0} = [(\hat{X}_{0|0} - X_0)(\hat{X}_{0|0} - X_0)^{\mathrm{T}}] \simeq Q, \tag{5.97}$$

$$\hat{X}^a_{0|0} = [X^a_0] = [\hat{X}^{\mathrm{T}}_{0|0} \quad \mathbf{0} \quad \mathbf{0}]^{\mathrm{T}}, \tag{5.98}$$

$$P^a_{0|0} = [(\hat{X}^a_{0|0} - X^a_0)(\hat{X}^a_{0|0} - X^a_0)^{\mathrm{T}}] = \begin{bmatrix} P_{0|0} & \mathbf{0} & \mathbf{0} \\ \mathbf{0} & Q & \mathbf{0} \\ \mathbf{0} & \mathbf{0} & R \end{bmatrix}, \tag{5.99}$$

where $\mathbf{0}$ is a zero matrix of appropriate size, and Q and R are the noise covariance matrices of the hidden process (in Equation (5.85) and observable process (in Equation (5.86), respectively. The prediction phase of the algorithm is:

Prediction:

$$\mathsf{X}^a_{k-1|k-1} = \left[\hat{X}^a_{k-1|k-1} \quad \hat{X}^a_{k-1|k-1} \pm \sqrt{(L+\lambda)P^a_{k-1|k-1}}\right], \tag{5.100}$$

$$\mathsf{X}^x_{k|k-1} = \mathcal{F}(\mathsf{X}^x_{k-1|k-1}, 0, \mathsf{X}^w_{k-1|k-1}), \tag{5.101}$$

$$\hat{X}_{k|k-1} = \sum_{i=0}^{2L} W_i^{(m)} \mathsf{X}^x_{k|k-1;i}, \tag{5.102}$$

$$P_{k|k-1} = \sum_{i=0}^{2L} W_i^{(c)} [\mathsf{X}^x_{k|k-1;i} - \hat{X}_{k|k-1}][\mathsf{X}^x_{k|k-1;i} - \hat{X}_{k|k-1}]^{\mathrm{T}}, \tag{5.103}$$

$$\mathsf{Z}_{k|k-1} = \mathcal{H}(\mathsf{X}^x_{k|k-1}, \mathsf{X}^v_{k|k-1}), \tag{5.104}$$

$$\hat{Z}_{k|k-1} = \sum_{i=0}^{2L} W_i^{(m)} \mathsf{Z}_{k|k-1;i}. \tag{5.105}$$

In Equation (5.100), $\mathsf{X}^a = [(\mathsf{X}^x)^{\mathrm{T}} \quad (\mathsf{X}^w)^{\mathrm{T}} \quad (\mathsf{X}^v)^{\mathrm{T}}]$ is the sigma points, that is, the points where the noise distributions are sampled. The notation $\sqrt{(L+\lambda)P^a_{k-1|k-1}}$ corresponds to the Cholesky factorization, and its summation with $\hat{X}^a_{k-1|k-1}$ corresponds to a columnwise sum. The parameters L and λ are the dimension of X and a scaling parameter, respectively; the latter is set to $\lambda = (10^3)^2 L - L$ as advised in Julier and Uhlmann (1997). In Equations (5.101) and (5.104), $\mathcal{F}$ and $\mathcal{H}$ are, respectively, the evolution model and the observation model as defined by Equations (5.85) and (5.86). In Equations (5.102) and (5.105),

the weights $W_i^{(m)}$ are calculated as in Julier and Uhlmann (1997), and the symbol $\mathsf{X}^x_{k|k-1;i}$ corresponds to column i of the matrix $\mathsf{X}^x_{k|k-1}$. The correction phase of the algorithm is:
Correction:

$$P_{z_k z_k} = \sum_{i=0}^{2L} W_i^{(c)} [\mathsf{Z}_{k|k-1;i} - \hat{Z}_{k|k-1}][\mathsf{Z}_{k|k-1;i} - \hat{Z}_{k|k-1}]^{\mathrm{T}}, \tag{5.106}$$

$$P_{x_k z_k} = \sum_{i=0}^{2L} W_i^{(c)} [\mathsf{X}_{k|k-1;i} - \hat{X}_{k|k-1}][\mathsf{Z}_{k|k-1;i} - \hat{Z}_{k|k-1}]^{\mathrm{T}}, \tag{5.107}$$

$$K_k = P_{x_k z_k} P^{-1}_{z_k z_k}, \tag{5.108}$$

$$\hat{X}_{k|k} = \hat{X}_{k|k-1} + K_k (Z_k - \hat{Z}_{k|k-1}), \tag{5.109}$$

$$P_{k|k} = P_{k|k-1} + K_k P_{z_k z_k} K_k^{\mathrm{T}}, \tag{5.110}$$

where the $W_i^{(c)}$, similarly to the $W_i^{(m)}$, are calculated as in Julier and Uhlmann (1997). The matrix $P_{z_k z_k}$ corresponds to the innovation covariance and $P_{x_k z_k}$ corresponds to the cross-correlations between the hidden and the observable process.

5.3.1.3 Convergence Issues

Iterative methods such as the KF have a particular aspect concerning the convergence of the estimators that cannot be disregarded. This aspect is connected to the number of steps of the various iterations that are performed. In the case of the KF, since each step is performed whenever fresh measurements are obtained, the convergence issue is tightly connected to the frequency of fresh measurements.

Let us assume a simple scenario where three access points (APs) measure the distance to a target that is moving through their coverage range. We consider two cases, where fresh measurements are available every time the target moves either 1 m or 5 m. We have run a filter for these two cases with the appropriate parameters.[5] Figure 5.2 shows the sequence of position estimators when the extended KF is used. As can be seen, more frequent measurements (Figure 5.2 (left) in comparison with Figure 5.2 (right)) means faster convergence to the true trajectory. The same conclusion can be drawn by calculating the Euclidean distance between the true position and the estimated position for every time step. Figure 5.3 shows that this Euclidean distance tends to lower the values as more measurements are available. At this point, however, it is important to mention that this tendency may not be necessarily true if there are non-Gaussian errors in the measurements or in the presence of colored noise.

Though the Euclidean distance between the true and the estimated position indicates the estimation error, the true trajectory is seldom known. For this reason, it is common to analyze the trace of the covariance matrix P. As Figure 5.4 shows, the trace of P tends to become lower and lower as the number of available measurements increases. This value gives a level

[5] Since we analyze the influence of measurement frequency and not the influence of filter configuration in this section, the details of the filter are not given here.

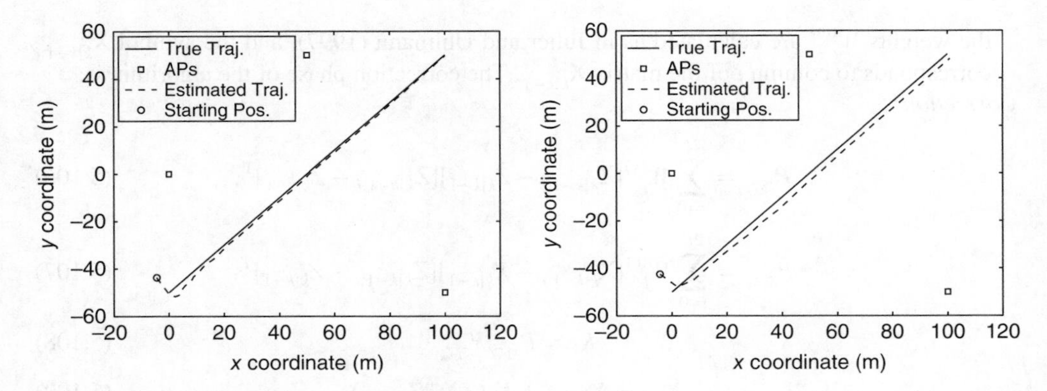

Figure 5.2 Example of an application where a wireless target is tracked using three access points capable of measuring the distance to the target. The left plot shows the tracking result when fresh measurements are available every time the target moves 1 m. The right plot shows the equivalent result when fresh measurements are available every time the target moves 5 m.

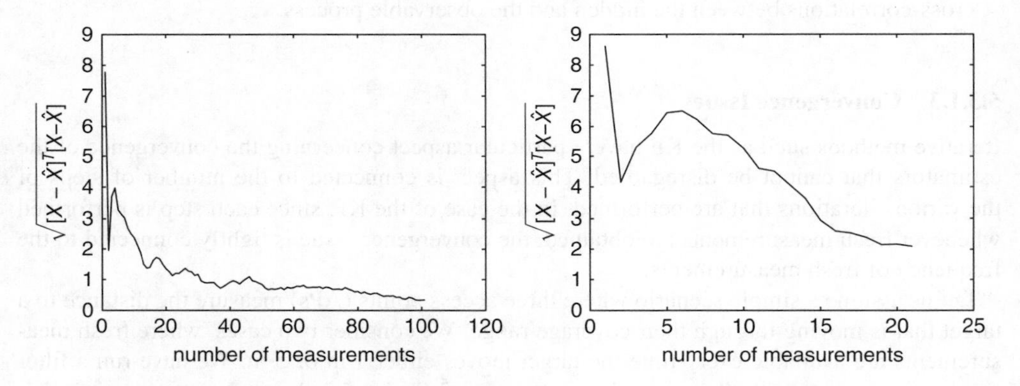

Figure 5.3 Evolution of the Euclidean distance between the true position and the estimated position at every time step. The left plot was obtained for a setup where fresh measurements are available every time the target moves 1 m. The right plot was obtained for a setup where measurements are available every time the target moves 5 m.

of confidence for the estimator of position. Thus, a common rule to define the convergence of estimators is to define a threshold below which the filter is said to converge.

5.3.2 *The Particle Filter*

The particle filter (PF) is the name given to a Monte Carlo implementation of the Bayesian framework. This filter is adequate for nonlinear and non-Gaussian estimation, but it suffers from the numerical problems that characterize Monte Carlo algorithms. Several different approaches are used, depending on characteristics of the noisy set of data. As stated by Arulampalam et al. (2002), the algorithms used have a basis in sequential importance

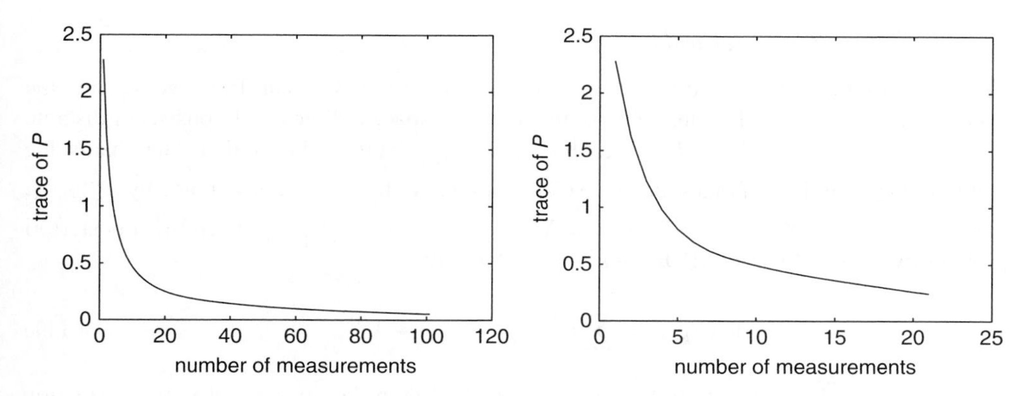

Figure 5.4 Evolution of the trace of the covariance matrix P at every time step. The left plot was obtained in a setup where fresh measurements are available every time the target moves 1 m. The right plot was obtained for a setup where measurements are available every time the target moves 5 m.

sampling. From the available solutions, we present here the "bootstrap filter" introduced by Gordon et al. (1993).

Suppose that at time t_{k-1}, M random samples $X^{(m)}_{k-1|k-1}$ $(1 \leq m \leq \mathsf{M})$, commonly referred to as particles, are available and that they are drawn from a probability distribution $\mathrm{f}(X_{k-1}|Z_{1:k})$, that is, the belief $\mathrm{Bel}(X_{k-1})$, as expressed in Equation (5.66). The bootstrap filter is an algorithm that propagates these samples through the system models (i.e., the models of Equations (5.85) and (5.86)) in order to obtain random samples $X^{(m)}_{k|k}$ that approximate the belief $\mathrm{Bel}(X_k)$ at time t_k. Similarly to any Bayesian filter, the bootstrap filter runs in a cyclic fashion:

Initialization:

$$X^{(m)}_{0|0} \sim \mathrm{f}(X_0), \tag{5.111}$$

Prediction:

$$X^{(m)}_{k-1|k-1} \sim \mathrm{Bel}(X_{k-1}), \tag{5.112}$$

$$X^{(m)}_{k|k-1} = \mathcal{F}(X^{(m)}_{k-1|k-1}, U_{k-1|k-1}, \zeta^{(m)}_{k-1}) \quad \zeta^{(m)}_{k-1} \sim \mathrm{f}(\zeta_{k-1}), \tag{5.113}$$

where the $\zeta^{(m)}_{k-1}$ are samples drawn from the noise component ζ in the system dynamics:
Correction:

$$W^{(m)}_k = \frac{\mathrm{f}(Z_k|X^{(m)}_{k|k-1})}{\sum_{m=1}^{\mathsf{M}} \mathrm{f}(Z_k|X^{(m)}_{k|k-1})}, \quad \mathrm{f}(Z_k|X^{(m)}_{k|k-1}) = \int \delta(Z_k - \mathcal{H}(X^{(m)}_{k|k-1}, \eta_k))\mathrm{f}(\eta_k)\,\mathrm{d}\eta_k, \tag{5.114}$$

$$\hat{X}_{k|k} = \sum_{m=1}^{\mathsf{M}} W^{(m)}_k X^{(m)}_{k|k-1}, \tag{5.115}$$

$$\mathrm{Bel}(X_k) = \sum_{m=1}^{\mathsf{M}} W^{(m)}_k \delta(X_k - X^{(m)}_{k|k-1}), \tag{5.116}$$

where $\hat{X}_{k|k}$ is the estimator of $X_{k|k}$ at time t_k given measurements until time t_k.

5.3.3 Grid-based Methods

The grid-based methods are optimal implementations of the Bayesian framework when the state space is discrete and finite. Suppose that the state space at time $k-1$ consists of discrete states $X^{(i)}_{k-1|k-1}$, with $i = 1, \ldots, l_s$. For each state $X^{(i)}_{k-1|k-1}$, let the belief of that state, that is, the conditional probability of that state given observation until time $k-1$, be denoted by $w^{(i)}_{k-1|k-1}$, that is, $\text{Bel}(X_{k-1} = X^{(i)}_{k-1|k-1}) = \Pr(X_{k-1} = X^{(i)}_{k-1|k-1}|Z_{1:k-1}) = w^{(i)}_{k-1|k-1}$. Then the a posteriori probability density function (PDF) at $k-1$ can be written as:

$$\text{Bel}(X_{k-1}) = \sum_{i=1}^{l_s} w^{(i)}_{k-1|k-1}\delta(X_{k-1} - X^{(i)}_{k-1|k-1}), \tag{5.117}$$

where $\delta()$ is the Dirac function. Using Equation (5.117) in the update and in the correction function, it is possible to define the grid-based filter by the following equations:

Initialization:

$$\hat{X}_{0|0} = \arg\max_{X^{(i)}_{0|0}}(w^{(i)}_{0|0}), \tag{5.118}$$

Prediction:

$$w^{(i)}_{k|k-1} = \sum_{j=1}^{l_s} w^{(j)}_{k-1|k-1} P(X^{(i)}_{k|k}|X^{(j)}_{k-1|k-1}), \tag{5.119}$$

$$\text{Bel}^{-}(X_k) = \sum_{i=1}^{l_s} w^{(i)}_{k|k-1}\delta(X_k - X^{(i)}_{k|k}), \tag{5.120}$$

Correction:

$$w^{(i)}_{k|k} = \frac{w^{(i)}_{k|k-1} P(Z_k|X^{(i)}_{k|k})}{\sum_{j=1}^{l_s} w^{(j)}_{k|k-1} P(Z_k|X^{(j)}_{k|k})}, \tag{5.121}$$

$$\text{Bel}(X_k) = \sum_{i=1}^{l_s} w^{(i)}_{k|k}\delta(X_k - X^{(i)}_{k|k}), \tag{5.122}$$

$$\hat{X}_{k|k} = \arg\max_{X^{(i)}_{k|k}}(w^{(i)}_{k|k}). \tag{5.123}$$

The above equations require the distributions $p(X^{(i)}_{k|k}|X^{(j)}_{k-1|k-1})$ and $p(Z_k|X^{(j)}_{k|k})$ to be known, although they are not constrained to be any particular form of distribution. Note that in Equations (5.118) and (5.123), the estimated state is defined as the maximum likelihood of the discrete belief.

5.4 Estimating Model Parameters and Biases in Observations

In wireless positioning and in general in any estimation process, it can happen that some of the model parameters are unknown or even that the models are unknown. In this section, we present a few typical methods for estimating such unknown parameters.

5.4.1 Precalibration

A very simple but time-consuming method for determining model parameters and biases in the measurements is to run a precalibration phase. In this phase, typical MS movements are performed within a test area in order to obtain actual measurements of the positioning system. For instance, in a WLAN positioning solution, one would need to obtain RSS measurements. While the true position of the MS gives us the so-called ground truth, the measurements give us data to estimate the target position. Combined analysis of the two types of data gives us information on the missing parameters.

A typical, generic method of precalibration might run as follows:

- Typical target trajectories are defined, and both true position data and corresponding timestamps are recorded. This data defines the ground truth.
- During the movement of the target, the system obtains measurements from the radio links that are available and are being calibrated. The timestamps of the measurements are also stored.
- By using appropriate models, the data measured from the radio links is translated into position information. To accomplish this, the unknown parameters of the models must be chosen on a rule-of-thumb basis.
- By comparing the estimated position information and ground truth, the parameters are determined either by trial-and-error tests or by using more advanced mathematical approaches.

Note, however, that the number of parameters to be estimated must guarantee the observability of the system. Only in this case can a single solution be found for each parameter.

Let us assume that the ground truth position at time k is given by X_k^{true}. We assume also that at that same time k, the measurement Z_k was obtained and that the measurements have a linear relation with the parameters ξ to be estimated:

$$Z = H\xi + \eta, \tag{5.124}$$

where each H_k is a function of the ground truth X_k^{true}. Using the LS estimator, the parameters can be determined from:

$$\hat{\xi} = (H^{\mathrm{T}}H)^{-1}H^{\mathrm{T}}Z, \tag{5.125}$$

where Z and H stack all the measurements Z_k and terms H_k along time into column vectors. Then, based on the estimator $\hat{\xi}$, it is possible to determine the covariance matrix of the parameters from:

$$P = (H^{\mathrm{T}}R^{-1}H)^{-1}, \tag{5.126}$$

with $R = [\{Z - H\xi - [Z - H\hat{\xi}]\}\{Z - H\xi - [Z - H\hat{\xi}]\}^{\mathrm{T}}] = [\eta\eta^{\mathrm{T}}]$ with η zero mean distributed.

5.4.2 Joint Parameter and State Estimation

In joint estimation, both the state space and the parameters are estimated together. In order to actually implement such a concept, it is necessary to augment Equations (5.69) and (5.70) with the parameters:

$$\begin{bmatrix} X_k \\ \hline \xi_k \end{bmatrix} = \begin{bmatrix} A & | & A_\xi \\ \hline 0 & | & 1 \end{bmatrix} \begin{bmatrix} X_{k-1} \\ \hline \xi_{k-1} \end{bmatrix} + \begin{bmatrix} B \\ \hline 0 \end{bmatrix} U_{k-1} + \zeta_{k-1},$$
$$\zeta \sim \mathrm{Norm}\left(0, \begin{bmatrix} Q & | & 0 \\ \hline 0 & | & Q_\xi \end{bmatrix}\right), \tag{5.127}$$

$$Z_k = \begin{bmatrix} H & | & 0 \end{bmatrix} \begin{bmatrix} X_k \\ \hline \xi_k \end{bmatrix} + \eta_k \quad \text{and} \quad \eta \sim \mathrm{Norm}\left(0, \begin{bmatrix} R & | & 0 \\ \hline 0 & | & R_\xi \end{bmatrix}\right), \tag{5.128}$$

where the parameters ξ are considered independent of time. Using the new definition of the state space and observation vector, the desired estimation algorithm can be used.

5.5 Alternative Approaches

Although LS algorithms and Bayesian filtering techniques are some of the most commonly used techniques to circumvent errors in the measurements obtained from radio channels, several other options exist in the literature.

5.5.1 Fingerprinting

Let us assume that for localizing an MS, measurements from the communication channel are obtained by several BSs identified by an index i. Let us further assume a general case where a precalibration phase is performed. The precalibration is done in such a way that for scattered positions X over the entire scenario, the distribution of the observed measurements Z at BS_i is determined as $\mathrm{f}(Z^{(i)}|X)$. This information is considered as the calibration content of a database. Let us now suppose that new data $Z^{(i)}_{\mathrm{new}}$ is available in order to perform positioning:

$$\mathrm{f}(X|Z^{(i)}_{\mathrm{new}}) = \frac{\mathrm{f}(Z^{(i)}_{\mathrm{new}}|X)\mathrm{f}(X)}{\mathrm{f}(Z^{(i)}_{\mathrm{new}})}. \tag{5.129}$$

Given that $\mathrm{f}(Z^{(i)}_{\mathrm{new}})$ is not straightforward to determine,

$$\mathrm{f}(X|Z^{(i)}_{\mathrm{new}}) = \frac{\mathrm{f}(Z^{(i)}_{\mathrm{new}}|X)\mathrm{f}(X)}{\int_X \mathrm{f}(Z^{(i)}_{\mathrm{new}}|X)\mathrm{f}(X)\,\mathrm{d}X}. \tag{5.130}$$

Assuming that the measurements from the various APs are i.i.d. when conditioned to X,

$$\mathrm{f}(X|Z_{\mathrm{new}}) = \frac{\mathrm{f}(Z^{(i)}_{\mathrm{new}}|X)\mathrm{f}(X)}{\int_X \mathrm{f}(Z^{(i)}_{\mathrm{new}}|X)\mathrm{f}(X)\,\mathrm{d}X}. \tag{5.131}$$

Note that in Equation (5.131) the density function $\mathrm{f}(Z^{(i)}_{\mathrm{new}}|X)$ is obtained from the calibration information contained in the database, that is, the density value is obtained as $\mathrm{f}(Z^{(i)} = Z^{(i)}_{\mathrm{new}}|X)$. Given Equation (5.131), it is possible to estimate the position of the MS using the maximum-likelihood estimator:

$$\hat{X} = \arg\max_{X|P}(\mathrm{f}(X|Z_{\text{new}})). \tag{5.132}$$

As an example, a simulation where three APs measure received-power information has been performed. The MS was assumed to be static, continuously communicating with all the APs, and the measurements of power were assumed to be corrupted by Gaussian noise. Since the purpose of this simulation was not to provide details of the actual implementation and positioning accuracy but instead to illustrate the methodologies analyzed in this section, the values of the parameters will not be mentioned here. The calibration phase was performed over the entire 10 m × 10 m area under consideration using a grid with a granularity of 5 cm. Figure 5.5 shows the expected value of $\mathrm{f}(Z^{(i)}|X)$ and the corresponding 1σ confidence intervals for each AP. After the calibration, the MS was placed in a random position and new measurements of received power were obtained. Using the new measurements and the calibration data, Equations (5.130) and (5.131) were applied, and the results are plotted in Figures 5.6 and 5.7, respectively. As can be seen in Figure 5.6, the positions with equal density of probability are arranged in a circular shape. This fact is a consequence of the dependency that received-power measurements have on the distance between the AP and the MS. In Figure 5.7 it is possible to see that the

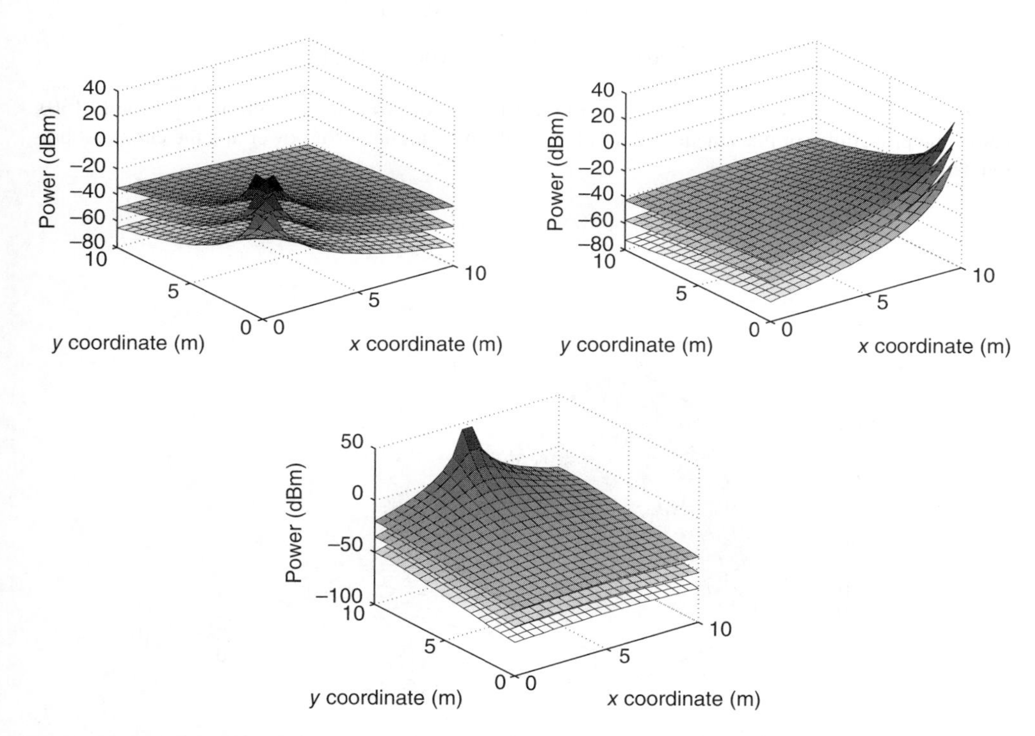

Figure 5.5 Each of these three plots shows the average and the 1σ confidence interval as a function of position. The top curve shows the top limit of the 1σ confidence interval, the middle curve shows the mean value and the bottom curve shows the bottom limit of the 1σ confidence interval. Each plot corresponds to a different AP.

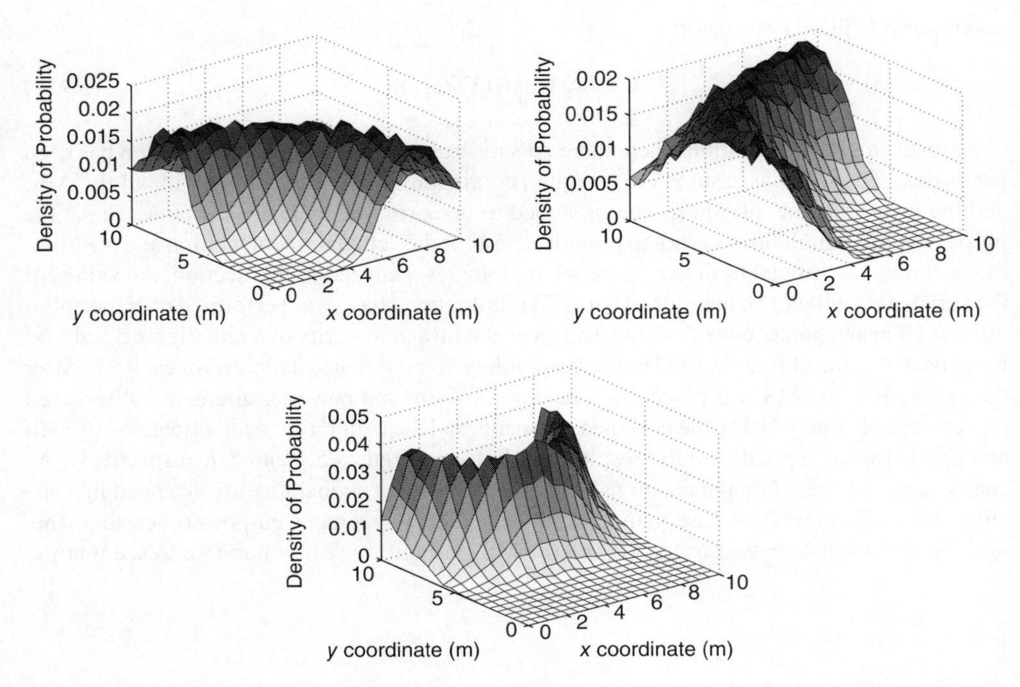

Figure 5.6 These plots show, for each AP, the distribution of probability given by Equation (5.130), that is, the probability of placement of the MS given the measurements of power and the entire calibration data.

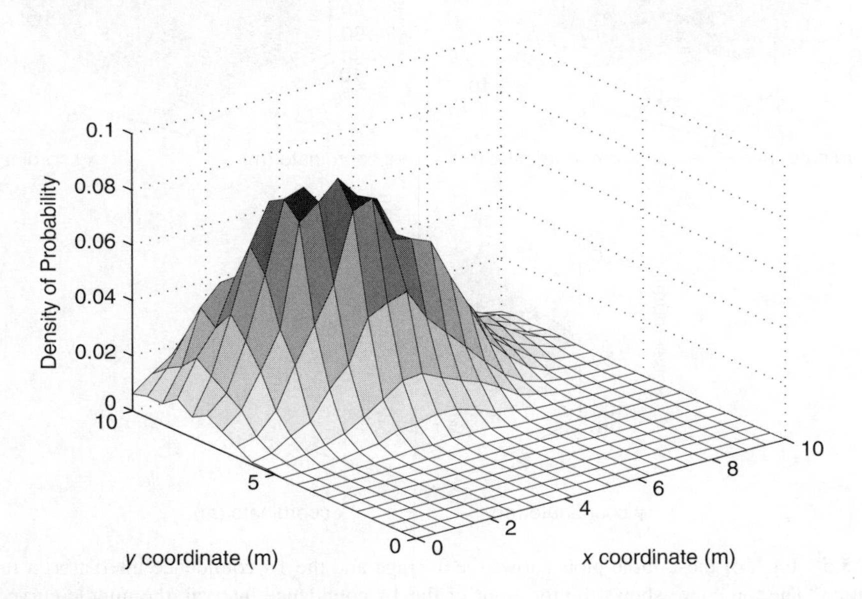

Figure 5.7 Probability density function of placement of the MS with respect to the measurements obtained.

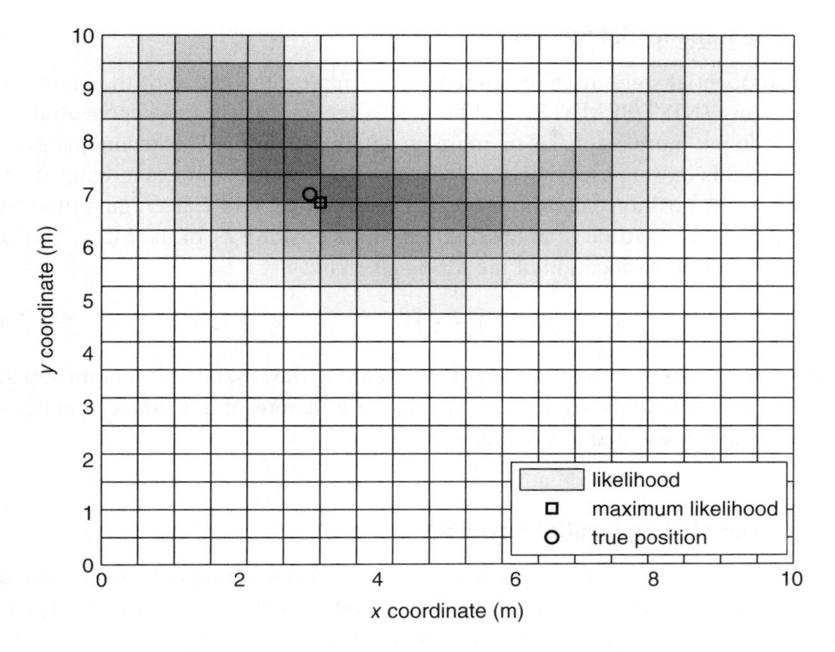

Figure 5.8 Heat map representing the likelihood of the MS position. The darker the map, the more likely is the placement of the MS.

density of probability given by Equation (5.131), which takes into account all APs, shows a cone-like shape. This shape is a consequence of the product of the distributions in Figure 5.6. Note, however, that as stated in Section 4.6.2 the distribution shown in Figure 5.7 is highly dependent on the placement of the APs. The estimation of position resulting from Equation (5.132) is shown in Figure 5.8. In contrast to the LS and Bayesian approaches, fingerprinting techniques are typically more accurate in scenarios with several physical obstructions in the communication links (e.g., indoor scenarios). The disadvantage of fingerprinting techniques is that they depend strongly on the preacquired data, resulting in weakness when the propagation conditions change.

5.5.2 Time Series Data

Exponential smoothing techniques are often used in the context of business and economics (Box and Jenkins 1976; NIST/SEMATECH 2007). Their application in the context of positioning is possible and has been used in previous work (Aloi and Korniyenko 2007; LaViola 2003). Time series techniques suit applications where data points, corrupted by noise, have an internal structure such as trends or seasonal variations.

This type of technique is simpler to implement than the techniques described above, but they are mostly suitable for systems where the MS position has been estimated beforehand. In this case, time series are appropriate tools for smoothing the variations in a sequence of position estimations.

5.5.2.1 Single Exponential Smoothing

The single exponential smoothing technique is the simplest technique of the family of exponential smoothing (NIST/SEMATECH 2007). This technique uses an exponential weighted averaging law to put more emphasis on more recent observations. This technique assumes no trends, and thus it is most appropriate for localizing static nodes or nodes moving at velocities low enough that the position can be considered constant over all the averaging time window.

Thus, supposing that at time t_k an observation of the position X_k of the MS is available, the smoothed estimator of the location of the MS is given by:

$$S_k = \gamma_s X_k + (1 - \gamma_s) S_{k-1}, \quad 0 \leq \gamma_s \leq 1, \tag{5.133}$$

where the variable γ_s is the smoothing forgetting factor. In the absence of other information, the initial condition is commonly set as $S_1 = X_1$. The observations of position X_k can be obtained by using the models presented in Chapter 4.

5.5.2.2 The Double Exponential Smoother

When the data points show a trend, a second exponential smoothing equation can be used for estimating that trend. The idea is to consider a second equation that estimates the constant increment that each data point is subject to with time:

$$\begin{cases} S_k = \gamma_s X_k + (1 - \gamma_s)(S_{k-1} + B_{k-1}), & 0 \leq \gamma_s \leq 1, \\ B_k = \gamma_b (S_k - S_{k-1}) + (1 - \gamma_b) B_{k-1}, & 0 \leq \gamma_b \leq 1, \end{cases} \tag{5.134}$$

where the parameter γ_b is the forgetting factor of the incremental trend B_k. In practice, S_k is the smoothed estimator of the position data points and B_k is the smoothed estimator of the increment that the data points are subject to at each time step. It is important to note that the first line of Equation (5.134) includes the trend B_{k-1} at a previous time, meaning that it takes into account the time lag between two consecutive data points. The initial conditions of Equation (5.134) are commonly set as:

$$\begin{cases} S_1 = X_1, \\ B_1 = X_2 - X_1. \end{cases} \tag{5.135}$$

Forecasting the data points into the future is possible using the following equation:

$$S^{\mathrm{f}}_{k+m} = S_k + mB_t, \tag{5.136}$$

where m is the number of time steps that the forecast is looking at.

5.6 Conclusions

This chapter focused on data fusion techniques for positioning applications. The traditional methods for data fusion were described in the context of wireless positioning. The LS methods were described as simple mathematical formulations to be used primarily in static scenarios. Owing to the necessity for tracking moving users, the Bayesian framework was analyzed.

Within this framework, the main focus was given to KF, with less focus on particle filters and grid-based methods. KFs are studied as convenient algorithms for tracking MSs in the presence of white Gaussian noise. In contrast, particle filters are suitable for any kind of noise, but the computational complexity may be considerably higher compared with KF methods. Grid-based methods are used when the state space, that is, the possible positions for finding the MS, is discrete and finite. We explained how data fusion algorithms can be used for estimating unknown parameters in a system, such as the path loss parameters for an RSS-based positioning system. In the closing section of the some alternative approaches were presented, the first using a fingerprinting technique and the second using a time series approach.

6

Fundamentals of Tracking

João Figueiras
Aalborg University, Aalborg, Denmark

6.1 Introduction

In the context of wireless positioning, one of the typical problems is to track a moving mobile station (MS) with time. The simplest approach would be to completely disregard any structural information contained in the sequence of observations and continuously run the estimation algorithms independently every time new observations were available. As was shown in Chapter 5, there are several tools that consider time history by propagating past knowledge into the present and future. As was previously seen, this approach provides the foundations for tracking moving devices.

This chapter starts with a conceptual explanation of the fundamental differences and the increment in complexity related to the movement of MSs. This explanation is then extended to the introduction of cooperative approaches. While taking movement into account constrains the position estimators by forcing a propagation of information along time, cooperative positioning influences the position estimators by constraining the estimation of the position of an MS to the position of neighboring MSs.

In order to guide the reader through the field of tracking applications, the present chapter will also introduce several of the most common mobility models. The chapter explains conventional models, geographically constrained models, group mobility and cooperative mobility. These models are commonly the initial step for any further analysis concerning tracking algorithms.

As a generalized case of the tracking of wireless devices, the first step concerns the treatment of the raw measurements. In this chapter, a simple mitigation algorithm is explained as a solution for excluding outliers in the observations. More complex models are introduced in Chapter 7. The theory behind tracking both nonmaneuvering and maneuvering MSs is then explained. Finally, as the last step of a tracking application, it is possible to have a methodology for learning about typical user movements. This information can, for instance, be used to analyze further activities of the MSs.

Mobile Positioning and Tracking: From Conventional to Cooperative Techniques, Second Edition.
Simone Frattasi and Francescantonio Della Rosa.

6.2 Impact of User Mobility on Positioning

This section highlights the important concepts that arise when mobility and cooperation are introduced into positioning applications. As is conceptually explained below, these concepts increase the complexity of the problem.

6.2.1 *Localizing Static Devices*

Positioning static devices is, conceptually, the simplest scenario. Since the device is static, the mobility model does not depend on the time component. This fact constrains the problem of positioning a wireless device to the problem of finding its time-independent spatial coordinates. It is often possible to determine these coordinates based on indirect observations, that is, physical phenomena that can be measured and have a relation to those coordinates (see Chapter 4). Mathematically, this means:

$$Z = \mathcal{H}(X), \tag{6.1}$$

where Z is the observable physical phenomena, X is the time-independent position coordinates and $\mathcal{H}(\bullet)$ is a function that relates the two variables. From Equation (6.1) it is possible to see that given two observations Z and two independent linear functions $\mathcal{H}(\bullet)$, the two coordinates (or three, depending on whether 2D or 3D positioning is being performed) can be calculated from:

$$X = \mathcal{H}^{-1}(Z). \tag{6.2}$$

It is important to note, however, that if $\mathcal{H}(\bullet)$ is nonlinear, two independent functions may not suffice.

6.2.2 *Added Complexity in Tracking*

When we assume movement of the wireless devices, the problem gets considerably more complex. Contrarily to the case in Section 6.2.1, the derivatives of the coordinates with respect to time are not zero, meaning that it is necessary to consider the derivatives of the space coordinates in Equation (6.1), such as

$$Z = \mathcal{H}(X, \dot{X}), \tag{6.3}$$

which implies that at least twice the number of measurements is necessary to determine the position coordinates and their derivatives compared with the static case. Note that Equation (6.3) could also be augmented with higher-order derivatives, but due to the noise existing in typical wireless positioning systems, only the first derivative is considered. Higher orders are commonly difficult to determine.

6.2.3 *Additional Knowledge in Cooperative Environments*

An advanced and more recent concept in wireless positioning is the use of cooperative schemes among several users. These schemes can equally well be used in the positioning of either static or moving devices. To simplify this concept, this section assumes static devices, since cooperative positioning of moving devices may introduce problems due to the proximity among the terminals.

Cooperative schemes assume that wireless devices are able to obtain measurements of some kind among them, that is, they are able to relatively position themselves. A typical example is short-range communications in ad hoc mode. Cooperation permits the augmentation of the system of equations in Equation (6.1), for example in a system with two users, we have:

$$\begin{cases} Z_1 = \mathcal{H}_1(X_1), \\ Z_2 = \mathcal{H}_2(X_2), \\ Z_c = \mathcal{H}_c(X_1, X_2), \end{cases} \tag{6.4}$$

where the indices 1 and 2 identify the two wireless devices and the index c identifies the cooperative part. As Equation (6.4) shows, the cooperative factor augments the range of physical phenomena used for positioning wireless devices compared with the individualistic approach of Section 6.2.1.

One of the advantages of cooperative schemes is that by correlating the positions of the wireless devices, it is possible to minimize propagation effects such as shadowing. As Chapter 3 has explained, the shadowing effect can introduce bias into the positioning estimators. Since this bias is dependent on the position of the user, correlating the users tends to minimize the effect of this bias.

6.3 Mobility Models

A particulary important aspect of the tracking of moving devices concerns mobility models. By definition, a mobility model is a mathematical or conceptual formulation of rules that dictate and resemble the movement of the MSs depending on the scenario considered. In this section, several models for individual mobility, group mobility and social-based mobility are described.

6.3.1 Conventional Models

Conventional models are models that present a deterministic behavior. These models are commonly given by the basic laws of physics, namely, Newton's laws of motion and equations of motion derived from them, available widely in the literature. For instance, the equations:

$$\dot{r} = \dot{r}_0 + \ddot{r}t \quad \text{and} \quad r = r_0 + \dot{r}_0 t + \frac{1}{2}\ddot{r}t^2 \tag{6.5}$$

represent uniformly accelerated linear motion. In Equation (6.5), r represents the movement in a certain direction, and t the time elapsed since time t_0, the initial state of the movement.

6.3.2 Models Based on Stochastic Processes

6.3.2.1 Brownian-motion Model

The Brownian-motion model is a random-movement model that describes a randomized movement of particles in a certain space. This model is often used in areas as diverse as the mechanics of fluids and finance, and in positioning applications. The model is simple and it is often used because of its mathematical convenience rather than its accuracy.

Consider the example of a million marbles inside a box where they all touch the bottom of the box in such a way that there is still plenty of space for marbles to move. Assume now that the box starts shaking and that all marbles start moving randomly due to the movement of the box and clash with neighboring marbles. If we focus on a single marble, it is possible to see that it describes a randomized path. This path can be seen as the result of a Brownian motion.

The process that dictates a Brownian motion is equivalent to a Wiener process. A Wiener process is a continuous-time stochastic process that can be considered as the *limit* scenario of a random walk (see Section 6.3.2.2). A Wiener process W_t is characterized by three facts:

- $W_0 = 0$.
- W_t is almost surely[1] continuous.
- W_t has independent increments with distribution $W_t - W_s \sim \text{Norm}(0, t - s)$, for $0 \leq s < t$.

The condition that it has independent increments means that if $0 \leq s_1 \leq t_1 \leq s_2 \leq t_2$ then $W_{t_1} - W_{s_1}$ and $W_{t_2} - W_{s_2}$ are independent random variables, and a similar condition holds for n increments.

A Wiener process has the following properties:

$$f_{W_t - W_s}(x) = \frac{1}{\sqrt{2\pi(t-s)}} e^{-x^2/(2(t-s))}, \tag{6.6}$$

$$\text{E}[W_t - W_s] = 0, \tag{6.7}$$

$$\text{E}[(W_t - W_s)^2] - \text{E}^2[W_t - W_s] = t - s, \tag{6.8}$$

$$\text{cov}(W_{t_1} - W_{s_1}, W_{t_2} - W_{s_2}) = \min(t_1 - s_1, t_2 - s_2) \tag{6.9}$$

and

$$\text{corr}(W_{t_1} - W_{s_1}, W_{t_2} - W_{s_2}) = \frac{\min(t_1 - s_1, t_2 - s_2)}{\sqrt{(t_1 - s_1)(t_2 - s_2)}}. \tag{6.10}$$

In physical terms, Equation (6.7) says that whatever path the MS follows, the average position is always the starting point. Figure 6.1 illustrates an example of a Brownian-motion movement, which shows a chaotic behavior.

6.3.2.2 Random Walk Model

The random walk model is a discrete motion model which evolves with time in constant steps. At each time step, the model defines the next movement by randomly choosing a direction in which the next constant step is taken. When the step size tends to zero, the random walk is equivalent to the Brownian motion. If the step size s tends to 0, one needs to walk L/s^2 steps in order to *approximate* a Brownian motion of length L. Depending on the context, a random walk is often considered as a Markov chain (see Section 6.3.2.5), where the model evolves in time according to state transitions defined in the chain. Other approaches of interest consider random walks on graphs, on lines, in planes, on maps, in higher-order dimensions, in groups or in sets.

[1] We say "almost surely" when a certain event has probability close to 1 to happen.

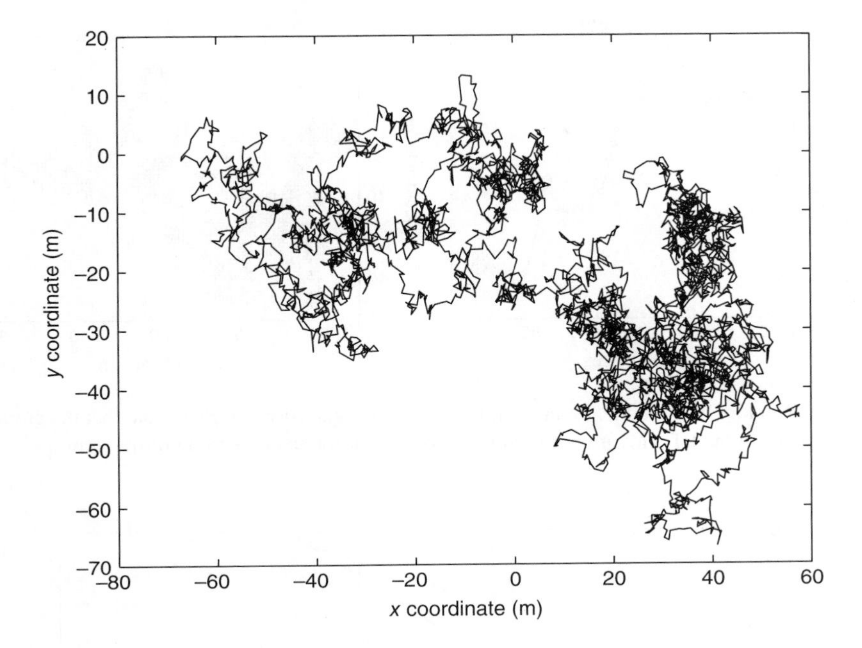

Figure 6.1 Example of a Brownian motion.

The random walk can be defined as follows:

- $W_0 = 0$.
- W_k is discrete.
- W_k has independent increments of either $-s$ or s, each outcome with probability $1/2$, where s defines the step size.

A one-dimensional random walk can be defined as follows:

- $W_k = 0$ for $k = 0$.
- $W_k = \sum_{k=1}^{n} W_{k|k-1}$, where $W_{k|k-1}$ is either $-s$ or s, each outcome with probability $1/2$.

Figure 6.2 shows a random walk in one dimension (left plot) and in two dimensions (right plot), both with unit step size. The random walk presented in Figure 6.2 (left) is commonly called the simple random walk on $\mathbb{Z}$.

6.3.2.3 Random Waypoint Walk

The random waypoint walk model is a model often used as a benchmark in simulations of ad hoc networks when it is necessary to consider moving nodes. The model considers a predetermined area within which the nodes are allowed to move. A node, starting at a random position, selects the next destination, called a waypoint, within the simulation area based on a uniform distribution. Then, a sample velocity is drawn from a uniform distribution between a

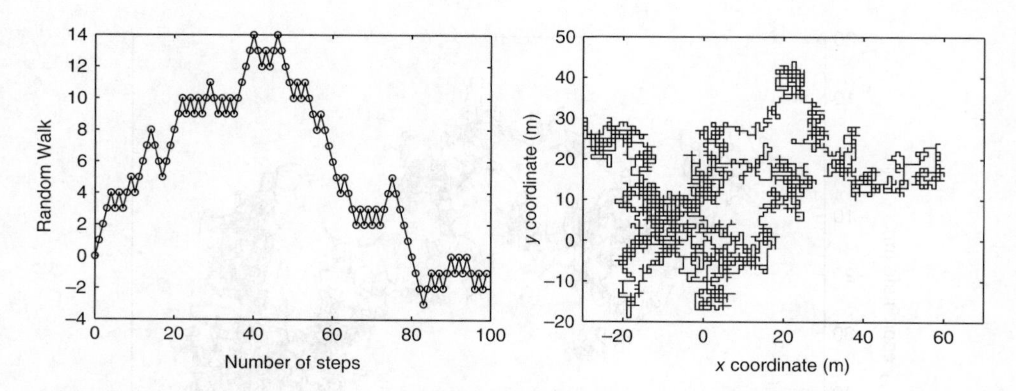

Figure 6.2 Random walk in one dimension (left) and two dimensions (right). Note that the circles that identify the steps in the left plot are not used in the right plot for easier reading of the figure.

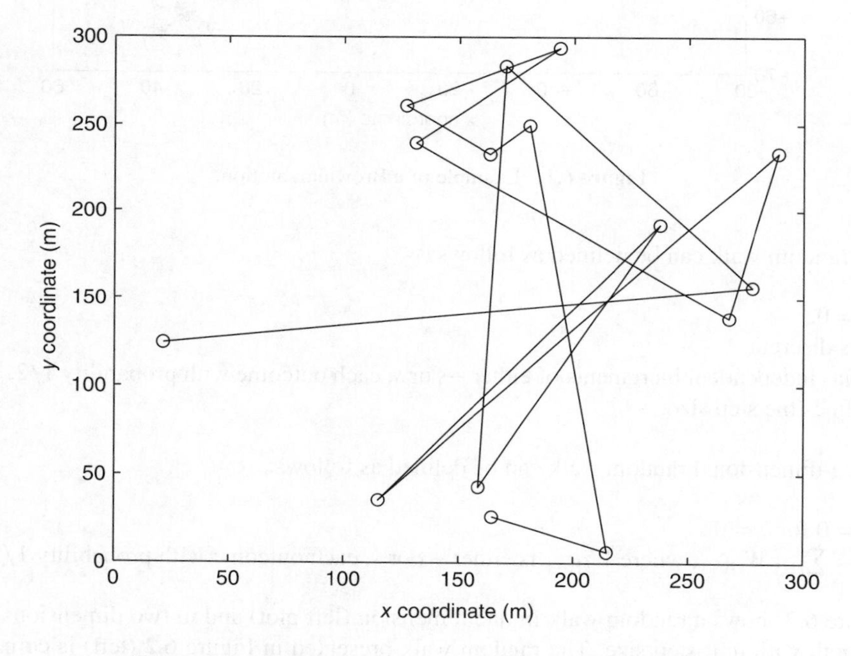

Figure 6.3 Example of a random waypoint walk movement.

minimum and a maximum speed, and the node moves to the chosen destination at that speed. Optionally, the model can assume pause times when the node arrives at a waypoint. This pause time can also be drawn from a uniform distribution. Figure 6.3 exemplifies the random waypoint walk model. Although this model is used in a wide range of research work, one should be careful in using it. As detailed in the work of Yoon et al. (2003), this model fails to provide a steady state. This means that in certain simulations, where a steady state is necessary, such as in routing studies, the algorithm can lead to wrong conclusions. The problem is related to the randomness of the waypoints and the velocity at which the node travels until it gets to those waypoints. When waypoints are placed long distances apart and the velocity is relatively low, the time to get to a waypoint may be too large compared with the simulation time. When the velocity tends to zero, the node takes an infinite time to arrive at the waypoint.

6.3.2.4 Gauss–Markov Model

Another motion model that can appear in various formulations is the Gauss–Markov model. This model permits adaptation of the randomness of the mobility pattern by using a parameter γ that weights the random part of the model against the deterministic part. At fixed time steps, the movement is updated based on three parameters: (1) the previous position, (2) the mean speed and direction, and (3) a random variable. This updating is given by the following equations:

$$s_n = \gamma s_{n-1} + (1-\gamma)\bar{s} + \sqrt{(1-\gamma^2)}\, s_{x_{n-1}} \tag{6.11}$$

and

$$d_n = \gamma d_{n-1} + (1-\gamma)\bar{d} + \sqrt{(1-\gamma^2)}\, d_{x_{n-1}}, \tag{6.12}$$

where $0 \leq \gamma \leq 1$ is the tuning factor, s_n and d_n are the velocity and direction at time t_n, $\bar{s}$ and $\bar{d}$ are the mean velocity and direction, and $s_{x_{n-1}}$ and $d_{x_{n-1}}$ are random variables from a Gaussian distribution. Given Equations (6.11) and (6.12), it is possible to obtain the Cartesian coordinates as:

$$x_n = x_{n-1} + s_{n-1}\cos(d_{n-1}), \tag{6.13}$$

$$y_n = y_{n-1} + s_{n-1}\sin(d_{n-1}), \tag{6.14}$$

where (x_n, y_n) are the Cartesian coordinates x and y where the mobile device is placed at time t_n. By inspecting the above equations it is possible to see that when γ tends to 0, the Gauss–Markov model approximates the Brownian-motion model. On the other hand, when γ tends to 1, the model approximates a linear movement.

Figure 6.4 shows two different examples of the Gauss–Markov model with two different values of γ. By comparing the two plots in Figure 6.4, it is possible to see that the higher

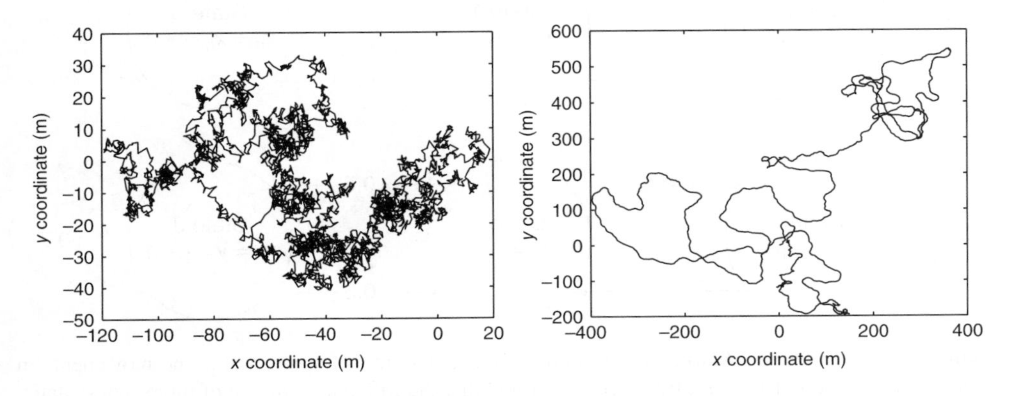

Figure 6.4 Example of a Gauss–Markov movement with $\gamma = 0.1$ (left) and $\gamma = 0.99$ (right).

the parameter γ, the higher is the tendency to maintain the direction of movement. Thus, by dynamically varying the parameter γ depending on the position obtained from the model, it is possible to differentiate between, for instance, areas with great flexibility of movement and areas with low flexibility of movement. For instance, in indoor scenarios, one may want movements in rooms to be more flexible with respect to changes in direction, while in corridors one may want movements to be more constrained.

6.3.2.5 Models Based on Markov Chains

Markov chains are stochastic processes that comply with the Markov property, that is, future states depend only on the present state and a fixed number m of past states, and not on states from the more distant past. The Markov chain in this case is said to be of order m:

$$\mathrm{P}(X_n|X_{n-1}, X_{n-2}, \ldots, X_1) = \mathrm{P}(X_n|X_{n-1}, X_{n-2}, \ldots, X_{n-m}). \tag{6.15}$$

In practice, a Markov chain is a system represented by a set of possible states with different transition probabilities between them. The system can either evolve by changing state or stay in the same state, depending on the probabilities associated with the state transitions. The random walk described in Section 6.3.2.2 is an example of a Markov chain.

In this section we show another example, obtained from Chiang (1998). The movement model is a simple Markov chain where the Cartesian coordinates are treated independently and, for each coordinate, there are only three possible states: the coordinate value decreases by one unit, maintains its value or increases by one unit. This chain, shown in Figure 6.5, has its states represented by circles, while possible transitions between states are represented by directed arrows. The transition probabilities between states are given by the values associated with the arrows. Using this chain, it is possible to sample different movement paths. Figure 6.6 shows a possible sampling of a movement path using the Markov chain shown in Figure 6.5. As it is possible to see, this model favors linear movements rather than the chaotic movement that is characterized by Brownian motion. This linear-movement behavior may resemble the movement of people, as people tend to walk in segmentwise linear trajectories.

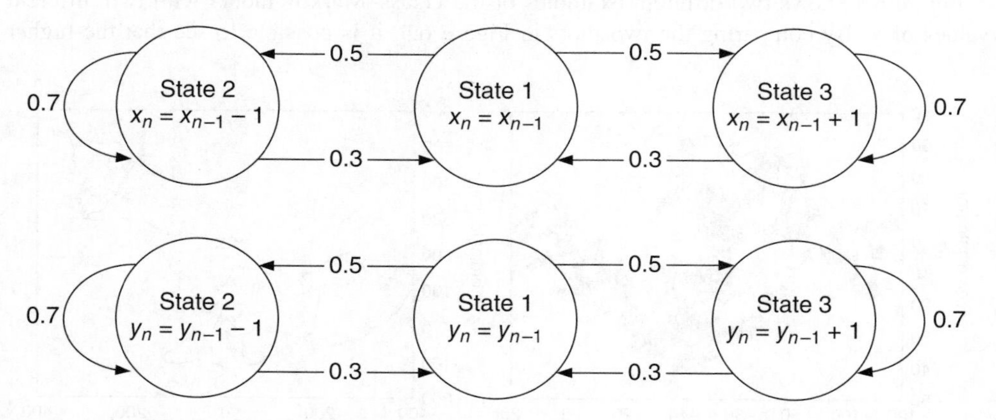

Figure 6.5 Example of a Markov chain motion model. The model assumes that the movements in each coordinate are independent from each other and that the MS can be in one of three states: static, decreasing by one unit or increasing by one unit.

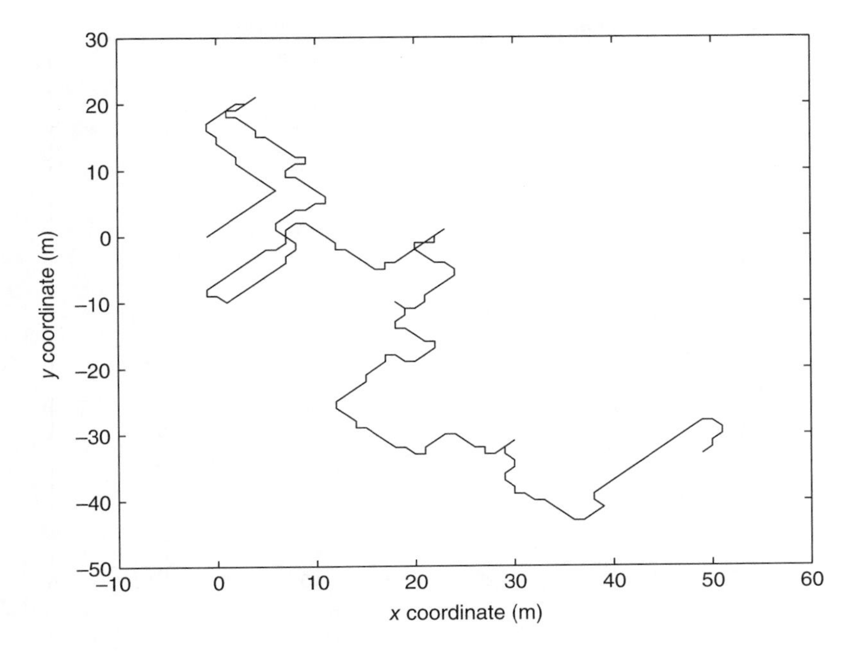

Figure 6.6 Example path sampled from the Markov model shown in Figure 6.5.

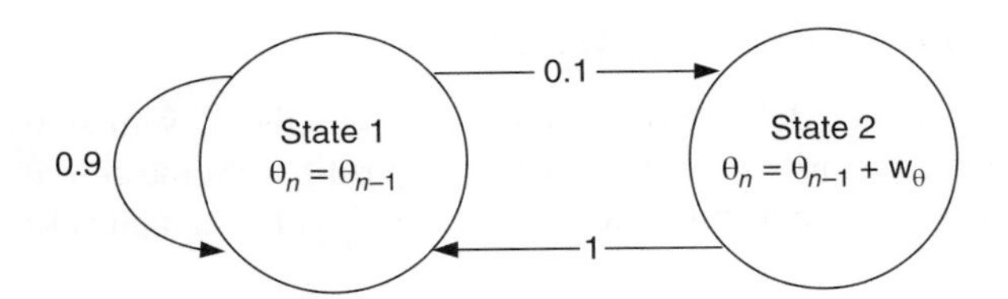

Figure 6.7 Example of a Markov chain motion model. The velocity is assumed constant and the direction to be variable. Each change in direction in state 2 follows a Gaussian distribution.

Another example of a Markov chain is shown in Figure 6.7, where only two states are possible. This is a simple model that could surely be implemented in several ways, but it is still a simple and good example of the use of Markov chains. The model assumes a constant velocity of movement and the direction of movement θ is being governed by a Markov chain. Inspecting Figure 6.7, it is possible to see that once the process is in state 1, it has a 90% probability of maintaining the same direction and a 10% probability of changing direction. In order for the direction to change, the system must be in state 2, where a Gaussian-distributed random variable w_θ is added to the current direction. Once in this state, the system evolves towards state 1 with 100% probability. Figure 6.8 shows one sampling of a possible path obtained from this model. As in the case of the trajectory in Figure 6.6, this model also resembles the segmentwise linear trajectories followed by people.

Markov chains have great flexibility for modeling a wide variety of movement patterns. For instance, if the states represent actual positions in space, the chain can be designed to model a geographical-restriction model.

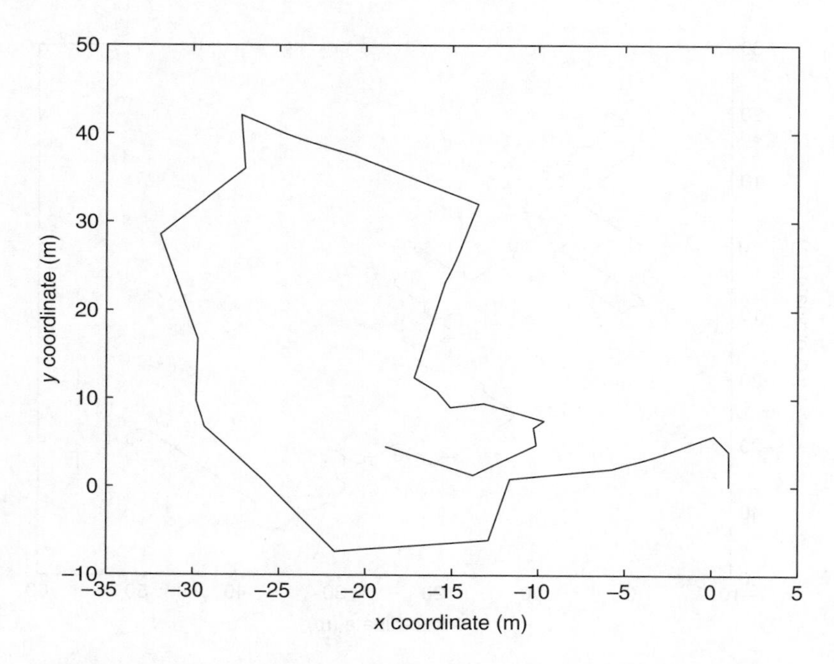

Figure 6.8 Example path sampled from the Markov model shown in Figure 6.7.

6.3.3 *Geographical-restriction Models*

Geographical-restriction models are models that constrain the movement to a certain portion of space, to discrete trajectories or to discrete sets of positions. Some algorithms for modeling movement according to map constraints are good examples of geographical-restriction models.

6.3.3.1 Pathway Mobility Model

One example of the pathway mobility model that is widely present in the literature is the Manhattan mobility model proposed by the Third Generation Partnership Project (3GPP) (ETSI 1998). The Manhattan model has its widest application in the field of intervehicle communications. The model assumes a map with the typical shape of the streets of Manhattan. This urban map, pictured in Figure 6.9, permits users to move along the streets with a predefined probability of changing direction at every crossing. At every crossing, the probability of continuing in the same direction is 0.5, while the probability of turning either left or right is 0.25. Turning back has zero probability. Although the 3GPP proposed a model with defined speed limits and well-defined discretization, more generalized models can be derived. Figure 6.10 shows a single run of the Manhattan model.

Another example is the freeway model (Bai et al. 2003). This model emulates the mobility behavior of vehicles on a freeway. The model assumes that there are several freeways, where each freeway has several lanes in each direction. The mobility is restricted to the paths of the freeways with variable movements. The model assumes that several nodes move on a freeway and defines a safety distance between nodes, that is, the mobility model must take into account neighboring nodes and avoid exceeding the velocity of the preceding node.

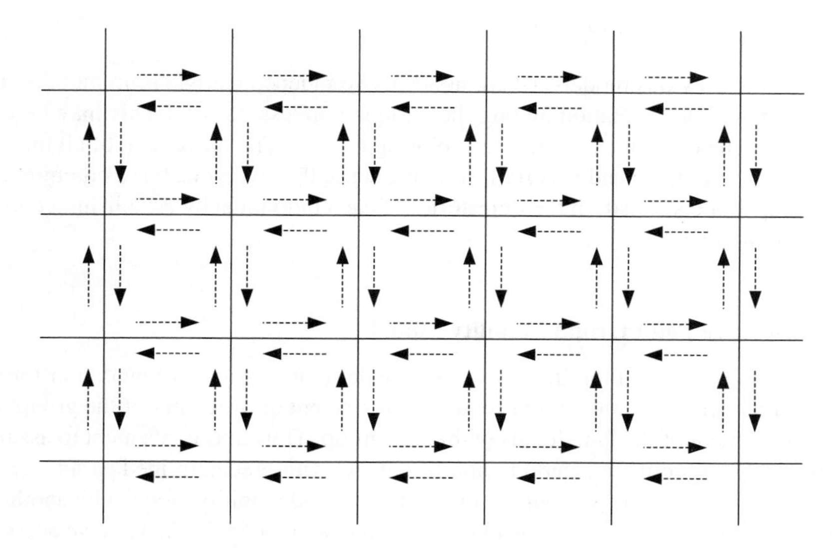

Figure 6.9 Example of a map used by the Manhattan model.

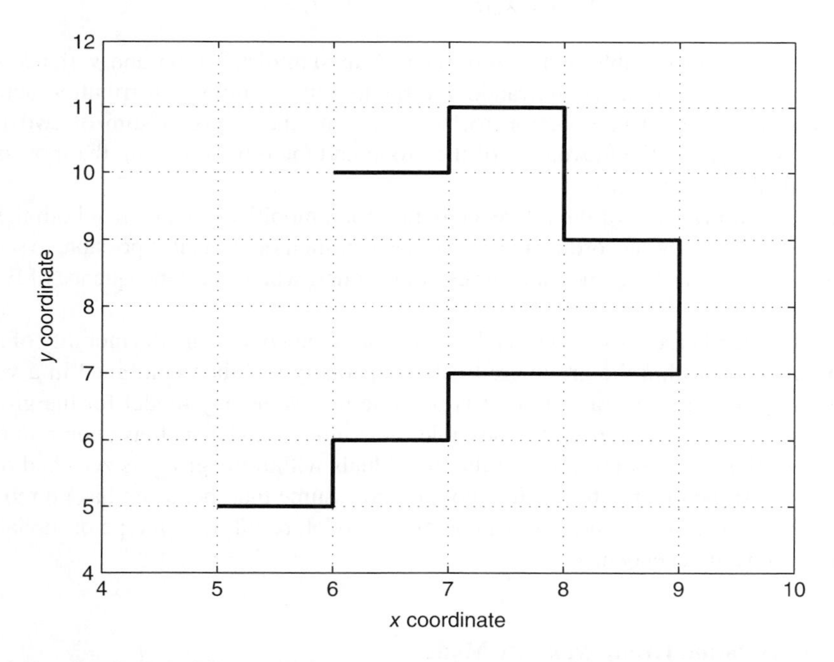

Figure 6.10 Single run of the Manhattan model. At every crossing, the probability of continuing straight is 0.5, of turning either left or right is 0.25, and of turning back is 0.

6.3.4 Group Mobility Models

In contrast to the previous models, where nodes move independently, group mobility models assume that there is a correlation among the various wireless devices. This may be the case when patrol, rescue or military teams are exploring an area. The movement of all the individuals is commonly strategic and is correlated in the sense that everyone has a common goal. In practice, group models constrain the trajectories of each individual by establishing correlation properties among them.

6.3.4.1 Reference Point Group Mobility Model

The reference point group mobility model is a model that provides a foundation for several other derived models. This model represents the movement of the center of the group in addition to the movement of the individuals within the group. Thus, it is convenient to assume that the group moves according to a certain mobility model with predetermined parameters, while the individuals move within the group according to a second mobility model with another set of parameters. For instance, assuming that both the group and the individuals move according to a waypoint random model, the position of each individual i at a certain time t is determined by:

$$X_i(t) = X_g(t) + r_i, \qquad X_g(t) = r_g, \tag{6.16}$$

where r_i is a random variable with a zero mean Gaussian distribution and variance σ_i, and $X_g(t) = r_g$, with r_g being a second random variable with a uniform distribution across the entire scenario. In practice, Equation (6.16) represents the vectorial sum of two random variables, one dictating the movement of the group and the other dictating the movement of each individual within the group.

Figure 6.11 shows a run of the reference point group mobility model when both the group and each individual move according to a random waypoint model. As it is possible to see, there is a strong correlation among the trajectories of the nodes, which is a consequence of Equation (6.16).

In a more general case, this model can be seen as a framework where the mobility of a group and of individuals within the group are treated separately and then combined in a vectorial summation. Given this generalization, it is possible to assume any model for the group and for the individuals. For instance, it is possible to assume a model such as people moving in a bus, that is, the relative movement of the individuals within the group is zero and only the group moves. At the other extreme, it is possible to assume that the group has no movement and the individuals move according to some given model, resulting in a type of model where individuals move independently.

6.3.4.2 Correlation Group Mobility Model

Another type of mobility model considers group mobility as a correlation among several devices. One device is taken as a reference and all the others are bound to the trajectory of this reference according to a certain correlation factor. The correlation factor can be seen as a variable that defines the affinity of a particular individual to the entire group. Thus, assuming that $X_1(t) = r_1$ (where r_1 is a random variable with a uniform distribution, taking into account

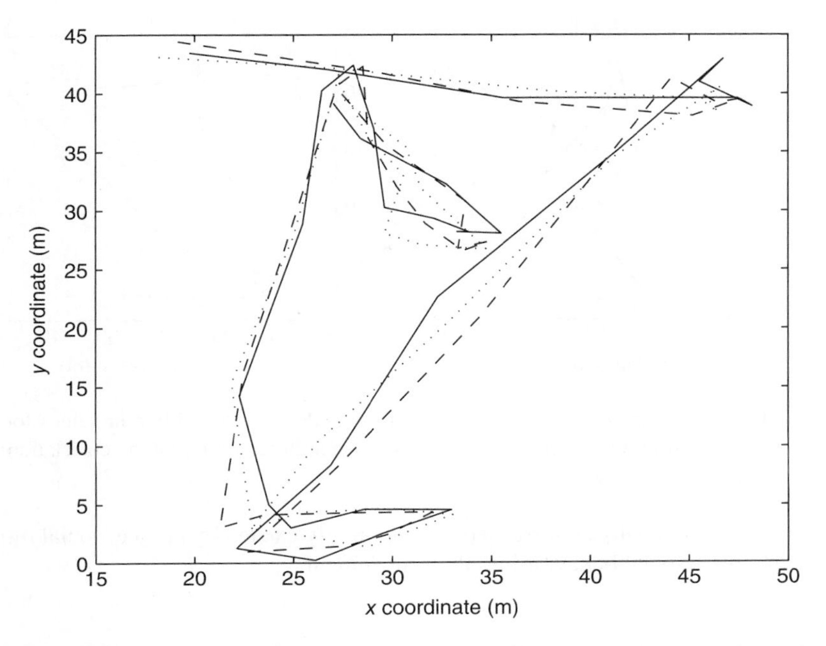

Figure 6.11 Sample pattern obtained by a simulation of a reference point group mobility model. Each line represents the trajectory of a different individual.

the entire simulation area), the positions $X_i(t)$ of the other individuals are defined as:

$$X_i(t) = \rho_i X_1(t) + \sqrt{1 - \rho_i^2} r_i, \tag{6.17}$$

where ρ_i is the correlation factor between the individual i and the reference, and r_i is a random variable with a uniform distribution. In general, r_i is taken as a generalized distribution on the entire simulation area and is the same for every device. In contrast, the correlation factor ρ_i can be considered individual-dependent. This allows one to define different affinities between each device and the entire group. Figure 6.12 shows two different runs of this group mobility model. In the left plot, where the correlation factor is 0, it is possible to see that the trajectories do not look alike among the different individuals. In fact, this is an extreme case, where nodes have independent trajectories and thus the model results in individual mobility instead of group mobility. In the right plot, with a correlation factor of 0.999, it is possible to see that the trajectories of the individuals have structural similarities, where individuals do move as a group.

6.3.5 *Social-based Models*

Social-based mobility models started to be investigated during the boom in social networking a few years ago. In contrast to group mobility models, which simulate the movement of individuals both within a group and as a group, social-based models have an extra dimension which takes into account the social behavior of individuals and groups with their peers. As

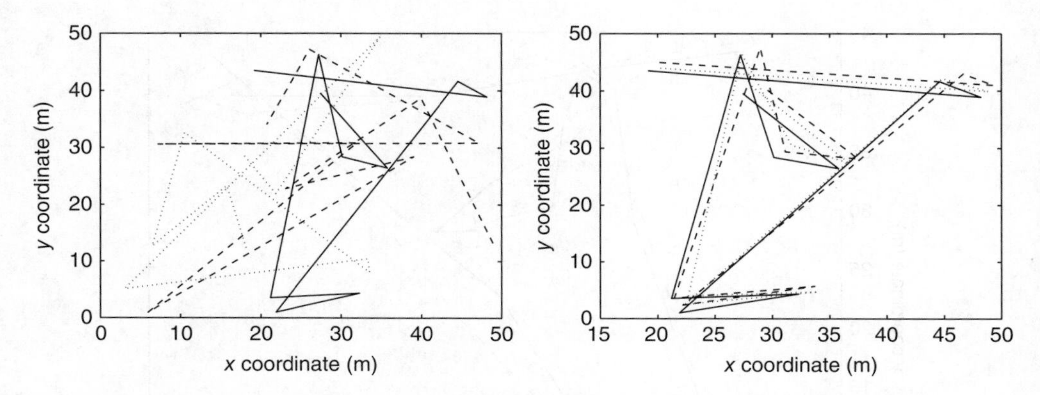

Figure 6.12 Example of the correlation group mobility model with two different values for the correlation factor. In the left plot the correlation factor is 0, while in the right plot the correlation factor is 0.999.

wireless devices to be localized or tracked are commonly carried by people, social models try to integrate people's social behavior into their movement.

6.3.5.1 Model Based on a Sociability Factor

This model, proposed by Musolesi et al. (2004), is based on statistical generation of social relationships. Social behavior is represented by a weighted graph, where nodes are wireless devices and edges represent the social affinity between the persons carrying those wireless devices. This social affinity between the nodes is defined by an interaction indicator m, which is then randomly chosen between 0 and 1, where 0 indicates no interaction and 1 represents strong interaction. The generation of the values of m can be done using a uniform distribution or any other distribution that is intended to simulate the desired society.

The interaction matrix M representing the social graph is then defined as:

$$M = \begin{bmatrix} 1 & m_{1,2} & \cdots & m_{1,n} \\ m_{2,1} & 1 & \cdots & m_{2,n} \\ \vdots & \vdots & \ddots & \vdots \\ m_{n,1} & m_{n,2} & \cdots & 1 \end{bmatrix}, \tag{6.18}$$

where $m_{i,j}$ represents the interaction indicator between nodes i and j. The diagonal elements are conventionally set to 1 and the matrix is assumed to be symmetric, that is, the interaction between the two nodes of a pair is reciprocal.

Given the interaction matrix of Equation (6.18), the social factor S of each individual is calculated as:

$$S = \frac{\sum_{j=1,\, j\neq i, m_{i,j}>\xi}^{n} m_{i,j}}{\upsilon}, \tag{6.19}$$

where ξ is a connection threshold below which individuals are assumed to be socially disconnected, and υ is the total number of individuals, so that $m_{i,j} > \xi$. Once the social factor is known, the nodes are sorted inversely according to this factor. Then, the first node is placed in a random group. The following nodes i are placed in a randomly chosen group that has no nodes.

When each group has one node, the remaining nodes are placed into the groups according to their group attraction, which is calculated as:

$$A_{i,G} = \frac{\sum_{j=1, j \in G}^{n} m_{i,j}}{w}, \tag{6.20}$$

where G stands for group and w is the number of members in the group. As an alternative, Equation (6.20) was rewritten in Musolesi et al. (2004) in order to accommodate a relation between the attraction and the distance between the node and the group to which attraction is being calculated. Optionally, a predetermined number of nodes with the lowest group attraction can be placed randomly within the entire simulation area instead of assigning them to a group.

Once the placement of all the nodes is accomplished according to the aforementioned social rules, the movement is modeled as a discrete-event simulation. Then, any movement is simulated according to a social variant of the waypoint random walk model described in Section 6.3.2.3. The major difference is the introduction of decision points originating from social behavior. First, an entire group is assumed to be moving, then the movement of each individual is augmented by movement within the group and, finally, nodes outside the groups move within the entire simulation area. Once the nodes arrive at a goal given by the waypoint random walk model, social decisions are made based on a uniform distribution, that is, a node decides to move outside of any group if a sampling of the uniform distribution is lower than the social factor of that node. Otherwise, the node stays within a group, but not necessarily the group that it belonged to until then. The node will move to the group to which it has the biggest group attraction. If that is the same group, the node selects the next goal within that same group. Otherwise, it moves to the second group.

Figure 6.13 shows a single run of the social mobility model. The left plot shows the placement of the groups and the placement of five nodes within the simulation area. The two squares define the limits of each group. In the right plot of Figure 6.13, it is possible to see the outcome of two simulation steps. In this simulation, it was assumed that the goals set by each node were accomplished within a single simulation step. As an alternative, one could set a constant velocity and then assume that goals are reachable only when the entire distance required to get to that specific goal has been covered. This approach specifies that new goals are sampled only when the target gets to the previously defined goal. If in any simulation step the node is moving toward a certain goal, that movement must be respected.

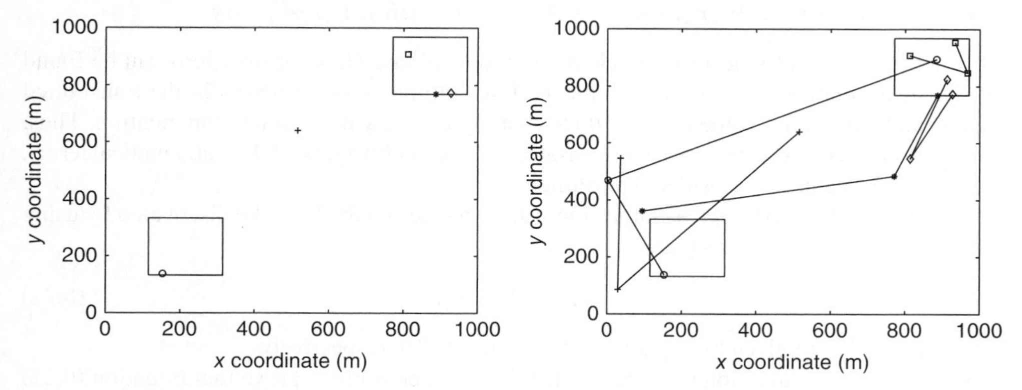

Figure 6.13 Simulation of the sociability factor model. The left plot shows the initial random placement of the nodes and the right plot shows a simulation of a movement where nodes move two steps.

6.4 Tracking Moving Devices

In both positioning and tracking applications, one of the first phases concerns the treatment of the measurements. This step regards the initial analysis that is necessary in order to identify some effects that can be canceled at a very early stage. For instance, detecting outliers and ignoring them is a type of analysis that can be done before the actual position estimation algorithm. Although Chapter 7 introduces some of these techniques, a technique for detecting outliers will be described in this section.

Regarding the positioning mechanism, the simplest assumption in the case of tracking moving MSs is that the kinetic laws governing the movement do not change with time, and that the position coordinates and their derivatives do not change with time. Instead, the set of equations governing the movement are the same along the entire trajectory. This type of movement can be defined as a nonmaneuverable movement.

A natural step ahead of tracking a moving MS is to detect maneuvers in the movement of the MS. As has been seen in previous chapters, the major problem is that the data available for estimating the position are typically corrupted by noise components with high variances compared with the magnitude of the expected values of the measurements. Moreover, the precise distributions of these noise components are unknown. Thus, it is common to approximate the noise distributions by known distributions. This approach results in estimation algorithms that cannot, in a simple fashion, track MSs with frequent maneuvers. The term "frequent" is defined here, loosely and fuzzily, as the inverse of the time that an algorithm needs to "understand" that changes have occurred in the MS movement rules. If these maneuvers can be detected, the common strategy is to analyze the sequence of observations and detect systematic changes in the expected value so that it is possible to infer changes in movement laws.

On top of these solutions for wireless positioning, it is possible to consider mechanisms that are able to learn further information from the estimated positions or trajectories, for example clustering of MSs or trajectory-learning methods. These type of algorithms, as will be seen in Chapter 12, provide a foundation for more complex cooperative schemes. This section shows in addition some of the algorithms used for postprocessing the results of the positioning or tracking algorithms.

6.4.1 Mitigating Obstructions in the Propagation Conditions

Obstructions frequently occur in wireless communications. These obstructions can be found by statistical treatment of the measurements. For example, an obstruction in the transmitted signal can be observed at the receiver through a higher delay or a stronger attenuation. Thus, this type of effect can be detected, compensated for or possibly ignored. In mathematical terms, these measurements are considered as outliers.

An outlier is detected when a "large" innovation happens, which can be discovered by using the normalized innovations squared:

$$\varepsilon_k = \tilde{Z}_k^{\mathrm{T}} S_k^{-1} \tilde{Z}_k, \tag{6.21}$$

where $\tilde{Z}_k$ and S_k are given by Equations (5.78) and (5.79), respectively.

Since $\tilde{Z}_k$ is by assumption Gaussian distributed, it is possible to prove that Equation (6.21) is a chi-squared random variable with l degrees of freedom (where l is the dimension of the observation vector Z).

Based on the assumption that ε_k is chi-squared, a possible solution for detecting an outlier would be to check when ε_k exceeds a predetermined threshold. Thus, by ignoring the measurements that exceed this predetermined threshold, obstructions can be handled, reducing the effect on the estimated positioning information.

6.4.2 Tracking Nonmaneuvering Targets

Positioning nonmaneuverable MSs in wireless communications is typically accomplished by assuming a kinetic law given by the following linear differential equation:

$$\dot{X}(t) = FX(t) + G(t), \tag{6.22}$$

where

$$X(t) = \begin{bmatrix} x(t) \\ y(t) \\ \dot{x}(t) \\ \dot{y}(t) \end{bmatrix}, \qquad F = \begin{bmatrix} 0 & 0 & 1 & 0 \\ 0 & 0 & 0 & 1 \\ 0 & 0 & 0 & 0 \\ 0 & 0 & 0 & 0 \end{bmatrix}, \qquad G(t) = \begin{bmatrix} 0 \\ 0 \\ g_x(t) \\ g_y(t) \end{bmatrix}, \tag{6.23}$$

and where $g_x(t)$ and $g_y(t)$ are independent zero-mean Gaussian distributions with standard deviations σ_x and σ_y, respectively. The covariance matrix of $G(t)$ is defined as:

$$\mathrm{E}[G(\tau)G(\upsilon)^{\mathrm{T}}] = \begin{bmatrix} 0 & 0 & 0 & 0 \\ 0 & 0 & 0 & 0 \\ 0 & 0 & \sigma_x \delta(\tau - \upsilon) & 0 \\ 0 & 0 & 0 & \sigma_y \delta(\tau - \upsilon) \end{bmatrix}, \tag{6.24}$$

where $\delta()$ is the Dirac delta function. Note that these equations are restricted to the first derivative of the position, that is, the velocity. The problem can be easily augmented for the case where the acceleration is also of interest. A consequence of this augmentation is that the problem becomes more complex and the position estimators may diverge more quickly than in the case where the highest derivative of position considered is the velocity. In addition, it is worth mentioning that another problem is related to the fact that major wireless technologies obtain channel measurements with a frequency that may not be as large as necessary to have enough samples to estimate changes in the acceleration. For these reasons, the model is kept simple in the context of wireless positioning. Note as well that Equations (6.22) and (6.23) represent a straight-line movement, although several other approaches could be considered, which could be derived in an equivalent manner.

In order to obtain a recursive expression for the movement, it is necessary to solve Equation (6.22). The first step is to multiply each side of Equation (6.22) by e^{-Ft}:

$$e^{-Ft}\dot{X}(t) = e^{-Ft}FX(t) + e^{-Ft}G(t), \tag{6.25}$$

$$\frac{\mathrm{d}}{\mathrm{d}t} e^{-Ft}X(t) = e^{-Ft}G(t). \tag{6.26}$$

Then, integrating both sides over the interval $[t, t + \mathrm{t}]$, we get:

$$e^{-F(t+\mathrm{t})}X(t + \mathrm{t}) - e^{-Ft}X(t) = \int_t^{t+\mathrm{t}} e^{-F\tau}G(\tau)\,\mathrm{d}\tau, \tag{6.27}$$

$$X(t + \mathrm{t}) = e^{F\mathrm{t}}X(t) + \int_t^{t+\mathrm{t}} e^{F(t+\mathrm{t}-\tau)}G(\tau)\,\mathrm{d}\tau. \tag{6.28}$$

By evaluating the matrix exponential $e^{F\text{t}}$ by series expansion, it is possible to see that

$$e^{F\text{t}} = I + \text{t}F + \sum_{i=2}^{\infty} \frac{\text{t}^i F^i}{i!}, \tag{6.29}$$

where the last term equals the zero matrix, since F to a power larger than 1 also equals the zero matrix,

$$A \equiv e^{F\text{t}} = I + \text{t}F = \begin{bmatrix} 1 & 0 & \text{t} & 0 \\ 0 & 1 & 0 & \text{t} \\ 0 & 0 & 1 & 0 \\ 0 & 0 & 0 & 1 \end{bmatrix}. \tag{6.30}$$

Thus, the expected value of Equation (6.28) is given by:

$$\text{E}[X(t+\text{t})] = A\text{E}[X(t)], \tag{6.31}$$

with A given by Equation (6.30).

Regarding the covariance matrix of Equation (6.28), it is possible to see that

$$Q \equiv \text{E}[X(t+\text{t})X(t+\text{t})^{\text{T}}] \tag{6.32}$$

$$= E\left[\left(\int_t^{t+\text{t}} e^{F(t+\text{t}-\tau)} G(\tau)\,\text{d}\tau\right)\left(\int_t^{t+\text{t}} e^{F(t+\text{t}-\upsilon)} G(\upsilon)\,\text{d}\upsilon\right)^{\text{T}}\right] \tag{6.33}$$

$$= \int_t^{t+\text{t}} \int_t^{t+\text{t}} e^{F(t+\text{t}-\tau)} E[G(\tau)G^{\text{T}}(\upsilon)] e^{F(t+\text{t}-\upsilon)^{\text{T}}}\,\text{d}\tau\,\text{d}\upsilon \tag{6.34}$$

$$= \int_t^{t+\text{t}} e^{F(t+\text{t}-\tau)} \begin{bmatrix} 0 & 0 & 0 & 0 \\ 0 & 0 & 0 & 0 \\ 0 & 0 & \sigma_x^2 & 0 \\ 0 & 0 & 0 & \sigma_y^2 \end{bmatrix} e^{F(t+\text{t}-\tau)^{\text{T}}}\,\text{d}\tau, \tag{6.35}$$

where the step from Equation (6.34) to Equation (6.35) is a consequence of the Dirac delta function included in Equation (6.24), which is forced to be nonzero only when $\tau = \upsilon$. The solution for Equation (6.35) can be determined, after manipulation, to be:

$$Q = \text{t} \begin{bmatrix} \sigma_x^2\text{t}^2/3 & 0 & \sigma_x^2\text{t}/2 & 0 \\ 0 & \sigma_y^2\text{t}^2/3 & 0 & \sigma_y^2\text{t}/2 \\ \sigma_x^2\text{t}/2 & 0 & \sigma_x^2 & 0 \\ 0 & \sigma_y^2\text{t}/2 & 0 & \sigma_y^2 \end{bmatrix} \tag{6.36}$$

Thereby, the typical movement laws used in estimating the positions of MSs in wireless technologies are given by the expected value of X, as in Equation (6.31), and the covariance matrix of the noise component, as in Equation (6.36).

6.4.3 Tracking Maneuvering Targets

This section introduces some of the algorithms used for detecting maneuvers. These algorithms can be considered as an add-on to the approach previously presented for tracking nonmaneuverable targets. The concept is that when a maneuver is detected, the positioning system takes action in order to recalculate the new movement parameters. This action can, for instance, momentarily increase the covariance of the noise component in the positioning estimators, that is, increase the uncertainty in the position estimators.

6.4.3.1 Process Adaptation Using Maneuver Detection

This type of maneuver detection is closely related to the formulation of the Kalman filter. The strategy is to evaluate the innovation process as a function of time given by Equation (5.78). As mentioned in Bar-Shalom et al. (2001), one maneuver detection scheme suitable for use when the noise components are Gaussian is the "white noise model with adjustable level".

A maneuver is detected when a sequence of "large" innovations happens, which can be discovered by using the normalized innovations squared:

$$\varepsilon_k = \tilde{Z}_k^{\mathrm{T}} S_k^{-1} \tilde{Z}_k, \tag{6.37}$$

where $\tilde{Z}_k$ and S_k are given by Equations (5.78) and (5.79), respectively.

Since $\tilde{Z}_k$ is by assumption Gaussian-distributed, it is possible to prove that Equation (6.37) is a chi-squared random variable with l degrees of freedom (where l is the dimension of the observation vector Z). Based on the assumption that ε_k is chi-squared, a possible solution for detecting a maneuver would be to check when ε_k exceeds a predetermined threshold. However, owing to the stochastic properties of ε_k, maneuvers can be incorrectly detected. To minimize this problem, an exponential discounted average can be used with a forgetting factor γ:

$$\varepsilon_k^{\gamma} = \gamma \varepsilon_{k-1}^{\gamma} + \varepsilon_k, \qquad 0 < \gamma < 1, \qquad \varepsilon_0^{\gamma} = 0, \tag{6.38}$$

where ε_k^{γ} is by the first-moment approximation chi-squared distributed with $l/(1-\gamma)$ degrees of freedom. If $l/(1-\gamma)$ is not an integer, the gamma function should be used. For additional details, see Bar-Shalom et al. (2001).

When a maneuver is estimated, that is, ε_k^{γ} has exceeded the predefined threshold, the level of process noise is momentarily increased, meaning that the uncertainty in the estimation increases. Conversely, when ε_k^{γ} goes below the threshold, a lower process noise is used again.

6.4.3.2 Multiple-model Approaches

An alternative approach to tracking maneuvering users is to use a bank of r different models, running in parallel. Then a decision is required on how they can be combined. The r models can be considered, for example, as modeling different movement dynamics of the MS, for example static vs. moving, forward movement vs. sideways movement or horizontal movement vs. vertical movement.

The simplest multiple-model approach assumes that the model that the system obeys is fixed, but unknown. Thus, the problem is to decide which one of the r models is the one that produces the output which is observed. This solution is suitable when the model does not change with time. The final estimation can be considered as a decision about which model is the most probable filter.

In a more advanced approach, the model that describes the system can undergo a switching process with time. These systems, also known as jump-linear systems or system-jumping processes, can be modeled by:

$$X_k = A(M_k)X_{k-1} + V(M_k)\zeta_k, \tag{6.39}$$

$$Z_k = H(M_k)X_k + W(M_k)\eta_k, \tag{6.40}$$

where M_k denotes the model at time k. The model that is active at time t_k is assumed to be one of the possible r models (or filters).

Assuming that the r filters are running in parallel, a decision about combining the filters can be taken by a switch variable Y_k. This variable specifies which movement pattern to use at each time instant t_k. In the case of Kalman filters, this switching variable decides which $\{A, Q, H, R\}$ matrices to use. The combination of the models is based on a weighted scheme where weights are given by $Pr(Y_k = i|Z_{1:k})$, that is, the probability of using model i given measurements $Z_{1:k}$ from time t_1 until time t_k. This is called a "soft switching" scheme.

The lth model history (or sequence of models) through time k is denoted by:

$$M^{k,l} = \{M_{i_{1,l}}, \dots M_{i_{k,l}}\}, \qquad l = 1, r^k, \tag{6.41}$$

where $i_{\kappa,l}$ is the model index at time κ from history l, and

$$1 \leq i_{\kappa,l} \leq r, \qquad \kappa = 1, \dots, k. \tag{6.42}$$

Figure 6.14 shows a schematic representation of a multiple model running with two filters in parallel.

A fundamental problem of this multimodel framework is that running r filters in parallel represents a geometric complexity cost (see Figure 6.14). The reason is that at each time step, every single estimator output of the parallel filters must be input into all r filters.

For simplification and easier tractability of the mathematical formulation, it is common to assume that the mode switching has Markovian properties of the first order:

$$m_{ij} \equiv \mathrm{P}\{M(k) = M_j | M(k-1) = M_i\}. \tag{6.43}$$

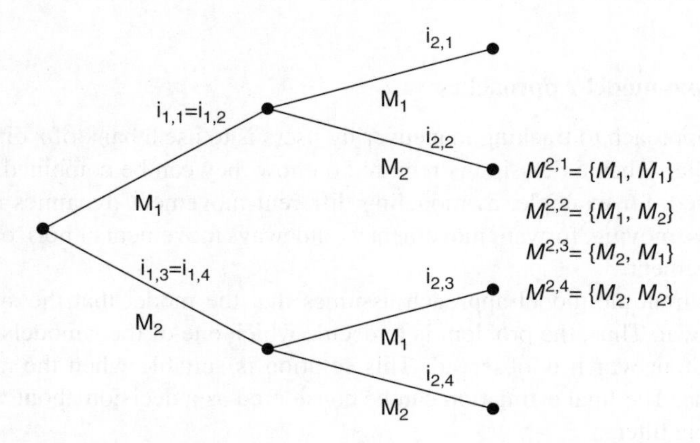

Figure 6.14 Example of the execution of the multiple-model framework. Two filters run in parallel for two units of time. The result is four histories.

The system of equations defined by Equations (6.39), (6.40) and (6.43) is a generalized hidden Markov model. Given this system, it is possible to obtain the probability for each history (Bar-Shalom et al. 2001):

$$\mu^{k,l} = \frac{1}{c} \mathrm{P}\{Z_k | M^{k,l}, Z_{1:k-1}\} m_{ij} \mu^{k-1,s}, \tag{6.44}$$

where $i = s_{k-1}$ is the last model of the parent sequence s and c is the normalization constant. As shown in Equation (6.44), there is still the necessity to condition on the entire history, even though the switching model has Markovian properties.

Taking into account the entire history means manipulating an algorithm that grows exponentially in time. This behavior is clearly intractable after a few time steps, even for the smallest system with only two models. In order to solve this scalability problem, there are several approaches. The generalized pseudo-Bayesian (GPB) (Bar-Shalom et al. 2001) approach is an algorithm that keeps track of the recent history, collapsing the older part of the history into a single estimator. Thus, the first-order GPB approach collapses the history for time $k - 1$ when the estimation is at time k. The second-order GPB approach collapses the history until time $k - 2$. Correspondingly, the first- and second-order approaches use r and r^2 filters running in parallel. An alternative solution to the second-order GPB approach is the interacting-multiple-model approach, which reduces the number of parallel filters from r^2 to r (Murphy 1988).

6.4.4 *Learning Position and Trajectory Patterns*

The learning of position patterns is related to artificial intelligence and machine learning. This is an enormous area with an extensive literature applied to a wide range of different applications. The reason why machine learning is such a wide area is related to the core problem that it addresses. In simple terms, the problem can be defined as: Given a set of inputs and, optionally, the corresponding set of outputs, what is the function and/or parameters that relate the input data to the output data? If the output data is not available, machine learning attempts to do the same job based on predefined rules.

Based on the methodology for learning patterns, machine-learning algorithms can be classified into several groups. Most of the algorithms fall into one of the following groups:

- **Supervised** learning aims at estimating the relationship between the inputs and outputs of a system. By using the estimated relationship, these techniques are able to predict future outputs based on future inputs. As an example, one can think of the problem of determining the propagation parameters in a certain scenario, given a set of propagation measurements and the set of target positions that produced those measurements.
- **Unsupervised** learning aims at automatically discovering structures, groups, features or representations, given only the inputs in a system. An example might be the problem of clustering wireless devices in a network.
- **Reinforced** learning, given inputs, real-world actions and coefficients of reward or punishment, aims at learning to select a sequence of actions resulting from future inputs in a way that maximizes the expected reward. A typical example is the type of learning used by robots.

6.4.4.1 The Expectation Maximization Algorithm

The expectation maximization (EM) algorithm is an efficient iterative procedure to compute the maximum-likelihood estimator (MLE) in the presence of a hidden random variable. Each iteration of the algorithm consists of two steps: the expectation step (E-step) and the maximization step (M-step). The E-step consists of estimating the missing variables given the observed data and the current estimators for the model parameters. The M-step consists of maximizing the likelihood function under the assumption that the hidden variables are known.

Ignoring at this point the hidden variables, the problem is to find the parameters θ such that $\mathrm{P}(X|\theta)$ is a maximum. In order to estimate θ, it is common to introduce the log-likelihood function, which is defined as:

$$L(\theta|X) = \ln \mathrm{P}(X|\theta). \tag{6.45}$$

The foundation of Equation (6.45) is that the parameter θ that maximizes $\mathrm{P}(X|\theta)$ also maximizes $L(\theta|X)$, owing to the monotonic property of the logarithm. The problem is that maximizing Equation (6.45) is typically intractable. The EM algorithm provides a framework which by including hidden variables makes the problem computationally simpler. These hidden variables can be either nonmeasurable variables of the system, as in the tracking application (see below), or purely a mechanism to ensure tractability of the problem:

$$L(\theta|X, W) = \ln \int_w \mathrm{P}(X, w|\theta), \tag{6.46}$$

where W is the hidden variable. Furthermore, it is possible to prove that maximizing Equation (6.46) is equivalent to iteratively maximizing (Borman 2004):

$$\mathrm{E}_{W|X,\theta}[L(\theta|X, W)]. \tag{6.47}$$

Based on Equation (6.47), is now possible to define the algorithm. The algorithm iteratively executes the following steps:

E-step: Determine the conditional expected value of Equation (6.47) given that θ is known:

$$\mathrm{E}_{W|X,\theta^{(t)}}[L(\theta|X, W)]. \tag{6.48}$$

M-step: Maximize Equation (6.47) with respect to θ:

$$\theta^{(t+1)} = \arg\max_{\theta^{(t)}} \{\mathrm{E}_{W|X,\theta^{(t)}}[L(\theta|X, W)]\}. \tag{6.49}$$

The initial value of $\theta^{(0)}$ is set manually to an expected value.

The EM algorithm permits several tracking applications, such as methods for learning user position and mobility. For instance, a simple example concerns clustering of position data. In the specific context of group mobility and cooperative positioning, clustering is an important matter. In a scenario where no prior clustering data is available, it is necessary to perform some kind of unsupervised learning in order to define a cluster. We assume that data points in clusters are formed according to a mixture of three Gaussian random variables, which are defined by:

$$\mathrm{f}(X|\theta) = \sum_{j=1}^{3} \frac{\tau_j}{(2\pi)|\Sigma_j|^{1/2}} \exp\left(-\frac{1}{2}(X-\mu_j)^{\mathrm{T}}\Sigma_j^{-1}(X-\mu_j)\right), \tag{6.50}$$

where $\theta = \{\mu_1, \mu_2, \mu_3, \Sigma_1, \Sigma_2, \Sigma_3, \tau_1, \tau_2, \tau_3\}$ is the parameter set, $\tau_j = \mathrm{P}(W = j)$, with W as the hidden variable that identifies each single Gaussian distribution, and μ_j and Σ_j are the mean and covariance of the Gaussian distribution j.

From Equation (6.50), it is possible to obtain the log-likelihood function, which is given by:

$$L(\theta|X, W) = \ln\left[\prod_{i=1}^{n}\sum_{j=1}^{3}\frac{\delta(W_i = j)\tau_j}{(2\pi)|\Sigma_j|^{1/2}}\exp\left(-\frac{1}{2}(X_i - \mu_j)^{\mathrm{T}}\Sigma_j^{-1}(X_i - \mu_j)\right)\right], \tag{6.51}$$

where $\{X_1, \dots, X_n\}$ are n input data points for the random variable X and $\{W_1, \dots, W_n\}$ represent realizations of the random variable W. Note that the latter can be seen as an artifact introduced in order to ease the algorithm because these variables are not known from input data. By solving Equation (6.51), it is possible to obtain:

$$L(\theta|X, W) = \sum_{i=1}^{n}\sum_{j=1}^{3}\delta(W_i = j)\left[\ln(\tau_j) - \frac{1}{2}\ln|\Sigma_j| - \frac{1}{2}(X_i - \mu_j)^{\mathrm{T}}\Sigma_j^{-1}(X_i - \mu_j) - \ln(2\pi)\right]. \tag{6.52}$$

Note that in Equation (6.52) the ln() function is passed into the summation because $\delta(W_i = j)$ forces the summation in the variable j to have only one nonzero term. Applying Equation (6.52) in Equation (6.47), it is possible to obtain:

$$\mathrm{E}[L(\theta|X, W)] = \sum_{i=1}^{n}\sum_{j=1}^{3}\xi_{j,i}\left[\ln(\tau_j) - \frac{1}{2}\ln|\Sigma_j| - \frac{1}{2}(X_i - \mu_j)^{\mathrm{T}}\Sigma_j^{-1}(X_i - \mu_j) - \ln(2\pi)\right], \tag{6.53}$$

where Equation (6.53) corresponds to the E-step, and $\xi_{j,i}$ is given by:

$$\begin{aligned}\xi_{j,i} &= \mathrm{P}(W|X, \theta) \\ &= \frac{(\tau_j/((2\pi)|\Sigma_j|^{1/2}))\exp(-(1/2)(X_i - \mu_j)^{\mathrm{T}}\Sigma_j^{-1}(X_i - \mu_j))}{\sum_{k=1}^{3}(\tau_k/((2\pi)|\Sigma_k|^{1/2}))\exp(-1/2)(X_i - \mu_k)^{\mathrm{T}}\Sigma_k^{-1}(X_i - \mu_k))}.\end{aligned} \tag{6.54}$$

At this point the M-step can be calculated by setting to zero the gradient of Equation (6.53) with respect to the variables θ and finding their zeros. The result is:

$$\tau_j^{(t+1)} = \frac{1}{n}\sum_{i=1}^{n}\xi_{j,i}^{(t)}, \tag{6.55}$$

$$\mu_j^{(t+1)} = \frac{\sum_{i=1}^{n}\xi_{j,i}^{(t)}X_i}{\sum_{i=1}^{n}\xi_{j,i}^{(t)}}, \tag{6.56}$$

$$\Sigma_j^{(t+1)} = \frac{\sum_{i=1}^{n}\xi_{j,i}^{(t)}(X_i - \mu_j^{(t)})^{\mathrm{T}}(X_i - \mu_j^{(t)})}{\sum_{i=1}^{n}\xi_{j,i}^{(t)}}. \tag{6.57}$$

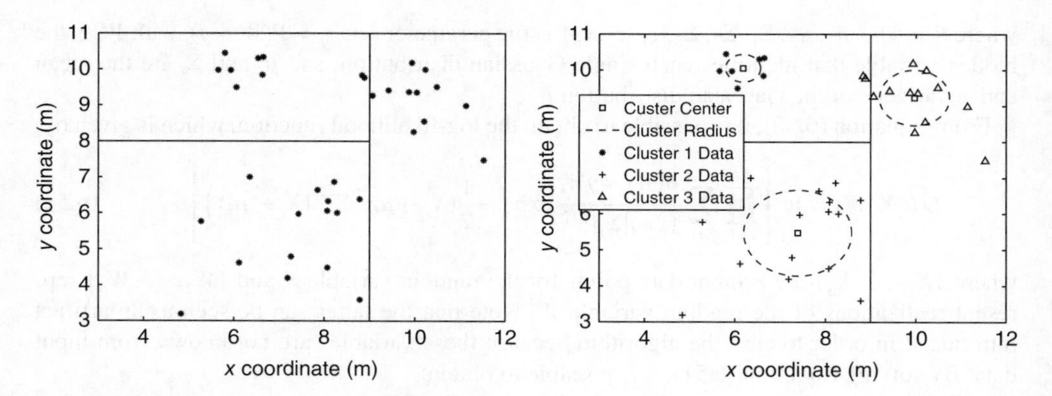

Figure 6.15 Clustering positioning data in order to identify different areas. The right plot shows the data points in a hypothetical room. The left plot shows the execution of the EM algorithm and the classification of each positioning estimator.

As an example, assume that a possible sampling of the distribution of Equation (6.50) is the one shown in Figure 6.15 (left). This figure shows a hypothetical scenario of a room with three divisions. The data shows a hypothetical set of Cartesian points obtained by a positioning method. Suppose that the purpose is to group data points in order to infer further information, such as scenario mapping. Given these samples as inputs, the EM algorithm is run in order to cluster the data. The EM algorithm is composed of the iterative sequence of Equation (6.54) and Equations (6.55)–(6.57). The result of the EM algorithm for a mixture of Gaussian distributions is given in Figure 6.15 (right). The initial values for the parameters were rather close to the real values. However, the higher the number of data points X_i, the more robust is the EM algorithm for handling the initial values properly. As the figure shows, the algorithm was able to clusterize the data given the sampling data. This is an unsupervised method for determining groups of data.

6.4.4.2 The *k*-means Algorithm

The *k*-means algorithm defines a methodology to partition an entire set of elements into a predefined number of groups. Each element belongs to the group for which it has the lowest Euclidean distance to the average of all elements in that same group.

To formulate the algorithm, let us assume that a set of n elements $\{X_1, \ldots, X_n\}$ is to be partitioned into k groups $\{G_1, \ldots, G_k\}$ with $k < n$. The k-means algorithm *defines* that each element belongs to the group which satisfies the condition of minimum within-cluster sum of squares:

$$\hat{G} = \arg\min_{G} \sum_{i=1}^{k} \sum_{X_j \in G_i} \| X_j - \mu_i \|^2, \tag{6.58}$$

where μ_i is the mean value of the elements of G_i.

As can be seen, Equation (6.58) is intractable because of the computational complexity of the optimization procedure. Thus, it is common to approach the problem in an iterative fashion in which we assign elements to groups and then determine the mean value of the elements in the group. Given a set of initial mean values for each group $\{m_1^{(0)}, \dots, m_k^{(0)}\}$ (the superscript index identifies the iteration step) defined in a random or heuristic manner, the iterative k-means algorithm can be defined as follows.

Assigning: The Euclidean distance is calculated between all group mean values and each element, so that each element is assigned to the group to which it has the shortest Euclidean distance. This decomposition is known as the Voronoi diagram:

$$X_j \in \hat{G}_i^{(l)} = \arg\min_{G_i^{(l)}} \|X_j, m_i^{(l)}\|, \tag{6.59}$$

where l is the iteration step.

Updating: The mean values for each group at time step $l + 1$ are updated to $\{m_1^{(l+1)}, \dots, m_k^{(l+1)}\}$ according to the average of all the elements in the group:

$$m_i^{(l+1)} = \frac{1}{n_i^{(l)}} \sum_{X_j \in G_i^{(l)}} X_j, \tag{6.60}$$

where $n_i^{(l)}$ is the number of elements in group $G_i^{(l)}$ at time step l.

Once the assignment of elements no longer changes in subsequent time steps, the algorithm terminates. As in any other optimization algorithm, convergence to an absolute minimum is not assured.

The k-means algorithm can be used, for instance, to estimate average routes obtained from several tracking estimations performed in a predetermined area. Let us assume, for example, a scenario with three different routes. For each route, there are five estimated paths, supposedly obtained from a tracking algorithm. The actual tracking algorithm is unimportant in this example, since our interest is focused on the k-means algorithm for grouping estimated trajectories. Let us suppose that these 15 different estimated paths are those given in Figure 6.16 (left). As one can see, the estimated paths may be corrupted by noise. Each of the 15 estimated trajectories is composed of a sequence of position coordinates, which typically differ in

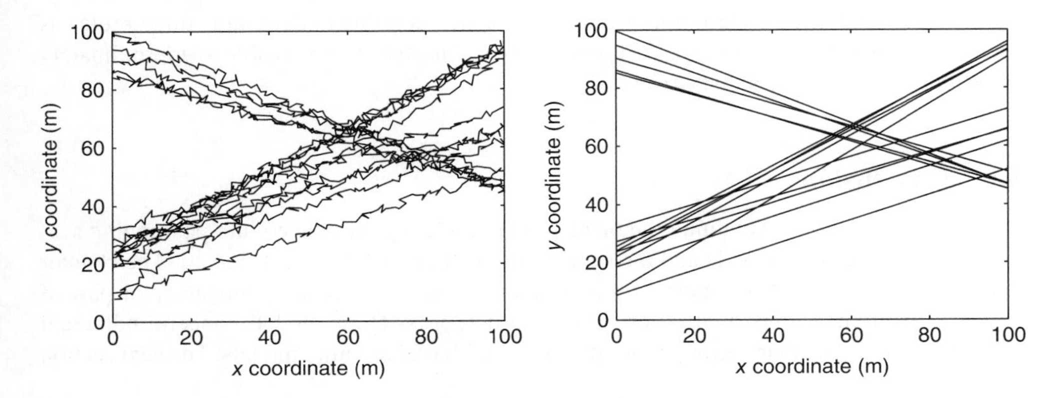

Figure 6.16 Example of 15 noisy routes (left) and the corresponding linear regression (right).

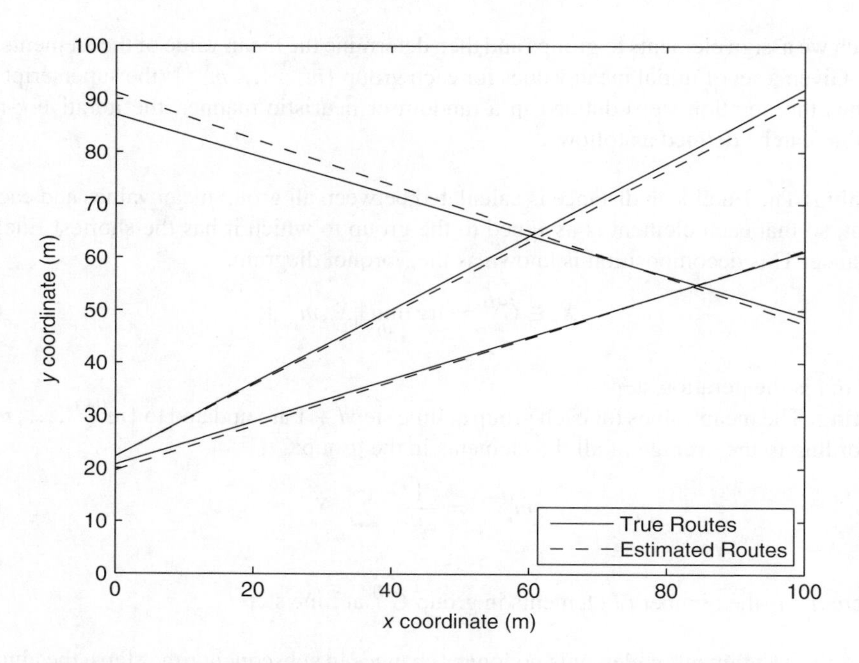

Figure 6.17 True routes and routes estimated using the *k*-means algorithm.

number from another estimated trajectory. This problem makes it complex to directly establish a definition of a distance between two trajectories. Hence, we have determined for each trajectory the corresponding linear regression, which is plotted in Figure 6.16 (right). This regression permits the approximation of the trajectory by a known expression so that it is possible to calculate distances between different estimated trajectories. In order to calculate the distances between pairs of trajectories, all the expressions obtained from the regression were sampled at equidistant points, which were used to calculate the mean square error. This metric was defined as the distance between trajectories and was used in the assignment step to calculate the distance to the center of the group, which itself had an initially random trajectory. By applying the *k*-means algorithm given above, it was possible to determine three routes as given in Figure 6.17. As Figure 6.17 shows, the *k*-means algorithm was able to approximately determine the three routes that existed.

6.5 Conclusions

The chapter started by explaining the increase in complexity introduced into positioning systems when movement of MSs and then cooperation among MSs are considered. The chapter then described some of the most common mobility models used as a fundamental part of wireless positioning, in particular for simulation purposes. These models concern individual mobility, mobility on maps, group mobility and social-based mobility models. The next section presented the various steps involved in a tracking system. In an initial step, the measurements are processed in order to filter out some useless measurements. Then algorithms for nonmaneuvering and maneuvering MSs are applied. Finally, envisioning cooperative schemes in wireless positioning, we described some methods and techniques for learning position and path patterns.

7

Error Mitigation Techniques

Ismail Guvenc
Wireless Access Laboratory, DOCOMO Communications Laboratories, Palo Alto, CA, USA

7.1 Introduction

The previous chapters have introduced several algorithms and mechanisms to perform positioning and tracking of mobile stations (MS)s in wireless networks. The common characteristic of all those methods is that all the approaches considered use raw measurements of the wireless channel, that is, no preprocessing of the measurements is done to detect possible destructive effects existing in the measured data. Thus, this chapter describes the mechanisms that can mitigate and circumvent some of these destructive effects present in raw measurements.

The most common effect affecting position accuracy is the non-line-of-sight (NLOS) effect. NLOS scenarios occur when there is an obstruction between the transmitter (TX) and the receiver (RX), and are commonly encountered in modern wireless system deployment in both indoor environments (e.g., residential and office buildings shopping malls) and outdoor environments (e.g., metropolitan and urban areas). As illustrated in Figure 7.1, the signal transmitted by the TX arrives at the RX through multiple paths. In an NLOS scenario, the LOS path is blocked by an obstruction and hence does not reach the RX.[1] Therefore, even if the first path arriving at the RX can be correctly identified, the distance estimate calculated using this path includes a positive bias, which, if not handled properly in the localization algorithm, may considerably degrade the localization accuracy.

Several previous studies reported in the literature (e.g., Gezici 2008; Gustafsson and Gunnarsson 2005; Liu et al. 2007; Patwari et al. 2005; Sayed et al. 2005b) provide extensive reviews of localization techniques using angle of arrival (AOA), time of arrival (TOA), time difference of arrival (TDOA) and received signal strength (RSS). However, the impact of NLOS propagation and how this can be mitigated has not been investigated in detail. In Caffery

[1] In some scenarios, a signal following this path may also arrive at the receiver after being attenuated and delayed by the obstruction (Allen et al. 2007).

Mobile Positioning and Tracking: From Conventional to Cooperative Techniques, Second Edition.
Simone Frattasi and Francescantonio Della Rosa.

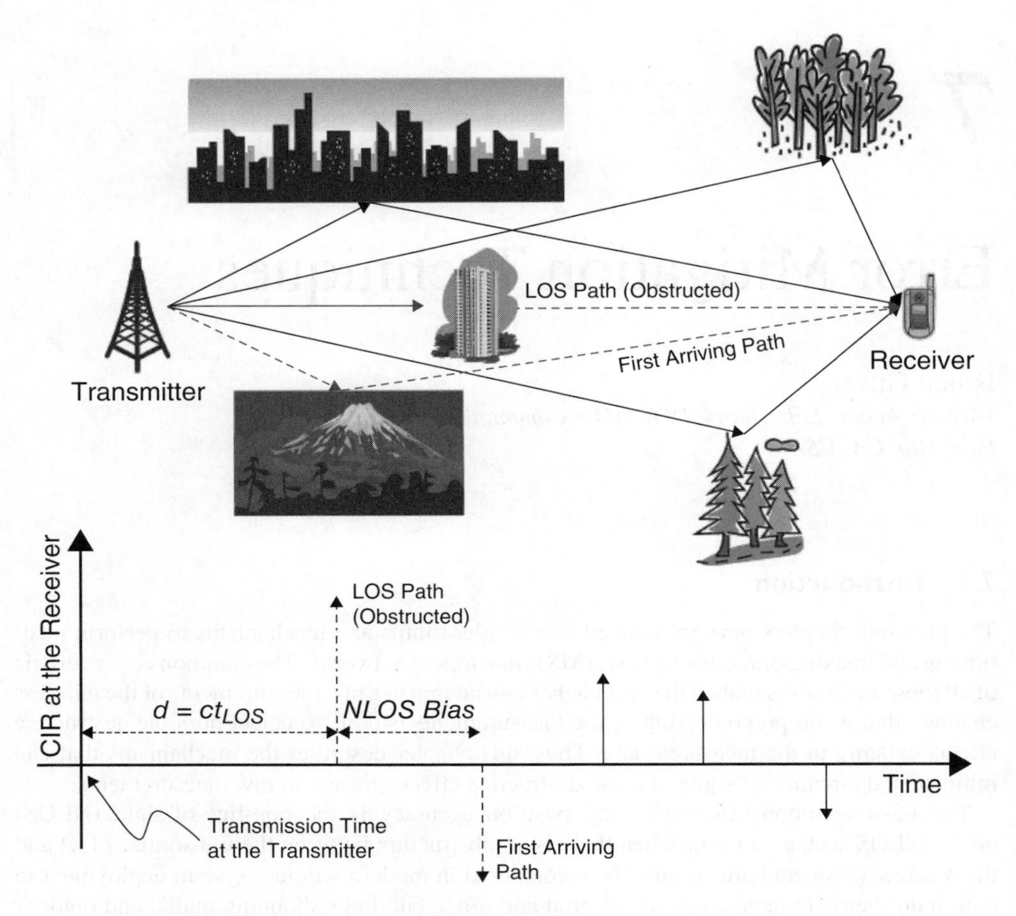

Figure 7.1 Illustration of the NLOS problem in wireless localization.

(1999), an analysis of NLOS effects in code division multiple access (CDMA) cellular networks was provided and two particular NLOS mitigation techniques were summarized: LOS reconstruction and NLOS measurement weighting. However, several other NLOS mitigation techniques have been proposed in the literature subsequently. The goal of the current chapter is to provide a comprehensive and up-to-date survey of NLOS mitigation techniques for wireless location estimation. While this chapter will focus on TOA-based approaches for the sake of a unified analysis, several of the NLOS mitigation techniques reviewed may also be applied to TDOA-, AOA- and RSS-based systems after some simple extensions.

The chapter is organized as follows. Section 7.2 extends previous chapters by outlining the TOA-based localization scenario and presenting the system model and the problem definition to be used throughout the chapter. A brief overview of the fundamental lower bounds in LOS scenarios is provided, and some of the key techniques based on the maximum likelihood (ML) method, which can achieve accuracies close to these fundamental bounds, are summarized. Section 7.3 provides lower bounds and some ML techniques for localization in NLOS scenarios. Section 7.4 is on least-squares (LS) techniques, Section 7.5 is on constraint-based techniques, Section 7.6 is on robust estimators and Section 7.7 is a review of the identify and discard type of techniques, all designed for improving the localization accuracy in NLOS scenarios. Finally, Section 7.8 briefly summarizes the techniques discussed and provides some concluding remarks.

7.2 System Model

Consider a wireless network as shown in Figure 7.2 where there are n fixed reference points (FRP)s,[2] $\hat{X} = [\hat{x}\ \hat{y}]^{\mathrm{T}}$ is the estimate of the MS location, and $\hat{d}_i$ is the measured distance between the MS and the ith FRP:

$$\hat{d}_i = d_i + b_i + \eta_i = ct_i, \quad i = 1, 2, \ldots, n, \tag{7.1}$$

where t_i is the TOA of the signal at the ith FRP, c is the speed of light, d_i is the real distance between the MS and the ith FRP, $\eta_i \sim \mathrm{Norm}(0, \sigma_i^2)$ is the additive white Gaussian noise (AWGN), with variance σ_i^2, and b_i is a positive distance bias introduced owing to the blockage of direct path, given by

$$b_i = \begin{cases} 0 & \text{if the } i\text{th FRP is in LOS,} \\ \psi_i & \text{if the } i\text{th FRP is in NLOS.} \end{cases} \tag{7.2}$$

For NLOS FRPs, the bias term ψ_i has been modeled in different ways in the literature depending on the wireless propagation channel environment; it may be exponentially distributed

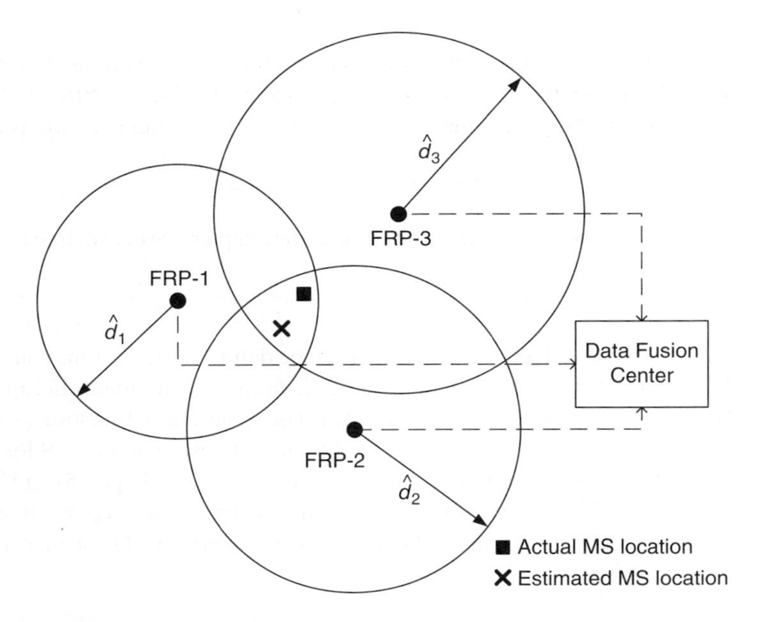

Figure 7.2 Illustration of a simple scenario for wireless localization.

[2] The FRPs are usually base stations (BSs) in cellular networks, anchor nodes in sensor networks, and access points (APs) in wireless local area networks (WLANs).

(Chen 1999; Gezici and Sahinoglu 2004), uniformly distributed (Jourdan and Roy 2006; Venkatesh and Buehrer 2006), Gaussian distributed (Dizdarevic and Witrisal 2006); constant in a time window (Riba and Urruela 2004) or based on an empirical model derived from measurements (Denis et al. 2006; Jourdan et al. 2006).

Let

$$D = D(X) = [d_1, d_2, \ldots, d_n]^{\mathrm{T}} \tag{7.3}$$

be a vector of true distances between the MS and the n FRPs, let

$$\hat{D} = [\hat{d}_1, \hat{d}_2, \ldots, \hat{d}_n]^{\mathrm{T}} \tag{7.4}$$

be a vector of measured distances and let

$$B = [b_1, b_2, \ldots, b_n]^{\mathrm{T}} \tag{7.5}$$

be a bias vector. Also, let

$$Q = \mathrm{E}[\eta\eta^{\mathrm{T}}] = \mathrm{diag}[\sigma_1^2, \sigma_2^2, \ldots, \sigma_n^2]^{\mathrm{T}} \tag{7.6}$$

denote the covariance of the noise vector $\eta = [\eta_1, \eta_2, \ldots, \eta_n]^{\mathrm{T}}$ with the assumption that all of the noise terms are zero-mean, independent Gaussian random variables.

In the absence of noise and NLOS bias, the true distance d_i between the MS and the ith FRP defines a circle around the ith FRP corresponding to possible MS locations:

$$(x - x_i)^2 + (y - y_i)^2 = d_i^2, \quad i = 1, 2, \ldots, n \tag{7.7}$$

where all the circles intersect at the same point, and solving these equations jointly gives the exact MS location. However, the noisy measurements and NLOS bias at different FRPs yield circles which do not intersect at the same point, resulting in inconsistent equations as follows:

$$(x - x_i)^2 + (y - y_i)^2 = \hat{d}_i^2, \quad i = 1, 2, \ldots, n. \tag{7.8}$$

In order to have more compact expressions throughout the chapter, we also define the following terms:

$$s = x^2 + y^2 \quad \text{and} \quad k_i = x_i^2 + y_i^2. \tag{7.9}$$

The problem of TOA-based location estimation can be defined as the estimation of the location X of an MS from the noisy (and possibly biased) distance measurements in Equation (7.4) given the FRP locations X_i; in other words, given the set of equations in Equation (7.8). Various localization techniques have been proposed in the literature to estimate the MS location from Equation (7.8) in LOS scenarios (which have also been discussed in Chapter 5) and NLOS scenarios. Before reviewing the various NLOS mitigation algorithms, first we will briefly review ML-based localization techniques and fundamental lower bounds in LOS scenarios.

7.2.1 Maximum-likelihood Algorithm for LOS Scenarios

In the absence of NLOS bias (i.e., $b_i = 0$ for all i), the conditional probability density function (PDF) of $\hat{D}$ in Equation (7.4) given X can be expressed as follows (Chan et al. 2006a; Cheng and Sahai 2004):

$$f(\hat{D}|X) = \prod_{i=1}^{n} \frac{1}{\sqrt{2\pi\sigma_i^2}} \exp\left\{ -\frac{(\hat{d}_i - d_i)^2}{2\sigma_i^2} \right\} \tag{7.10}$$

$$= \frac{1}{\sqrt{(2\pi)^n \det(Q)}} \exp\left\{ -\frac{J}{2} \right\}, \tag{7.11}$$

where

$$J = [\hat{D} - D(X)]^{\mathrm{T}} Q^{-1} [\hat{D} - D(X)], \tag{7.12}$$

with Q as given in Equation (7.6). Then, the ML solution for X is the one that maximizes $f(\hat{D}|X)$, that is,

$$\hat{X} = \arg\max_{X} f(\hat{D}|X). \tag{7.13}$$

Note that solving for X from Equation (7.13) requires a search over possible MS locations, which is computationally intensive.

For the special case of $\sigma_i^2 = \sigma^2$ for all i, the ML solution of Equation (7.13) is equivalent to minimizing J. In order to find the minimum value of J, the gradient of J with respect to X is equated to zero, yielding (Chan et al. 2006a)

$$\sum_{i=1}^{n} \frac{(d_i - \hat{d}_i)(x - x_i)}{d_i} = 0 \tag{7.14}$$

and

$$\sum_{i=1}^{n} \frac{(d_i - \hat{d}_i)(y - y_i)}{d_i} = 0, \tag{7.15}$$

which are nonlinear equations. Hence, X cannot be solved for in closed form from Equations (7.14) and (7.15) using a linear LS algorithm. Also, both Equation (7.14) and Equation (7.15) depend on the d_i, which are unknown. However, even though a closed-form ML solution is not possible, iterative ML and approximate ML techniques can be derived as discussed in Chan et al. (2006a) and Chan and Ho (1994), respectively, which may asymptotically achieve the Cramér-Rao lower bound (CRLB).

7.2.2 Cramér–Rao Lower Bounds for LOS Scenarios

Given the conditional PDF of $\hat{D}$ as in Equation (7.10), we may derive the CRLB for TOA-based location estimation (see Section 4.6.3). The CRLB, in general, can be defined as the theoretical lower bound on the variance of any unbiased estimator of an unknown parameter. The CRLB for TOA-based location estimation depends mainly on the following factors: (1) the positions of the FRPs (X_i), (2) the true position of the MS (X) and (3) the measurement noise variances (σ_i^2). It is calculated using the Fisher information matrix (FIM), whose elements are defined as

$$[I(X)]_{ij} = -\mathrm{E}\left[\frac{\partial^2 \ln f(\hat{D}|X)}{\partial X_i\, \partial X_j} \right]. \tag{7.16}$$

Then, using the PDF given in Equation (7.10), the FIM can be calculated as (Chan et al. 2006b; Cheng and Sahai 2004)

$$I(X) = \begin{bmatrix} \sum_{i=1}^{n} \frac{(x - x_i)^2}{\sigma_i^2 d_i^2} & \sum_{i=1}^{n} \frac{(x - x_i)(y - y_i)}{\sigma_i^2 d_i^2} \\ \sum_{i=1}^{n} \frac{(x - x_i)(y - y_i)}{\sigma_i^2 d_i^2} & \sum_{i=1}^{n} \frac{(y - y_i)^2}{\sigma_i^2 d_i^2} \end{bmatrix} \tag{7.17}$$

$$= \begin{bmatrix} \sum_{i=1}^{n} \frac{\cos^2(\theta_i)}{\sigma_i^2} & \sum_{i=1}^{n} \frac{\cos(\theta_i)\sin(\theta_i)}{\sigma_i^2} \\ \sum_{i=1}^{n} \frac{\cos(\theta_i)\sin(\theta_i)}{\sigma_i^2} & \sum_{i=1}^{n} \frac{\sin^2(\theta_i)}{\sigma_i^2} \end{bmatrix}, \tag{7.18}$$

where θ_i defines the angle from the ith FRP to the MS, and the CRLB is given by $I^{-1}(X)$. Thus, for an estimate $\hat{X}$ of the MS location obtained with any unbiased estimator, we have

$$\mathrm{E}_X[(\hat{X} - X)(\hat{X} - X)^{\mathrm{T}}] \geq I^{-1}(X). \tag{7.19}$$

The CRLB can be related to another important measurement metric, referred to as the geometric dilution of precision (GDOP) in Section 4.6.2. For identical noise variances $\sigma_i^2 = \sigma^2$ at different FRPs, the GDOP can be defined as

$$\mathrm{GDOP} = \frac{\mathrm{RMSE_{loc}}}{\mathrm{RMSE_{range}}} = \frac{\sigma_{\mathrm{loc}}}{\sigma}, \tag{7.20}$$

where $\mathrm{RMSE_{loc}}$ and $\mathrm{RMSE_{range}}$ are the root mean square error (RMSE) of the location estimate and the range estimate, respectively, and σ_{loc} is the standard deviation of the location estimate. Similarly to the CRLB, the GDOP also depends on the positions of the FRP and the MS, but it is independent of the measurement noise variance σ_i^2. GDOP values smaller than three are usually preferable, and values larger than six may imply a very bad geometry of the FRPs. If the location estimator employed can achieve the CRLB, the GDOP is given by

$$\mathrm{GDOP} = \frac{\sqrt{\mathrm{trace}[I^{-1}(X)]}}{\sigma} \tag{7.21}$$

$$= \sqrt{\mathrm{trace}[\tilde{I}^{-1}(X)]}, \tag{7.22}$$

where

$$\tilde{I}(X) = \begin{bmatrix} \sum_{i=1}^{n} \cos^2(\theta_i) & \sum_{i=1}^{n} \cos(\theta_i)\sin(\theta_i) \\ \sum_{i=1}^{n} \cos(\theta_i)\sin(\theta_i) & \sum_{i=1}^{n} \sin^2(\theta_i) \end{bmatrix}. \tag{7.23}$$

The relation between the achievable localization accuracy and the geometry between the locations of the MS and the FRPs is evident from Equation (7.22).

Example 7.1 In order to see some typical values of the GDOP in wireless location estimation, some simple node topologies were simulated, as shown in Figure 7.3. Three MS locations were

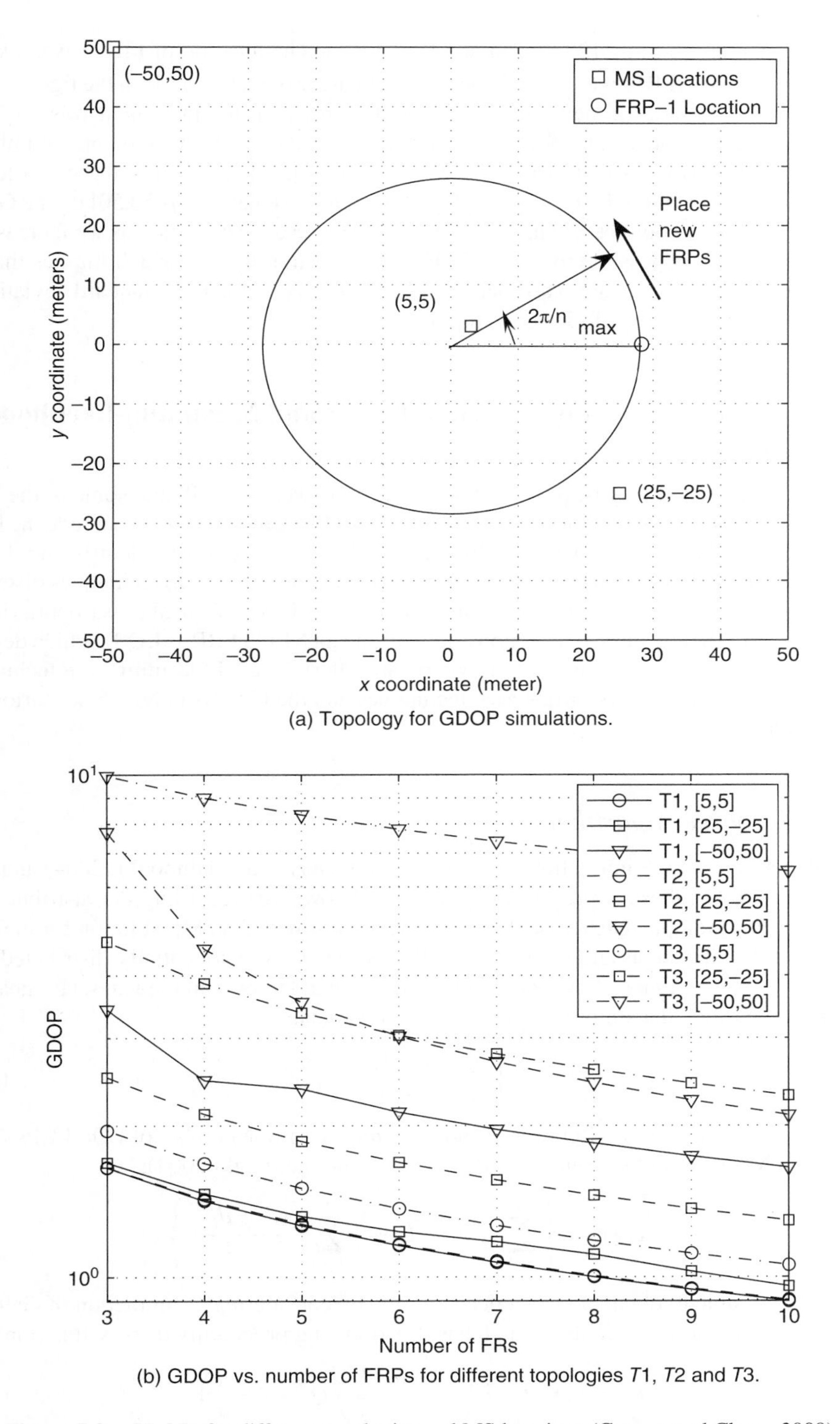

(a) Topology for GDOP simulations.

(b) GDOP vs. number of FRPs for different topologies $T1$, $T2$ and $T3$.

Figure 7.3 GDOPs for different topologies and MS locations (Guvenc and Chong 2009).

considered, namely [5,5] m, [25,-25] m and [-50,50] m. The location of FRP-1 was fixed at $[20\sqrt{2}, 0]$ m and new FRPs were added counterclockwise around the circle in the figure. Three topologies were considered, namely T1, T2 and T3, in which the increment was $2\pi/n_{\max}$, $\pi/n_{\max}$ and $\pi/2n_{\max}$ radians, respectively, where $n_{\max} = 10$ denotes the maximum number of FRPs. The GDOP was less than two for an MS located at [5,5] m for all of these topologies, and became better the more FRPs were deployed. For an MS located at [–50,50] m, the GDOP was the worst, and it could be as large as 10 for T3. The results show that, as we increase the number of FRPs, it is possible to have GDOP values smaller than 1, which implies that the standard deviation of the location estimate may become smaller than the standard deviation of the distance estimates for well-designed topologies.

7.3 NLOS Scenarios: Fundamental Limits and Maximum-likelihood Solutions

In realistic situations, it may be possible that the LOS between the MS and some of the FRPs may be obstructed (i.e., $b_i > 0$ for some i). These NLOS FRPs may seriously degrade the localization accuracy. The simplest way to mitigate the effects of NLOS is to identify and discard the NLOS FRPs and estimating the MS location by using one of the LOS techniques discussed in the previous section. However, there is always the possibility of false alarms (identifying an LOS FRP as NLOS) and missed detections (identifying an NLOS FRP as LOS), which degrade the localization accuracy. In this section, we review alternative NLOS mitigation techniques reported in the literature. First, ML-based techniques and the CRLBs in NLOS scenarios will be discussed.

7.3.1 ML-based Algorithms

ML approaches to NLOS mitigation were discussed in Gezici and Sahinoglu (2004) and Riba and Urruela (2004); these approaches require prior knowledge regarding the distribution of NLOS bias. For example, Gezici and Sahinoglu (2004) provided an ML solution for the position of the MS with the assumption that the NLOS bias b_i is exponentially distributed with parameter λ_i. Since in most cases the NLOS bias is much larger than the Gaussian measurement error, the following simplifying assumption was made:

$$\hat{d}_i = d_i + \begin{cases} b_i, & i = 1, 2, \dots, n_{\mathrm{NL}}, \\ \eta_i, & i = n_{\mathrm{NL}} + 1, \dots, n, \end{cases} \tag{7.24}$$

where n_{NL} is the number of NLOS FRPs and $n_{\mathrm{L}} = n - n_{\mathrm{NL}}$ is the number of LOS FRPs. Thus, a simplified ML solution is given as follows (Gezici and Sahinoglu 2004):

$$\hat{X} = \arg\min_{X} \left\{ \sum_{i=1}^{n_{\mathrm{NL}}} \lambda_i(\hat{d}_i - d_i) + \sum_{i=n_{\mathrm{NL}}+1}^{n} \frac{(\hat{d}_i - d_i)^2}{2\sigma_i^2} \right\}. \tag{7.25}$$

It is also possible to obtain an exact decision rule by considering a summation of Gaussian and exponential random variables, which has the following probability density function:

$$\mathrm{f}(u) = \lambda \exp\left(-\lambda\left(u - \frac{\lambda\sigma^2}{2}\right)\right) Q\left(\lambda\sigma - \frac{u}{\sigma}\right), \tag{7.26}$$

where $Q(a) = (1/\sqrt{2\pi}) \int_a^{\infty} \exp(-u^2/2)\, du$ denotes the Q-function, and the exact ML solution becomes (Gezici and Sahinoglu 2004)

$$\hat{X} = \arg\min_X \left\{ \sum_{i=1}^{n_{\mathrm{NL}}} \lambda_i \left(\hat{d}_i - d_i - \frac{\lambda_i \sigma_i^2}{2} \right) \right.$$
$$\left. - \sum_{i=1}^{n_{\mathrm{NL}}} \log \left[Q\left(\lambda_i \sigma_i - \frac{\hat{d}_i - d_i}{\sigma_i} \right) \right] + \sum_{i=n_{\mathrm{NL}}+1}^{n} \frac{(\hat{d}_i - d_i)^2}{2\sigma_i^2} \right\}. \tag{7.27}$$

Example 7.2 Consider a simple cellular wireless localization scenario with seven FRPs, as illustrated in Figure 7.4. An MS is located at (1,1.5) km and the locations of the FRPs are as illustrated in the figure. Each of the cells has a radius of 3 km and it is assumed that the transmitted MS signal is successfully received at all seven FRPs and utilized for location estimation. The distance estimate at each of the FRPs is subject to zero-mean AWGN with a variance of 0.2 km^2. Moreover, for distance estimates at FRP-1 an exponentially distributed NLOS bias exists, which is assumed to have a mean equivalent to 10% of the true measurement. Given this scenario, the metrics for the approximate ML in Equation (7.25) and exact ML in Equation (7.27) are plotted in Figures 7.4(a) and 7.4(b), respectively, for one realization of the noise and NLOS bias. It is observed that both metrics are minimized close to the true location of the MS.

Another NLOS mitigation technique for TOA-based systems based on the ML approach was introduced by Riba and Urruela (2004). In this technique, several hypotheses about the set of FRPs are considered, and then, utilizing the ML principle, the best set (which is assumed to be composed of LOS FRPs) is selected for location estimation. The hypothesis index estimate for the best FRP set is derived as (Riba and Urruela 2004)

$$\hat{\zeta} = \arg\min_i \left[\ln \gamma^{-1}(i) + \sum_{j \in S_i^{\mathrm{LOS}}} \frac{n_{\mathrm{trn}}}{2} \frac{\hat{\sigma}_j^2(i)}{\sigma_j^2} \right], \tag{7.28}$$

where S_i^{LOS} denotes the ith set of FRPs which are hypothesized to be in LOS, the $\gamma(i)$ are assigned according to the a priori probability of each hypothesis (equivalent to 1, if no information is available),

$$\hat{\sigma}_j^2(i) = \frac{1}{n_{\mathrm{trn}}} \sum_{k=1}^{n_{\mathrm{trn}}} \left(t_j^{(k)} - \frac{\|\hat{X}(i) - X_k\|}{c} \right)^2 \tag{7.29}$$

denotes the estimated variance of the n_{trn} TOA measurements associated with the jth FRP under the ith hypothesis, and $t_j^{(k)}$ denotes the kth TOA measurement at the jth FRP. Note that the above approach requires buffering of n_{trn} TOA measurements for the purpose of obtaining the noise statistics for a particular FRP. Once the set of LOS FRPs has been selected using the ML principle, the MS location is estimated using only these FRPs and the ML algorithm. Simulation results (Riba and Urruela 2004) show that this yields better accuracy than the residual-weighting (Rwgh) technique introduced by Chen (1999) (see Section 7.4.2), and slightly worse than when only the true LOS FRPs are used for localization. Also, at low signal-to-noise ratio (SNR) and small NLOS bias values, simulation results show that it may be better not to employ any NLOS mitigation, in order not to degrade the accuracy.

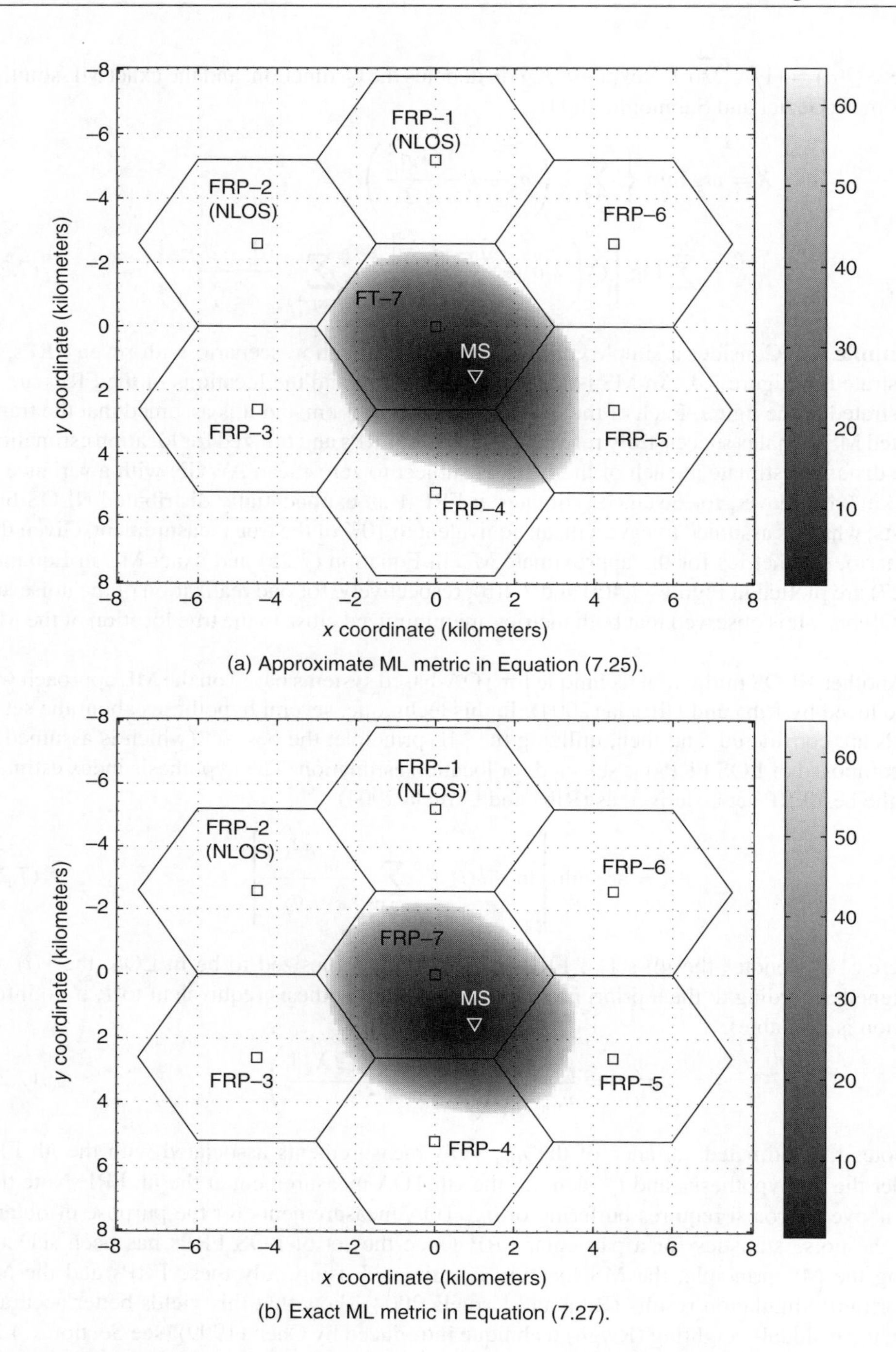

Figure 7.4 Simulations of exact and approximate ML techniques for NLOS mitigation.

7.3.2 Cramér–Rao Lower Bound

In NLOS scenarios, the CRLB depends on whether there is any prior information available about the NLOS bias. First, let us assume that there is no prior information about the NLOS bias, but that we know the FRPs that are in NLOS. Then, an extended version of the FIM in Equation (7.19) is given by (Qi et al. 2006)

$$I(X_b) = \tilde{A} I(D) \tilde{A}^{\mathrm{T}}, \tag{7.30}$$

where

$$X_b = [x, y, b_1, b_2, \dots, b_{n_{\mathrm{NL}}}]^{\mathrm{T}} \tag{7.31}$$

is an $(n_{\mathrm{NL}} + 2) \times 1$ vector of unknown parameters incorporating the NLOS bias values, and

$$\tilde{A} = \frac{\partial D}{\partial X_b} = \begin{bmatrix} \frac{\partial d_1}{\partial x} & \frac{\partial d_2}{\partial x} & \cdots & \frac{\partial d_{n_{\mathrm{NL}}}}{\partial x} & \cdots & \frac{\partial d_n}{\partial x} \\ \frac{\partial d_1}{\partial y} & \frac{\partial d_2}{\partial y} & \cdots & \frac{\partial d_{n_{\mathrm{NL}}}}{\partial y} & \cdots & \frac{\partial d_n}{\partial y} \\ \frac{\partial d_1}{\partial b_1} & \frac{\partial d_2}{\partial b_1} & \cdots & \frac{\partial d_{n_{\mathrm{NL}}}}{\partial b_1} & \cdots & \frac{\partial d_n}{\partial b_1} \\ \vdots & \vdots & \ddots & \vdots & \ddots & \vdots \\ \frac{\partial d_1}{\partial b_{n_{\mathrm{NL}}}} & \frac{\partial d_2}{\partial b_{n_{\mathrm{NL}}}} & \cdots & \frac{\partial d_{n_{\mathrm{NL}}}}{\partial b_{n_{\mathrm{NL}}}} & \cdots & \frac{\partial d_n}{\partial b_{n_{\mathrm{NL}}}} \end{bmatrix}, \tag{7.32}$$

and

$$I(D) = \mathrm{E}_D \left[\frac{\partial \ln f(\hat{D}|D)}{\partial D} \left(\frac{\partial \ln f(\hat{D}|D)}{\partial D} \right)^{\mathrm{T}} \right]. \tag{7.33}$$

As discussed in Qi et al. (2006), $\tilde{A}$ can be written in terms of its LOS and NLOS components as

$$\tilde{A} = \begin{bmatrix} \tilde{A}_{\mathrm{NL}} & \tilde{A}_{\mathrm{L}} \\ I_{n_{\mathrm{NL}}} & \mathbf{0}_{n_{\mathrm{NL}}, n_{\mathrm{L}}} \end{bmatrix}, \tag{7.34}$$

where $I_{n_{\mathrm{NL}}}$ and $\mathbf{0}_{n_{\mathrm{NL}}, n_{\mathrm{L}}}$ are an identity matrix of size $n_{\mathrm{NL}} \times n_{\mathrm{NL}}$ and a zero matrix of size $n_{\mathrm{NL}} \times n_L$, respectively, and

$$\tilde{A}_{\mathrm{NL}} = \begin{bmatrix} \cos\theta_1 & \cos\theta_2 & \dots & \cos\theta_{n_{\mathrm{NL}}} \\ \sin\theta_1 & \sin\theta_2 & \dots & \sin\theta_{n_{\mathrm{NL}}} \end{bmatrix} \tag{7.35}$$

and

$$\tilde{A}_{\mathrm{L}} = \begin{bmatrix} \cos\theta_{n_{\mathrm{NL}}+1} & \cos\theta_{n_{\mathrm{NL}}+2} & \dots & \cos\theta_n \\ \sin\theta_{n_{\mathrm{NL}}+1} & \sin\theta_{n_{\mathrm{NL}}+2} & \dots & \sin\theta_n \end{bmatrix}. \tag{7.36}$$

Similarly, $I(D)$ can be written in terms of its NLOS and LOS components as (Qi et al. 2006)

$$I(D) = \begin{bmatrix} \Lambda_{\mathrm{NL}} & \mathbf{0} \\ \mathbf{0} & \Lambda_{\mathrm{L}} \end{bmatrix}, \tag{7.37}$$

where $\Lambda_{\rm NL} = {\rm diag}(\sigma_1^{-2}, \ldots, \sigma_{n_{\rm NL}}^{-2})$ and $\Lambda_{\rm L} = {\rm diag}(\sigma_{n_{\rm NL}+1}^{-2}, \ldots, \sigma_n^{-2})$. After some manipulations, $I(X_b)$ can be obtained as (Qi et al. 2006)

$$I(X_b) = \begin{bmatrix} \tilde{A}_{\rm NL}\Lambda_{\rm NL}\tilde{A}_{\rm NL}^{\rm T} + \tilde{A}_{\rm L}\Lambda_{\rm L}\tilde{A}_L^{\rm T} & \tilde{A}_{\rm NL}\Lambda_{\rm NL} \\ \Lambda_{\rm NL}\tilde{A}_{\rm NL}^{\rm T} & \Lambda_{\rm NL} \end{bmatrix}. \tag{7.38}$$

Note that Equation (7.38) depends on both NLOS and LOS signals. However, it was proven in Qi et al. (2006) that the CRLB for the MS location is given by

$$\mathrm{E}[(\hat{X}_b - X_b)(\hat{X}_b - X_b)^{\rm T}] \geq I^{-1}(X_b) = (\tilde{A}_{\rm L}\Lambda_{\rm L}\tilde{A}_L^{\rm T})^{-1}. \tag{7.39}$$

In other words, the CRLB depends exclusively on the LOS signals if the NLOS FRPs can be accurately identified. Hence, an ML estimator that can achieve the CRLB in NLOS scenarios first identifies the NLOS FRPs and discards their measurements, and then obtains the location estimate using the LOS FRPs (Figure 7.5(a)).

If there is further side information related to the statistics of the NLOS bias vector B, a better positioning accuracy can be obtained. Then, the generalized Cramer–Rao lower bound

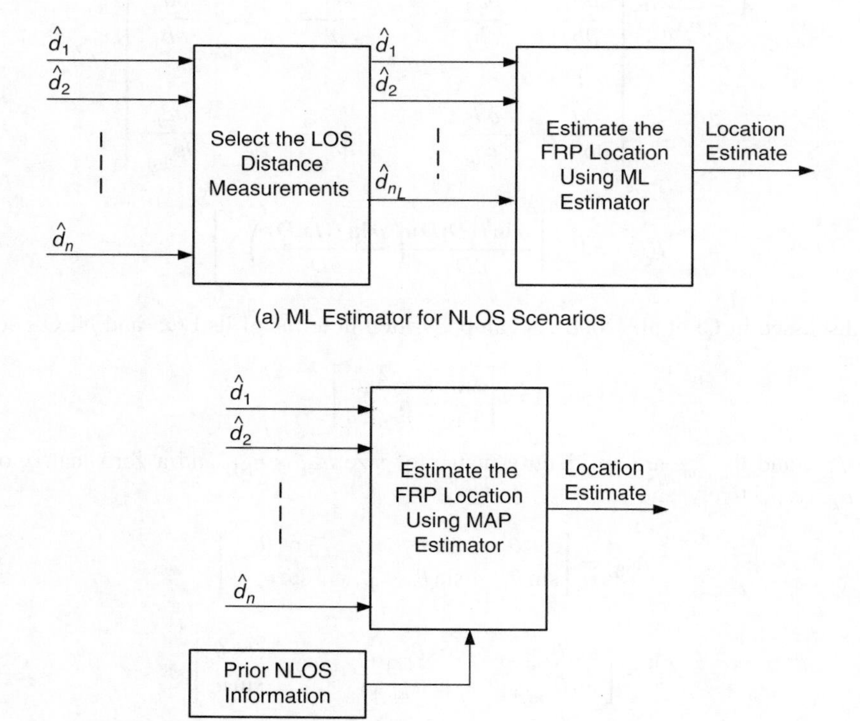

Figure 7.5 Block diagrams for (a) ML estimator and (b) MAP estimator for NLOS scenarios. In (a), without loss of generality, it is assumed that the first $n_{\rm L}$ measurements are the LOS measurements (Guvenc and Chong 2009).

(G-CRLB) can be written as (Qi et al. 2006)

$$\mathrm{E}[(\hat{X}_b - X_b)(\hat{X}_b - X_b)^{\mathrm{T}}] \geq \left(I(X_b) + \begin{bmatrix} \mathbf{0} & \mathbf{0} \\ \mathbf{0} & \mathbf{\Omega} \end{bmatrix} \right)^{-1} \tag{7.40}$$

$$= \left(\begin{bmatrix} \tilde{A}_{\mathrm{NL}}\Lambda_{\mathrm{NL}}\tilde{A}_{\mathrm{NL}}^{\mathrm{T}} + \tilde{A}_L\Lambda_{\mathrm{L}}\tilde{A}_{\mathrm{L}}^{\mathrm{T}} & \tilde{A}_{\mathrm{NL}}\Lambda_{\mathrm{NL}} \\ \Lambda_{\mathrm{NL}}\tilde{A}_{\mathrm{NL}}^{\mathrm{T}} & \Lambda_{\mathrm{NL}} + \mathbf{\Omega} \end{bmatrix} \right)^{-1}, \tag{7.41}$$

where $\mathbf{\Omega} = \mathrm{diag}(\tilde{\sigma}_1^2, \ldots, \tilde{\sigma}_{n_{\mathrm{NL}}}^2)$, and $\tilde{\sigma}_i^2$ can be interpreted as the variance[3] of b_i. As an upper bound on the G-CRLB, when the variances $\tilde{\sigma}_i^2$ are infinitely large, the G-CRLB is reduced to the CRLB since there is practically no information available about b_i. The estimator that asymptotically achieves the G-CRLB is given by the maximum a posteriori (MAP) estimator, and it employs the statistics of the NLOS biases (Figure 7.5(b)).

7.4 Least-squares Techniques for NLOS Localization

The LS techniques for location estimation can be tuned to suppress the effects of NLOS bias, for example through some appropriate weighting. In this section, weighted LS approaches and the residual-weighting algorithm will be briefly reviewed.

7.4.1 Weighted Least Squares

A simple way to mitigate the effects of NLOS FRPs is to give less emphasis to the corresponding NLOS terms in the LS solution. In Caffery and Stüber (1998b) and Gezici and Sahinoglu (2004), with the assumption that the variances of the distance measurements are larger for NLOS FRPs, the inverses of these variances are used as a reliability metric. From the ML algorithm, the location estimate is given by

$$X_{\mathrm{ML}} = \arg\max_X \mathrm{p}(\hat{D}|X), \tag{7.42}$$

where

$$\mathrm{f}(\hat{D}|X) = \mathrm{f_n}(\hat{D} - D|X). \tag{7.43}$$

If the noise is Gaussian distributed, we have

$$\mathrm{f_n}(\eta) = \frac{1}{\sqrt{2\pi}\sigma_i} \exp\left(-\frac{\eta^2}{2\sigma_i^2} \right). \tag{7.44}$$

Then, the joint probability function becomes

$$\mathrm{f}(\hat{D}|X) = \frac{1}{(2\pi)^{n/2}\prod_{i=1}^n \sigma_i} \exp\left(-\sum_{i=1}^n \frac{(\hat{d}_i - \|X - X_i\|)^2}{2\sigma_i^2} \right). \tag{7.45}$$

On further manipulation of Equation (7.45), the ML solution becomes equivalent to

$$\hat{X}_{\mathrm{ML}} = \arg\min_X \sum_{i=1}^n \frac{(\hat{d}_i - \|X - X_i\|)^2}{\sigma_i^2}, \tag{7.46}$$

[3] For a Gaussian-distributed NLOS bias, this is strictly the variance of the NLOS bias.

which is in essence a weighted LS solution with weights set to $\beta_i = 1/\sigma_i^2$. However, for a static MS, the variance of the TOA measurements may not be significantly different for LOS and NLOS FRPs. Still, the bias in the NLOS distance measurements may degrade the localization accuracy. Hence, in Guvenc et al. (2007, 2008), an alternative weighting technique was proposed, which uses certain statistics of the multipath components of the received signals. In particular, the kurtosis, mean excess delay and root mean square (RMS) delay spread of the received signal are used to evaluate the likelihood of the received signal to be in LOS. The likelihood values are then used to evaluate the weighting parameters β_i.

7.4.2 Residual-weighting Algorithm

The Rwgh algorithm proposed by Chen (1999) is based on the observation that the residual error is typically larger if NLOS FRPs are used when estimating the MS location, where the residual error is defined as

$$r_{\mathrm{es}}(X) = \sum_{i=1}^{n} (\hat{d}_i - \|X - X_i\|)^2. \tag{7.47}$$

Assuming that there are more than three FRPs available, Rwgh estimates the MS location as follows (Chen 1999). (1) Form $n_{\mathrm{cb}} = \sum_{i=3}^{n} \binom{n}{i}$ range measurement combinations, where $\binom{n}{i}$ denotes the total number of combinations of i FRPs selected from a total of n FRPs. Also, let $\{S_k | k = 1, 2, \ldots, n_{\mathrm{cb}}\}$ denote the set of FRPs for the kth combination. (2) For each set of combinations S_k, compute an intermediate LS location estimate as follows:

$$\hat{X}_k = \arg\min_{X} \{r_{\mathrm{es}}(X; S_k)\}, \tag{7.48}$$

where $r_{\mathrm{es}}(X; S_k)$ is the residual error when only the FRPs in set S_k are used for calculating the MS location. Also, we define the normalized residual

$$\bar{r}_{\mathrm{es}}(\hat{X}_k; S_k) = \frac{r_{\mathrm{es}}(\hat{X}_k; S_k)}{|S_k|}, \tag{7.49}$$

where $|S_k|$ denotes the number of elements of S_k. (3) Find the final location estimate by weighting the intermediate location estimates with their corresponding normalized residual errors:

$$\hat{X} = \frac{\sum_{k=1}^{n_{\mathrm{cb}}} \hat{X}_k [\bar{r}_{\mathrm{es}}(\hat{X}_k; S_k)]^{-1}}{\sum_{k=1}^{n_{\mathrm{cb}}} [\bar{r}_{\mathrm{es}}(\hat{X}_k; S_k)]^{-1}}. \tag{7.50}$$

The main idea behind Rwgh is that it calculates a weighted average of all candidate location estimates, where the location estimates with larger residual errors (i.e., location estimates that are possibly impacted by NLOS propagation) are given smaller weights.

It was shown (Chen 1999) through simulations that Rwgh performs better than choosing the location estimate with the minimum residual error. A suboptimal version of the Rwgh algorithm that has lower computational complexity was proposed by Li (2006). In the latter, instead of considering all combinations of the FRPs (which may be very large in number if n is large), all combinations with $(n - 1)$ FRPs are used to calculate the intermediate location estimates and the corresponding residuals. Then, out of n different combinations, the FRP which is not employed in the best estimator (i.e., the estimator corresponding to the combination with the smallest residual error) is discarded. The process is iterated until a predetermined stopping rule is reached (such as when a minimum number of FRPs is reached or when the change in the residual error is small).

Example 7.3 Consider a cellular localization scenario with parameters as specified in Example 7.2. We have evaluated three different location estimators. The first one was a linear least-squares (LLS-ALL) estimator which uses all seven FRPs for location estimation. A second linear least-squares estimator assumed that the NLOS FRPs were accurately identified and used only the LOS FRPs for estimating the MS location (LLS-LOS). The third estimator was the Rwgh algorithm. For comparison purposes, the linear least-squares estimator when all the FRPs are in LOS, and the CRLB in LOS scenarios have also been plotted (Figure 7.6).

Two different NLOS settings have been considered: the first one assumes that only FRP-1 is in NLOS, while the second assumes that both FRP-1 and FRP-2 are in NLOS. The simulation results in Figure 7.6 show that, if the NLOS propagation effects are not handled, there is a significant degradation in the localization performance, as illustrated by the LLS-ALL results. If the NLOS FRPs can be accurately identified, those FRPs can be removed from the localization process to improve the accuracy (LLS-LOS). Since accurate identification of the NLOS FRPs may not always be possible, utilizing the Rwgh technique yields a reasonable accuracy improvement compared with LLS-ALL. Note that if the statistics of the NLOS bias are available, it may be possible to obtain accuracy levels closer to the CRLB, for example by utilizing techniques such as ML-based location estimation.

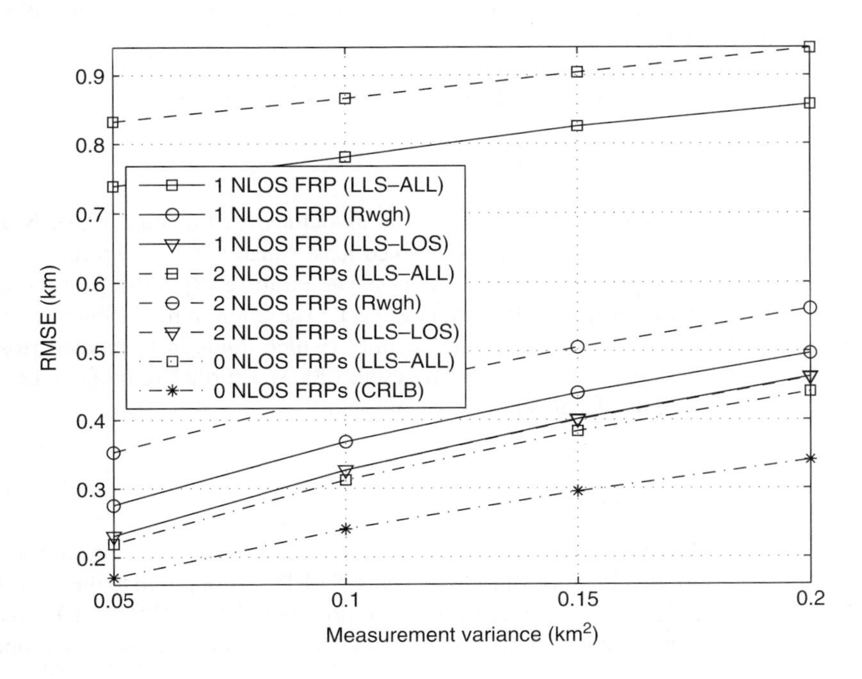

Figure 7.6 Simulations for several different NLOS mitigation techniques.

7.5 Constraint-based Techniques for NLOS Localization

In this section, a different class of NLOS mitigation algorithms that utilize some constraints related to the NLOS measurements will be briefly reviewed.

7.5.1 *Constrained LS Algorithm and Quadratic Programming*

The two-step ML algorithm discussed by Chan et al. (2006a) is an accurate localization technique in LOS settings, but it is not robust against NLOS. In Wang et al. (2003), a quadratic programming (QP) technique for NLOS environments was developed. First, Equation (7.8) is written in matrix form as

$$A_1 = \begin{bmatrix} x_1 & y_1 & -0.5 \\ x_2 & y_2 & -0.5 \\ \vdots & \vdots & \\ x_n & y_n & -0.5 \end{bmatrix} \tag{7.51}$$

and

$$\boldsymbol{\theta} = \begin{bmatrix} x \\ y \\ s \end{bmatrix}, \quad \mathbf{p}_1 = \begin{bmatrix} k_1 - \hat{d}_1^2 \\ k_2 - \hat{d}_2^2 \\ \vdots \\ k_n - \hat{d}_n^2 \end{bmatrix}, \tag{7.52}$$

with $s = x^2 + y^2$ being a part of the vector of unknown variables. Then, the mathematical programming is formulated as follows:

$$\hat{\boldsymbol{\theta}}_{cw} = \arg\min_{\boldsymbol{\theta}} (A_1\boldsymbol{\theta} - \mathbf{p}_1)^{\mathrm{T}} \boldsymbol{\Psi}^{-1} (A_1\boldsymbol{\theta} - \mathbf{p}_1), \tag{7.53}$$

$$\text{such that } A_1\boldsymbol{\theta} \leq \frac{1}{2}\mathbf{p}_1, \tag{7.54}$$

where $\boldsymbol{\Psi} = BQB$, $B = \mathrm{diag}(2d_1, \ldots, 2d_n)$ and Q is as defined in Equation (7.6). Note that Equations (7.53) and (7.54) constitute a constrained least squares (CLS) algorithm that can be solved using quadratic programming techniques. The intuitive explanation of the CLS is that Equation (7.53) finds a weighted least squares (WLS) solution to the MS location, while the constraint in Equation (7.54) relaxes the equality (which holds in LOS scenarios) into an inequality for the NLOS scenarios. In Wang et al. (2003) a further refining stage is also introduced to incorporate the dependency between s and X.

7.5.2 *Linear Programming*

In Venkatesh and Buehrer (2006, 2007), a linear programming approach was introduced which assumes perfect a priori identification of LOS and NLOS FRPs. As opposed to the identify and discard (IAD) type of algorithms (see Section 7.7), it does not discard NLOS FRPs, but uses them to construct a linear feasible region for the MS location. The location estimate is obtained using a linear programming technique that employs only the LOS FRPs, with the constraint that it has to be within the feasible region obtained.

If the ith FRP is identified to be in NLOS, we have the nonlinear constraint

$$||X - X_i|| \leq \hat{d}_i, \quad (7.55)$$

which is used in Wang et al. (2003) to formulate a quadratic programming technique for the two-step ML algorithm (without the assumption that the NLOS FRPs are accurately identified). In order to linearize the constraints in Equation (7.55), we can relax them to (Larsson 2004)

$$x - x_i \leq \hat{d}_i, \quad -x + x_i \leq \hat{d}_i, \quad (7.56)$$

$$y - y_i \leq \hat{d}_i, \quad -y + y_i \leq \hat{d}_i, \quad (7.57)$$

where $i = 1, 2, \ldots, n$. In essence, the above is equivalent to relaxing the circular constraints into rectangular (i.e., square) constraints for the purpose of linearizing the equations (see Figure 7.7). By defining some slack variables,[4] Equations (7.56) and (7.57) can be converted to equalities that define the feasible region. Then, the MS location is estimated by using only the LOS FRPs for minimizing an objective function, with the constraint that the location estimate should be within the feasible region obtained using both the NLOS and LOS FRPs.

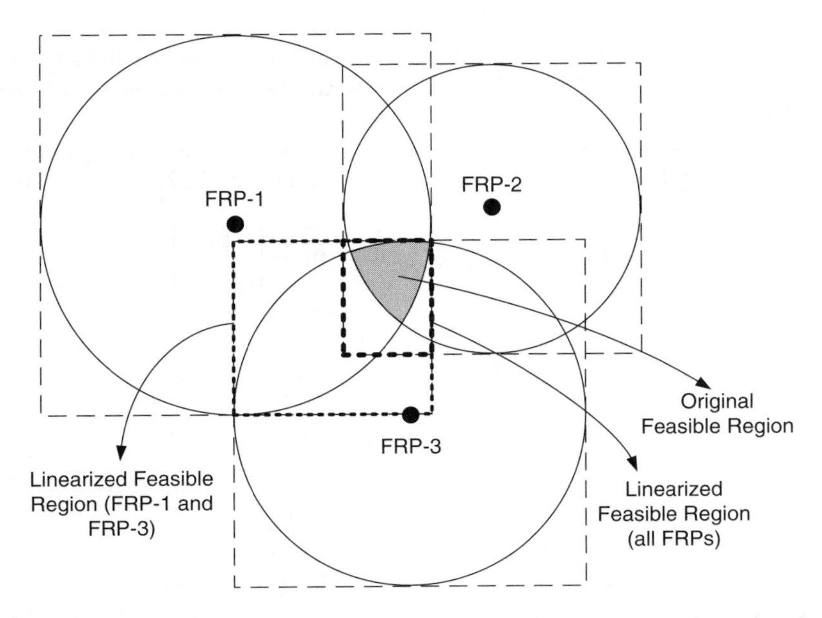

Figure 7.7 Illustration of the CLS technique. The NLOS FRPs are used to determine the feasible region. For the original (nonlinear) model, the feasible region is obtained from the intersections of the circles. For the linear model, the feasible region is obtained from the intersections of the squares (Guvenc and Chong 2009).

[4] A slack variable is a variable which is used to turn an inequality into an equality (e.g., $x < 5 \Rightarrow x + s = 5$).

7.5.3 Geometry-constrained Location Estimation

In Chen and Feng (2005b) a geometry-constrained location estimation method was proposed, which uses the two-step ML technique of Chan and Ho (1994) with some additional parameters to incorporate the geometry of the FRPs (only a scenario with three FRPs was considered). Let the intersection points of the three circles in Equation (7.8) be $X_A = [x_A, y_A]^T$, $X_B = [x_B, y_B]^T$ and $X_C = [x_C, y_C]^T$. Then, a constrained cost function, which is referred to as the *virtual distance*, is defined as (Chen and Feng 2005b)

$$\gamma = \sqrt{\frac{1}{3}[\|X - X_A\|^2 + \|X - X_B\|^2 + \|X - X_C\|^2]}. \tag{7.58}$$

For an expected position X_e of the MS, we can calculate the *expected virtual distance* using Equation (7.58) as $\gamma_e = \gamma + n_e$, with n_e denoting the noise in the expected virtual distance. The coordinates for X_e are chosen as (Chen and Feng 2005b)

$$X_e = \omega_1 X_A + \omega_2 X_B + \omega_3 X_C, \tag{7.59}$$

where the weights are obtained as

$$\omega_i = \frac{\sigma_i^2}{\sigma_1^2 + \sigma_2^2 + \sigma_3^2}, \quad i = 1, 2, 3. \tag{7.60}$$

Basically, it is assumed that for NLOS FRPs, the measurement variance will be larger and the weights defined in Equation (7.60) will move X_e towards the center of the NLOS FRP circle, yielding a better location estimate.

These geometric constraints are then incorporated into the two-step ML algorithm of Chan and Ho (1994) by updating A_1 and $\mathbf{p}_1$ in Equations (7.51) and (7.52) as follows (Chen and Feng 2005b):

$$\tilde{A}_1 = \begin{bmatrix} & A_1 & \\ \gamma_x & \gamma_y & -0.5 \end{bmatrix} \quad \text{and} \quad \tilde{\mathbf{p}}_1 = \begin{bmatrix} \mathbf{p}_1 \\ \gamma_k - \gamma_e^2 \end{bmatrix}, \tag{7.61}$$

where

$$\gamma_x = \frac{1}{3}(x_A + x_B + x_C), \tag{7.62}$$

$$\gamma_y = \frac{1}{3}(y_A + y_B + y_C) \tag{7.63}$$

and

$$\gamma_k = \frac{1}{3}(x_A^2 + x_B^2 + x_C^2 + y_A^2 + y_B^2 + y_C^2). \tag{7.64}$$

The geometric constraints are also incorporated into other variables as

$$\tilde{B} = \text{diag}(d_1, d_2, d_3, \gamma) \tag{7.65}$$

and

$$\tilde{\eta} = [\eta_1, \eta_2, \eta_3, \eta_\gamma]^T. \tag{7.66}$$

Then, the two-step ML algorithm given in Chan and Ho (1994) is employed to solve for the MS location using the variables updated with the geometric constraints, thus improving the performance in NLOS scenarios.

7.5.4 Interior-point Optimization

In Kim et al. (2006) an interior-point optimization (IPO) method was proposed for finding the optimum location estimate in the presence of NLOS bias. By linearizing the system using a Taylor series approximation, the nonlinear function $D(X)$ in Equation (7.3) can be linearized around a reference point X_0. Then, neglecting the higher-order terms, we can write

$$D(X) \approx D(X_0) + H_0(X - X_0), \tag{7.67}$$

where H_0 is the Jacobian matrix of $D(X)$ around X_0, which is defined as

$$H_0 = \begin{bmatrix} \frac{\partial d_1}{\partial x} & \frac{\partial d_2}{\partial x} & \cdots & \frac{\partial d_n}{\partial x} \\ \frac{\partial d_1}{\partial y} & \frac{\partial d_2}{\partial y} & \cdots & \frac{\partial d_n}{\partial y} \end{bmatrix}^{\mathrm{T}}_{X=X_0}. \tag{7.68}$$

Note that the reference point X_0 must be chosen sufficiently close to the true location in order for Equation (7.67) to be valid. By substituting Equation (7.67) into Equation (7.46) and by means of some manipulations, we may obtain a linear system which can be written in a matrix form and solved using an LS estimator. A more accurate iterative technique could use this LS estimate as an intermediate estimate, plug it into Equation (7.67) to relinearize the system around it and iterate until convergence (Caffery and Stüber 1998a).

By using Equation (7.67), a linearized measurement vector is defined as follows (Kim et al. 2006):

$$Z = H_0X + B + \eta. \tag{7.69}$$

Then, if NLOS bias is neglected, the bias-free position estimate is given by (see Section 5.2.1)

$$\hat{X} = (H_0^{\mathrm{T}}Q^{-1}H_0)^{-1}H_0^{\mathrm{T}}Q^{-1}Z. \tag{7.70}$$

On the other hand, if the bias vector B is known, a more accurate bias-free location estimate can be derived (Kim et al. 2006):

$$\hat{X}^* = \hat{X} + VB, \tag{7.71}$$

where

$$V = -(H_0^{\mathrm{T}}Q^{-1}H_0)^{-1}H_0^{\mathrm{T}}Q^{-1} \tag{7.72}$$

is a bias correction matrix. However, in reality, B is unknown and has to be estimated.

In order to estimate B from Equation (7.69), the observed bias metric is defined as (Kim et al. 2006)

$$Y = Z - H_0\tilde{X}, \tag{7.73}$$

which can be simplified to $Y = SB + W$, where $S = I + H_0V$ and the *bias noise* is given by

$$W = H_0(X - \hat{X}) - \eta. \tag{7.74}$$

Then the following constrained optimization problem is defined to estimate the NLOS bias errors (Kim et al. 2006):

$$\hat{B} = \arg\min_{B} (Y - SB)^{\mathrm{T}}Q_w^{-1}(Y - SB), \tag{7.75}$$

$$\text{such that } b_i \in B_i, \quad i = 1, 2, \ldots, n, \tag{7.76}$$

where $B_i = [l_i, u_i]$ is the a priori information for the range of b_i lower-bounded by $l_i \geq 0$ and upper-bounded by u_i, and Q_w is the covariance matrix of W.

In order to solve the constrained optimization problem in Equations (7.75) and (7.76), an IPO technique was used in Kim et al. (2006). In particular, Equations (7.75) and (7.76) were modified to

$$\hat{B} = \arg\min_{B} (Y - SB)^{\mathrm{T}} Q_w^{-1} (Y - SB), \tag{7.77}$$

$$\text{such that } g_i(b_i) - s_i = 0 \quad \text{and} \quad s_i > 0, i = 1, \ldots, n, \tag{7.78}$$

where s_i is a slack variable and $g_i(b_i)$ is a barrier function that satisfies $g_i(b_i) > 0, \forall b_i \in [l_i, u_i]$. A smooth second-order function that is generally used and which satisfies this requirement is $g_i(b_i) = (u_i - b_i)/(b_i - l_i)$. Equations (7.77) and (7.78) are then solved by minimizing the following Lagrangian (Kim et al. 2006):

$$\mathcal{L}(B, \lambda, \mathbf{s}) = (Y - SB)^{\mathrm{T}} Q_w^{-1} (Y - SB) - \mu \sum_{i=1}^{n} \ln s_i - \lambda^{\mathrm{T}}(\mathbf{g}(b) - \mathbf{s}), \tag{7.79}$$

where λ is a vector of Lagrange multipliers and $\mathbf{g}(b)$ and $\mathbf{s}$ are obtained upon stacking $g_i(b_i)$ and s_i into $n \times 1$ vectors. Note that the logarithmic barrier function, with μ as a positive bias parameter

$$\mu \sum_{i=1}^{n} \ln s_i, \tag{7.80}$$

ensures that $s_i > 0$ and the bias error is always within $[l_i, u_i]$.

The solution to Equation (7.79) can be obtained by differentiating Equation (7.79) with respect to B, λ and $\mathbf{s}$, and solving the resulting equations together to obtain B. Once an estimate of the bias vector B is obtained, the bias correction matrix in Equation (7.72) is used to calculate the bias-free location using Equation (7.71).[5] The simulation results reported in Kim et al. (2006) show that better accuracies can be obtained with IPO than with Rwgh or iterative LS algorithms in NLOS scenarios.

7.6 Robust Estimators for NLOS Localization

Robust estimators are commonly used to suppress the impact of outliers in a given set of data, and several different classes of robust estimators have already been used in the literature for NLOS mitigation purposes. Below, a few of the popular robust estimators considered for NLOS mitigation are briefly reviewed.

7.6.1 Huber M*-estimator*

The *M*-estimators, which are of the ML type of estimators, are one class of robust estimators that have been considered for NLOS mitigation purposes. As discussed in the previous

[5] This is similar to the LOS construction algorithm discussed in Caffery (1999).

sections, the ML algorithm tries to maximize a function of the form

$$\prod_{i=1}^{n} \mathrm{f}(x_i), \tag{7.81}$$

which is equivalent to minimizing $\sum_{i=1}^{n} - \log \mathrm{f}(x_i)$. In the presence of outliers,[6] the ML algorithm fails to yield accurate results. A generalized form of the ML algorithm referred to as the M-estimator, which was introduced in Huber (1964), aims at minimizing

$$\sum_{i=1}^{n} \rho(x_i), \tag{7.82}$$

where $\rho(.)$ is a convex function. For the Huber M-estimator, $\rho(\cdot)$ is defined as (Petrus 1999)

$$\rho(v) = \begin{cases} v^2/2 & |v| \leq \xi, \\ \xi|v| - \xi^2/2 & |v| > \xi, \end{cases} \tag{7.83}$$

which is not strictly convex and, therefore, minimization of the objective function yields multiple solutions (Petrus 1999).

In Sun and Guo (2004), the M-estimator, used to estimate the MS location in the presence of NLOS bias, was defined as

$$\hat{X} = \arg\min_{X} \left\{ \sum_{i=1}^{n} \rho((\hat{d}_i - \|X - X_i\|)/\sigma_i) \right\}. \tag{7.84}$$

Simulation results (Sun and Guo 2004) show that the M-estimator outperforms the conventional LS estimator, especially for large NLOS bias errors. If the bootstrapping technique[7] is used in conjunction with the M-estimator, the accuracy can be improved even further (Sun and Guo 2004).

7.6.2 Least Median Squares

In Li et al. (2005), a least median of squares (LMS) technique was proposed for NLOS mitigation; this is one of the most commonly used robust fitting algorithms. It can tolerate a proportion of outliers as high as 50% in the absence of noise. The location estimate of the MS using the LMS solution is given by (Li et al. 2005)

$$\hat{X} = \arg\min_{X} \{\mathrm{med}_i(\hat{d}_i - \|X - X_i\|)^2\}, \tag{7.85}$$

where $\mathrm{med}_i(\theta(i))$ is the median of $\theta(i)$ over all possible values of i. Since calculation of Equation (7.85) is computationally intensive, Li et al. (2005) proposed a lower-complexity implementation that uses random subsets of X_i to obtain several candidate $\hat{X}$, and the one with the least median of residue is selected as the solution. It should be noted that Li et al. (2005)

[6] An outlying observation, or outlier, is one observation that appears to deviate markedly from other members of the sample in which it occurs (Grubbs 1969).

[7] Bootstrapping can be defined as obtaining the statistics of an estimator by measuring those statistics by sampling from an approximating distribution, such as the empirical distribution of the observed data (Chernick 1999).

uses the LMS algorithm in the context of security, where some of the measurements may be outliers in the presence of certain attacks against the localization system. A parallel approach that uses the LMS algorithm for NLOS mitigation purposes has been reported in Casas et al. (2006).

7.6.3 Other Robust Estimation Options

Besides the M-estimation and the LMS techniques discussed above, there are some other robust estimation techniques (Ripley 2004) which, to the best of our knowledge, have not been considered in detail for NLOS mitigation purposes. For example, the least-trimmed squares (LTS) technique aims at minimizing the sum of squares for the smallest n of the residuals. On the other hand, the S-estimator, which attempts to find a line minimizing a robust estimate of the scale of the residuals, is quite robust to outliers, but it is not as efficient as the M-estimator. Finally, the MM-estimator can be considered as an M-estimator which starts at the coefficients given by the S-estimator (Knight and Wang 2009). With more computational complexity compared with the other two techniques, the MM-estimator has both good robustness and good efficiency.

There are also some other parametric approaches for robust regression, where the normal distribution is replaced with a heavy-tailed distribution in order to model the outliers. For example, the t-distribution can be a good distribution for modeling the outliers in a Bayesian robust regression algorithm (Taylor 2002). Alternatively, a mixture of zero-mean Gaussian distributions (i.e., a contaminated normal distribution) with different variances can be considered to model the noise in the presence of NLOS bias:

$$\eta_i \sim (1-\epsilon)\text{Norm}(0, \sigma^2) + \epsilon\ \text{Norm}(0, c_{nlos}\sigma^2), \tag{7.86}$$

where the second Gaussian distribution is intended to capture the outliers, and $\epsilon < 1$ is a small number (typically smaller than 0.1) that characterizes the impact of the outliers together with the constant term c_{nlos}.

7.7 Identify and Discard Techniques for NLOS Localization

As discussed at the beginning of Section 7.3, one of the simplest techniques to mitigate NLOS effects is to identify the NLOS FRPs and discard them during localization (i.e., to find the MS location using only the LOS FRPs). In fact, the ML estimator of Qi et al. (2006) illustrated in Figure 7.5(a) and the ML estimator of Riba and Urruela (2004) discussed in Section 7.3.1 are also estimators of the IAD type. Moreover, the WLS technique discussed in Section 7.4.1 also becomes an IAD type of technique if the weights are set to 0 for NLOS FRPs and 1 for LOS FRPs. A common problem in all of these approaches is the accurate identification of the NLOS FRPs. In this section, we will review another IAD-based technique, which uses a residual test algorithm to identify the NLOS FRPs.

7.7.1 Residual Test Algorithm

The response time (RT) algorithm proposed in Chan et al. (2006b) falls into the group of algorithms where the MS is localized using only the LOS FRPs. Hence, the NLOS FRPs have to be correctly identified and discarded, which is achieved as follows. First, by using the approximate ML algorithm discussed in Chan et al. (2006a) and by employing different combinations of FRPs,[8] different location estimates $\hat{X}_m$ are computed, and we define

$$S_0 = \sum_{i=3}^{n} \binom{n}{i}. \tag{7.87}$$

For each m, the squares of the normalized residuals[9] are computed as

$$\chi_x^2(m) = \frac{[\hat{x}_m - \hat{x}_{S_0}]^2}{I_x(m)} \quad \text{and} \quad \chi_y^2(m) = \frac{[\hat{y}_m - \hat{y}_{S_0}]^2}{I_y(m)}, \tag{7.88}$$

where $I_x(m)$ and $I_y(m)$ are obtained from Equation (7.17) and are the CRLBs[10] for x and y dimensions and for the mth hypothesis. Without loss of generality, S_0 indexes the case when all the FRPs are used in localization.

If all the FRPs in the mth hypothesis are in LOS, we have $\chi_x(m) \sim \text{Norm}(0, 1)$ and $\chi_y(m) \sim \text{Norm}(0, 1)$. Hence, $\chi_x^2(m)$ and $\chi_y^2(m)$ have centralized chi-square distributions with one degree of freedom. On the other hand, if there is at least one NLOS FRP in the mth hypothesis, both random variables have noncentralized chi-square distributions with noncentrality parameters depending on the NLOS bias. This observation suggests that if the PDFs of $\chi_x^2(m)$ and $\chi_y^2(m)$ (which should ideally be the same) can be identified correctly, it allows one to determine whether all the FRPs are in LOS or not. This can be simply achieved by a threshold test. For example, for a case with seven FRPs, as in Chan et al. (2006b), an appropriate threshold to characterize the chi-square PDF for the LOS case has been determined to be 2.71. If the area under the PDF to the right of this threshold is larger than 0.1, the random variable is identified as having a noncentralized chi-square distribution. If the PDF is determined to be a noncentralized chi-square distribution, this implies that there is at least one NLOS FRP. Then, the algorithm forms $\binom{n}{n-1}=n$ sets of FRPs, with $(n-1)$ FRPs in each set. For each of these n sets,

$$\sum_{i=1}^{n-1} \binom{n-1}{i} \tag{7.89}$$

estimates of X_m are obtained. If any of these sets is found to be distributed according to a centralized chi-square distribution using the threshold test, then the number of LOS FRPs is $(n-1)$, and the FRPs within that particular set are used to estimate the MS location. Otherwise, the algorithm iterates until there are at least three LOS FRPs. Since three FRPs do not have a sufficient number of realizations for a reliable application of the RT algorithm, the delta test procedure has been proposed, which takes two FRPs first and then combines them with one of the rest of the FRPs to check if all three of those FRPs are in LOS. Simulation results show that the proposed technique outperforms the Rwgh (Chen 1999) and CLS (Wang et al. 2003) algorithms, and can achieve the CRLB if the number of LOS FRPs is larger than half of the total number of FRPs.

[8] The summation in Equation (7.87) starts from 3 since at least three FRPs are required for location estimation in 2D.
[9] Note that the definition of "residual" in the RT algorithm is different from that in the Rwgh algorithm.
[10] The computation of the CRLB requires the MS location X. Since that is not available, X_{S_0} is used as an approximation for X to compute an approximate CRLB.

Table 7.1 Overview of TOA-based localization algorithms for use in LOS and NLOS scenarios.

	Algorithm name	References	Summary	Complexity, a priori knowledge etc.
ML-type algorithms	ML algorithm utilizing NLOS statistics	Gezici and Sahinoglu (2004); Qi et al. (2006)	In Gezici and Sahinoglu (2004), the X that maximizes the joint probability density function of the observations in NLOS scenarios is selected. A MAP estimator utilizing the NLOS bias statistics is introduced in Qi et al. (2006).	The probability density function of the NLOS bias and the distance measurements are assumed known, and the method requires a search over possible MS locations.
	IAD-based ML algorithm	Qi et al. (2006); Riba and Urruela (2004)	Uses the ML principle to discard the NLOS FRPs. Then, only the LOS FRPs are used in location estimation.	Need to collect n_{trn} TOA measurements for each FRP to capture the noise statistics (Riba and Urruela 2004). There is always be a possibility of misidentification of the LOS FRPs.
LS algorithms	Weighted LS	Guvenc et al. (2007, 2008)	Uses some appropriate weights (e.g., using the variance of the distance measurements or the statistics of the multipath components) to assign less reliability to NLOS FRPs.	For a static MS, the variance information may not be very different for LOS and NLOS FRPs.
	Residual-weighting algorithm	Chen (1999)	Different possible combinations of FRPs are considered. Then, the corresponding location estimates are weighted with the inverses of the residual errors to obtain the final location estimate.	Need to solve for $n_{\text{cb}} = \sum_{i=3}^{n} \binom{n}{i}$ location estimates for different hypotheses before weighting them.
Constrained localization techniques	Constrained LS with QP	Wang et al. (2003)	A two-step ML technique is used to obtain an estimate of the MS, with a quadratic constraint as given in Equation (7.54).	May have high computational complexity.

	Constrained LS with LP	Venkatesh and Buehrer (2006, 2007)	The NLOS FRPs are used to obtain a feasible region composed of squares. Then, the LOS FRPs are used to solve for the MS location via the LLS-1 technique so that the solution is within the feasible region.	Linear constraints yield a less complex (but coarser) solution compared with the quadratic constraints.
	Geometry-constrained localization	Chen and Feng (2005b)	A constraint related to the intersection points of circles is incorporated into the two-step ML algorithm.	Slightly more complex than the two-step ML algorithm.
	Interior-point optimization	Kim et al. (2006)	First, the NLOS bias values are estimated with IPO. Then, the NLOS bias estimates are used in a WLS solution (linearized using a Taylor series approximation).	Bias estimation through IPO may be computationally complex.
Robust estimators	*M*-estimators	Sun and Guo (2004)	Employs a convex function $\rho(v)$ to capture the effects of NLOS bias values in an ML-type of estimator.	Need to tune $\rho(v)$ appropriately. Better alternatives such as *S*-estimators (more robust) and *MM*-estimators (both robust and efficient) are available.
	Least median of squares	Li et al. (2005)	The location that minimizes the LMS of the residual is selected as a location estimate.	Robust up to 50% of outliers. More computationally complex than the NLS method.
Identify and discard techniques	Residual test algorithm	Chan et al. (2006b)	Identifies and discards the NLOS FRPs. The residual errors are normalized by the CRLBs, and the resulting variables are checked to find if they have a centralized or noncentralized chi-square distribution.	Computationally complex owing to testing numerous hypotheses, Delta test etc.

7.8 Conclusions

In this chapter, an extensive survey of TOA-based localization and NLOS mitigation techniques was presented. While some techniques discussed in this chapter can perform close to the CRLB, they may require high computational complexity and the availability of various types of prior information. For example, prior information regarding the NLOS bias statistics may not be available in many scenarios, forcing the use of techniques such as the residual-weighting algorithm. In Table 7.1 we provide a brief summary of the various NLOS mitigation methods and of their complexity and requirements.

Several of the NLOS mitigation techniques reviewed may also be extended to TDOA-, AOA- and RSS-based systems after some simple modifications. For example, robust estimation techniques similar to those discussed in Section 7.6 may also be utilized for handling outliers in TDOA, AOA or RSS measurements. Using techniques similar to those discussed in Section 7.4, appropriate weights may be used to suppress TDOA, AOA or RSS measurements that are subject to NLOS bias. Extensions of the maximum-likelihood and constrained-based techniques as discussed in Sections 7.3 and 7.5 to other localization scenarios are also possible. All in all, the achievement of practical, efficient, low-complexity localization techniques in the presence of NLOS propagation still requires further research. The authors hope that this chapter will serve as a valuable resource for researchers for evaluating the merits and trade-offs of the various techniques available in the present state of the art with the aim of developing more efficient and practical NLOS mitigation methods.

8

Positioning Systems and Technologies

Andreas Waadt, Guido Bruck and Peter Jung
Lehrstuhl für KommunikationsTechnik, Universität Duisburg-Essen, Germany

8.1 Introduction

Applications requiring positioning in mobile networks have gained importance in recent years. Examples include not only location-based services (LBSs), but also enhanced emergency call services such as the North American E911 and the newly defined emergency call (eCall), which is planned to become a European standard as part of the eSafety initiative of the European Commission (European Commission 2005). Mobile network providers already offer LBSs along with positioning, although the accuracy and reliability of the present techniques do not yet meet the requirements of such services in all cases.

One of the most popular positioning systems in navigation is the Global Positioning System (GPS), which is satellite based, widely available and quite accurate, as long as the receiver to be localized and the satellites have a line-of-sight (LOS) connection. Satellite positioning systems are discussed in Section 8.2.

Since most common mobile phones do not support the GPS, network providers evaluate the position of the network's base stations (BSs) which have connections to the mobile stations (MSs) to be located. The accuracy of the present position estimation techniques can be increased by evaluating additional protocol data from the cellular network, for example in the case of the Global System for Mobile Communications (GSM). Some additional information which can be evaluated is measured by the MS and available only there. This leads to the approach of mobile-assisted position estimation techniques (Waadt et al. 2008). Section 8.3 focuses on positioning in cellular networks.

With less area-wide coverage but nevertheless with interesting positioning capabilities and applications, wireless local area networks (WLANs) are increasingly prospering. Examples include Wi-Fi and ultra-wideband (UWB) systems, the latter allowing precise real-time

Mobile Positioning and Tracking: From Conventional to Cooperative Techniques, Second Edition.
Simone Frattasi and Francescantonio Della Rosa.

localization and tracking (see Section 8.4). Positioning is also discussed in that section in the context of dedicated systems, where the solutions include infrared, radio frequency identification (RFID) and ultrasound systems.

Positioning in ad hoc networks has been considered an important asset for their performance improvement. This topic is briefly addressed in Section 8.5.

Finally, hybrid positioning solutions, which combine two or more different communication concepts, are being considered in order to facilitate improved ease of use in terms of low latency of position estimation. At present the combination of cellular and WLAN positioning and also the Assisted Global Navigation Satellite System (A-GNSS), for example the Assisted Global Positioning System (A-GPS), is of interest. These concepts will be concisely presented in Section 8.6. Section 8.7 concludes this chapter.

8.2 Satellite Positioning

8.2.1 Overview

Positioning methods are generally based on the determination of distances, directions or angles between known points and unknown points. In geodesy (from the Greek *geodaisia*, division of the Earth), which is the science of measuring the Earth and primarily concerns positioning, the known points are often trigonometric points, whose locations have been determined in the past. The first and most fundamental reference point of a geodetic coordinate system is the geodetic reference datum. Together with the reference ellipsoid, a mathematical approximation of the Earth's shape, the geodetic datum defines the coordinate system. A popular reference ellipsoid is the European Bessel Ellipsoid, which is often used in combination with the Potsdam datum as a reference datum; in contrast, the World Geodetic System 1984 (WGS84) uses a reference frame of 12 fundamental reference points, called the geodetic data, which are distributed over the Earth. The principle of using trigonometric points as references for positioning requires a broad and reliable network of known trigonometric points, which can be seen from every point whose location could potentially be determined. Our ancestors used known stars and constellations as a reference for navigation. One essential advantage of this archaic method is that it provides global coverage, since the stars can be seen from nearly every location on Earth, as long as there is a clear view of the night sky.

Nowadays, known stars and constellations have been replaced by the satellites of global navigation satellite systems (GNSSs). For positioning, it is sufficient to receive the radio signals of at least four satellites whose positions are known. Mentionable GNSSs include the American Transit and GPS, the Russian Globalnaya Navigationnaya Sputnikovaya Sistema (GLONASS), the European GALILEO and the Chinese Compass. The basic principles and the mathematical treatment are given in Section 8.2.2.

A historical review of GNSSs forms part of Section 8.2.3. Satellite-based augmentation systems (SBASs) also include stationary transmitters, which transmit correction signals to improve the accuracy of the positioning. Some examples of SBASs are the American wide area augmentation system (WAAS), the Japanese Multifunctional Satellite Augmentation System (MSAS), the Indian GPS-aided Geo-augmented Navigation System (GAGAN) and the European Geostationary Navigation Overlay Service (EGNOS). Finally, the future of GNSSs, in particular GPS III and GALILEO, will be touched upon.

Accuracy and reliability are considered in Section 8.2.4, leading to the identification of the drawbacks of satellite positioning when applied to mobile positioning (Section 8.2.5).

8.2.2 Basic Principles

GNSSs make use of satellites which broadcast their positions and clock times. To estimate its location, a receiver must receive the signals of at least four satellites at the same time. It then calculates the time of flight (TOF) and, from that, its distance from each satellite. With such distances and the position of the satellites, the location of the receiver can be calculated, including longitude, latitude and altitude.

The basic principle of the localization procedure, previously described in Section 4.4.2.1, is also illustrated by the following example. The measured TOF and the resulting calculated distance to a satellite define a sphere around the satellite itself (Figure 8.1). In the case of error-free measurements and calculations, the receiver must be located on the surface of this sphere. The measurement of the signal of a second satellite defines a second sphere. The two spheres intersect in a circle, limiting the possible positions where the receiver may be located. When a third sphere is evaluated by exploiting the signal from a third satellite, the three spheres will intersect in two points, one at the location of the receiver and one at an ambiguous point in distant space. This solution should actually be sufficient to determine the position of the receiver when it is known a priori that the receiver is located on the Earth. However, the TOF measurements require that the satellites and the receiver use accurately synchronized clocks. The satellites do in fact use accurately synchronized atomic clocks, but the clock of a receiver may not be sufficiently synchronized to allow accurate TOF measurements. This would result in enormous positioning errors if not compensated for. The unknown time difference between the receiver's clock and the satellites' clock is a fourth unknown variable to be determined, requiring a fourth equation and therefore a fourth satellite, which must be seen by the receiver. The localization of a receiver requires therefore at least four satellites, assuming that the positions of the satellites are accurately known.

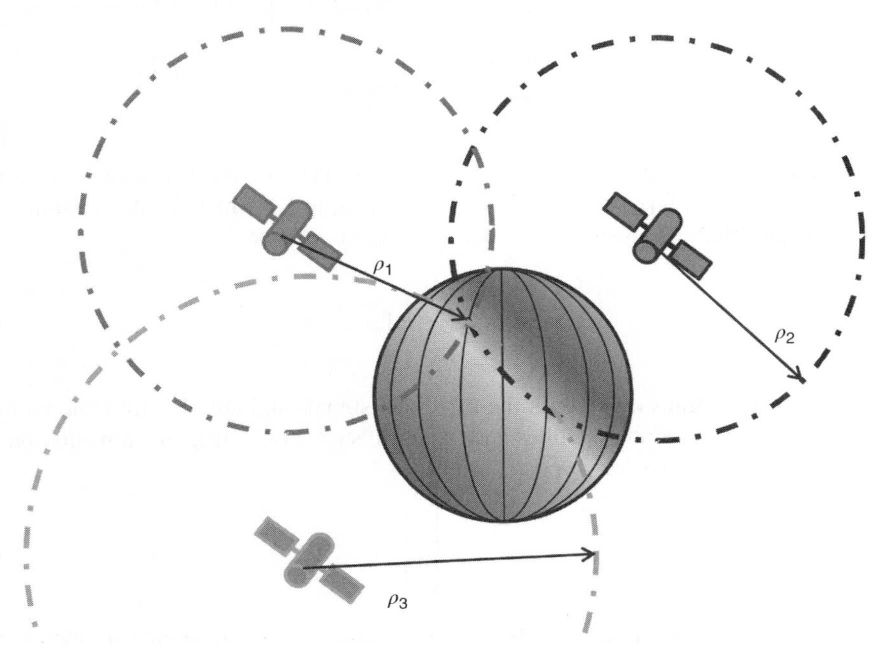

Figure 8.1 Distance measurements to satellites define orbits intersecting in a point.

In practice, GNSS receivers usually evaluate signals from more than four transmitters in order to minimize the drawback of error-prone distance measurements to the satellites. The reasons for measurement errors include changes in the travel time of the radio signal in the ionosphere (see Section 4.5.1.6), and noise and quantization errors in the receiver (Xu 2003). Owing to these inaccuracies, the spheres resulting from the measurements of the satellites' signals will generally not intersect in one or two locations, and the equation system will not result in a unique solution. The receiver's position must consequently be estimated, for instance with WLS estimators or Kalman filtering (see Chapter 5).

8.2.2.1 Mathematical Background

Let us assume that a receiver receives n signals from n satellites and extracts timestamps when a signal is received. The signal transmitted by the ith satellite contains the timestamp t_i^{tx}, $i \in \{1, \dots, n\}$, which is the point in time of transmission from the satellite. The clocks of the n satellites are assumed to be accurately synchronized, and all timestamps provided by the satellites are in terms of the same satellite reference time. In the satellite reference time, the signal reaches the receiver at a point in time t_i. However, since the receiver is not perfectly synchronized with the satellites, the timestamp

$$\tilde{t}_i = t_i + \tau, \quad i \in \{1, \dots, n\}, \tag{8.1}$$

extracted by the receiver is a pseudo-time or rough estimate of the true time of reception, t_i, plus a deterministic time measurement error, τ, which is the offset between the satellite reference time and the receiver's clock. Stochastic components resulting from propagation conditions are not considered at this point. The receiver can calculate the pseudo-distances

$$\hat{d}_i = (\tilde{t}_i - t_i^{(\mathrm{tx})})c = \underbrace{(t_i - t_i^{(\mathrm{tx})})c}_{d_i} + \underbrace{\tau c}_{b}, \quad i \in \{1, \dots, n\}, \tag{8.2}$$

where c is the speed of light and $\tilde{t}_i$ are the timestamps. The pseudo-distances $\hat{d}_i$ contain the true distances d_i, equal to $(t_i - t_i^{(\mathrm{tx})})c$, and a bias b, equal to τc, which results from the clock offset in Equation (8.2) and does not depend on i. The coordinates

$$X_i = \begin{bmatrix} x_i \\ y_i \\ z_i \end{bmatrix}, \quad i \in \{1, \dots, n\}, \tag{8.3}$$

of the n satellites are transmitted with the satellites' signals and are therefore known by the receiver. The movement of the satellites and the Earth's rotation during the transmission have not yet been considered. The receiver's coordinates

$$X = \begin{bmatrix} x \\ y \\ z \end{bmatrix} \tag{8.4}$$

then need to be determined. Equations (8.2) and (8.4) contain four unknown variables, namely the receiver's coordinates x, y, z and the bias b. Hence, the calculation of the receiver's position requires measurements of at least four satellites' signals, leading to the equation

$$(x_i - x)^2 + (y_i - y)^2 + (z_i - z)^2 = d_i^2 = (\hat{d}_i - b)^2, \quad i \in \{1, \dots, n\}, \tag{8.5}$$

with $n \geq 4$. If $n = 4$, Equation (8.5) yields a four-dimensional system of quadratic equations with four unknown variables and a unique solution. If the receiver sees more than four satellites, the method of least squares (see Section 5.2) can be used to obtain the optimal solution.

In the following, we illustrate a possible way to solve the above equation. By subtracting the ith equation from the first equation, the quadratic components are eliminated and we obtain the linear equation

$$\kappa_i - \kappa_1 - 2x_i^* x - 2y_i^* y - 2z_i^* z = \hat{d}_i^2 - \hat{d}_1^2 - 2\hat{d}_i^* b, \tag{8.6}$$

where

$$\kappa_i = x_i^2 + y_i^2 + z_i^2 \tag{8.7}$$

and

$$w_i^* = w_i - w_1, \quad w \in \{x, y, z, \hat{d}\}. \tag{8.8}$$

Separating the known terms from those with unknown variables in Equation (8.6), that is, the receiver's coordinates x, y, z and the bias b, we obtain

$$x_i^* x + y_i^* y + z_i^* z - \hat{d}_i^* b = \frac{1}{2}[\kappa_i - \kappa_1 - (\hat{d}_i^2 - \hat{d}_1^2)], \quad i \in \{1, \dots, n\}. \tag{8.9}$$

Using the system matrix

$$H = \begin{bmatrix} x_1^* & y_1^* & z_1^* & -\hat{d}_1^* \\ x_2^* & y_2^* & z_2^* & -\hat{d}_2^* \\ \vdots & \vdots & \vdots & \vdots \\ x_n^* & y_n^* & z_n^* & -\hat{d}_n^* \end{bmatrix}, \tag{8.10}$$

and the vectors

$$\mathsf{X} = \begin{bmatrix} X \\ b \end{bmatrix} = \begin{bmatrix} x \\ y \\ z \\ b \end{bmatrix} \tag{8.11}$$

and

$$C = \frac{1}{2} \begin{bmatrix} \kappa_2 - \kappa_1 - (\hat{d}_2^2 - \hat{d}_1^2) \\ \kappa_3 - \kappa_1 - (\hat{d}_3^2 - \hat{d}_1^2) \\ \vdots \\ \kappa_n - \kappa_1 - (\hat{d}_n^2 - \hat{d}_1^2) \end{bmatrix}, \tag{8.12}$$

the $n - 1$ equations in Equation (8.9) can be written in matrix vector notation:

$$H\mathsf{X} = C. \tag{8.13}$$

When $n > 4$, the receiver sees more than the minimum of four satellites required. Then, the matrix H is not square anymore and cannot be directly inverted. Rather, a pseudo-inverse $H^\dagger$ needs to be determined, for example the Moore–Penrose pseudo-inverse (Penrose 1955),

$$H^\dagger = (H^{\mathrm{T}} H)^{-1} H^{\mathrm{T}}. \tag{8.14}$$

In Equation (8.14), H^{T} denotes the transposed version of the matrix H, and $(H^{\mathrm{T}}H)^{-1}$ is the inverted matrix of $(H^{\mathrm{T}}H)$. Applying $H^{\dagger}$ to Equation (8.13), we obtain

$$\mathrm{X} = (H^{\mathrm{T}}H)^{-1}H^{\mathrm{T}}C. \tag{8.15}$$

An elegant solution of Equation (8.15) can be implemented by starting out from

$$(H^{\mathrm{T}}H)\mathrm{X} = H^{\mathrm{T}}C \tag{8.16}$$

and using the two-iteration-step approach illustrated in Jung (1997), which is based on Cholesky and Schur decomposition.

The offset of the receiver's clock with respect to the satellites' reference clock is actually not the only bias. Other systematic sources of error include the effects of the atmosphere, the Earth's rotation and Einstein's relativity. The large motion velocities and the gravitational potential differences between the satellites and receiver, for instance, lead to nonnegligible relativistic effects. These include a frequency offset between the signals transmitted by the satellites and the signals observed by the receiver, and path range effects caused by time dilation. For details of the sources of errors, the reader is referred to Kaplan and Hegarty (2006) and Xu (2003).

8.2.3 Satellite Positioning Systems

8.2.3.1 Introductory Remarks

The first GNSS and the predecessor of the well-known GPS was originally called the Navy Navigation Satellite System (NNSS), later called Transit. Starting in 1958, Transit was developed by the US Navy to provide accurate location information to ballistic missiles and submarines. From 1967 it was also used for civil applications. Its accuracy was between 500 and 15 m. Transit was shut down in 1996.

GLONASS is a satellite navigation system developed by the Russian Space Forces. It was deployed during the Cold War as a competitor to the GPS, starting in 1976.

The Chinese Beidou navigation system is a project for the development of an independent satellite navigation system. It is named after the Big Dipper constellation, which is called *Beidou* in Chinese. It is an experimental system of only four satellites, but the final GNSS with global coverage will include 35 satellites and be called Compass.

8.2.3.2 The Global Positioning System

The GPS, officially the navigational satellite timing and ranging (NAVSTAR) GPS, is the successor of Transit and probably the best known and most popular satellite positioning system for localization and navigation purposes. Developed from the 1970s by the US Department of Defense, it replaced Transit in 1985.

A constellation of at least 24 satellites transmits signals with a data rate of 50 bit/s and a frame period of 30 s on two frequencies, the L1 signal at 1.57542 GHz and the L2 signal at 1.2276 GHz (additional signals on other frequencies have been proposed and are intended to be deployed). Code division multiple access (CDMA) is used to separate the signals of different satellites. The good autocorrelation properties of the CDMA codes used allow accurate synchronization and therefore accurate determination of the time of arrival (TOA), that is, the point in time of the radio reception, which is necessary to estimate the TOF. GPS uses two codes, a public coarse/acquisition code for civil applications and an encrypted precise code for military use (Parkinson and Spilker 1996).

8.2.3.3 Augmentation Systems

For further improvement of the accuracy of GNSSs, there are several augmentation systems in use. Most of them are designed to cooperate with the GPS. Examples include the Differential Global Positioning System (DGPS), the US WAAS, the European EGNOS, the Japanese MSAS and the Indian GAGAN system. Augmentation systems provide additional information to the receiver. This may include information about the speed of a vehicle, which can be used advantageously if the positioning system is integrated into a car or an aircraft, for instance. In the DGPS, a network of fixed BS receivers is used as a reference for the localization of MS receivers. After a comparison of TOF measurements with the expected results, a BS receiver can estimate sources of errors, such as ionospheric delays. Since those error sources are nearly constant within small areas, they can be compensated for by estimations made at the mobile receiver side.

8.2.3.4 GPS III and GALILEO

The improvement of GNSSs is an ongoing process. In the GPS, for instance, the satellites of the first generation have been shut down and a third generation, GPS III, is to be deployed (Luba et al. 2005). GPS III will allow improved accuracy and reliability, making it more robust against jamming and spoofing. Although similar in design, the European satellite navigation system GALILEO is, in contrast to the GPS, designed only for civil applications and is without a separately encrypted CDMA channel for military use. Furthermore, it is not under military control. As of today, GALILEO (Antonini et al. 2004) is still under construction and will be compatible with GPS III, leading to better coverage and accuracy if used in combination.

8.2.4 Accuracy and Reliability

The accuracy of satellite positioning is usually in the range of few tens of meters; this is the case, for example, for GPS and GLONASS. In reality, the accuracy depends on the weather conditions and the sensitivity of the receiver. When augmentation systems are used, such as the DGPS, for example, the accuracy can be improved to the range of meters or centimeters, depending on the distance from the BS receivers. A disadvantage of satellite positioning is the LOS constraint between the receiver and the satellites, which results in satellite positioning not being fully reliable. For details on the accuracy and reliability of GNSSs, the reader is referred to David (2009).

8.2.5 Drawbacks When Applied to Mobile Positioning

Satellite positioning systems are dedicated positioning systems designed for accurate, global navigation, and the accuracy and global availability of satellite positioning are their major strengths compared with other positioning techniques. Furthermore, satellite positioning supports fast tracking of mobile users. However, two requirements of satellite positioning systems constrain their usability when applied to mobile positioning: (1) the need for an LOS connection between the satellites and the receiver, and (2) knowledge of the orbital elements, locations and ephemerides of the satellites. The first requirement leads to a limited capability of satellite positioning systems in indoor environments and bad weather conditions. The second requirement implies that the orbital elements must be downloaded before the first position fix is acquired. Owing to the small data rate in satellite positioning systems, it can take minutes to localize the receiver the first time. As a consequence, the receiver must stay on line in order to keep the orbital elements up to date. This fact, together with the energy management approach that satellite positioning systems implement while in normal operational mode, increases the power consumption, which is a critical parameter in mobile applications. This makes satellite positioning systems only conditionally applicable to mobile positioning.

8.3 Cellular Positioning

8.3.1 Overview

The localization of mobile terminals in cellular networks has been discussed over the past ten years, usually focusing on GSM-based solutions (Drane et al. 1998; Kyammaka and Jobmann 2005). It is particularly noteworthy that the techniques discussed in Kyammaka and Jobmann (2005) influenced the standardization of universal mobile telecommunications systems (UMTSs). For example, one specification (3GPP 2008a) specifies the positioning methods to be supported in UMTS.

Nowadays, many manufacturers offer numerous models of GPS-enabled cell phones. However, the increased cost of GPS-enabled mobiles, as well as their drawbacks when applied to mobile positioning, is reducing the speed at which they are spreading on the market. Owing to the fact that most of the currently circulating cell phones are still without GPS, the following will focus on nonsatellite-based, or GPS-free, positioning, which is achieved by exploiting information from the cellular network. Generally, we distinguish between mobile-assisted GPS-free positioning and mobile- or network-based GPS-free positioning. In mobile-assisted GPS-free positioning, the mobile calculates its position using signals received from base transceiver stations (BTSs), whereas in network-based GPS-free positioning the position of a mobile is determined on a server in the network. Network-based GPS-free positioning is quite common, whereas mobile-assisted GPS-free positioning has not yet been widely introduced.

One simple network-based GPS-free positioning method is based on cell coverage, by means of evaluating the cell identification (ID). This scheme is commonly deployed by mobile network operators. The position of a mobile connected to a particular BTS, which is identified by its cell ID, is determined by the location of the BS itself. A more advanced network-based GPS-free positioning method using the cell coverage determines the location of a mobile connected to a multiantenna BTS by evaluating the center of gravity of the sector that the mobile belongs to, thus using that location as the estimated position of the mobile.

Recently, network-based GPS-free positioning techniques have also been using recursive Bayesian filtering (RBF) based on models of the radio environment, that is, data from network planning and radio propagation prediction in the vicinity of the mobile, and predictions of the user's mobility. Although these techniques facilitate position estimation accuracies of the order of those achieved with GNSSs, the creation and maintenance of the required radio propagation databases and the estimation of mobility are extremely cumbersome. Since positioning services might have only a minor market importance compared with multimedia services, network operators are not much interested in bearing the burden of creating these databases. More elaborate positioning methods evaluate the TOA and time difference of arrival (TDOA) of radio signals. These methods require reasonably accurate synchronization among the BSs, which is usually cumbersome.

Several cellular positioning techniques have already been commercially deployed for emergency call location and for mobile tracking services from mobile network operators, which allow a mobile to be localized using a Web interface, for instance. The Google Maps Mobile Application also provides GPS-free cellular positioning for navigation purposes (Google 2006).

8.3.2 GSM

8.3.2.1 Cell ID

One simple network-based GPS-free positioning method is cell coverage positioning by means of evaluating the cell ID. This scheme is commonly deployed by mobile network operators. The position of a mobile that is connected to a particular BTS, identified by its cell ID, is determined by the location of the BS itself. This scheme is referred to as "BTS connected". This is a very rough, quite inaccurate localization technique, since it does not take the cell's geometry into account. Owing to their directional characteristics, BTS antennas are most often located at the borders of the cell sectors, and this leads generally to large localization errors when the BTS connected method is used. If, besides the cell ID, no additional information is known, then the estimation error can be minimized by estimating the MS's location as the center of gravity of the cell. The determination of such a center of gravity requires a knowledge of the cell's geometry.

A more advanced network-based GPS-free positioning method using the cell coverage determines the location of a mobile connected to a BTS sector by first evaluating the center of gravity of the sector and using the location of that as the estimated position of the mobile. A simple but reasonably accurate way of determining a virtual center of gravity is the following: (1) determine the distances between the connected BTS and the BTSs in the neighborhood, (2) select the smallest distance value d, (3) divide the cell into sectors according to the directivity of the BTS antennas, (4) the virtual center of gravity is in the middle of the sector of the connected antenna, at a distance $\rho_{\mathrm{vc}} = d/3$ from the connected BTS (Figure 8.2). This advanced cell coverage positioning scheme is referred to as the "virtual center" scheme. BTS connected and virtual center methods can also be implemented in a mobile-assisted GPS-free positioning fashion.

The virtual center method assumes a symmetric network geometry, which is usually not the case. An accurate determination of the cell's geometry can be achieved by measurements of the radio signal power from the reference BTS and the neighboring BTSs. This empirical method, though, is cumbersome. The cell geometry can be determined roughly in a simplified way without the need for radio measurements. Let us assume free-space propagation, and an equivalent radiated power (ERP) that is assumed to be the same for all BTSs. In this case, the cell geometry becomes the Voronoi diagram of the BTSs (Baert and Seme 2004).

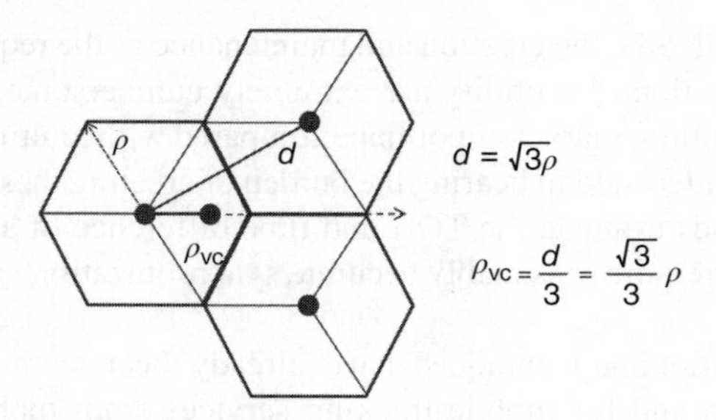

Figure 8.2 Virtual center in a symmetric cellular network.

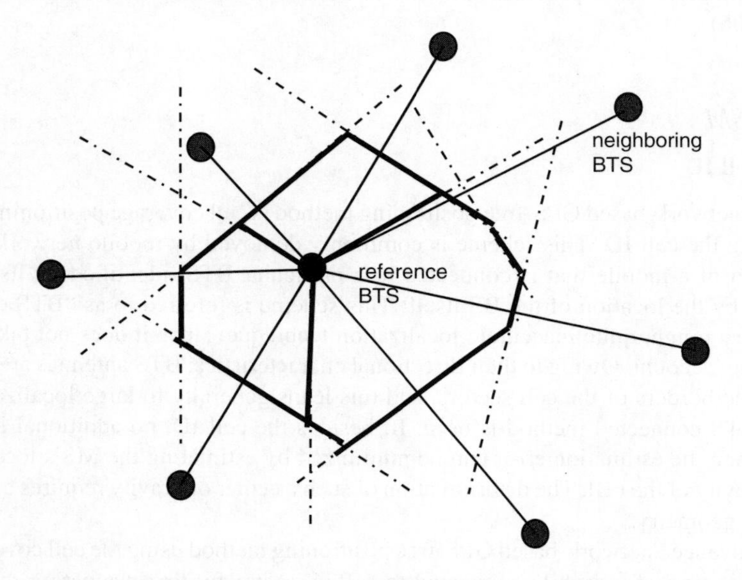

Figure 8.3 Construction of the cell geometry.

Figure 8.3 illustrates the construction of the cell geometry for an example where there is a reference BTS surrounded by seven neighboring BTSs. The boundaries of the cells are constructed from the perpendicular bisectors of the line segments between the reference BTS and the neighboring BTSs. If the reference BTS is composed of several antennas with different directional characteristics, then the cell must be divided into appropriate cell sectors. Figure 8.3 shows an example with three cell sectors from three BTSs, one pointing to the north (0°), one pointing to the southeast (120°) and one pointing to the southwest (240°). Assume that the corners of the polygonal cell in Figure 8.3 are numbered from 1 to n and their corresponding positions are given by

$$X_i = \begin{bmatrix} x_i \\ y_i \end{bmatrix}, \quad i \in \{1, \ldots, (n+1)\}, \quad X_1 = X_{n+1}, \tag{8.17}$$

where x_i and y_i are the coordinates of the ith cell corner in two dimensions and X_{n+1} is equal to X_1. From geometrical theory for the calculation of the area and the centroid of a polygon, it is possible to obtain the area

$$A = \frac{1}{2} \sum_{i=1}^{n} (x_i y_{i+1} - x_{i+1} y_i) \tag{8.18}$$

and the coordinates of the centroid

$$X = \frac{1}{6A} \begin{bmatrix} \sum_{i=1}^{n} (x_i + x_{i+1})(x_i y_{i+1} - x_{i+1} y_i) \\ \sum_{i=1}^{n} (y_i + y_{i+1})(x_i y_{i+1} - x_{i+1} y_i) \end{bmatrix}. \tag{8.19}$$

Figure 8.4 shows an example of a computer-constructed cell geometry of a cellular GSM network in Duisburg, Germany. The black marker in the middle of the figure represents the location of a reference BTS. The other six markers near the centers of the three cell sectors represent the centers of gravity of these sectors and the true locations of the test mobiles to be located.

The estimation error d_{err} of the above positioning technique has been determined by the authors from real-life measurements; d_{err} was the distance between the position estimated by the cell ID positioning method and the position obtained from a GPS measurement, which served as a reference and was assumed to be error free. Figure 8.5 shows the cumulative distribution function (CDF) of d_{err}, which we refer to as the estimation accuracy. According to Figure 8.5, 67% of the estimated position coordinates are less than 447 m away from the real position and 95% are less than 1000 m away. The median corresponds to an estimation accuracy of 356 m.

Since the cell ID is also known on the mobile network side, this method can be used in network-based positioning methods. The MSs know the cell IDs from evaluating the broadcast

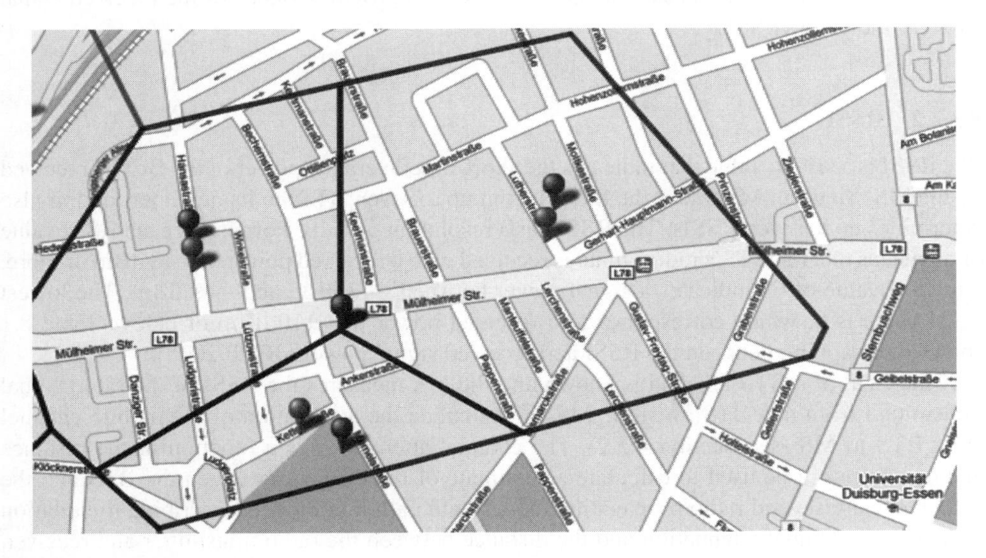

Figure 8.4 Cell geometry of a GSM network in Duisburg, Germany (© 2009 Google – Map data © 2009 Tele Atlas).

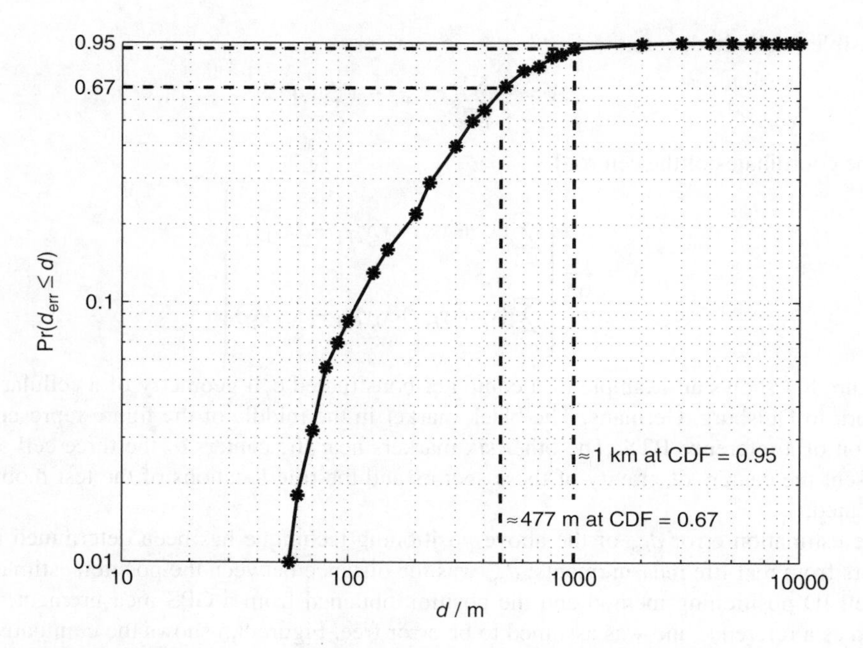

Figure 8.5 Localization accuracy measured in the area of Duisburg, using the cell ID positioning method.

control channels (BCCHs) of the BTSs. Every MS tracks the BCCHs of up to seven BTSs in its neighborhood. This is usually done to allow appropriate preparations for handover. Besides the cell IDs, the tracking and measurement of the BCCHs provides values of the received signal strength indicator (RSSI).

8.3.2.2 RSSI

The RSSI is a six-bit value that indicates the short-time-averaged power of the BCCH received by the MS. Since an MS tracks the BCCH from up to seven BTSs in its neighborhood, it also determines up to seven RSSIs. The RSSI has a resolution of 1 dB. For instance, an RSSI value of 63 means that the BTS's radio signal is received with a received power of −48 dBm or more. An RSSI value of 62 indicates a signal power between −48 dBm and −49 dBm. The lowest RSSI value is 0, which corresponds to a received power of −110 dBm or below. Table 8.1 shows the mapping between the RSSI and received signal power (3GPP 2005).

If the ERP of a BTS antenna is known and the MS measures the RSSI of the radio signal transmitted from that BTS, then the MS can calculate the attenuation of the mobile channel from BTS to MS (see Section 4.3.2). The channel attenuation is a function of the distance and can therefore be used to calculate an estimate of the distance between the BTS and the MS. There are several radio propagation models and path loss models describing the relation between the channel attenuation and the distance between the radio transmitter and receiver. Common models which can be applied to cellular networks are the Okumura Hata model, the

Table 8.1 Mapping between the RSSI parameter in GSM and the actual value of received signal strength in dBm.

RSSI	Received signal power, p/dBm
0	$p < -110$
1	$-110 \leq p < -109$
⋮	⋮
62	$-49 \leq p < -48$
63	$-48 \leq p$

COST Hata model and the COST Walfish Ikegami model (see Section 3.2.1 or, for more detail, CCIR (1989)).

Owing to the channel impairments discussed in Section 3.2.1 and the consequent statistical properties of the noise, RSS measurements are unreliable for performing accurate positioning in GSM. Generally, the mobile-channel attenuation is assumed to be log-normal distributed (CCIR 1989). With ERP in dB and RSS in dBm, the mean logarithmic attenuation a in dB can be estimated from

$$a(d) = \text{ERP} - \text{RSSI}(d) - 110.5\ \text{dB}, \tag{8.20}$$

d being the distance between the MS and the BTS.

Figure 8.6 shows the estimated attenuation, calculated after RSSI measurements, as a function of the distance d in the area of Duisburg. The solid line shows the logarithmic approximation to the relation between distance and attenuation which minimizes the absolute error of the estimated distance d. The mean error of this approximation is about 238 m. This estimate can be used to improve the localization when several RSSIs from different BTSs are known.

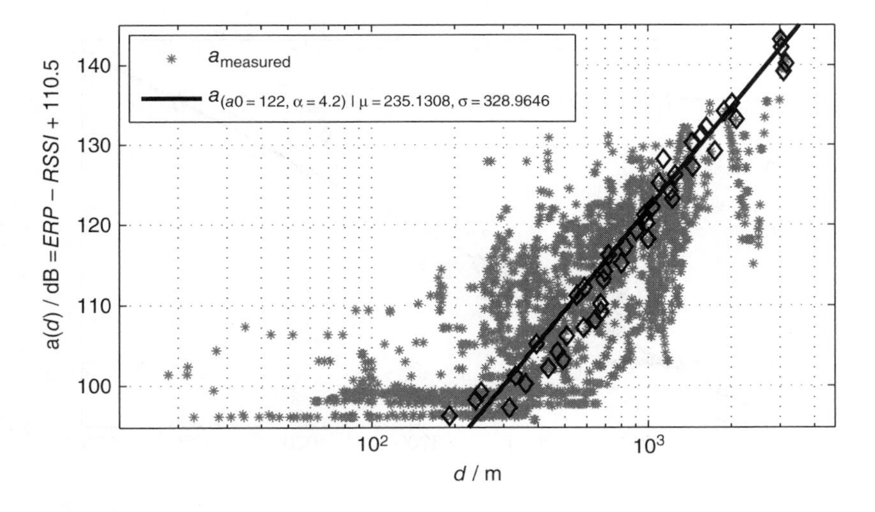

Figure 8.6 Channel attenuation obtained from RSSI measurements in the area of Duisburg.

8.3.2.3 Mobile-Assisted TOA

Since GSM uses time division multiple access (TDMA), the radio signals from the MSs must reach the BTS in certain time slots. To allow accurate synchronization, the MSs must know the signal propagation delay of the mobile channel from MS to BTS. The timing advance (TA) is a six-bit value that indicates the signal propagation delay from the MS to the BTS and back – the so-called round-trip time. The latter is quantized in bit periods, that is, the signal propagation time from BTS to MS and back is TA bit periods. In GSM, the bit period is (3GPP 2009)

$$t_{\mathrm{b}} = \frac{48}{13}\ \mu\mathrm{s} \approx 3.69\ \mu\mathrm{s}. \tag{8.21}$$

If we assume free-space propagation or LOS between the MS and the BTS, the distance d_{TA} between the MS and the BTS can be estimated as

$$d_{\mathrm{TA}} = \frac{t_{\mathrm{b}}c}{2} \cdot \mathrm{TA} \approx 554\ \mathrm{m} \cdot \mathrm{TA}, \tag{8.22}$$

where c is the speed of light. However, in most cases non-line-of-sight (NLOS) channels must be considered (see Chapter 7). This is usually the case in urban areas, where the radio signals often reach the receivers after reflections and scattering. This situation leads to an increased signal propagation delay and, consequently, to an overestimated distance d.

Along with cell IDs, the authors also measured the TA with a mobile phone in the area of Duisburg. The coordinates of the measurement points were determined by using a GPS-enabled device. The distances between the measurement points and the BTSs were calculated and served as reference distances to calculate the TA-based distance estimation error. Figure 8.7 illustrates the relation between the measured TA and the distance d. The solid line shows the linear approximation which minimizes the absolute estimation error. In field measurements, the average of the absolute distance estimation error was about 217 m when the linear approximation in Figure 8.7 was used.

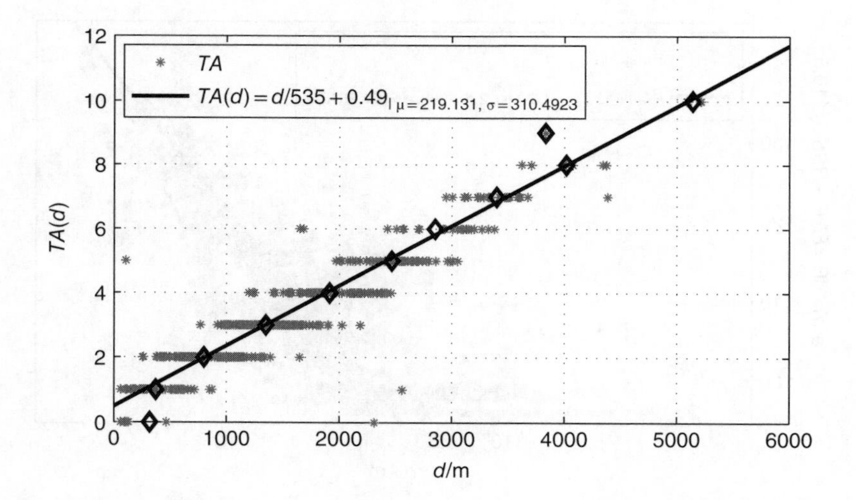

Figure 8.7 Measured timing advance and measured distance in the city of Duisburg.

The TA can generally be used to increase the localization accuracy. However, since it can be measured only when a dedicated channel is allocated, for example when a call has been initiated, the TA method is only conditionally applicable to positioning applications. On the other hand, the cell IDs and the RSSIs can always be queried from the mobile, and allow localization with increased accuracy compared with network-based methods without mobile assistance. The accuracy can be further improved by averaging consecutive measurements. Figure 8.8 shows the cumulative distribution function of the localization error d_{err} for mobile-assisted localization. Similarly to Figure 8.5, Figure 8.8 shows the CDF of d_{err}. According to Figure 8.8, 67% of the estimated position coordinates are less than 150 m away from the real position and 95% are less than 300 m away. The median corresponds to an estimation accuracy of 117 m.

8.3.2.4 Accuracy and Reliability

Selected results obtained by the authors are shown in Figures 8.9, 8.10 and 8.11. Figure 8.9 refers to measurements in Duisburg, and Figures 8.10 and 8.11 present a comparison of the positioning accuracy between network-based GPS-free positioning and mobile-assisted GPS-free positioning.

Figure 8.9 shows the map of the measurement area in Duisburg. The solid line shows the route taken and the positions estimated using GPS positioning. The 23 markers refer to virtual center positioning. The speed of the mobile was around 45 km/h and no temporal averaging of the measured GSM data was used. The quality of the measurements was improved in a second step by taking account of low mobility, for example at typical pedestrian speeds, and

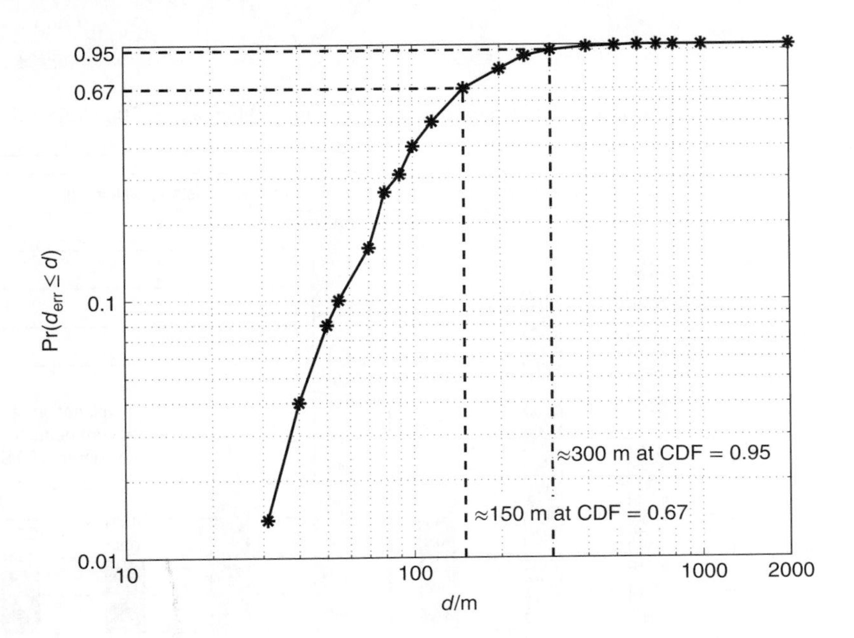

Figure 8.8 Localization accuracy measured in the area of Duisburg, using mobile-assisted TOA positioning.

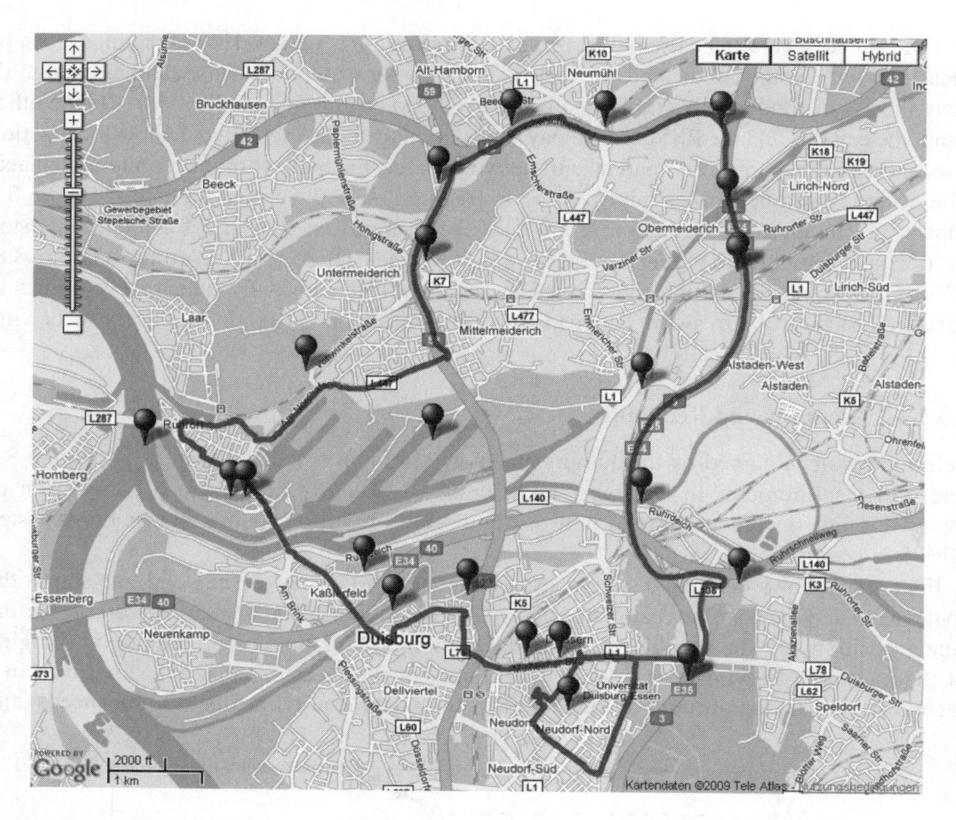

Figure 8.9 Measurement route in Duisburg (© 2009 Google – Map data © 2009 Tele Atlas).

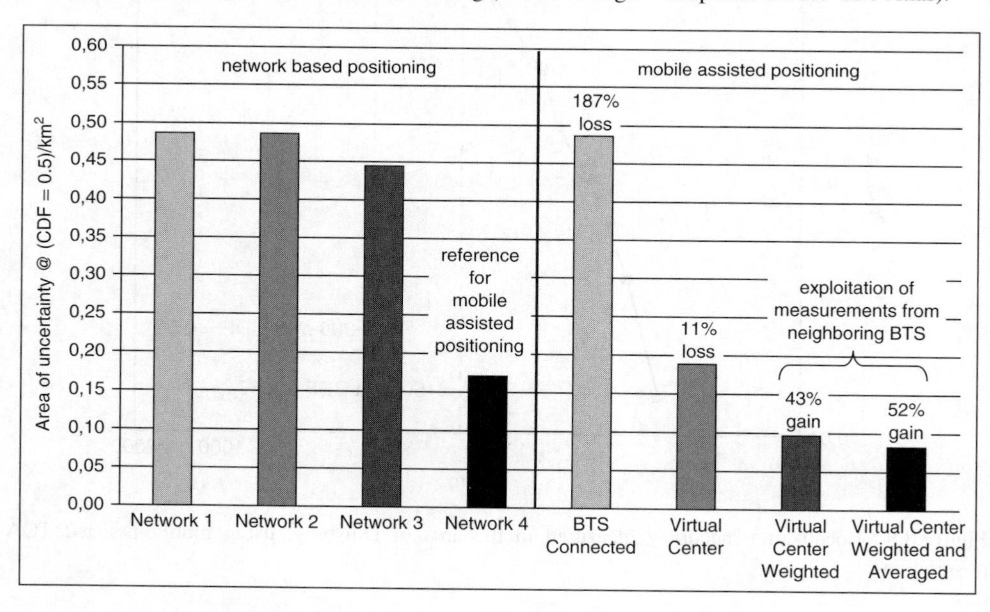

Figure 8.10 Median area of uncertainty (urban area).

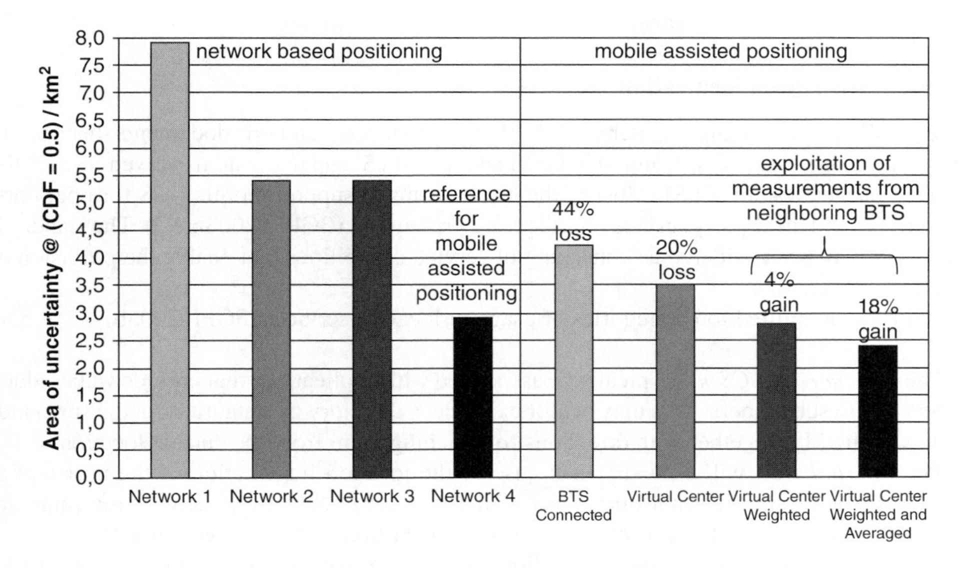

Figure 8.11 Median area of uncertainty (rural area).

by taking the measurements of the candidate BTSs into account, for example by using linear estimation techniques. Let X_i be the position of the virtual center associated with the ith BTS and let $\text{RSSI}_i^{\text{BTS}}$ be the received power value associated with the ith BTS and measured at the mobile; then the position of the mobile can be determined from

$$X = \frac{1}{\sum_{\forall i}\text{RSSI}_i^{\text{BTS}}} \cdot \sum_{\forall i} \text{RSSI}_i^{\text{BTS}} \cdot X_i. \tag{8.23}$$

This scheme is referred to as the "virtual center weighted" scheme. A further improvement can often be obtained by temporal averaging of consecutive positioning results. The latter scheme is termed the "virtual center weighted and averaged" scheme (Waadt et al. 2008).

Figures 8.10 and 8.11 show results for the positioning of mobiles in GSM networks in urban and rural areas. Only cell ID and RSSI measurements were used. The measurements were taken in Duisburg and in rural areas between Duisburg and the Dutch border. The results are depicted as the area of uncertainty, that is, the area of a circle around the estimated location of a mobile with a radius representing half of the maximum positioning error. The positioning error is equal to the difference between the GPS coordinates of the mobile and the estimated coordinates of the mobile. Mobile-network-based positioning by four network operators was considered (Network 1 to Network 4). It was found that mobile-assisted positioning was often superior to mobile-network-based positioning and increased the localization accuracy by 52% in urban areas and 18% in rural areas.

Simple mobile-assisted GPS-free positioning is possible at a reasonably low implementation complexity and viable at the same time, when we consider the results in terms of positioning errors. Such mobile-assisted GPS-free positioning might be considered as an interesting alternative to mobile-network-based GPS-free positioning.

8.3.3 UMTS

8.3.3.1 3GPP Standardization

Since LBSs are increasingly in demand, 3GPP has produced standards documents that specify the location services (LCS) featured in GSM and in UMTS, and the system evolved from it, the evolved packet system (EPS). Most of the mechanisms to support mobile LCS for operators, subscribers and third-party service providers are presented in 3GPP (2008b,c,d). The standardized LCS may be considered as consisting of service capabilities that enable the provision of location applications.

3GPP has identified four categories of usage of location services (3GPP 2008b):

- The *Commercial LCS* will typically be associated with applications that provide value-added services to subscribers. This may be, for example, a directory of restaurants in the surroundings of the MS, together with directions for reaching them from the current location.
- The *Internal LCS* will typically make use of the location information of the user equipment for accessing internal operations of the network. These may include, for example, location-assisted handover, traffic and coverage measurements (see Section 2.4).
- The *Emergency LCS* will typically allow subscribers who place emergency calls to be located. This service may become mandatory in some jurisdictions. In the USA, for example, this service is already mandatory for all mobile voice subscribers.
- The *Lawful Intercept LCS* will use location information to support legally required or sanctioned services, for example localization for law enforcement agencies, as required by the local jurisdiction.

As discussed in the earlier sections of this chapter, the capability to determine the geographic location of an MS is provided by making use of the radio signals. The location information may be requested by and reported to a client application associated with the MS, or by a client within or attached to the core network. Furthermore, the location information request may ask for the speed of the MS as part of the positioning information. The latter is reported in accordance with the specifications in 3GPP (2008d).

The LCS feature will be based on the following positioning methods (3GPP 2008a): (1) cell ID, (2) observed time difference of arrival-idle period downlink (OTDOA-IPDL), (3) uplink–time difference of arrival (U-TDOA), and (4) network-assisted GNSS methods, also termed A-GNSS-based positioning. Since the cell-coverage-based positioning method has already been presented in Section 8.3.2, only the OTDOA-IPDL, the U-TDOA and the A-GNSS-based positioning methods will be presented in the following subsections.

8.3.3.2 OTDOA-IPDL

The OTDOA-IPDL positioning method (3GPP 2008a) involves measurements made between the MS and the location measurement unit (LMU) of the UMTS terrestrial radio access network (UTRAN). The simplest case of OTDOA-IPDL positioning is the one without idle periods. In this case, the method can be referred to as simply OTDOA positioning. A node B (e.g., a BS) may provide idle periods in the downlink in order to potentially improve the hearability of neighboring nodes B.

The primary standard OTDOA measurement is the observed time difference between two system frame numbers (SFNs), which is observed by the MS. These measurements, together with other information concerning the surveyed geographic position of the transmitters and the relative time difference (RTD) of the actual transmissions of the downlink signals, may be used to calculate an estimate of the position and, optionally, the speed of the MS. Each OTDOA measurement for a pair of downlink transmissions describes a line of constant TOA difference, yielding a hyperboloid in three dimensions. The MS's position is determined by the intersection of the resulting lines for at least two pairs of nodes B. The accuracy of the position estimates made with this technique depends on the precision of the timing measurements and the relative position of the nodes B involved, and it is also subject to the effects of multipath radio propagation. The best results are obtained when the nodes B equally surround the MS. This is illustrated in Figure 8.12 (3GPP 2008a).

8.3.3.3 U-TDOA

The U-TDOA positioning method (3GPP 2008a) is based on network measurements of the TOA of a known signal sent from the MS and received by at least four LMUs. The signal propagation time from an MS to an LMU is proportional to the distance between the MS and the LMU. The difference between the TOAs at two LMUs defines a hyperbola. The MS's location can be estimated from the intersections of these hyperbolas. An advantage of this method is that it does not require knowledge of the time that the MS transmits, nor does it require any new functionality in the MS.

Figure 8.13 illustrates how the time delay is determined (3GPP 2008a). Two receivers, RX1 and RX2, represent the fixed locations of two U-TDOA-capable LMUs in a network. The pairs of intersecting circles centered around RX1 and RX2 have radii that represent the times t^{tr} it takes a signal to travel from an MS transmitter to each U-TDOA-capable LMU. The difference between the radii, however, is a constant equal to the difference between t_1^{tr} for the first U-TDOA-capable LMU and t_2^{tr} for the second LMU. This difference, $(t_1^{tr} - t_2^{tr})$, is the value

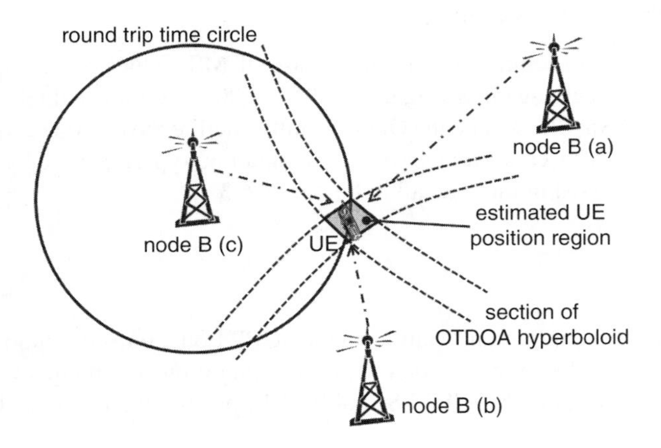

Figure 8.12 OTDOA positioning. UE, user equipment.

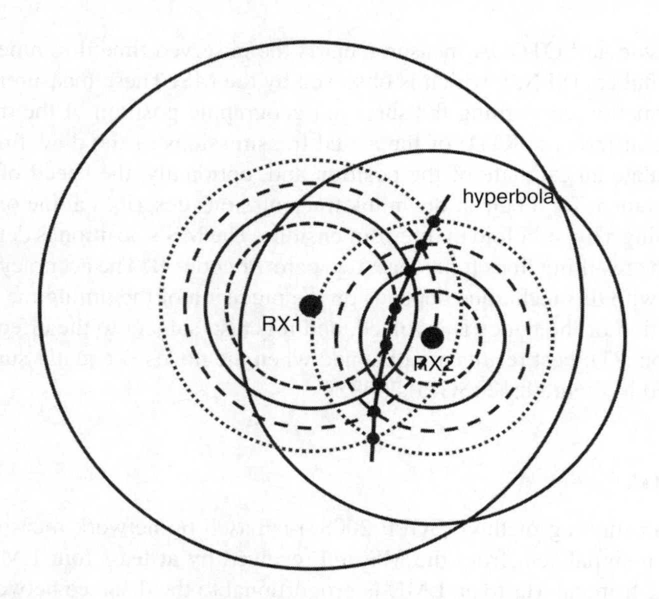

Figure 8.13 Deriving the hyperbolic function for the time delay.

which the U-TDOA system measures. When the points of intersection of the pairs of circles are plotted and connected, the resulting shape is a hyperbola. The latter, defined by a reference site in comparison with another site, constitutes a baseline. If one constructs a similar hyperbola for another pair of U-TDOA-capable LMUs receiving transmissions from the same MS, that hyperbola will intersect with the first hyperbola at two points, yielding two possible locations for the MS. A third hyperbola will yield a unique location for the MS, so that a minimum of four reception sites is needed to obtain a unique location estimate.

8.3.3.4 A-GNSS-based Positioning

The A-GNSS-based positioning method makes use of MSs which are equipped with radio receivers capable of receiving GNSS signals (3GPP 2008a). Examples of GNSS include GPS, GALILEO, GLONASS, SBASs and the Quasi-Zenith Satellite System (QZSS). In this context, different GNSSs can be used separately or in combination to perform the location of an MS. Assisted GPS is discussed in more detail in Section 8.6.3.

8.3.4 LTE[1]

Positioning is one of the important features in the 3GPP Long Term Evolution (LTE) system. The first LTE release 8 does not provide LTE positioning protocol and this was introduced in LTE release 9 (3GPP TS 36.355; 3GPP TS 36.455). The positioning feature in LTE is mainly to

[1] Basuki Priyanto, Sony Mobile Communications, Sweden

fulfil the localization requirement of an emergency call as required by FCC e911 (FCC Rules 2015). The requirements for user equipment (UE)-based and UE-assisted positioning are:

- 50 m accuracy – within 50 m for 67% of all calls measured at country level
- 150 m accuracy – within 150 m for 95% of all calls measured at county level.

The Federal Communications Commission (FCC) as the regulatory body mandates operators in the USA to fulfill these requirements, therefore positioning techniques have been added in 3GPP LTE specifications starting from release 9.

The positioning techniques in LTE can be classified into two groups:

- radio access technique (RAT) dependent
- RAT independent.

The positioning for the RAT-dependent group is the positioning technique that is based on LTE signals. In RAT-independent group, the positioning is utilizing other signals, such as the GPS from satellites. The positioning techniques in LTE are summarized in Figure 8.14. A-GNSS was introduced so that LTE system can provide assisted information to significantly improve the time-to-fix of the GNSS positioning method. E-CID is a positioning technique that uses cell ID information where the UE is clamped on. Other measurements, such as timing advance (TA), round trip time (RTT) and angle of arrival (AOA), can be used to enhance CID positioning.

The observed time difference of arrival (OTDOA) is essentially the same method as described in the previous section (UMTS positioning). However, the LTE OTDOA uses LTE specific positioning signals transmitted from the BS to determine the UE position. In LTE release 11, Uplink TDOA (UTDOA) is introduced. This is another form of OTDOA that utilizes the uplink signals from the UE. The UE position is determined based on the positioning measurement at multiple BSs.

Support of the BeiDou navigation satellite system (BDS) in LTE was introduced in release 12. BDS is a Chinese satellite navigation system. In LTE release 13, positioning enhancement that particularly addresses indoor positioning was studied and resulted in many new positioning techniques that are suitable for indoor scenarios.

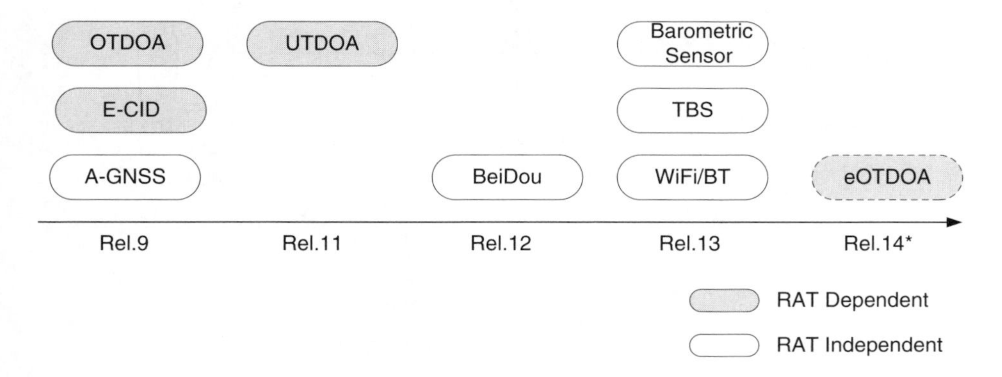

Figure 8.14 Positioning techniques in LTE.

8.3.4.1 OTDOA in LTE

OTDOA is one of the most common RAT-dependent positioning techniques in LTE and is illustrated in Figure 8.15. First, the location server (LS) checks the UE capability for performing OTDOA positioning. Then, the LS provides assistance data. The assistance data contains the reference cell and neighboring cell information. The BSs, also known as eNB in LTE, transmit positioning reference symbols (PRS) to the UE so that the UE can perform positioning measurements. The PRS signal generation and resource mapping in the LTE resource grid are described in 3GPP TS 36.211. The UE measurement is known as the reference signal time difference measurement (RSTD) (3GPP TS 36.214). The RSTD is a relative time difference between two cells (reference and neighbouring cell). The UE reports the RSTD measurements back to the LS. The LS determines the geographical position of the UE based on the received measurement results (e.g., using the triangulation method).

8.3.4.2 Enhancements of LTE Positioning Techniques

The FCC has identified that the positioning requirements for indoor scenarios cannot be met by most operators, thus new and tighter requirements have been released (FCC Rules 2015):

- horizontal accuracy – 50 m accuracy for 40%, 50%, 70% and 80% of emergency calls within 2, 3, 5 and 6 years, respectively
- vertical accuracy – uncompensated barometric data is provided and a z-axis accuracy metric should be developed within 3 years.

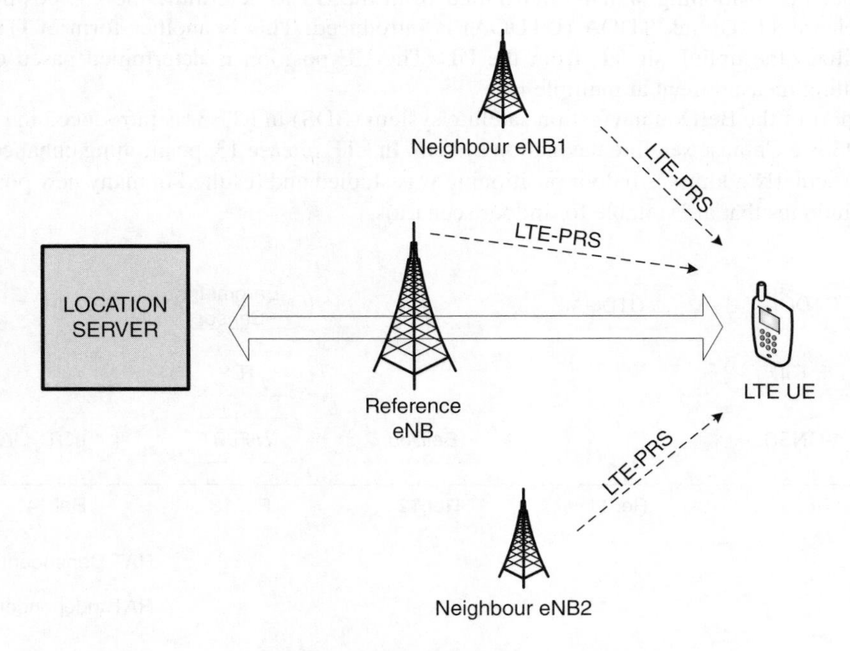

Figure 8.15 OTDOA techniques in LTE.

These requirements initiated the LTE release 13 indoor positioning study item (3GPP TS 37.857). This study focused on the following positioning enhancements:

- OTDOA enhancements
- E-CID enhancements
- WLAN/Bluetooth-based positioning
- barometric pressure sensor positioning
- terrestrial beacon systems (PRS beacons and metropolitan beacon systems)
- D2D-aided positioning.

Because of the limited time constraint in LTE release 13, some parts of WLAN/Bluetooth positioning, barometric pressure sensor positioning and TBS are included in the LTE specifications. The remaining techniques and other potential techniques will be considered in the next LTE releases. One potential big challenge in the upcoming LTE releases is to identify the positioning techniques that are suitable for machine-type communications (MTC) and the narrow-band Internet of Things (NB-IoT) communications as both have relatively narrow bandwidths of 1.4 MHz and 200 kHz, respectively.

8.3.4.3 Positioning Techniques Outlook in 5G

In the previous section, the key factor and requirements for the positioning techniques were the requirements defined by the regulatory body (the FCC). In 5G, where the study has just been initiated (3GPP TSG SA WG1 SP-150818), it can be observed that the positioning requirements are no longer defined by the regulatory body for positioning the emergency call. The positioning is more driven by the expected use-cases in 5G.

The report on the positioning accuracy and the use-cases proposed by several organizations and companies involved in 3GPP are summarized in Table 8.2 (ETRI S1-160014). The expected use-cases are, for example, tracking vehicles, car-to-car communications, tracking of a high number of devices (IoT) and drone (UAV) communications.

Based on the proposals in Table 8.2, a positioning accuracy of 0.5–1 m is expected. Moreover, some use-cases use a narrow bandwidth communication system, which creates many challenges and stimulates research activities in the positioning technique area.

8.3.5 Emergency Applications in Cellular Networks

A classical application of cellular positioning is the localization of the origin of an emergency call. The purpose of this application is to allow the mobile network to route the emergency call to the closest public safety answering point (PSAP) and to immediately inform the PSAP call-taker about the location of the emergency caller and where exactly to send emergency services. The facility that a network offers to localize the origin of an emergency call is often referred to as the caller location. In North America, this service is also called enhanced 9-1-1 (or E911).

In wired telephone networks, caller location has been employed since the late 1960s, with the result that there is only a small delay between the reception of an emergency call at the PSAP and the arrival of emergency services at the location of the emergency caller. Owing to the lack of caller location services for cellular networks, this delay has increased significantly

Table 8.2 5G Positioning Accuracy and Use-Cases.

Proposed by		Accuracy level and use-cases
US FCC	Accuracy	50 m (horizontal) × 3 m (vertical) accuracy in 80% of indoor calls
	Reason	Enhancing the ability of a public safety answering point to accurately identify the location of wireless 911 callers
NGMN	Accuracy	From 10 m to <1 m, on 80% of occasions better than 1 m indoors
	Reason	Tracking high-speed vehicles, providing instant personalized service, network-based 3D positioning
5G Forum	Accuracy	1 m in urban and indoor scenarios
	Reason	To avoid collisions between unmanned objects moving at 100 km/h
Nokia	Accuracy	Vehicular 1 m, robot 10 cm
	Reason	To support quick positioning in 10–15 ms for cars moving at 280 km/h, robots at 40 km/h
ZTE	Accuracy	3 m to <1 m on 80% of occasions
	Reason	Auto-parking needs <1 m accuracy, to support enhanced indoor LBSs 3 m to <1 m accuracy is needed
ETRI	Accuracy	1 m in 95% of urban and indoor scenarios
	Reason	For car crash avoidance at intersections, emergency rescue in buildings, 1 m accuracy is necessary in tunnels, car-parks, indoor, etc.
Qualcomm	Accuracy	0.1–0.5 m for vehicular and 10 cm for UAVs
	Reason	Car-to-car distance in platooning requires 0.5 m accuracy and side-by-side accuracy of 0.1 m
		0.1 m accuracy required for UAV control and crash avoidance between group flying UAVs
ALU, Huawei	Accuracy	Less than 0.5 m indoor inventory and outdoor
	Reason	To support tracking of millions of (internet purchased) items per km^2 shipped by trucks moving at 100 km/h
GM	Accuracy	0.1 m for high-speed vehicles moving at 100 km/h (relative speed is 200 km/h)
	Reason	To support multi-lane highway vehicles of density 10,000 per km^2

with the spread of mobile phones since the 1990s. Mobile networks and PSAPs provide the caller location service in a fully automated fashion, that is, the location of the emergency caller is automatically displayed to the PSAP call-taker. However, this service has not been implemented in all networks and PSAPs. The eCall project in the eSafety initiative of the European Commission is intended to provide a unique, fully automatic caller location service for all European streets (European Commission 2005).

Figure 8.16 shows a typical system framework for a caller location service. The emergency call, originating at an MS, is recognized by the operator of a public land mobile network (PLMN). Today, PLMN operators typically use network-based cellular positioning techniques to determine the location of an emergency call. The network operator then routes the call through the PLMN and the public switched telephone network (PSTN) to the closest PSAP and provides the caller location information (CLI). The evaluation of the location information

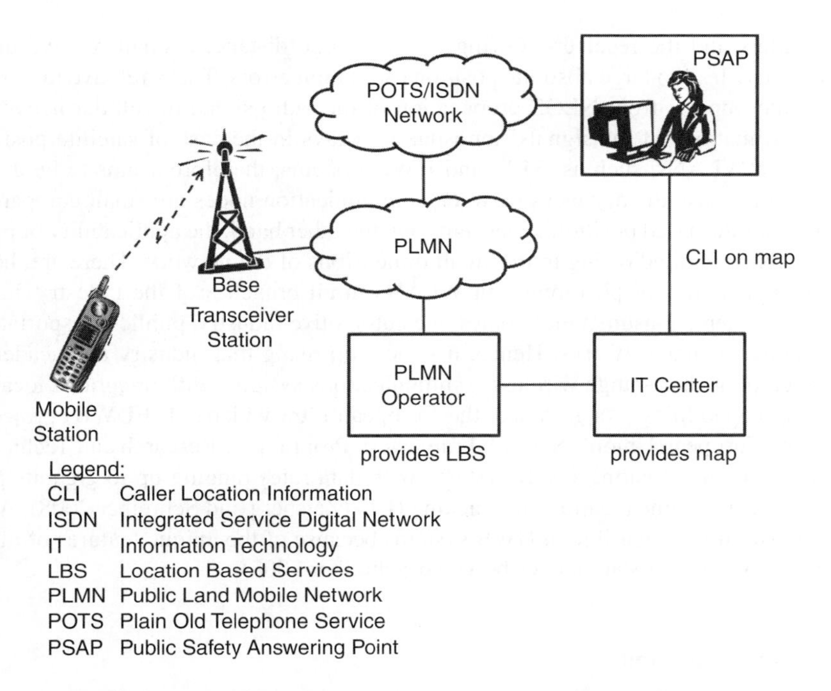

Figure 8.16 Emergency caller localization system in a mobile network.

and the presentation of the mobile subscriber's location on the map are often accomplished by a third-party service provider, illustrated as the information technology (IT) center in Figure 8.16.

8.3.6 Drawbacks When Applied to Mobile Positioning

Mobile positioning in cellular systems has many advantages. These include a short time to the first position fix, good coverage and low cost owing to the fact that the information to be exploited is gathered in the mobile network itself. However, there are also some disadvantages. A major drawback is the limited accuracy. It has been shown that the accuracy is typically not better than approximately 100 m at a reasonable complexity. Furthermore, the accuracy varies greatly with the density of nodes B: in the case of urban areas with a dense network, the accuracy is reasonably good, whereas in rural areas, with a much lower number of nodes B, the absolute accuracy decreases. The results obtained in the previous sections show that the relative accuracy can achieve values of the order of approximately 15–20% of the cell radius.

8.4 Wireless Local/Personal Area Network Positioning

8.4.1 Solutions on Top of Wireless Local Networks

One reason for the limited accuracy and reliability of localization in satellite and cellular positioning systems lies in the large distances between the communication nodes, that is, the transmitters and the receivers. Owing to these large distances, small relative measurement errors may lead to large absolute position estimation errors. These relative measurement errors include impacts from NLOS propagation of the radio signal in cellular networks and erroneous estimations of the signals' ionospheric delays in the case of satellite positioning. Positioning in WLANs, such as Wi-Fi and UWB systems, therefore seems to be a promising approach because the distances between communication nodes are small compared with cellular or satellite-based positioning systems. On the other hand, the applicability of positioning in WLANs is limited owing to the small dimensions of the network. There are, however, numerous applications of positioning in WLANs in all branches of the industry, including home entertainment, consumer electronics, the automotive industry, public transportation and heterogeneous cellular networks. Hence, it is not surprising that industry and academia are jointly developing short-range wireless communications systems with integrated localization and tracking capabilities. For instance, the European ultra-wideband (EUWB) project, supported by the European Union's Seventh Framework Program for Research and Technological Development, is investigating UWB systems with data rates ranging up to gigabits per second and precise real-time location and tracking (LT) (Zeisberg and Schreiber 2008). Accurate real-time positioning is feasible in UWB systems because of the unique features of ultrawide radio frequency band allocation (Zeisberg and Schreiber 2008).

8.4.1.1 Ultra-Wideband

UWB is a short-range communication system that uses a large frequency bandwidth at a low spectral power density. One goal of UWB is to transmit in a way which does not interfere with narrowband communication systems in the same frequency band. There are various standardization and industry associations specifying UWB in the physical and MAC layers of communication systems. Examples of UWB applications include high-data-rate services over short distances, such as indoor video streaming and wireless personal area networks (WPANs) (Heidari 2008), and low-power, down-market devices that allow moderate data rates and positioning (IEEE 2005).

Interesting scenarios for the application of LT in the EUWB project include both small-scale environments and spacious indoor environments such as shopping malls, train stations, airports, exhibition centers and sports stadiums. Providing data transmission and LT capabilities simultaneously, a single UWB network can be used to interconnect several different sensors and to provide location information to mobile users. In the same manner as in automotive navigation systems, users can locate themselves on their mobile device and find their way to a shop, a check-in desk or a seat on an aircraft. Furthermore, users can also get location-related information through the UWB network, such as points of interest. The network can evaluate knowledge about users' locations; when this is sent to a cellular network, it can improve RRM performance (see Section 2.4). In small-scale environments, indoor or in-cabin, LT can be used to localize a wireless key, for instance, or to identify whether a particular seat on an aircraft has been occupied or not.

Cross-layer functionality is required to accomplish LT applications. An adequate communications protocol comprises three steps: (1) synchronization of the nodes, (2) localization of a mobile device and (3) triangulation and mapping.

Accurate positioning in WLANs is based on TOF or TOA measurements. Similarly to cellular and satellite positioning, these measurements first require synchronization of the nodes, including the infrastructure, that is, the access points (APs), and the MS. After synchronization, the measurement and calculation of the distances between the APs and the MS can be accomplished. Finally, the location of the MS can be computed by an AP and conveyed to the other nodes, if required.

The accuracy of the aforementioned LT procedure has been simulated for a scenario with $n = 3$, $n = 4$ or $n = 5$ APs equally spaced on a circle with a radius of 1 m. The MS was located in the middle of the configuration. Figure 8.17 shows the root mean square (RMS) error of the position estimation; it is in the range of centimeters and depends on the standard deviation σ_t of the TOA measurements.

8.4.1.2 Bluetooth

Bluetooth is a short-range technology in which two different approaches are used for positioning. One approach assumes that positioning or tracking is to be performed on a large scale, that is, a large area covered by several APs, where the granularity of the positioning system is given by the positions of the APs (Gonzalez-Castano and Garcia-Reinoso 2002; Pels et al. 2005). Another approach aims at estimating position on a subcell-scale level, that is, the granularity is finer than the AP coverage (Bandara et al. 2004; Feldmann et al. 2003; Figueiras and Schwefel 2008; Figueiras et al. 2005).

On large scales, the typical information used to localize an MS is cell ID data. The use of this cell ID technique limits such systems to a granularity given by the coverage of the APs. Consequently, the magnitude of the achieved accuracy is given by the cell range in the Bluetooth communications, which depends on the class of the device: class 1 devices have a typical range of 100 m, whereas the class 2 devices have a typical range of tens of meters.

For subcell localization, the RSSI information defined by the Bluetooth specifications is the key to estimating positioning information. The RSSI is an octet defined in the Bluetooth specifications and ranges from -128 to 127 in single integers. The value 0 specifies the so-called golden receive power range, above which the RSSI values are positive and below which the

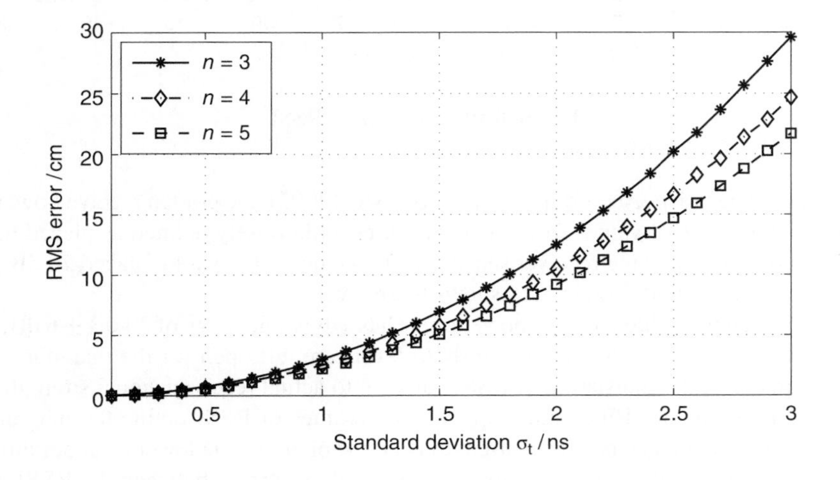

Figure 8.17 RMS localization error in UWB scenarios.

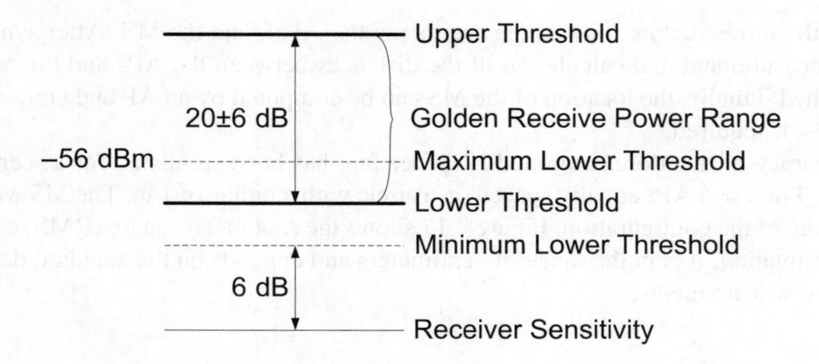

Figure 8.18 RSSI thresholds in Bluetooth.

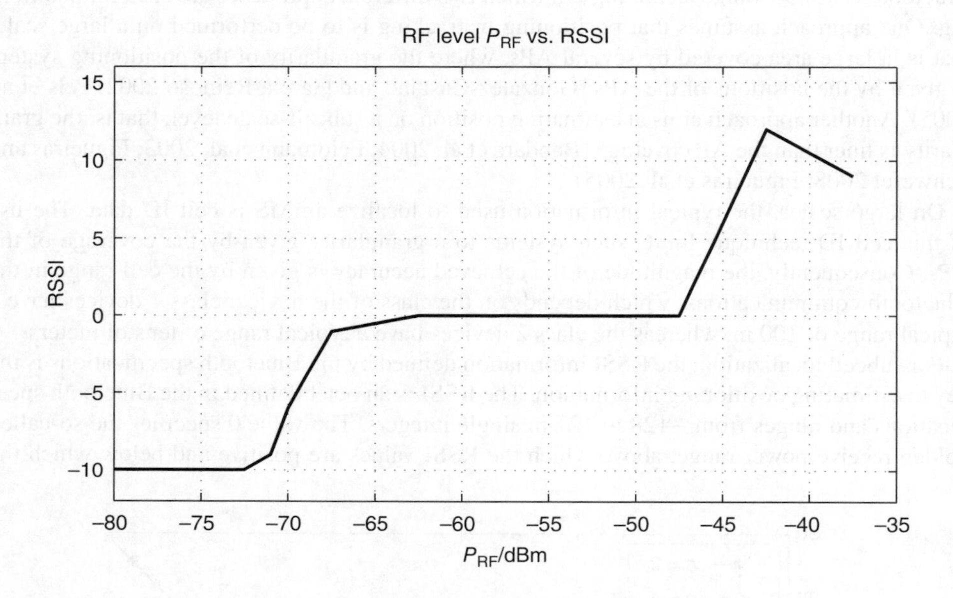

Figure 8.19 RF level vs. RSSI.

RSSI values are negative. According to Bluetooth SIG (2003), the golden receive power range is defined as the range of values from the lower threshold, loosely defined as placed between −56 dBm and 6 dB above the receiver sensitivity, to the upper threshold, placed 20 dB ± 6 dB above the lower threshold. Figure 8.18 shows these levels.

This means that the golden receive power range has a dynamic range of 20 dB ± 6 dB, within which the RSSI value is 0. Since the threshold levels are left open for the manufacturers of the Bluetooth chips, the translation of RSSI values into actual received signal strength values depends on the particular Bluetooth chip. Nonzero values of RSSI define the magnitude in decibels by which the received signal strength is below or above the lower or upper threshold, respectively. Figure 8.19 shows an example of a possible relation between the RSSI and the received signal strength.

Referring to the Bluetooth specifications, the RSSI value is tightly connected to the power control mechanism that adapts the link conditions in a Bluetooth communication. Careful analysis reveals that when power control is active, the transmission power used by the transmitter is not necessarily known at the receiver, meaning that the estimation of path loss may be complex to perform. A common solution is to run the positioning mechanisms when the Bluetooth device is in the inquiry state[2] because during the inquiry state the power control mechanism is inactive and the transmission is done at the maximum transmission power specified by the Bluetooth standards.

8.4.1.3 WLAN (Wi-Fi)

There was already a commercially available positioning system for WLAN called Loki in early 2007 (Skyhook 2007), which works similarly to the cell ID method. Instead of the cell IDs of BTSs, it uses the MAC addresses of nearby WLAN APs to estimate a position. Since the WLAN cell size is much smaller than the cell size in a cellular mobile network and a WLAN client can observe several MAC addresses, the WLAN positioning service outperforms the cell ID method with respect to its accuracy. Loki requires access to a database of registered WLAN APs and their locations. Therefore, it is a requirement that every user of this service registers their own WLAN AP with this publicly available database.

8.4.2 Dedicated Solutions

The positioning methods discussed above make use of communication systems whose main purpose is data transmission rather than positioning. Dedicated positioning solutions, in contrast, are expressly designed for positioning applications and require specific hardware for this purpose. The only signals exchanged by the transmitter and receiver are the positioning signals.

8.4.2.1 RFID

An old but newly reemerged technology is RFID, using so-called RFID tags for identification and tracking. RFID tags can also be used to increase the accuracy and reliability of popular positioning systems such as the GPS. For instance, we could have a number of fixed installations on a roadway. Every installation includes an RFID tag along with information about its accurate position. The mobile receiver contains a RFID reader to read this location information in order to provide more accurate positioning, especially in tunnels and in downtown areas, where satellites are often obstructed.

The system described in Chon et al. (2004) was proposed as a complementary outdoor system to GNSSs, and can be summarized as follows. RFID tags need to be installed on a road in a way that maximizes the coverage and the accuracy of positioning (see Figure 8.20). During installation, necessary information, such as the coordinates of the location where the tag is

[2] The inquiry state is the initial state that a Bluetooth device enters in order to discover other Bluetooth devices with the intention of establishing a communication link.

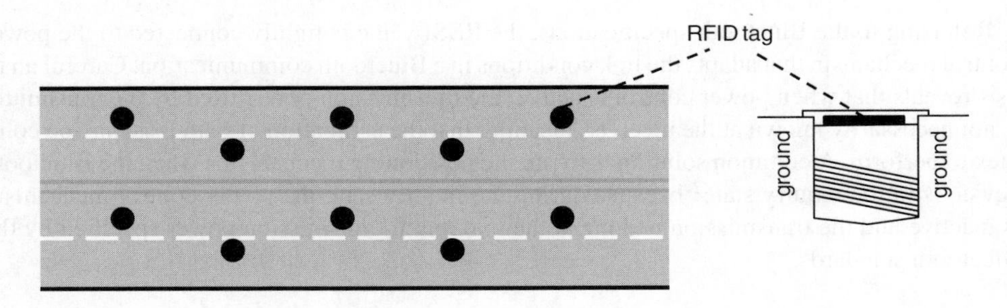

Figure 8.20 Road-mounted RFID tags.

installed, needs to be written into each tag. The accuracy of this position information is critical for this technique to be successful. The position information can be acquired by using, for example, the DGPS. Vehicles then need to be equipped with an RFID reader that can communicate with the tags on a road. No matter how accurate the RFID positioning is, the system provides the positions where the tags are (Chon et al. 2004). Therefore the vehicles need also to be equipped with a GPS receiver and inertial sensors such as a gyroscope for positioning when there are no tags around. While being driven, the vehicles constantly monitor for the presence of a tag. On detection, the reader retrieves the information from the tag, including its coordinates. Figure 8.20 shows RFID tags deployed on a road. The circles on each lane represent a tag and the tag itself is enclosed in a special-purpose reflector, a so-called "cat's eye."

8.4.2.2 Infrared

In 1977 the Institute of Electrical and Electronics Engineers (IEEE) redefined radar as "an electromagnetic means for target location and tracking," which includes electro-optical devices such as laser radars (lidars) and laser rangefinders in general. Relative to microwave and millimeter-wave systems, electro-optical sensors are characterized by extremely short wavelengths, thus affording much higher resolution but suffering greater attenuation due to atmospheric conditions. The optical region of the electromagnetic spectrum is broken up into the ultraviolet, visible and infrared domains. This subsection concentrates on positioning in the infrared domain.

A typical model of an electro-optical ranging system comprises five main components: (1) the optical source, typically a light-emitting diode or a laser diode, (2) a mechanism to modulate the optical output from the source with the signal to be transmitted, (3) the transmission medium, (4) a photodetector, which converts the received optical power back into an electrical waveform, and (5) electronic amplification and signal processing to recover the signal and present it in a suitable form.

Infrared ranging is performed by using continuous wave signals, narrowband signals (Eltaher 2009) or pulse-train-based transmission, which uses, for example, short impulses or impulse compression concepts (Zimmermann 1991). Typically, the range of such systems is limited to a few tens of meters, but the accuracy can be as low as a few tens of micrometers. For a survey of infrared technology, the reader is referred to Eltaher (2009).

Figure 8.21 shows an infrared positioning scenario in the case of a small indoor environment. Each room is equipped with an infrared sensor. These infrared sensors are connected to

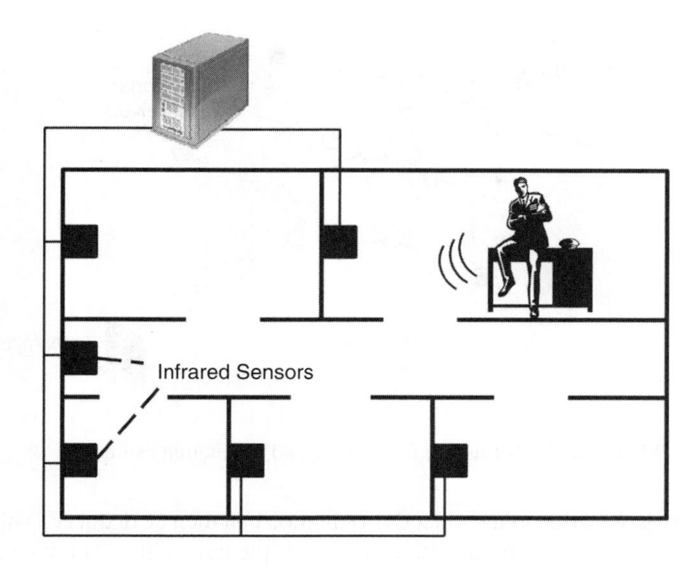

Figure 8.21 Infrared positioning.

a server that carries out the location estimation. A mobile infrared device, for example worn by a human, transmits infrared signals, which are detected by at least one particular sensor in the vicinity of the mobile infrared device. The detected signal is then used to evaluate the position of the mobile infrared device.

8.4.2.3 Ultrasound

For indoor applications, ultrasonic positioning systems have emerged which make use of a combination of RF and ultrasound technologies. A number of fixed nodes, installed on walls and ceilings, transmit their location information via RF signals and at the same time ultrasonic pulses, which are synchronized with the RF signals. A mobile node correlates the received RF signal and the ultrasonic pulses. By taking the difference between the propagation speeds of RF (the speed of light) and ultrasound (the speed of sound) into account, it can estimate the distances to different fixed nodes and then its own position. Examples of commercial solutions have been presented by Intersense (Wormell et al. 2007) and Olivetti (Ward et al. 1997).

A specialty of the work on ultrasonic positioning has been the estimation of the orientation of objects (Ward et al. 1997). If three MSs are placed on a rigid body at known noncollinear points, then, by calculating the positions of the devices, the orientation of the object can be found. This technique is most suitable for stationary or slow-moving objects because the three devices will be located at different times – any intervening movement of the object will introduce errors to the calculated orientation. Alternatively, an object's orientation can be obtained using knowledge of the set of receivers that have detected ultrasonic signals from an MS (Ward et al. 1997). The directional transmission pattern of the ultrasonic pulse is known to be hemispherical. With this knowledge in combination with the known position of the MS, the set of sensors which receive the MS's ultrasonic pulse can be used to estimate the MS's orientation (Figure 8.22). If the device is rigidly affixed to an object at a known point and in a known

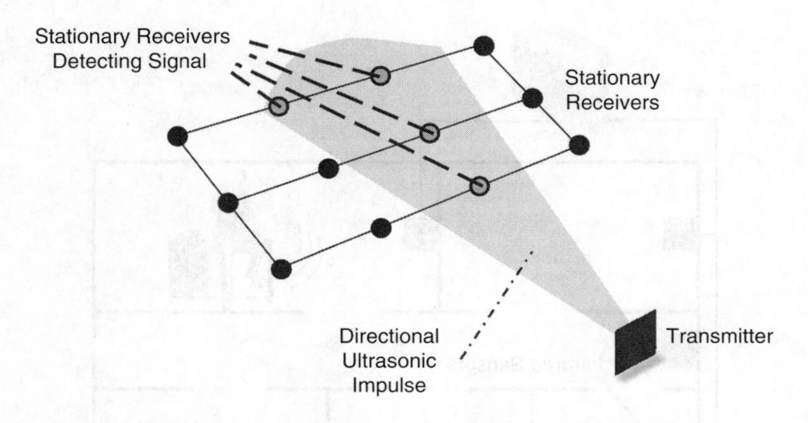

Figure 8.22 Ultrasonic positioning and orientation estimation.

orientation, an approximation to the object's orientation can then be deduced. This approach is most suitable when the tagging of the object with multiple transmitters could be cumbersome.

8.5 Ad hoc Positioning

In large isotropic networks, where the routing effort does not depend on the routing direction, the availability of position information enables routing without the use of large routing tables. This application is often referred to as the primary positioning application in ad hoc networks, and the GPS is often used (Ko and Vaidya 1998). In Section 8.3, we discussed positioning methods that are also applicable to ad hoc positioning. These rely on the nodes' capability to measure TOA, TDOA and RSS, and on a knowledge of the BSs' locations. Decentralized, self-organized, scalable ad hoc networks do not contain fixed BSs. Locations must therefore be determined in a relative coordinate system. Algorithms for relative localization can be found in Capkun et al. (2002). These are based on the detection of the nodes in the ad hoc network and on the TOA method to estimate the distances between neighboring nodes. In a decentralized approach, every node measures the distances to its neighboring nodes. The collected information about the network topology and the distances between the nodes is then communicated to all nodes in the network. Based on this knowledge, the relative positions can be estimated.

8.6 Hybrid Positioning

8.6.1 Heterogeneous Positioning

The coexistence of different wireless access technologies creates a highly dynamic environment, which can be viewed as a heterogeneous network. To establish appropriate interoperability, heterogeneous positioning systems can be created to facilitate increased positioning accuracy and better coverage for LBSs. In this context, two different kinds of heterogeneous positioning can be distinguished.

The aim of the first kind of heterogeneous positioning system is to extend conventional indoor positioning systems by introducing a homogeneous interface between them and outdoor positioning systems such as the GPS. In addition, when cellular networks, which have wide area coverage, high mobility, small bandwidth and, usually, moderate data rates, are combined with short-range systems such as WLANs, which typically have low coverage, low mobility, high bandwidth and, usually, high data rates, the positioning accuracy can be improved. This aspect is treated further in Section 8.6.2.

The aim of the second kind of heterogeneous positioning systems is to extend conventional outdoor positioning systems and to reduce the position acquisition times. For instance, this can be achieved by combining cellular positioning with GNSSs. In this case, the positioning coverage of GNSSs can be improved in cases where the LOS to the satellites is obstructed, for example. This aspect is covered further in Section 8.6.3.

8.6.2 Cellular and WLAN

Recently, the combination of cellular and WLAN radio interfaces in handheld devices has been introduced. Essentially, the position estimation techniques deployed in WLANs are the same as for cellular systems and can therefore be reused in combined cellular and WLAN positioning.

Seamless LBS provisioning for users moving between outdoor and indoor environments requires a complex interaction between the different positioning systems. Zuendt et al. (2005) provided a concept that facilitates an open homogeneous location architecture that delivers the best possible positioning information to an MS. A conceptual architecture of a combined GSM and WLAN positioning scheme is depicted in Figure 8.23. This figure illustrates the case of a multistandard-capable MS which supports both GSM and WLAN radio interfaces. Positioning is carried out by enhancing the LBS data obtained from the GSM by using the WLAN connectivity of the MS. In practice, the MS requests improved positioning via the WLAN AP, which conveys this request to the location server. The latter provides both cellular and WLAN positioning features and is connected to the WLAN and GSM infrastructure databases. These databases hold information about the location of BTSs and WLAN APs. The location server computes the position of the MS by exploiting the data in the databases and conveys this piece of information back to the MS. In addition, the location server could also provide a link to a map showing the location of the MS. The accuracy of this scheme is limited mainly by the coverage of the WLAN APs. Zuendt et al. (2005) reported a typical estimation accuracy as low as 2 m. The drawback of this scheme lies in the fact that the establishment of a location database of APs is cumbersome and thus at present limits the deployment of WLAN positioning.

8.6.3 Assisted GPS

When a GPS system is designed to interoperate with cellular networks, the network assists the receiver to improve the performance with respect to the startup time, sensitivity and power consumption. In order to perform localization, the receiver must first know the orbital elements of the satellites. These data are transmitted with the GPS broadcast signal. The data rate of the GPS broadcast signal is only 50 bits per second. That is why the provision of a first position fix can take up to 12.5 minutes. Alternatively, the satellites' orbital elements can be transmitted via a cellular network connection, thus reducing the startup and acquisition times to a few seconds.

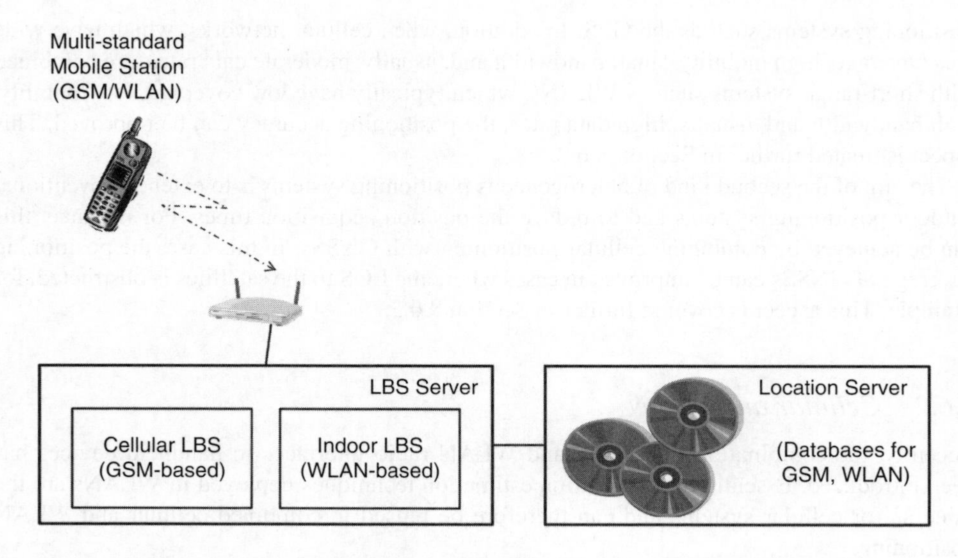

Figure 8.23 Conceptual architecture of combined cellular and WLAN positioning.

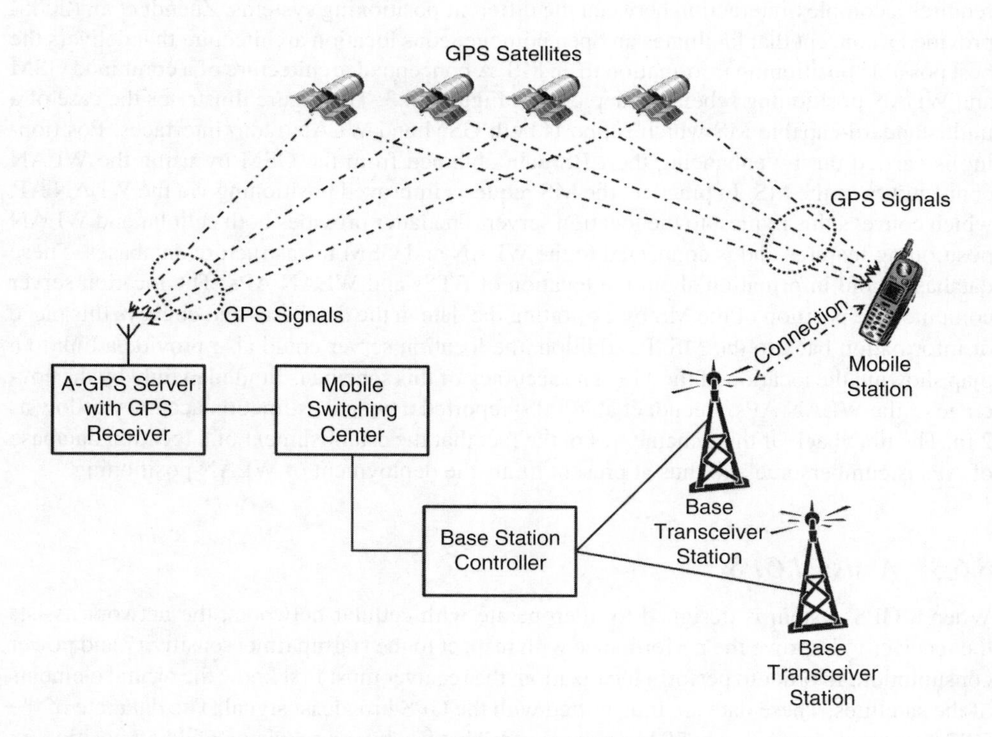

Figure 8.24 A-GPS conceptual architecture.

A-GPS is considered to be the first A-GNSS. It was initially proposed in the early 1980s. In A-GPS, the GPS receiver is provided with information that helps to reduce the time it takes the receiver to calculate its position. Also, the sensitivity can be improved by using this method when the receiver is unable to demodulate the GPS signals in low signal-to-noise ratio (SNR) situations. Moreover, the rapid startup time allows the receiver to stay in idle mode when positioning is temporally not needed. This reduces the receiver's power consumption.

The A-GPS concept is shown in Figure 8.24. The MS has a GPS receiver integrated into it. The A-GPS server has a reference GPS receiver, which is in LOS conditions and monitors the same satellites as are seen by the MS. Each A-GPS server provides information for various BTSs and has exact knowledge about the GPS signals that are available at the MSs. The A-GPS server is connected to the mobile switching center. The A-GPS server sends information to the MS by means of the cellular radio network. This includes information about the satellites, which can be received in the corresponding area. The MS utilizes this information and as a result improves its positioning capability. A-GPS became truly relevant and needed only after the wireless E911 mandate was issued FCC (1999).

8.7 Conclusions

GNSSs such as the GPS can provide precise positioning with global coverage, as long as the receivers have an LOS connection to at least four satellites. This constraint is the main disadvantage of the GPS, leading to limited reliability. Cellular positioning techniques have excellent reliability in urban areas. However, they are usually significantly less accurate than GNSSs. Nevertheless, they can potentially be improved by advanced mobile-assisted positioning. In particular, combination with the upcoming WLAN positioning leads to improvement in the position estimation accuracy, and thus makes them competitive with GNSSs.

Acknowledgements

The authors gratefully acknowledge the support of Björn Steiger Stiftung Service GmbH and of Vodafone. Furthermore, the authors wish to thank their colleagues at the Lehrstuhl für KommunikationsTechnik, Universität Duisburg-Essen, for fruitful discussions, hints and helpful suggestions.

9

Ultra-wideband Positioning and Tracking

Davide Dardari
University of Bologna, Italy

9.1 Introduction

The previous chapters have discussed how to infer the position of a mobile node by measuring the features of radio signals which are position-dependent, such as RSS, TOA, AOA or the carrier phase. Such measurements introduce geometric constraints that allow the location engine to estimate the node's position according to some method. As a consequence, the obtained accuracy of position estimation strongly depends on the quality of the measurements in combination with the geometric configuration of the network.

The exploitation of RSS measurements has received particular attention because they are available in every radio device regardless of whether or not it has been explicitly designed for positioning. Unfortunately, RSS might be weakly correlated with position/distance due to the random nature of radio propagation, especially in indoor environments and in NLOS conditions, thus leading to poor localization performance. Although for some applications the accuracy provided by RSS-based positioning systems is sufficient, there are several other potential applications in which sub-meter accuracy and a higher reliability have to be guaranteed. Examples can be found in logistics, augmented reality, and indoor autonomous navigation of robots and drones.

Time-based measurements are in general better suited when high-accuracy positioning is required. In fact, as will be explained in this chapter, the larger the signal bandwidth, the higher the resolution of time measurements. This justifies the adoption of wideband signals in the GPS and the interest in UWB signals in indoor positioning applications. Besides active positioning, thanks to its peculiarities UWB technology also enables other applications, such as multi-static radar for non-collaborative localization, life signs detection systems, and through-wall and underground imaging.

Mobile Positioning and Tracking: From Conventional to Cooperative Techniques, Second Edition.
Simone Frattasi and Francescantonio Della Rosa.

In this chapter, the fundamentals of UWB technology, the main issues and the methods of obtaining high-resolution ranging and localization are presented and discussed.

9.2 UWB Technology

9.2.1 History and Definitions

The term *ultra-wideband* (*UWB*) originated with the US Defense Advanced Research Projects in a study undertaken in 1990 and referred to the transmission of very short-duration pulses to be used for accurate ranging in radar applications (Win et al. 2009). However, the idea of emitting very short pulses was not new. In fact we could say the radio was born with UWB considering that the Italian inventor Guglielmo Marconi realized his early wireless communications using spark gap transmissions in the late 1890s. Despite of this UWB "dawn" after a few years wireless communication abandoned impulse transmissions and became "tuned" by employing modulated signals on sinusoidal carriers and tuned circuits to allow multi-user coexistence on the shared radio spectrum.

Impulsive transmission was rediscovered and studied using time-domain electromagnetic theory in the early 1960s whereas implemented systems for radar applications appeared in the 1970s thanks to the introduction of short-pulse generators using tunnel diodes.

The first studies of the possible application of UWB to communication systems were published during the 1990s in the pioneering works of Scholtz and Win (Scholtz et al. 2005; Win and Scholtz 1998a) that conceived the transmission of information by associating multiple pulses to the same bit according to some position or amplitude spreading code to allow multi-user spectrum sharing. Starting from these early works, the research activity on UWB went beyond the boundary of electromagnetism and involved studies and experimental characterizations for communication as well as indoor positioning (Lee and Scholtz 2002; Win and Scholtz 2002). The real boost to UWB came in February 2002 when the US FCC adopted the First Report and Order that permitted the marketing and operation of certain types of UWB products. A few years later the EU and other countries introduced similar regulatory frameworks. According to the FCC, a UWB signal is defined as a signal that has fractional (relative) bandwidth larger than 20% or an absolute bandwidth of at least 500 MHz Fed (2002).

9.2.2 Theory

According to FCC definition, each signal characterized by a bandwidth larger than 500 MHz is considered as UWB regardless of the way it is generated. Conventional carrier-based modulation schemes such as OFDM or CDMA can be used to generate UWB signals by broadening their bandwidth, even though the classical and simplest way to obtain UWB signals is by means of *impulse radio UWB* (*IR-UWB*). In an IR-UWB system, a number of short pulses (typically with duration less than 1ns) are transmitted per information bit in order to collect more energy per bit and allow multi-user access. With reference to Figure 9.1, in a typical IR-UWB communication system, a symbol of duration T_s is divided into N_s frames of duration T_{f}, which are further decomposed into N_{h} smaller time slots T_{c} (chips). The frame time T_{f} is usually chosen to be greater than the maximum multipath delay to avoid inter-symbol interference. The information is usually conveyed by pulse polarity, leading to pulse amplitude modulation (PAM), equivalent to binary phase shift keying (BPSK) signaling or pulse position, thus obtaining a

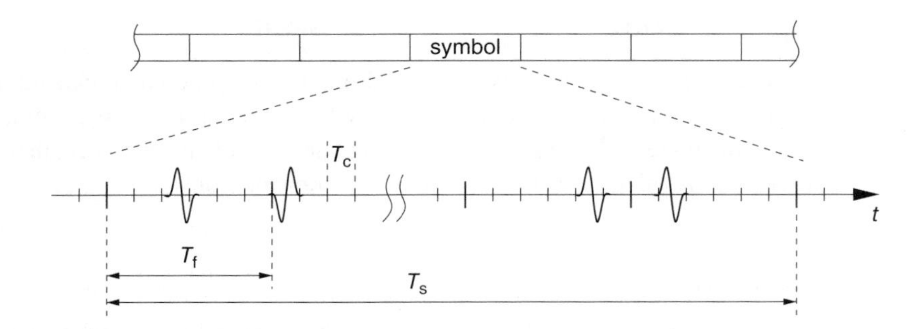

Figure 9.1 Example of a UWB modulated signal.

pulse position modulation (PPM) signaling scheme. To allow for multi-user access the short UWB pulse $p(t)$, with duration $T_p < T_c$, is transmitted in each frame in a chip position specified by a user-specific pseudo-random time-hopping code $\{c_n^{(k)}\} \in \{0, 1, \ldots, N_h - 1\}$ and/or a direct sequence code $\{a_n^{(k)}\} \in \{-1, 1\}$ having period N_s (Win and Scholtz 1998a). The general expression of the UWB signal modulated by the kth user is

$$s^{(k)}(t) = \sum_i \sum_{n=0}^{N_s-1} d_{\text{pam}_i}^{(k)} \, a_n^{(k)} \, p\left(t - d_{\text{ppm}_i}^{(k)} \delta - c_n^{(k)} \, T_c - n \, T_f - i \, T_s\right) \tag{9.1}$$

where $d_{\text{pam}_i}^{(k)} \in \{-1, 1\}$ is the information sequence associated to the PAM signaling (if any) and $d_{\text{ppm}_i}^{(k)} \in \{-1, 1\}$ is the information sequence associated to the PPM signaling (if any) of the kth user. Parameter δ determines the depth of the PPM modulation.

The simplest UWB receiver is a correlation receiver where the received signal is correlated with a local replica (template) of the transmitted pulse $p(t)$ or, equivalently, is passed through a filter matched to $p(t)$. Typical indoor environments often exhibit the presence of a dense multipath with delay spread much larger than the resolution capability of the signal being employed. The transmission of ultra-short pulses can potentially resolve an extremely large number of paths experienced by the received signal, thus eliminating significant multipath fading. This may considerably reduce the fading margin in link budgets and may allow low power transmissions. In addition, rich multipath diversity can be collected through the adoption of rake receivers which combine the signals coming over resolvable propagation paths in a way that maximizes the SNR (Win and Scholtz 1998b).

IR-UWB transmitters can be extremely simple and energy efficient as no oscillators and mixers are required in principle, even though commercial implementations usually prefer to sacrifice energy efficiency in favor of more conventional super-heterodyne schemes. In addition to communication, IR UWB systems can provide accurate ranging and positioning due to high time resolution of UWB pulses.

While communication-related applications of UWB have not had significant success, mainly due to the introduction on the market of high data rate and backward compatible IEEE 802.11x standards, there are some emerging applications in which the UWB technology is demonstrating its potential, such as indoor positioning, medical (e.g., wireless body area networking), military (personnel tracking, through-wall vision), automatic inventory and rescue scenarios (e.g., vital sign detectors).

9.2.3 Regulations

Regulation and standardization are major issues for the industrial and commercial future of wireless technologies. Since UWB signals occupy a very large portion in the spectrum, they must coexist with the incumbent systems without causing considerable interference, therefore a set of very conservative regulations is imposed on UWB transmissions.

9.2.3.1 US Regulations

The FCC Part 15 rules in the US limit UWB systems to transmit below certain power levels in order not to cause significant interference to the legacy systems in the same frequency spectrum. For indoor communications the average power spectral density (PSD) cannot exceed −41.3 dBm/MHz over the frequency band from 3.1 to 10.6 GHz, and it must be even lower outside this band, depending on the specific application (Fed 2002). This corresponds to a small fraction of a milliwatt of transmitted power over several gigahertz of bandwidth. Similarly, outdoor communications are allowed in the same band with a higher constraint on out-of-band emissions. As can be noted, FCC limits ensure that UWB emission levels are exceedingly small below spurious emission limits for all radios or below the lowest limits on unintentional emissions ever applied by the FCC to any system. For comparison, PSD limits for 2.4 GHz ISM and 5 GHz U-NII bands are 40 dB higher.

9.2.3.2 European UWB Regulations

The convoluted story of regulation in Europe has resulted in decisions which allow UWB "generic" emissions in certain bands, within a rather complicated spectrum mask (The Commission of the European Communities Official Journal of the European Union, 2007/131/EC, Feb. 23, 2007). Two bands are relevant for real-time locating system (RTLS), which are 3.1–4.8 GHz on the one hand and 6–8.5 GHz on the other.

The frequency band 6–8.5 GHz has been identified in Europe for long-term UWB operation with a maximum mean effective isotropic radiated power (EIRP) spectral density of −41.3 dBm/MHz and a maximum peak EIRP of 0 dBm measured in a 50 MHz bandwidth without the requirement for additional mitigation. The lower band 3.1–4.8 GHz is limited to −70 dBm/MHz at most, unless specific interference limitation techniques are used. These techniques are *detection and avoidance* (*DAA*), in the general case, or *low-duty cycle* (*LDC*) for devices emitting on rare occasions. They are already mentioned in the ECC decision (The Commission of the European Communities Official Journal of the European Union, 2007/131/EC, Feb. 23, 2007), but have been further investigated by CEPT and ETSI in order to standardize them.

DAA requires the UWB device to sense the use of the band by primary services, such as broadband wireless access. in the range 3.4 GHz to 3.8 GHz or radars signals in the range 3.1 GHz to 3.4 GHz and 8.5 GHz to 9 GHz, and apply spectrum shaping and transmit power control if necessary to protect incumbent services.

Alternatively, LDC transmissions are allowed without channel sensing and power control provided that the constraints listed in Table 9.1 are met, where T_{on} is the duration of a transmission burst and T_{off} is the time interval between two consecutive transmission bursts.

Table 9.1 LDC parameters.

T_{on} max	5 ms
T_{off} mean	$\geq$ 38 ms averaged over 1 s
$\sum T_{on}$	< 5% per second
$\sum T_{on}$	< 0.5% per hour
$\sum T_{off}$	> 950 ms per second

9.2.3.3 Worldwide UWB Regulation

Similar regulations for UWB transmissions have been fixed worldwide. For instance, in Japan the first emission task was announced in September 2005 and Established on Dec. 2006 by the Ministry of Internal Affairs and Communications. A power spectral density of −41.3 dBm/MHz is allowed in the band between 7.25 GHz and 10.25 GHz. The band 3.4–4.2 GHz is licensed for satellite (downlink) communication operators, and requires DAA. The regulation could be reviewed when satellite links are replaced by fibers. The band 4.2–4.8 GHz is reserved for 4G use in Japan and requires DAA. Without DAA, the emission power is limited to −70 dBm/MHz or less.

In Korea, the UWB regulation done by the Ministry of Information and Communication allows a power spectral density of −41.3 dBm/MHz in the band 7.2–10.2 GHz. DAA is required in the 3.1–4.8 GHz band, otherwise the emission power is limited to −70 dBm/MHz or less. In Table 9.2 the main regulatory emission masks are summarized.

Table 9.2 Worldwide UWB emission masks.

Band (GHz)	EIRP (dBm/MHz)	Mitigation factor
Europe		
3.1–4.8	–41.3	LDC or DAA
2.7–3.4	–70	No mitigation
3.4–3.8	–80	No mitigation
3.4–6	–70	No mitigation
6–8.5	–41.3	No mitigation
8.5–9	–41.3	DAA
Japan		
3.4–4.8	–41.3	DAA
3.4–4.2	–70	No mitigation
4.2–4.8	–41.3	No mitigation
4.8–7.25	–70	No mitigation
Korea		
7.25–10.25	–41.3	No mitigation
3.1–4.8	–41.3	LDC or DAA
7.2–10.2	–41.3	DAA
China		
4.2–4.8	–41.3	DAA
3.6–6	–70	No mitigation
6–9	–41.3	No mitigation
USA		
3.1–10.6	–41.3	No mitigation

9.3 The UWB Radio Channel

The ultimate performance of any wireless communication system is limited by the radio channel; the same applies to positioning systems as they are based on the measurement of some position-dependent features of the radio signal. Indoor radio propagation is typically more challenging than outdoor propagation because of the larger number of signal reflections that cause multipath components in the received signal. In narrowband communications multipath determines the fading effect which has to be mitigated by introducing a suitable margin in the link budget. In RSS-based positioning systems fading might seriously compromise the correlation between the received power and the distance/position.

The transmission of UWB pulses can potentially resolve an extremely large number of paths experienced by the received signal. Such paths can be ideally discriminated at the receiver and coherently summed up using rake receivers to get significant diversity gain in communication (Win and Scholtz 2002). From the localization perspective, the possibility of detecting and isolating the first arriving path, whose delay (TOA) is proportional to the distance between the wireless nodes (if in LOS), allows for precise ranging.

Figure 9.2 shows a typical measured UWB channel response in a LOS receiving location. It can be noted that a dense multipath could be present even in strong LOS conditions. In Section 9.6, the problem of detecting the first arriving path when in the presence of multipath and possible solutions will be described.

In the last 15 years several measurement campaigns have taken place from which dedicated channel models have been derived, allowing researchers to characterize and compare the performance of UWB communication and localization systems.

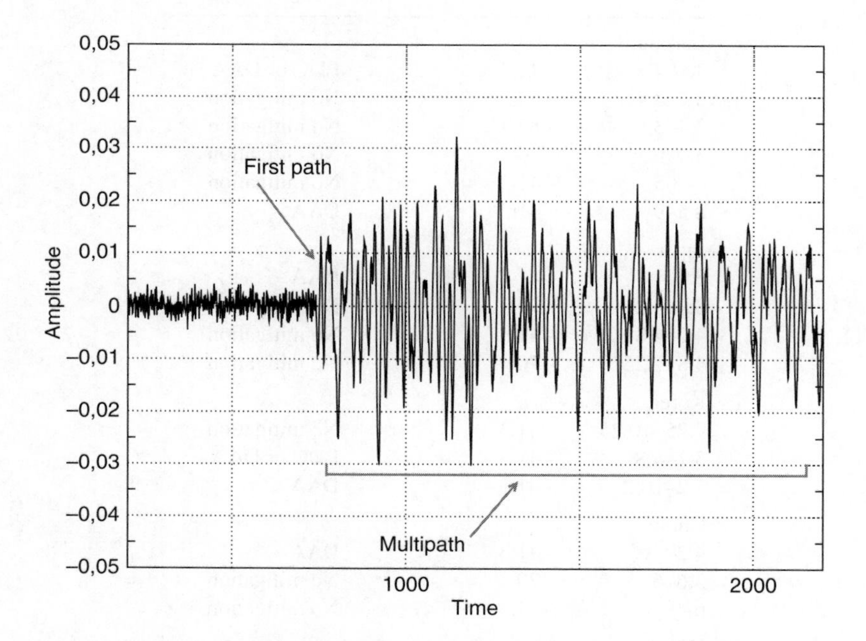

Figure 9.2 Example of measured UWB multipath profile in LOS condition.

A number of UWB channel models have been proposed depending on the frequency range and operating conditions (Cassioli et al. 2002; Molisch et al. 2006). The IEEE 802.15.3a Task Group developed the first standardized UWB channel model that is valid from 3 to 10 GHz for indoor residential and office environments with restricted distance (<10 m). A more general standard model valid for the high-frequency range (3–10 GHz) in a number of different environments, namely CM1-CM9, as well as for the frequency range below 1 GHz in office environments has been presented within the 802.15.4a Task Group (Molisch et al. 2005). More details are provided on the following sections.

9.3.1 *Path Loss*

In a wideband or UWB channel, the path loss is a function of frequency as well as of distance (Molisch et al. 2005)

$$L_0(f,d) = \Delta \left[\mathbb{E}_l \left\{ \mathbb{E}_s \left\{ \int_{f-\Delta/2}^{f+\Delta/2} |H(\tilde{f},d)|^2 \, d\tilde{f} \right\} \right\} \right]^{-1}, \tag{9.2}$$

where $H(f,d)$ is the transfer function from the transmitter antenna connector to the receiver antenna connector and Δ is chosen small enough so that electromagnetic (e.m.) characteristics of materials can be considered constant within the bandwidth. $\mathbb{E}_l\{\cdot\}$ and $\mathbb{E}_s\{\cdot\}$ denote the statistical expectation with respect to long-term fading (shadowing) and short-term fading, respectively.

In Molisch et al. (2006) the following frequency-dependent path-loss model is suggested

$$L_0(f,d) = 2\, k_0 \, \frac{(f/f_c)^{2(\kappa+1)}\, d^{\alpha}}{\eta_{TX}(f)\, \eta_{RX}(f)}, \tag{9.3}$$

where k_0 is the isotropic path-loss at the reference distance of 1 m and center frequency f_c, $\eta_{TX}(f)$ and $\eta_{RX}(f)$ are, respectively, the transmit and receive antenna efficiencies, and f_c is the center frequency, usually taken as 5 GHz. The coefficient 2 approximates the typical loss due to the presence of a person close to the antenna. Shadowing effects can be accounted for similarly to what is done in narrowband channel models by multiplying the path loss by a log-normal random variable (Molisch et al. 2006). In decibels it can be written as

$$L(f,d) = L_0(f,d) + S \tag{9.4}$$

where S is a zero-mean Gaussian random variable (RV) accounting for random fluctuations due to large-scale fading.

9.3.2 *Multipath*

The IEEE 802.15.4a UWB channel model is based on an extended version of the classical Saleh–Valenzuela (SV) indoor channel model, where multipath components arrive at the receiver in clusters following the Poisson point process. According to the SV model, the complex baseband channel impulse response is given by

$$h_0(t) = \sum_{k=0}^{K} \sum_{l=0}^{L} a_{k,l} e^{j\phi_{k,l}} \delta(t - T_k - \tau_{k,l}), \tag{9.5}$$

where $a_{k,l}$ is the amplitude of the lth path in the kth cluster, T_k is the delay of the kth cluster and $\tau_{k,l}$ is the delay of the lth path relative to the kth cluster arrival time T_k. The phases $\phi_{k,l}$ are uniformly distributed in the range $[0, 2\,\pi)$. The number $K + 1$ of clusters is assumed to be Poisson distributed. The classical SV model also considers a Poisson point process for the ray arrival times $\tau_{k,l}$. However, to better fit indoor and outdoor channel measurements, the IEEE 802.15.4a Task Group proposed to model the ray arrival times with mixtures of two Poisson processes. More details can be found in Molisch et al. (2006). The power delay profile (PDP) is assumed to be exponential negative within each cluster

$$\Lambda_{k,l} = \mathbb{E}\left\{|a_{k,l}|^2\right\} \propto \exp\,(-\tau_{k,l}/\epsilon_k), \tag{9.6}$$

where ϵ_k is the intra-cluster decay time constant.

Another widely adopted model is the dense multipath model with a single cluster composed of L independent equally spaced paths and exponential PDP. In this case the real impulse response is given by

$$h(t) = \sum_{l=1}^{L} a_l \, p_l \, \delta\,(t - \tau_l), \tag{9.7}$$

where a_l is the path amplitude and p_l is a RV which takes, with equal probability, the values $\{-1, +1\}$. The average path power gains Λ_l are given by

$$\Lambda_l = \mathbb{E}\left\{|a_l|^2\right\} = \frac{(e^{\Delta/\epsilon} - 1)e^{-\Delta(l-1)/\epsilon}}{e^{\Delta/\epsilon}(1 - e^{L\,\Delta/\epsilon})} \tag{9.8}$$

for $l = 1, \ldots, L$ (Cassioli et al. 2002). Parameter Δ is the resolvable time interval whereas ϵ describes the multipath spread of the channel (decay time constant). In most of these models the small-scale fading, which characterizes the path amplitudes a_l, follows a Nakagami-m distribution

$$f(x) = \frac{2}{\Gamma(m)}\left(\frac{m}{\Lambda}\right)^m x^{2m-1} \exp\,\left(-\frac{m}{\Lambda}\,x^2\right), \tag{9.9}$$

where m is the severity parameter, depending on the working conditions, and $\Gamma(\cdot)$ is the gamma function.

9.3.3 UWB Channel Models for Positioning

The previous channel models have been derived by having in mind communication systems and not positioning. While communication systems could in principle take advantage of the presence of a multipath, especially if resolvable, by collecting the energy carried by each path component through rake-based receivers, in positioning systems usually only the direct path (if it exists) is related to the distance between the transmitter and receiver, and the other paths represent nuisance components that could confuse the first path TOA estimator. In fact, it has been demonstrated theoretically that if the multipath components do not overlap with the direct

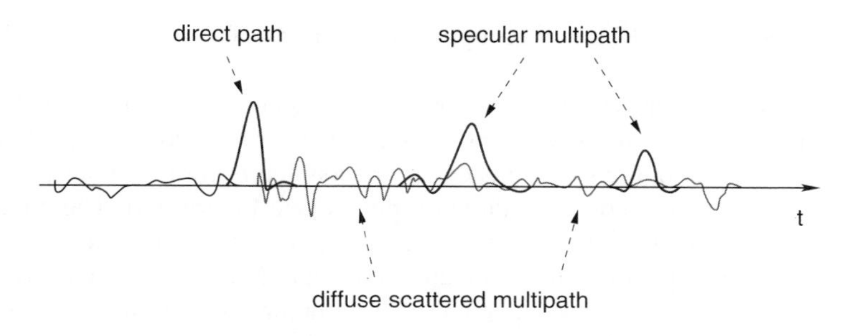

Figure 9.3 Specular and diffuse multipath components.

path and are statistically independent of each other they do not carry any information about the distance (Shen and Win 2010).

It is interesting to note, however, that if some a priori knowledge of the environment is available, it is possible to correlate the TOA of the dominant paths, corresponding to deterministic signal reflections to the mobile's position according to geometric relationships. Secondary paths coming from multiple reflections present weak or difficult-to-predict correlation with position and hence cannot be exploited for positioning purposes.

An interesting model that captures such a phenomenon is that proposed in Meissner and Witrisal (2012). According to this model, the baseband channel impulse response is given by

$$h_0(t, \mathbf{p}) = \sum_{l=1}^{L} a_l(\mathbf{p}) e^{j\phi_l(\mathbf{p})} \delta(t - \tau_l(\mathbf{p})) + \nu(t), \tag{9.10}$$

which consists of L multipath components caused by deterministic reflections with associated position-dependent amplitudes $a_l(\mathbf{p})$, phases $\phi_l(\mathbf{p})$ and delays $\tau_l(\mathbf{p})$, plus a Gaussian random process $\nu(t)$ modeling the diffuse scattered components (see Figure 9.3).

This model has been successfully adopted to investigate multipath-assisted positioning systems and validated using measurements in real scenarios (Witrisal et al. 2016).

9.4 UWB Standards

9.4.1 *IEEE 802.15.4a Standard*

After the FCC allowed the use of UWB signals in the USA, considerable amount of effort was put into the development and standardization of UWB systems (ECMA-368 2005; IEE 2007). This effort resulted in the publication of the IEEE 802.15.4a standard in 2007 (IEE 2007) as a physical layer (PHY) alternative based on UWB to the IEEE 802.15.4 standard (IEEE 2003). Compared to the existing IEEE 802.15.4 standard for wireless sensor networks (WSN), the main interest of the new PHY was in high-precision ranging (sub-meter accuracy) and ultra-low power consumption.

The baseline of the standard is constituted by two optional PHY layers consisting of a UWB impulse radio and a chirp spread interfaces system operating in an unlicensed 2.4 GHz spectrum (IEE 2007). Due to the larger bandwidth, the former is able to deliver high-precision time-based ranging in addition to communication. In particular, the UWB PHY supports a mandatory data rate of 851 Kbit/s, with optional data rates of 110 Kbit/s, 6.81 Mbit/s and 27.24 Mbit/s.

According to the scheme reported in Figure 9.4, input data is first coded through a Reed–Solomon encoder followed by a convolutional encoder with code rate $1/2$. The bits of the systematic and parity outputs of the encoder are coupled to form symbols. Each symbol is associated witha burst of short time duration pulses (see Figure 9.5). The duration of an individual pulse is nominally considered to be the length of a chip whose duration is equal to 2.02429 ns, which corresponds to a bandwidth larger than 500 MHz. A burst is formed by grouping 2^n, $n = 0, 1, 2, \ldots, 9$ consecutive chips. A combination of PPM or, more precisely, burst position modulation (BPM) and BPSK is considered. In particular, the two bits of information comprising each symbol determine, respectively, the position and the polarity (phase) of the burst within the symbol time. A symbol period is divided into two BPM intervals, and a burst occurs in one of the two intervals depending on the value of the first bit.

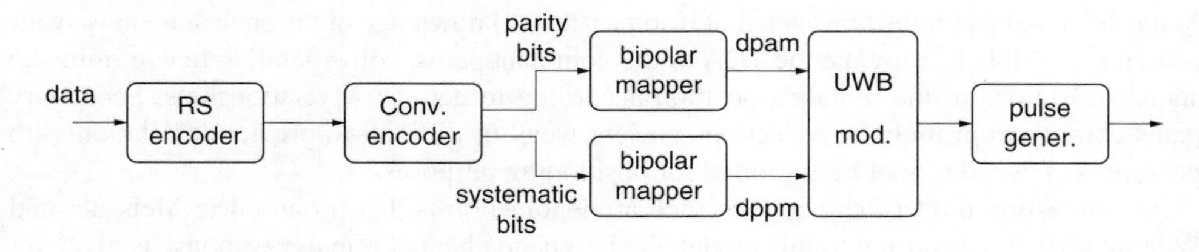

Figure 9.4 IEEE 802.15.4a transmission chain.

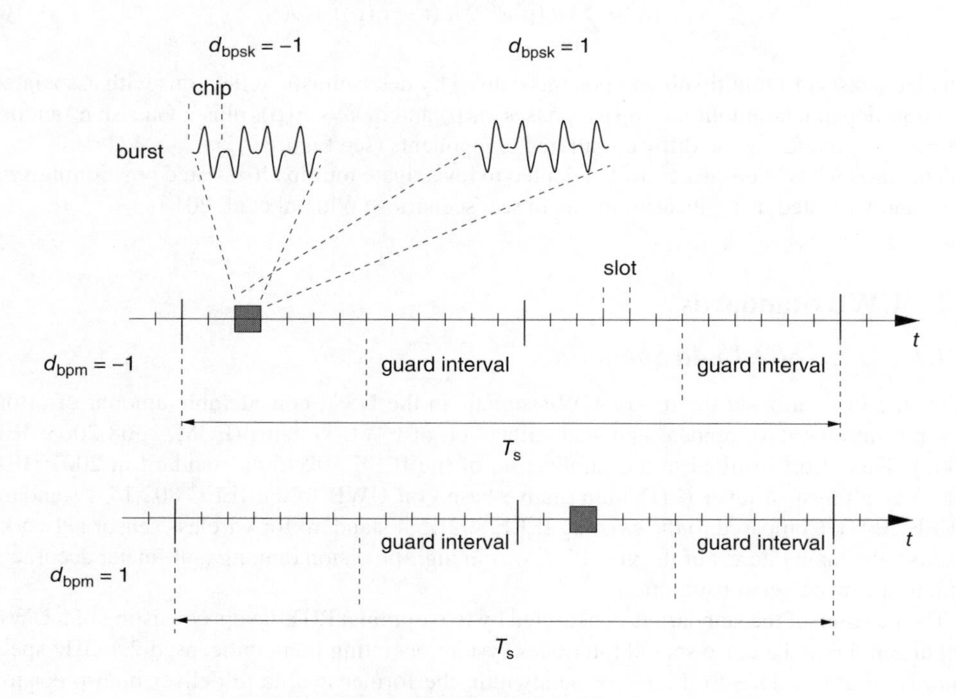

Figure 9.5 IEEE 802.15.4a signal structure.

Each symbol can have 8, 16, 32, 64 or 128 possible slots for a burst inside the BPM intervals. The UWB PHY provides interference suppression through scrambling and time-hopping. In fact, each symbol contains a single burst of pulses which are scrambled by a time-varying scrambling sequence, whereas each burst position in slots varies from one symbol to the successive one according to a time-hopping code. The guard interval prevents inter-symbol interference caused by the multipath.

The combination of both BPSK and BPM signaling allows for coherent and low complexity noncoherent receivers. In fact, while BPM signaling can be easily detected using low-cost energy detectors, the bits associated with BPSK signaling are lost. Nevertheless, the systematic part of the coded sequence of bits associated with the BPM signaling is sufficient to decode the information. Obviously, the performance results are degraded with respect to that obtainable using coherent receivers because of the absence of the convolutional coding gain.

At a higher level, the IEEE 802.15.4a packet structure shown in Figure 9.6 is exchanged between devices. Each packet consists of a synchronization header (SHR) preamble, a physical layer header (PHR) and a data field. The SHR preamble consists of a ranging preamble and a start of frame delimiter (SFD). The ranging preamble is used for acquisition, channel sounding and first path detection, whereas the SFD notifies the end of the preamble. The PHR contains data rate and frame length information, and the data field is the part that carries the communication data. The ranging preamble is the part that is used for range estimation in IEEE 802.15.4a. The main ranging protocol adopted by the standard is two-way ranging (TWR), but it also enables the use of TDOA or other combinations.

The preamble and SFD lengths are specified by the application according to channel conditions and receiver capabilities (coherent/noncoherent). The application bases its choice on figure of merit reports from PHY on how good the range measurement is; longer lengths are preferred for noncoherent receivers to improve the SNR via processing gain.

To optimize the ranging performance, each symbol of the SHR is composed of a ternary sequence with values in $\{-1, 0, +1\}$ taken from a set of eight sequences. Each ternary sequence, and its noncoherent version, has been designed to have an ideal periodic autocorrelation function so that what is observed at the receiver is only the channel response and the detection of the first path is facilitated (IEE 2007; Lei et al. 2006). An example of one out of the eight possible length-31 ternary sequences is the following: $[-000+0-0+++0+-000+-000+-+++00-+0-00]$.

9.4.2 *IEEE 802.15.4f Standard*

The IEEE 802.15.4f Active RFID System Task Group was chartered to define new wireless PHY layers and enhancements to the 802.15.4-2006 standard MAC layer to support active RFID system bi-directional and location determination applications. The IEEE 802.15.4f was finalized and published in 2012 (IEE 2012a). The applications in IEEE 802.15.4f are aimed at port and marine cargo, automotive, logistics, industrial and manufacturing applications.

SYNCH Preamble [16,64,1024,4096] symbols	Start of frame delimiter, SFD [8,64] symbols	PHY header (PHR)	Data field (PSDU)

Figure 9.6 The IEEE 802.15.4a packet structure.

This amendment defines a PHY layer, and only those MAC modifications required to support it, for active RFID (readers and tags). It allows for efficient communications with active RFID tags and sensor applications in an autonomous manner in a promiscuous network, using very low energy consumption (low-duty-cycle), and low transmitter power. The PHY parameters are flexible and configurable to provide optimized use in a variety of active RFID tag operations, including simplex and duplex transmission (reader-to-tags and tag-to-readers), multicast (reader to a select group of tags), uni-cast (reader to a single tag), tag-to-tag communication, and multi-hop capability. The PHY specification supports a large tag population (hundreds of thousands), which may consist of a number of densely populated (closely situated or packed) tags within a single reader field, and supports basic applications such as read and write with authentication and an accurate location determination capability. The communication reliability of the system is very high for applications such as active tag inventory counting or auditing. The active RFID device frequency band(s) used are available world-wide, with or without licensing, and the active RFID PHY is capable of avoiding, or operating in the presence of interference from other devices. The group achieved a common proposal on the PHY which is based on UWB, UHF at 433 MHz and 2.4 GHz.

9.4.2.1 IEEE 802.15.4f UWB Air Interface

The UWB PHY is based on on–off keying (OOK) modulation with 1 MHz pulse repetition frequency. Three symbol mapping modes are foreseen: *base mode* (one chip per symbol), *enhanced mode* (four chips per symbol) and *long-range mode* (64 chips per symbol). In the long-range mode the pulse repetition frequency is set at 2 MHz. The PHY structure is composed of:

- a preamble from 1024 to 8192 pulses, followed by base mode SFD at 1 chip per symbol, and by 16 to 64 symbols encoded as per data
- SFD at 64 chips per symbol encoded as per data
- PHR at 64 chips per symbol encoded as per data
- payload modulated at 64 chips per symbol, Manchester encoded.

A location-enabling information postamble is also defined.

9.4.2.2 IEEE 802.15.4f UHF Air Interface

The UHF interface is in the frequency range 433.05–434.79 MHz and 1.74 MHz width. Three channels centered at 433.34, 433.92 and 434.50 MHz are considered. The modulation is minimum shift keying (MSK) with data rates 250 and 31.25 Kb/s. RSS indicator is used as the location determination mechanism.

9.4.2.3 IEEE 802.15.4f 2.4 GHz Air Interface

The 2.4 GHz air interface is built on IEEE 802.15.4 PHY. It could be used alone for low-precision-RTLS RFID or to provide assistance to UWB PHY. The default channels

0xC5 (1 byte)	Seq. Num. (1 byte)	Tag IEEE ID (8 bytes)	CRC (2 bytes)

Figure 9.7 The ISO/IEC FDIS 24730 12 bytes minimal blink packet.

considered in IEEE 802.15.4f have been chosen in order to not affect Wi-Fi and Zigbee systems. The MSK modulation is used with a bit rate of 250 Kbps. This value is considered as a compromise between range, bandwidth and power consumption.

9.4.3 Other Standards

In 2005 ECMA International released two standards based on UWB, ECMA-368 and ECMA-369, with the purpose of specifying PHY and MAC protocols for high-speed and short-range WPANs (ECMA-368 2005). The transmission technique proposed is multi-band OFDM. Because of the low commercial success of this initiative and since these standards are not of particular interest for positioning, no further details are provided here.

For completeness, it is worth mentioning the IEEE 802.15.6 UWB-based standard, which is intended for short-range wireless communications in the vicinity of, or inside, a human body (IEE 2012b). Its PHY is optimized for medical applications with scalable data rates allowing for tradeoff of range versus data rate. It supports multiple frequency bands of operation and modulation modes. Among them the UWB mode foresees a PHY similar to that of the IEEE 802.15.4a standard.

At protocol level, a dedicated standard, namely ISO/IEC FDIS 24730-62, has recently been published (ISO 2013). It defines the packets format and protocol specifications for TDOA and TWR RTLS based on the IEEE 802.11.4a PHY layer. In Figure 9.7 an example of the minimal blink frame encoding a 64-bit IEEE ID for TDOA networks with only-transmitting tags is reported. The sequence number is an incremental number (modulo 256) which is followed by the 64-bit TAG ID. This field has an unique EUI-64 identifier that is typically stored in the device during manufacture. The default blink period is 3 seconds with random dithering to reduce permanent collisions between mobile nodes (tags). At the PHY layer the default operating mode is channel 5 at 850 Kbit/s.

9.5 Time-of-arrival Measurements

9.5.1 Two-way Ranging

Electromagnetic waves propagate in free space at the speed of light ($c \approx 3 \times 10^8$ m/s), so the distance information (ranging) between a couple of wireless devices can be obtained from the measurement of the propagation delay or time of flight (TOF) $\tau_\mathrm{f} = d/c$, where d is the distance between devices.

The basic method of performing ranging between two nodes A and B is *two-way ranging* (*TWR*). Specifically, node A (see Figure 9.8) emits a ranging request packet to node B, which answers by transmitting back a response packet to node A after a fixed response delay τ_d.

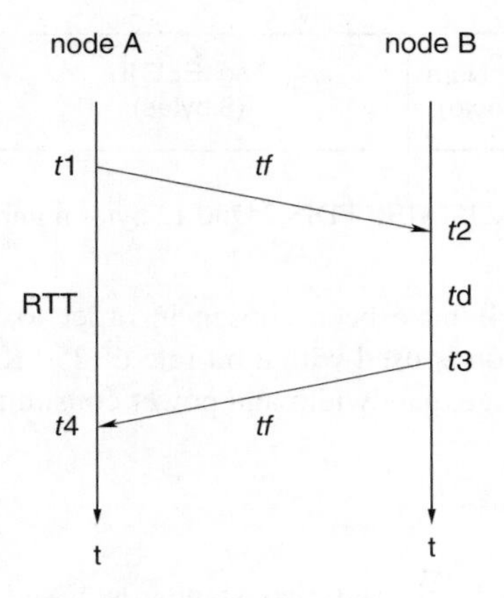

Figure 9.8 Two-way ranging.

Under ideal conditions (perfect estimates and clocks) the distance between A and B can be derived as

$$d = \tau_\mathrm{f}\ c = \frac{RTT - \tau_\mathrm{d}}{2} \tag{9.11}$$

where $RTT = t_4 - t_1$ is the round-trip time (RTT) measured by node A. If several reference nodes are present, multiple ranging measurements can be integrated by the localization algorithm to obtain the position estimate of the target node through multi-lateration (Dardari et al. 2015).

9.5.2 Time Difference of Arrival

In TDOA-based localization systems burst UWB signals are broadcast periodically by the mobile node and received by several reference nodes (anchors) placed in known positions. The anchors share their estimated TOA and compute the TDOA provided that they have a common reference clock. To calculate the position of the mobile node, at least three anchors with known position and two TDOA measurements are required. Each TDOA measurement can be geometrically interpreted as a hyperbola formed by a set of points with constant range-differences (time-differences) from two readers. TDOA localization has the main advantage of allowing only-transmitting mobile nodes thus minimizing the packet exchange and hence the energy consumption. The main disadvantage is that anchors must be kept tightly synchronized through a wired or wireless network with a consequent infrastructure cost increase.

9.5.3 Fundamental Limits in TOA Estimation

In real conditions, it is evident that TWR, TDOA and, in general, all time-based techniques require very accurate TOA estimation as well as precise clocks to mitigate the effect of drifts when measuring time intervals.

Before describing practical schemes to obtain an estimate of TOA, it is important to understand which are the main parameters that determine the fundamental limits in TOA estimation.

Consider the basic scenario where a generic unitary-energy signal $p(t)$, with duration T_p and spectrum $P(f)$, is transmitted and received undistorted with energy E_p through an additive white Gaussian noise (AWGN) channel. The received signal can be expressed as

$$r(t) = \sqrt{E_\mathrm{p}}\, p(t-\tau) + n(t), \tag{9.12}$$

where $n(t)$ is AWGN with zero mean and two-sided power spectral density $N_0/2$. The parameter τ describes the TOA to be estimated based on the received signal $r(t)$ observed over the interval $[0, T)$.

The maximum-likelihood (ML) TOA estimator, which is well-known to be asymptotically efficient, is given by

$$\hat{\tau}_{\mathrm{ML}} = \arg\max_{\tau} \int_0^T r(t)\, p(t-\tau)\, d\tau. \tag{9.13}$$

This equation suggests the typical implementation of the estimator shown in Figure 9.9 based on a filter matched to the transmitted signal $p(t)$ followed by a device which searches the time instant corresponding to the maximum peak of the output signal over the observation interval (Trees 2001).

For high SNR the ML estimate becomes unbiased with variance equal to the Cramér–Rao (CRB), which gives the lower bound on the estimation mean squared error (MSE) of any unbiased estimator $\hat{\tau}$ of τ, that is,

$$\mathrm{Var}\{\hat{\tau}\} = \mathbb{E}\{(\hat{\tau}-\tau)^2\} \geq CRB, \tag{9.14}$$

where

$$CRB = \frac{1}{8\, \pi^2\, SNR\, \beta^2} \tag{9.15}$$

with $SNR = E_\mathrm{p}/N_0$ and β the effective bandwidth of $p(t)$ given by

$$\beta \triangleq \sqrt{\frac{\int_{-\infty}^{\infty} f^2\, |P(f)|^2\, df}{E_\mathrm{p}}}. \tag{9.16}$$

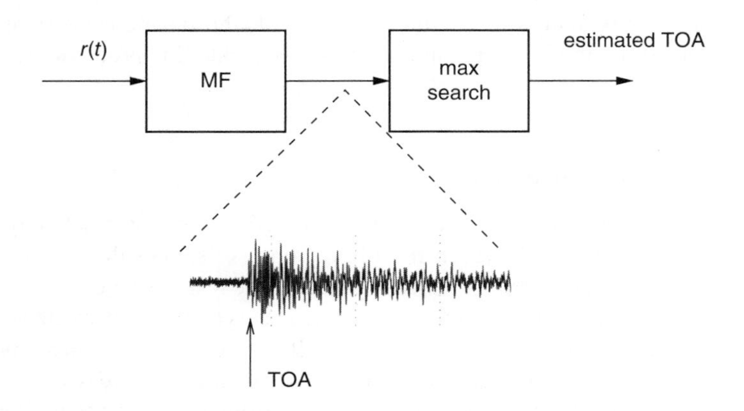

Figure 9.9 ML TOA estimator in AWGN.

Notice that the denominator of Equation (9.15) is proportional to the SNR whereas the proportionality constant β^2 depends on the shape of the pulse through Equation (9.16). Contrary to RSS-based ranging methods, in TOA-based ranging the performance depends on the signal structure. In particular, signals with (ultra-)wide bandwidth or high-frequency components (large β) provide accurate ranging. For instance, according to Equation (9.15) with $SNR = 10$ dB and a signal with effective bandwidth $\beta = 1$ GHz it is evident that centimeter-level accuracy can be theoretically achieved in AWGN. The investigation of the theoretical bounds in different working conditions can be found in Dardari and Win (2009), Decarli and Dardari (2014) and Mallat et al. (2014).

9.5.4 Main Issues in TOA Estimation

In real environments several sources of errors may impair TOA estimation accuracy and hence the localization performance. The impact of these errors has to be carefully examined, and proper methods to mitigate these effects have to be identified (Dardari et al. 2009).

9.5.4.1 Multipath Propagation

Multipath propagation is caused by the superimposition of several replicas of the transmitted signal caused by reflections that arrive at the receiver with different amplitudes and delays.

Without loss of generality consider a scenario in which a pulse $p(t)$ is transmitted through a channel with multipath and thermal noise. In the presence of L multipath components the received signal is

$$r(t) = \sum_{l=1}^{L} \alpha_l \, p(t - \tau_l) + n(t) \tag{9.17}$$

where α_l and τ_l denote the amplitude and delay of the lth path, respectively, and $n(t)$ is AWGN.

The problem is to estimate the TOA $\tau = \tau_1$ of the first path by observing the received signal $r(t)$ in the observation interval $[0, T)$, with T sufficiently large to accommodate all paths. This task can be challenging due to the presence of thermal noise and multipath components that could partially overlap the first path and whose parameters (amplitudes and delays) are unknown. Examples of possible methods to tackle this problem are provided in Section 9.6.

9.5.4.2 NLOS Channel Conditions

When the direct path (DP) between a couple of nodes is completely obstructed by an obstacle (e.g., a wall; Figure 9.10) then the receiver will only observe signals that arrive after having been reflected and/or diffracted by one or more objects. The result is an overestimation of the actual distance. Even when the DP exists and is detected but is partially obstructed by obstacles, a NLOS condition is still experienced and the DP is affected by the *excess delay* phenomenon where the propagation time of the signal depends not only on the travelled distance, but also on the encountered materials. Since the propagation of electromagnetic waves is slower in some

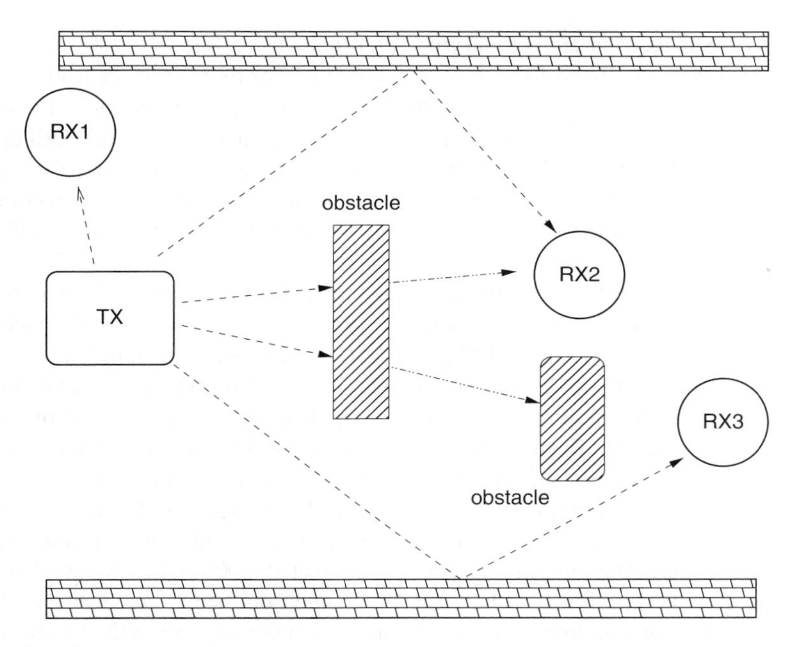

Figure 9.10 Possible LOS and NLOS configurations. RX1, LOS condition; RX2, NLOS condition, no DP blockage; RX3, NLOS condition, DP blockage.

materials compared to air, the signal arrives with excess delay (thus with a positive bias in the range estimate).

A common characteristic to these situations is that the measured range is larger than the true value (positive bias). Such a characteristic can be exploited by the localization algorithm to mitigate the effect of NLOS measurements, as will be described later.

The main approaches that can be pursued to deal with NLOS conditions are to design localization algorithms robust to ranging biases, or to detect the NLOS condition and apply some mitigation method.

9.5.4.2.1 Localization Algorithms Robust to Ranging Biases

Since NLOS measurements are always characterized by positive biases, geometric constraints can be added during the localization process (Venkatesh and Buehrer 2006; Wang et al. 2003).

In general, increasing the number of anchor nodes is beneficial as the constraints in the localization algorithm increase and the final position estimate becomes less sensitive to biased measurements (Jourdan et al. 2008). In a similar way, cooperation between mobile nodes is another option to introduce additional geometric constraints and hence improve the performance in the presence of NLOS conditions (Dardari, Conti, Lien and Win 2008; Mayorga et al. 2007; Pierrot et al. 2006; Wymeersch et al. 2009).

Another approach to mitigate the effect of NLOS measurements is to introduce memory into the ranging system through the adoption of tracking techniques based, for example, on Bayesian filtering (Dardari et al. 2015; Morelli et al. 2007).

9.5.4.2.2 NLOS Condition Detection and Mitigation

NLOS detection is generally performed by extracting one or more features from the received signal that might be correlated to the different channel conditions. Examples of features proposed are root mean square delay-spread, mean excess delay and kurtosis (Decarli et al. 2010; Guvenc et al. 2008; Maali et al. 2009). Once the features have been identified, the NLOS condition identification problem can be modeled according to a classical binary detection scheme and the classical decision theory can be applied by deriving the corresponding likelihood ratio test (LRT).

Nonparametric methods have also been proposed to tackle this problem, such as that studied in Gezici et al. (2003) where the PDF of ranging estimates between nodes is obtained from the measurements and compared to the LOS propagation's PDF. Another nonparametric solution based on least-squares support vector machines can be found in Maranò et al. (2009).

Once the LOS/NLOS condition has been properly detected, a large amount of NLOS mitigation algorithms available from the literature can be considered. Even if measurements in NLOS situations carry some useful information to improve positioning accuracy (Shen and Win 2010), in general it is difficult to exploit it, especially if no a priori knowledge of the environment is available. As a consequence the simplest way to consider the detected LOS/NLOS condition is just to discard NLOS measurements (Chan et al. 2006b). In any case, better results can be obtained by accounting somehow for NLOS measurements. For instance, the NLOS information can be exploited to weight the reliability of measurements in the localization algorithm (Guvenc et al. 2008; Venkatraman et al. 2004) or associate different statistical models with TOA estimates. Other methods can be found in Denis et al. (2003), Li et al. (2006), Chan et al. (2006b) and Schroeder et al. (2007).

9.5.5 Clock Drift

TOA estimation requires precise time measurements with errors in the order of 1 ns or less when centimeter accuracy is required.

In wireless devices, timing is derived from a clock reference provided by an internal oscillator. Numerous physical effects cause oscillators to experience independent frequency drifts, which result in large timing errors. These errors can be particularly significant in low-cost systems equipped with poor oscillators. The local time in a device can be expressed as a function $C(t)$ of the true time t, where $C(t) = t$ for a perfect clock. The device can base its measurements only on an estimate $\hat{t} = C(t)$ of the true time t. For short time intervals, $C(t)$ can be linearized as (Sivrikaya and Yener 2004)

$$C(t) = (1 + \delta)\, t + \mu, \tag{9.18}$$

where δ is the *clock drift* relative to the correct rate and μ is the *clock offset*. A typical measure of the performance of an oscillator is parts per million (ppm), which is defined as the maximum number of extra (or missed) clock counts over a total of 10^6 counts, that is $\text{ppm} = \delta \times 10^6$.

Clock drift affects time interval measurements. Suppose a node wants to measure the time interval τ elapsed between the true instants t_1 and t_2. According to Equation (9.18) its measurement $\hat{\tau}$ is

$$\hat{\tau} = C(t_2) - C(t_1) = \tau\, (1 + \delta). \tag{9.19}$$

Table 9.3 Ranging error with clock drift.

Response delay, τ_d	RTT estimation error	Ranging error
100 μs	0.1 ns	3 cm
1 ms	1 ns	30 cm
10 ms	10 ns	3 m

In a similar way, when the node attempts to generate a delay of τ_d, due to clock drifts it generates an actual delay of

$$\hat{\tau}_d = \frac{\tau_d}{1+\delta}. \tag{9.20}$$

9.5.5.1 Clock Drift in Two-way Ranging Protocols

In TWR protocols both accurate delay generation and measurement are important factors in accurate ranging. Here we analyze the effect of clock drift in the TWR protocol when a couple of nodes, A and B, have clock drifts δ_A and δ_B, respectively. According to Equation (9.20), the effective response delay $\hat{\tau}_d$ generated by node B in the presence of a clock drift δ_B is given by $\hat{\tau}_d = \tau_d/(1+\delta_B)$, whereas the estimated RTT denoted by $\hat{\tau}_{RT}$, according to node A's time-scale, is

$$\hat{\tau}_{RT} = 2\,\tau_f(1+\delta_A) + \frac{\tau_d(1+\delta_A)}{(1+\delta_B)}. \tag{9.21}$$

In the absence of other information, node A can estimate the propagation time $\hat{\tau}_f$ by equating Equation (9.21) with the supposed round-trip time $2\,\hat{\tau}_f + \tau_d$, leading to

$$\hat{\tau}_f = \tau_f\,(1+\delta_A) + \frac{\varepsilon\tau_d}{2(1+\delta_B)} \approx \tau_f + \frac{\varepsilon\tau_d}{2} \tag{9.22}$$

where $\varepsilon \triangleq \delta_A - \delta_B$ is the relative clock offset, having accounted for the fact that typically $\tau_f \ll \tau_d$.

In Table 9.3 the RTT estimation error and the associated error in ranging are reported for different values of the response delay τ_d by considering a relative clock offset of 1ppm corresponding to a high-performance oscillator. Note that a response delay larger than 1 ms, which is a typical value in a real protocol, can significantly compromise the ranging accuracy with respect to the theoretical performance.

RTT measurement accuracy can be improved by adopting high-precision oscillators (not feasible in low-cost devices) or by implementing rapid synchronization techniques at the PHY layer (Jiang and Leung 2007; Verdone et al. 2008; Zhen et al. 2007). Significant improvements can be obtained through the design of smart packet exchange protocols, such as those described in Section 9.6.2.

9.6 Ranging Algoritms in Real Conditions

9.6.1 ML TOA Estimation in the Presence of a Multipath

The transmission of a single pulse in real channels generates a sequence of echoes at the receiver, with different shapes and random delays over tens of nanoseconds, with the possibility of partial overlap that makes TOA estimation in UWB channels a challenging task. An intensive research activity has been carried out in this area.

The logical method to estimate the TOA in the presence of a multipath is to start from the the ML estimator derived for the AWGN scenario introduced in Section 9.5.3. In the presence of a multipath this estimator is no longer a ML estimator due to channel distortion and it has to be redesigned to account for channel parameters.

When channel parameters are unknown, TOA estimation in multipath environments is closely related to channel estimation (Lottici et al. 2002; Win and Scholtz 2002; Zhan et al. 2007b). In this case path amplitudes and TOA are jointly estimated and the ML estimate of the first path $\tau = \tau_1$ translates into a multi-dimensional correlation function maximization. In fact, the received signal $r(t)$ is correlated with test signal templates as follows

$$\hat{\tau} = \arg\max_{\tau} \max_{L,\{\tau_l\},\{\alpha_l\}} \int_0^{\infty} r(t) \sum_{l=1}^{L} a_l\, p(t - \tau_l)\, dt. \tag{9.23}$$

ML estimation can also be performed by adopting super-resolution techniques which operate in the frequency domain (Dumont et al., 1994; Guan et al. 2007; Li and Pahlavan 2004; Zhan et al. 2007). It has been already remarked that ML estimators are known to be asymptotically efficient, that is, their performance achieves the CRB in high SNR regions. Note that channel estimators concentrate their effort in estimating all multipath components. This does not correspond necessarily to the best estimate of the first path. Several sophisticated methods have been proposed, such as maximum entropy methods and spatial analysis techniques like the MUSIC (Schmidt 1986) and ESPRIT (Roy and Kailath 1989) algorithms. Unfortunately the computational complexity of ML or super-resolution estimators and the requirement of high sampling rates limit their implementation even though some solutions to reduce the complexity have been proposed (López-Salcedo and Vázquez 2005). To alleviate these problems several practical sub-optimal TOA estimators have been proposed in the literature. Some of them are illustrated in the following.

9.6.1.1 Threshold-based Estimator

The simplest TOA estimator is that based on threshold detection. It consists of comparing the absolute signal at the output of the matched filter (MF) in Figure 9.9 to a fixed threshold λ. The first threshold crossing event is taken as the estimation $\hat{\tau}$ of the TOA (Dardari, Chong and Win 2008). The choice of threshold is important. With a small threshold, the probability of detecting peaks due to noise, that is, false alarm, and thus estimating the position of an erroneous path that comes earlier than the actual direct path is high. With a large threshold, however, the probability of missing the direct path and thus estimating the position of an erroneous path that comes later, that is, missed detection, is high. Some criteria for properly setting λ are suggested in Dardari, Chong and Win (2008).

When the received signal experiences multipath components partially overlapped with the first path, thresholding approaches might fail due to signal cancellation. Because of this, more sophisticated schemes have been proposed. These algorithms essentially involve the detection of the strongest peak present at the output of a MF and the estimation of its parameters (TOA and amplitude). The estimated contribution of the strongest path is subtracted out from the received signal and a new observation signal is calculated from which the next largest peak is detected. The same process is repeated until the N strongest paths are found. At the end the smallest of the TOA values is selected as the estimate of the TOA of the first path. The parameter N has to be determined with an optimization process and affects the computational complexity of the estimator.

Experimental results have evidenced that, especially at medium and high SNRs, the actual advantage, when present, of sophisticated techniques over the simple thresholding method is modest and in any case TOA estimation errors less than 0.5 ns (<15 cm) can be reached even in the presence of a dense multipath (Falsi et al. 2006).

9.6.1.2 Sub-Nyquist TOA Estimation

A primary barrier to the implementation of ML or MF-based estimators is that they usually require implementation at the Nyquist sampling rate or higher, and these sampling rates can be difficult or costly to achieve due to the large bandwidth of UWB signals.

As an alternative, sub-sampling TOA estimation schemes based on *energy detectors* (*EDs*) have received significant attention in practical implementations (D'Amico et al. 2008; Dardari et al. 2009; Giorgetti and Chiani 2013; Guvenc and Arslan 2006; Guvenc and Sahinoglu 2005a,b,c,d; Guvenc et al. 2006; Rabbachin et al. 2006; Sahinoglu and Guvenc 2006). These TOA estimation techniques rely on the energy collected at sub-Nyquist rate over several time slots. Indeed, they do not require expensive pulse-shape estimation algorithms and represent a good solution for low-power and low-complexity systems.

A typical ED-based TOA estimator scheme is shown in Figure 9.11. The received signal is first passed through a band-pass zonal filter (BPZF) with bandwidth W and center frequency f_c around the signal band to eliminate the out-of-band noise.

The observed signal forms the input to the ED, whose output is sampled at every T_{int} seconds (integration time), thus $K = \lfloor T/T_{\text{int}} \rfloor$ samples (with indexes $[0, 1, \ldots, K-1]$ corresponding to K time slots) are collected in each sub-interval. The true TOA τ is contained in the time slot $n_{\text{TOA}} = \lfloor \tau/T_{\text{int}} \rfloor$. Note that the interval corresponding to the first n_{TOA} energy samples contains noise and possibly interference whereas the interval corresponding to the remaining $K - n_{\text{TOA}} + 1$ samples may also contain echoes of the useful signal, in addition to the noise and interference. Due to the presence of the ED, the integration time T_{int} determines the resolution in estimating the TOA and thus the minimum achievable RMSE on TOA estimation is given by $T_{\text{int}}/\sqrt{12}$.

In the following we describe some techniques proposed in the literature to detect the first path from the vector $\mathbf{z}$ containing the measured energies. For completeness, other techniques are referenced in the survey (Dardari et al. 2009).

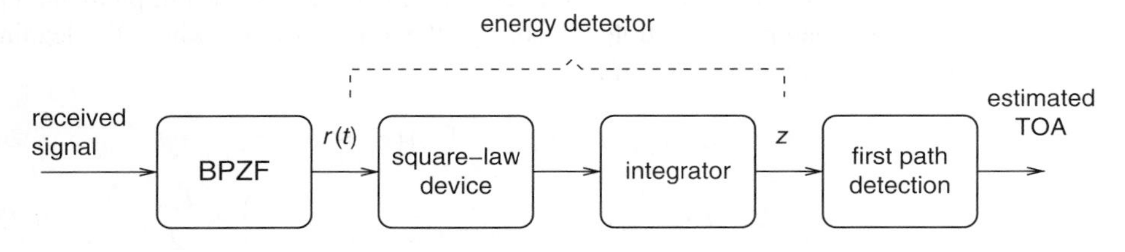

Figure 9.11 TOA estimator with ED scheme.

9.6.1.2.1 First Path Detection

1. Max
   The Max criteria is based on the selection of the strongest sample in **z** (Stoica et al. 2006). This criteria has the advantage of not requiring extra parameters to be tuned according to the received signal property (channel model and/or noise level). However, as will be shown later, it suffers from performance degradation when NLOS propagation reduces the energy of the first path so that the strongest path is not necessarily the first.
2. Peak-Max
   The Peak-Max (*P*-Max) criteria is based on the selection of the earliest sample among the *P* strongest in **z**. The TOA estimation performance depends on the parameter *P*, which has to be optimized according to channel characteristics.
3. Simple thresholding
   Simple thresholding takes an estimate of n_{TOA}, and hence τ, by comparing each element of **z** within the observation interval to a fixed threshold λ. Note that the threshold value must be optimized according to operating conditions and channel statistics. The main advantage of simple thresholding is that it can be implemented completely in analog hardware.
4. Jump back and search forward
   The jump back and search forward (JBSF) TOA estimation scheme is based on the detection of the strongest sample and a forward search procedure (Guvenc and Sahinoglu 2005b). In particular, it assumes that the receiver is synchronized to the strongest path and the leading edge of the signal is searched element-by-element in a window of length W_{sb} samples preceding the strongest one. The search begins from the sample in **z**, with index $k_{\max} - W_{\mathrm{sb}}$, and the search proceeds forward until the sample under test is above a threshold λ within the window (see Figure 9.12). Note that the optimal selection of W_{sb} and λ depend on channel characteristics as well as SNR, both of which are usually unknown a priori.
5. Serial backward search
   In Guvenc and Sahinoglu (2005b) other TOA estimation schemes based on the detection of the strongest sample and a search-back procedure are proposed. One scheme is the serial backward search (SBS), where the leading edge of the signal is backward searched element by element. The search begins from the strongest sample in **z**, with index $k_{\max}$, and the search proceeds sequentially backward until the sample under test goes below a threshold λ within a search back window of length W_{sb} (see Figure 9.12).
6. Serial backward search for multiple clusters
   The received multipath components in typical UWB channels usually arrive at the receiver in multiple clusters that are separated by noise-only samples. In this case, the SBS algorithm may lock to a sample that arrives later than the leading edge. The clustering problem may be handled by allowing a number of *D* consecutive occurrences of noise samples while continuing the backward search (Guvenc et al. 2005). This is the key idea implemented in the Serial backward search for multiple clusters (SBSMC) algorithm, where the leading edge estimation is then modified as follows

$$\hat{\tau} = T_{\mathrm{int}} \cdot \max \Big\{ k \in \{k_{\max}, \ldots, k_{\max} - W_{\mathrm{sb}}\} \big| \, z_k > \lambda \tag{9.24}$$

$$\cap \; \max\{z_{k-1}, \ldots, z_{\max\{k-D, k_{\max - W_{\mathrm{sb}}}\}}\} < \lambda \Big\} + \frac{T_{\mathrm{int}}}{2}. \tag{9.25}$$

In Guvenc et al. (2005) some criteria to calculate suitable values of λ, W_{sb} and *D* are presented.

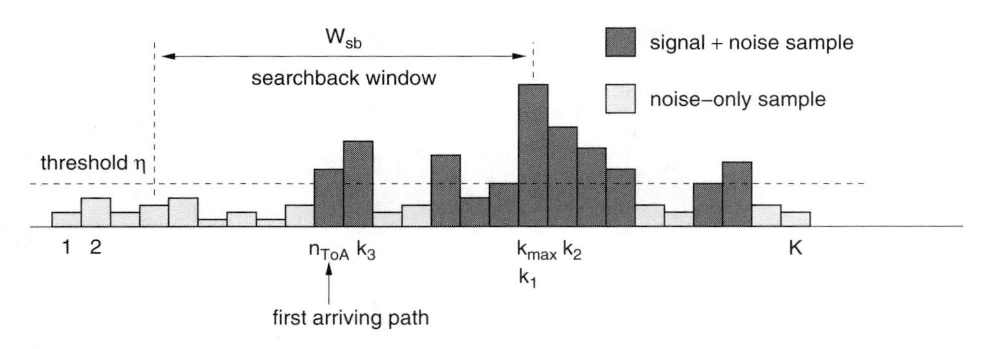

Figure 9.12 Illustration of the Max, *P*-Max, simple thresholding, JBSF, SBS and SBSMC algorithms (from Dardari et al. 2009).

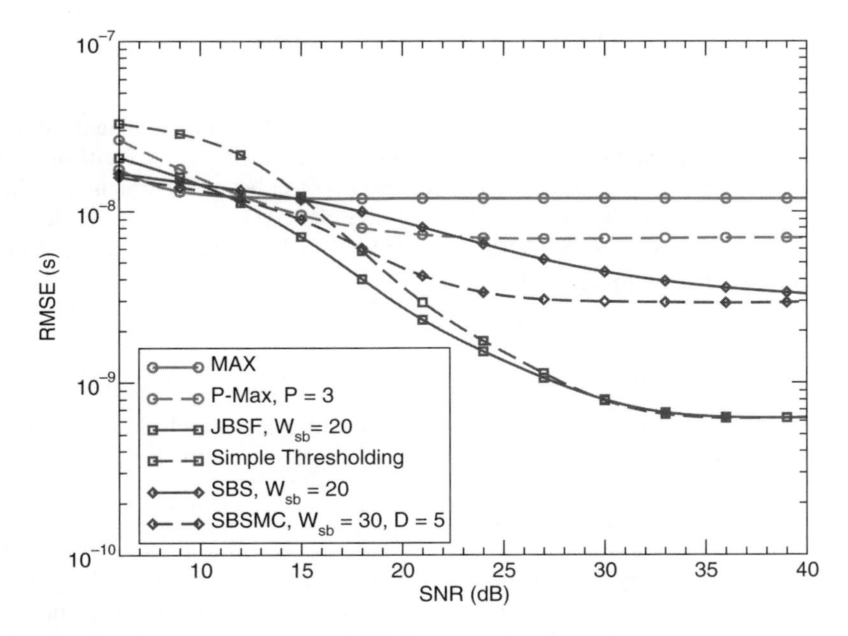

Figure 9.13 RMSE as a function of SNR for different ED-based TOA estimation schemes using the IEEE 802.15.4a channel model (from Dardari et al. 2009).

The performances of the illustrated ED-based TOA estimators are compared in Figure 9.13. We consider dense multipath channels based on the IEEE 802.15.4a channel models (Molisch et al. 2006) and root raised cosine (RRC) band-pass pulse with $f_c = 4$ GHz, and bandwidth $W = 1.6$ GHz. The other system parameters are $T_f = 128$ ns, $T = 120$ ns and $T_{int} = 2$ ns, and the RMSE is evaluated through Monte Carlo simulation. The corresponding RMSE in distance estimation can be obtained by multiplying that of TOA by the speed of light c. Figure 9.13 shows the RMSE as a function of the SNR for Max, *P*-Max, ST, JBSF, SBS and SBSMC. As already mentioned, all these techniques require the proper tuning of parameters (e.g., λ, W_{sb}, D or P) accordingly to channel conditions. In the simple thresholding scheme this problem can be avoided by adopting the sub-optimal criteria given in Dardari, Chong and Win (2008).

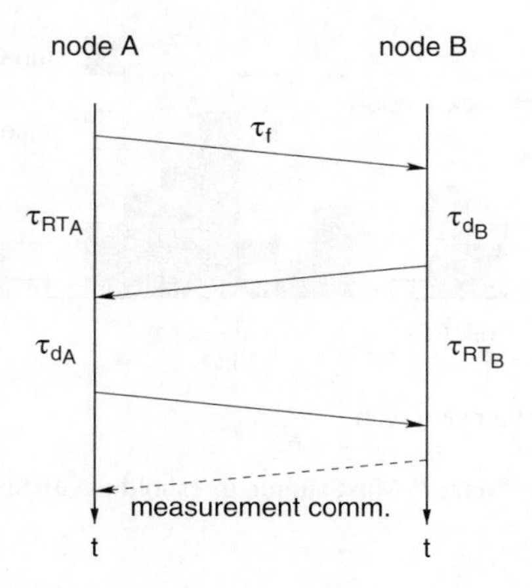

Figure 9.14 SDS ranging protocol.

Note that the Max and *P*-Max schemes exhibit an asymptotic floor higher than the other techniques. In addition, the maximum number of peaks *P* to be considered is a critical parameter to be tuned. The SBS and SBSMC estimation algorithms offer better performance in the intermediate SNR region but exhibit an error floor for high SNR values. The best performance is obtained with either the ST or JBSF scheme. Large SNRs achieve the theoretical performance limit for ED-based TOA estimators, that is, $T_{\rm int}/\sqrt{12} = 0.57$ ns.

9.6.2 Clock Drift Mitigation

From Equation (9.22) it turns out that in the presence of a relative clock drift ε between a couple of nodes A and B performing TWR the error in TOF estimation is

$$\hat{\tau}_{\rm f} - \tau_{\rm f} \approx \frac{\varepsilon\,\tau_{\rm d}}{2} \tag{9.26}$$

which could be significantly large if, for any practical reason, the response delay $\tau_{\rm d}$ cannot be kept at very low values. The sensibility of TOF estimation error to clock drift can be reduced by suitable design of the TWR protocol. In the following section two examples are provided.

9.6.2.1 Symmetric Double-sided TWR

The SDS-TWR consists of a symmetric exchange of TWR packets between nodes A and B, as shown in Figure 9.14 (Hach 2005). Specifically, node A sends a ranging request to B, which responds with an acknowledge (ACK) packet after $\tau_{\rm d_B}$ seconds. Node A does the same in response to the ACK packet after $\tau_{\rm d_A}$ seconds. Both nodes A and B measure the RTT, respectively $\tau_{\rm RT_A}$ and $\tau_{\rm RT_B}$, of each packet's exchange phase. An additional transmission is required to allow node B to communicate its RTT measurement $\tau_{\rm RT_B}$ to node A, so that at least four packets are required to calculate the TOF $\tau_{\rm f}$ at node A.

With perfect clocks it is

$$\begin{aligned}\tau_{\mathrm{RT_A}} &= 2\tau_\mathrm{f} + \tau_{\mathrm{d_B}} \\ \tau_{\mathrm{RT_B}} &= 2\tau_\mathrm{f} + \tau_{\mathrm{d_A}}\end{aligned} \tag{9.27}$$

and hence τ_f is

$$4\tau_\mathrm{f} = \tau_{\mathrm{RT_A}} - \tau_{\mathrm{d_A}} + \tau_{\mathrm{RT_B}} - \tau_{\mathrm{d_B}} \tag{9.28}$$

where the response delays are typically chosen equal, that is, $\tau_{\mathrm{d_A}} = \tau_{\mathrm{d_B}} = \tau_\mathrm{d}$.

In the presence of clock drifts δ_A and δ_B, the measured TOF is

$$4\hat{\tau}_\mathrm{f} = (\tau_{\mathrm{RT_A}} - \tau_{\mathrm{d_A}})(1+\delta_\mathrm{A}) + (\tau_{\mathrm{RT_B}} - \tau_{\mathrm{d_B}})(1+\delta_\mathrm{B}), \tag{9.29}$$

where, without loss of generality, we can write $\tau_{\mathrm{d_B}} = \tau_{\mathrm{d_A}} + \Delta\tau_\mathrm{d}$, being

$$\Delta\tau_\mathrm{d} = \tau_\mathrm{d}\left(\frac{1}{1+\delta_\mathrm{A}} - \frac{1}{1+\delta_\mathrm{B}}\right) \approx \tau_\mathrm{d}\frac{\varepsilon}{2}. \tag{9.30}$$

Therefore the TOF estimation error is

$$\hat{\tau}_\mathrm{f} - \tau_\mathrm{f} = \frac{1}{2}\tau_\mathrm{f}(\delta_\mathrm{A} + \delta_\mathrm{B}) + \frac{1}{4}\Delta\tau_\mathrm{d}\varepsilon \approx \frac{1}{8}\tau_\mathrm{d}\varepsilon^2. \tag{9.31}$$

It is evident that being $\varepsilon \ll 1$ the error in Equation (9.31) is much less than that of conventional TWR in Equation (9.26). For instance, by considering a relative clock drift $\varepsilon = 10^{-5}$ (10 ppm) and $\tau_\mathrm{d} = 5$ ms, which are common values in practical implementations, the corresponding TOF estimation errors are 25 ns (750 cm) and 6.25×10^{-5}ns (1.87 mm), respectively. An improved version of the SDS-TWR protocol can be found in Jiang and Leung (2007).

9.6.2.2 Double Response TWR

The DR-TWR protocol, proposed by Dardari, is reported in Figure 9.15, where node B transmits two consecutive response packets spaced $\tau_{\mathrm{d_B}}$ seconds apart. Node A measures the RTT t_1 from the transmission of the ranging request to the reception of the first response packet, and the time interval t_2 between the two response packets. With respect to the SDS-TWR protocol, here only three packets are sufficient and no additional packets containing the measurements are required to allow node A compute the TOF.

With perfect clocks we have

$$\tau_\mathrm{f} = \frac{t_1 - t_2}{2}. \tag{9.32}$$

In the presence of clock drifts δ_A and δ_B, the actual measured time intervals are

$$\begin{aligned}\hat{t}_1 &= 2\tau_\mathrm{f}(1+\delta_\mathrm{A}) + \tau_\mathrm{d}\frac{1+\delta_\mathrm{A}}{1+\delta_\mathrm{B}} \\ \hat{t}_2 &= \tau_\mathrm{d}\frac{1+\delta_\mathrm{A}}{1+\delta_\mathrm{B}}.\end{aligned} \tag{9.33}$$

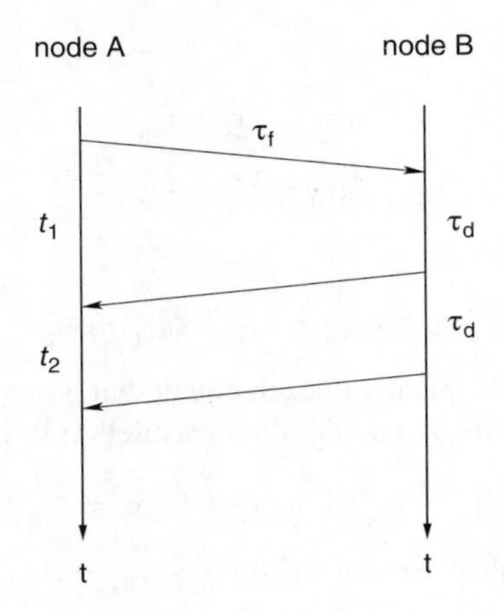

Figure 9.15 DR-TWR ranging protocol.

It turns out that the TOF estimation error is

$$\hat{\tau}_f - \tau_f = \frac{\hat{t}_1 - \hat{t}_2}{2} - \tau_f = \tau_f\, \delta_A. \tag{9.34}$$

Note that τ_f is typically in the order of nanoseconds and δ_A in the order of 10^{-5}–10^{-6}, therefore the TOF estimation error due to clock drifts is negligible and does not depend on the response delay τ_d.

9.6.3 Localization and Tracking with UWB

In this section we illustrate an experimental case study where UWB technology has been used to track a mobile node in a typical indoor environment.

Consider the problem of tracking a mobile node based on real-time UWB range measurements with K_a anchor nodes deployed at known fixed positions $\mathbf{p}_a^{(k)}$, $k = 1, 2, \ldots, K_a$. The state of the system at discrete time n is $\mathbf{x}_n = [\mathbf{p}_{u_n}, \mathbf{p}_a]$, where $\mathbf{p}_{u_n}$ are the positions of the mobile node. In tracking algorithms the objective is to find the marginal posterior distribution $p(\mathbf{x}_n|\mathbf{y}_{1:n})$ of the current state $\mathbf{x}_n$ given all past range measurements $\mathbf{y}_{1:n}$, and hence obtain a point estimate of $\mathbf{x}_{u_n}$.

A recursive and common approach for the evaluation of $p(\mathbf{x}_n|\mathbf{y}_{1:n})$ which requires a constant number of computations at each step is given by *Bayesian filtering*. Its formulation is as follows (Dardari et al. 2015):

- *Initialization*: The posterior marginal at time step 0 is set to the prior $p(\mathbf{x}_0)$.
- *Prediction step*: By exploiting the mobility model $p(\mathbf{x}_n|\mathbf{x}_{n-1})$, the predictive distribution of state $\mathbf{x}_n$ at time step n is given by

$$p(\mathbf{x}_n|\mathbf{y}_{1:n-1}) = \int p(\mathbf{x}_n|\mathbf{x}_{n-1})\, p(\mathbf{x}_{n-1}|\mathbf{y}_{1:n-1})\, d\mathbf{x}_{n-1}. \tag{9.35}$$

- *Update step*: The marginal posterior distribution of $\mathbf{x}_n$, given the new incoming measurement $\mathbf{y}_n$ at time step n (and all past measurements), can be computed using the Bayes' rule and the perception model $p(\mathbf{y}_n|\mathbf{x}_n)$ according to

$$p(\mathbf{x}_n|\mathbf{y}_{1:n}) = \frac{p(\mathbf{y}_n|\mathbf{x}_n)\, p(\mathbf{x}_n|\mathbf{y}_{1:n-1})}{\int p(\mathbf{y}_n|\mathbf{x}_n)\, p(\mathbf{x}_n|\mathbf{y}_{1:n-1})\, d\mathbf{x}_n}. \tag{9.36}$$

The recursive structure of the Bayesian filter is such that as new measurements are received, the previous posterior is propagated and updated using the mobility and perception models, respectively. Once the marginal posterior distribution $p(\mathbf{x}_n|\mathbf{y}_{1:n})$ of the current state $\mathbf{x}_n$ is computed at step n, we can obtain from it, for instance, the maximum a posteriori (MAP) position estimate

$$\hat{\mathbf{x}}_n = \arg\max_{\mathbf{x}_n} p(\mathbf{x}_n|\mathbf{y}_{1:n}). \tag{9.37}$$

We illustrate in the following tracking experimental results using a Bayesian filtering approach, with a PF implementation (Dardari et al. 2015). We also exploited prior knowledge from the mobility and perception models.

9.6.3.1 Mobility Models

A common way to model mobility uncertainties is by Gaussian distributions $p(\mathbf{x}_n|\mathbf{x}_{n-1}) = \mathcal{N}(\boldsymbol{\mu}_{n-1}, \sigma_m^2\mathbf{I})$, where $\mathbf{I}$ is an identity matrix, the variance σ_m^2 accounts for the uncertainty in the movements and $\boldsymbol{\mu}_{n-1}$ depends on $\mathbf{x}_{n-1}$, as will be shown in the following. To avoid forbidden locations (e.g., inside a wall or furniture), truncated distributions for $p(\mathbf{x}_n|\mathbf{x}_{n-1})$ should be considered. We present three examples of mobility models and assess the performance of the tracking scheme using experimental data (Conti et al. 2014).

- *[M0] No mobility*: When the position at each instant is estimated independently of the previous one, tracking reduces to a sequence of independent localization steps.
- *[M1] Mobility without speed information*: In this case, the Gaussian distribution is centered in the previously estimated position, that is, $\boldsymbol{\mu}_{n-1} = \mathbf{x}_{n-1}$.
- *[M2] Mobility with speed measurement*: In this model, the speed and direction of the motion are known because they are estimated from the inertial mobile unit (IMU) by integrating acceleration measurements. In this case, $\boldsymbol{\mu}_{n-1} = \mathbf{x}_{n-1} + \boldsymbol{v}_{n-1}\,\Delta T$, with $\boldsymbol{v}_{n-1}$ being the speed estimated at time $n-1$ and ΔT the time step. The standard deviation σ_m depends on the accuracy of the inertial device.
- *[M3] Mobility with speed learning*: Similarly to model $M2$, the speed is evaluated from previously estimated positions in a sliding window of length N_v, which leads to

$$v_{n-1} = \frac{1}{N_v\,\Delta T}\sum_{i=1}^{N_v}(\hat{\mathbf{x}}_{n-i} - \hat{\mathbf{x}}_{n-i-1}).$$

9.6.3.2 Perception Model

The perception model of the single range measure between the mobile node $k = 1$ and the ith anchor node follows a Gaussian distribution $p\left(y_n^{(i)}|\mathbf{x}_n\right) = \mathcal{N}\left(\left\|\mathbf{p}_{u_n} - \mathbf{p}_{a_n}^{(i)}\right\|, \sigma_\mathrm{p}^2\right)$. The standard deviation σ_p depends on the sensing technology and can be obtained from measurements. The simplest option is to consider σ_p constant.

9.6.3.3 Experimental Results

The experiment was performed in a typical office building located at the University of Bologna, Italy. All measurements were taken automatically by a small robot moving at controlled speeds along a predetermined trajectory. The map of the localization area, the trajectory of the robot, as well as the positions of the anchors are shown in Figure 9.16. The sensing devices that were used in the measurement campaign consisted of FCC-compliant UWB impulse radios, operating with 3.2 GHz bandwidth centered at 4.6 GHz with −12.8 dBm of equivalent isotropically radiated power and an accelerometer. Acceleration measurements composed of three acceleration values, one for each axis, were acquired every 2 ms. The UWB radios provided range estimates based on a TWR TOA protocol. The range measurements between a pair of devices were collected every $\Delta T = 500$ ms. For tracking we used the PF implementation with $N_s = 2000$ particles. The parameters of the mobility models were $\sigma_p = 0.26$ m, $\sigma_m = 0.056$ m (M1–M3 with variable speed) and $\sigma_m = 0.08$ m (M2 with inertial data). More details on the experimental setup can be found in Conti et al. (2014).

The impact of the mobility model in UWB positioning is displayed in Figure 9.17, where the estimated position error CDFs obtained with the mobility models M1–M3 are compared with that obtained when the mobility model was not used (model M0). The corresponding RMSE averaged over the trajectory were, respectively, 110, 85, 38 and 38 cm. The advantage of using the mobility model in tracking in comparison to performing independent localizations

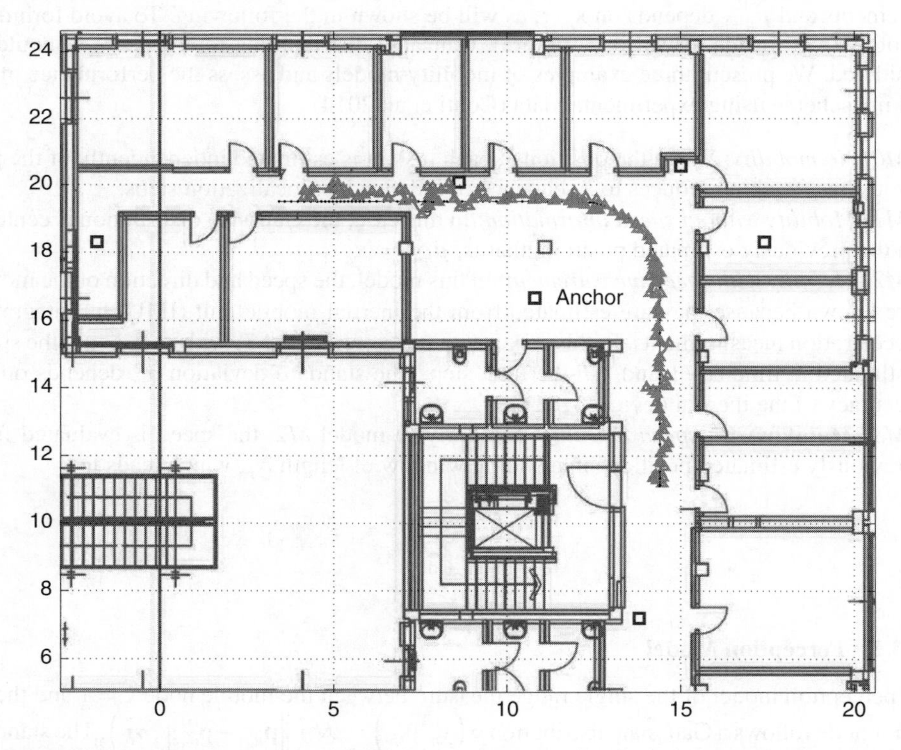

Figure 9.16 True trajectory of the mobile node (solid line), estimated positions of the mobile node (dots) and position of the anchors using M3.

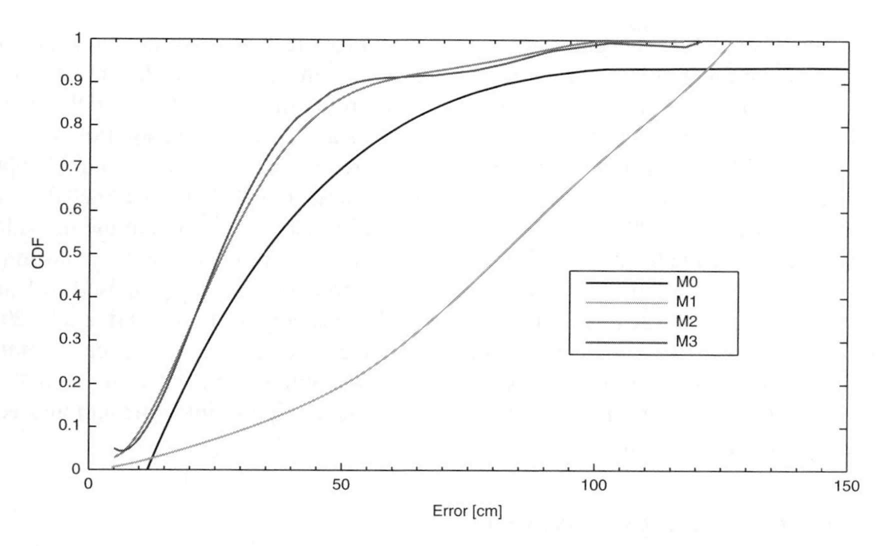

Figure 9.17 Estimated CDF of the position error for the four mobility models considered. $N_v = 18$ in model M3.

(no mobility) is apparent. Furthermore, the fusion of data from the IMU device in M2 and the velocity estimation from past positions allowed the mobility model to approximate the real movement of the node more accurately. A realization of the estimated trajectory with the M3 mobility model can be observed in Figure 9.16.

9.7 Passive UWB Localization

As anticipated in the introduction, besides classical localization involving active and collaborating mobile devices (tags), UWB technology offers the capability to track noncollaborative objects/persons that are not equipped with dedicated devices through UWB multi-static radars or wireless sensor radars (Sobhani et al. 2014). Recently, UWB has been coupled with the RFID technology to detect and track collaborative passive tags (Dardari et al. 2010) even in combination with multi-static radars (Decarli et al. 2014). In the following we illustrate some results related to the passive localization of energy autonomous tags using the UWB backscatter principle.

9.7.1 UWB-RFID

The adoption of RFID technology has experienced a tremendous increase in everyday life as, differently from bar codes, it does not require visibility conditions or any physical contact to work correctly. A typical RFID system consists of one or more readers and many tags applied to the objects to be identified. The reader interrogates the tags via a wireless link to obtain the ID or other data stored on them. Passive RFID tags are the most appealing for the mass market thanks to their low cost and energy consumption. They rely on backscatter modulation to communicate their identification information.

Future advanced RFID systems are expected to provide both identification and high-definition localization of objects with improved reliability and security while maintaining low power consumption and cost. Given the high-resolution ranging capability of UWB signals, the joint use of RFID and UWB technologies represents a very appealing solution to fuse the advantages of RTLS and RFID. Some works and research projects have analyzed this option, especially those considering active UWB-RFID schemes (Dardari 2004; Lazaro et al. 2013a). For example, in Dardari (2004) and Lazaro et al. (2013a) active reflectors at the tag side are exploited to reinforce the backscattered signal. The strict constraints on energy consumption make solutions based on chipless and passive tags more appealing, as proposed in Dardari et al. (2010), Arnitz et al. (2010), Zou et al. (2012), Ramos et al. (2013), Guidi et al. (2014a) and Guidi et al. (2014b). Due to the longer operating range and better management of multiple tags, tag-reader communication performed by the modulation of the backscatter signal at the tag side seems better suited for localization purposes than chipless implementations, as will be detailed in the next section.

9.7.1.1 The UWB Backscatter Principle

Backscatter modulation is based on the idea that the tag's antenna reflection properties are changed according to information data. This property is currently adopted in traditional passive ultra-high frequency (UHF) RFID tags based on continuous wave (CW) signals to carry information from the tag to the reader (Chawla and Ha 2007). The same idea has been proposed in Dardari et al. (2010) using UWB interrogation signals instead of CW signals to exploits the benefits offered by UWB in terms of TOA resolution.

With reference to Figure 9.18, when an UWB pulse is transmitted and encounters an antenna, it is partially reflected back depending on antenna configuration. The antenna scattering mechanism is composed of structural and antenna mode scattering which are plotted separately for convenience in the figure. The structural mode occurs owing to the antenna's given shape and material, and is independent from how the antenna is loaded. On the other side, antenna mode scattering is a function of the antenna load; when in short circuit the backscattered pulse is characterized by a sign inversion with respect to the open load condition (see Figure 9.19). This property can be exploited to establish a communication link between the tag and the reader. Moreover, by estimating the RTT of the reflected signal with respect to the interrogation signal timing it is possible to infer the distance between the reader and the tag with high accuracy. Unfortunately, signals scattered by the surrounding environment (clutter) as well as the antenna structural mode dominate the signal received by

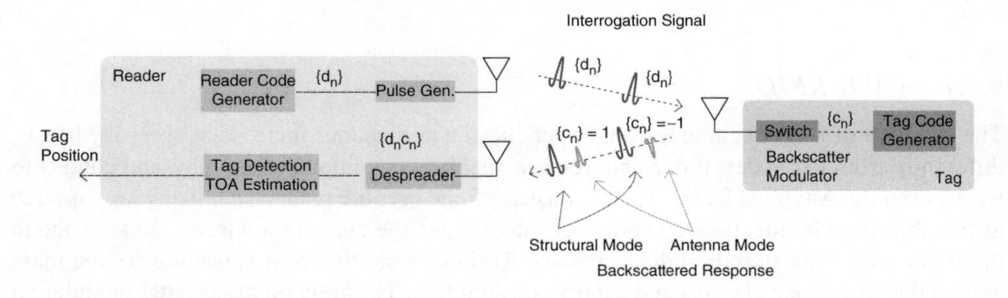

Figure 9.18 UWB-RFID backscattering scheme.

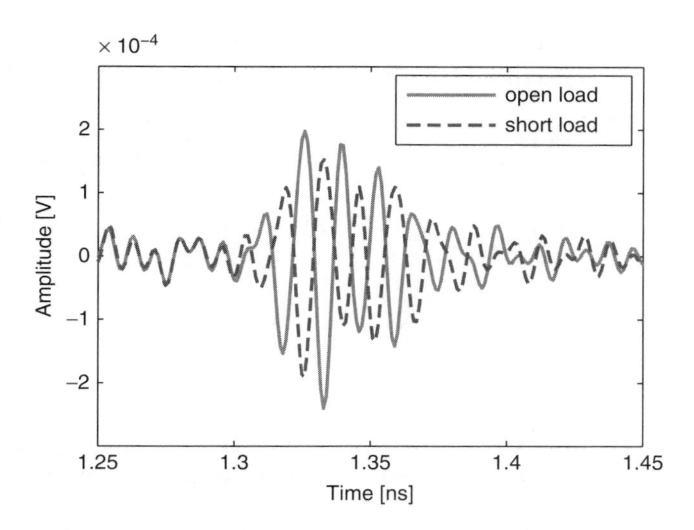

Figure 9.19 Example of measured backscattered signal for tag open and short circuit loads (only the antenna mode component is shown).

the reader, thus making the detection of the antenna mode component (which carries data) a main issue in passive UWB RFID. This requires a proper design of the signaling scheme, as suggested in Dardari et al. (2010). In particular, consider the following interrogation signal, composed of a coded sequence of UWB pulses $p(t)$, emitted by the mth reader

$$s(t) = \sum_{n=0}^{N_c-1} d_n^{(m)}\, p(t - n\, T_c), \tag{9.38}$$

where T_c is the chip time, or pulse repetition period, $\left\{ d_n^{(m)} \right\}$, $n = 0, 1, \ldots, N_c - 1$, is the (spreading) bipolar code identifying the mth reader with length N_c.

Each tag is designed to alternate its antenna load status at each chip time T_c, according to a bipolar code $\left\{ c_n^{(k)} \right\}$, for $n = 0, 1, \ldots, N_c - 1$. This ensures the creation of a unique backscattered signal signature allowing the discrimination of the kth tag among other tags (generating the multi-tag interference) and clutter.

For the sake of explanation, suppose the clock of the tag is synchronous with that of the reader, then the signal received back by the reader is

$$r(t) = \sum_{n=0}^{N_c-1} d_n^{(m)} c_n^{(k)}\, w(t - n\, T_c) + \sum_{n=0}^{N_c-1} d_n^{(m)}\, w_C(t - n\, T_c) + n(t) \tag{9.39}$$

where $n(t)$ is the AWGN, $w(t)$ is the reader-tag-reader channel response to $p(t)$ (antenna mode only), and $w_C(t)$ is the clutter obtained from each emitted pulse. As can be noticed in Equation (9.39), the first component appears as a UWB sequence spread using the combined code $\left\{ d_n^{(m)} c_n^{(k)} \right\}$, whereas the second term representing the clutter appears as a UWB sequence spread using the reader code $\left\{ d_n^{(m)} \right\}$. This difference allows for discrimination between the useful signal coming from tag k and the clutter (or the interference from other tags). In fact, if at the reader's receiver the signal is de-spread using the combined code $\left\{ d_n^{(m)} c_n^{(k)} \right\}$ and the tag's code $\left\{ c_n^{(k)} \right\}$ is designed so that it is balanced, then clutter can be (in principle) removed as evident from the following expression describing the de-spreading operation

$$
\begin{aligned}
Y &= \int_0^{(N_c-1)T_c} r(t) \cdot \sum_{n=0}^{N_c-1} d_n^{(m)} c_n^{(k)} \, w(t - n\, T_c) \, dt \\
&= N_c E_w + E_w \sum_{n=0}^{N_c-1} d_n^{(m)} c_n^{(k)} \, d_n^{(m)} + z(t) \\
&= N_c E_w + z(t)
\end{aligned}
\tag{9.40}
$$

since $d_n^{(m)} \, d_n^{(m)} = 1 \, \forall n$, $\sum_{n=0}^{N_c-1} c_n^{(k)} = 0$ by design, and where E_w is the energy of $w(t)$ and $z(t)$ is the thermal noise at the output of the de-spreader. From Equation (9.40) one can note that the presence of the kth tag can be detected if the intensity $N_c E_w$ of the useful signal at the output of the despreader is sufficiently high with respect to the thermal noise. This basic protocol can be easily extended by including data payload in the backscattered signal (Dardari et al. 2010). By combining range measurements from multiple readers it is possible to localize the tag.

9.7.1.2 Practical Implementations of UWB RFID Systems

The basic passive communication and ranging scheme illustrated in the previous section relies on some simplifying assumptions that are not met in real scenarios. In practical implementations some important issues have to be addressed.

First of all, because of the two-way link, the path loss can be very large even at short distances and only partially compensated for by increasing the transmission power because of the conservative regulation limitations in the UWB band. As can be seen in Equation (9.40), the SNR can be improved by increasing the number N_c of pulses associated with each interrogation signal at the expense of a longer identification time.

Another aspect is that tags are passive devices so no front-end is present inside to extract synchronization information from the UWB interrogation signal. The main consequence is that tags are completely asynchronous. This problem can be solved by a proper design of the spreading code and the introduction of side synchronization signals (e.g., in the UHF band) to provide a coarse synchronization (Decarli et al. 2016). To facilitate code acquisition at the receiver, in Guidi et al. (2014b) (quasi)-orthogonal codes allowing multiple tags operating simultaneously with limited reciprocal interference have been investigated.

Finally, although UWB backscattering is characterized by an extremely low power consumption since only the UWB switch is required to perform signal modulation, the circuitry at tag side (control logic in addition to the UWB switch) must be properly powered so that some limited source of energy is needed (e.g., energy harvesting). Due to the impossibility of extracting significant energy from the UWB link because of regulatory constraints, a solution is to exploit the UHF band to transfer the required energy to the tag as done in Gen.2 RFID (Decarli et al. 2016).

These issues have been tackled in some research projects, leading to prototypes providing proof of concept of the UWB-RFID technology. An example of the implemented UWB-RFID system realized in the context of the EU project SELECT for precise luggage sorting in airports is depicted in Figure 9.20 (Guidi et al. 2016). To simplify the processing at the reader side, a ED-based receiver has been designed as a trade-off between complexity and ranging performance. In this set up it was demonstrated that it is possible to detect and sort tags up to distances of 4.5 m, with errors within 30 cm.

In the context of the Italian project GRETA, the energy efficiency and eco-compatibility aspects of UWB-RFID systems have been investigated (Decarli et al. 2015, 2016). Coexisting UHF and UWB interrogation signals have been considered to transfer the energy to the tag and provide some synchronization reference. In Figure 9.21 an example of tag architecture is

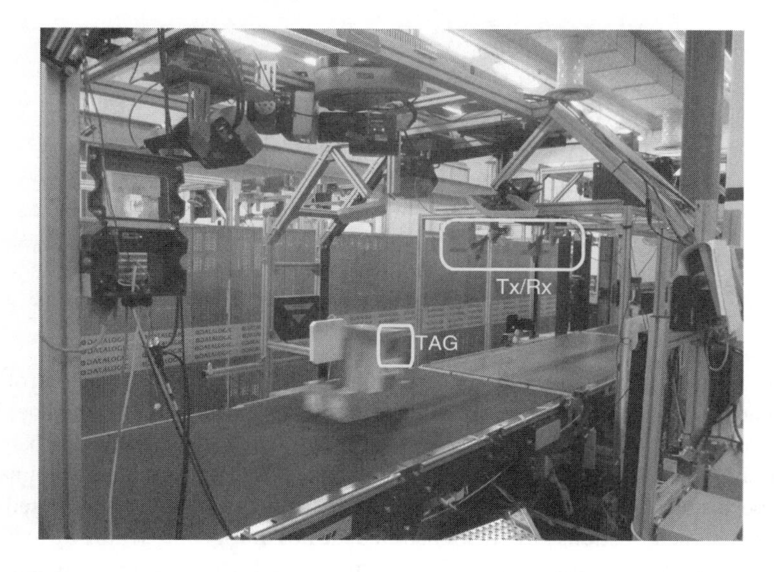

Figure 9.20 Luggage sorting on a conveyor belt using UWB-RFID tags (Guidi et al. 2016).

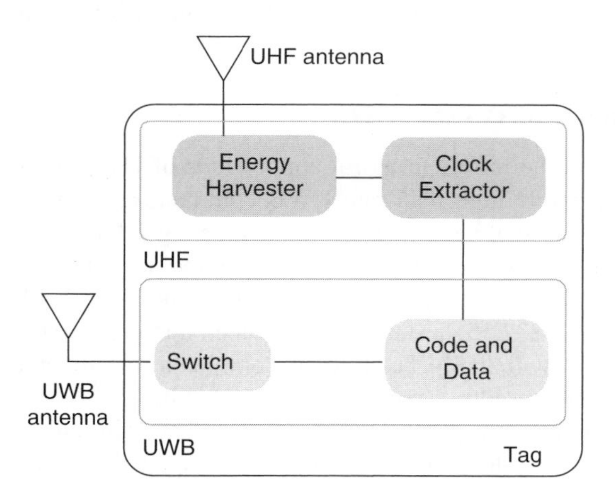

Figure 9.21 UWB-RFID tag with energy harvesting in the UHF band.

Figure 9.22 Single-port UHF/UWB antenna prototype on paper substrate (Fantuzzi et al. 2015).

reported in which the energy necessary to wake up the tag, and power the control logic and the UWB backscatter modulator is harvested from the UHF link. Due to the coexistence of UHF and UWB bands and to reduce the tag's dimensions, a dedicated dual- band antenna has been designed, as shown in Figure 9.22 (Fantuzzi et al. 2015). Such an antenna provides quite good performance also in the UWB band considering it is built on paper-based substrate. Experimental characterizations have demonstrated the possibility of reaching up to 6 m with 15 cm ranging accuracy. In Decarli et al. (2016) alternative architectures are discussed, including those in which the UWB link is exploited only for ranging and localization purposes while the UHF section is composed of a standard Gen.2 UHF tag for backward compatibility.

9.8 Conclusions and Perspectives

We have discussed that the most successful applications of UWB are those related to positioning and tracking, while high data rate UWB communication has not had a significant commercial impact because of better performing competing solutions. After initial slow market penetration, mainly caused by the high cost of proprietary devices and the fragmented power emission mask worldwide regulations, the availability since 2014 of low-cost chip sets compliant with the IEEE 802.15.4a standard has shaken up the market.

In the future, UWB will likely be used in combination with other technologies. The UWB-RFID solution represents a recent example of this, and is expected to take on an important role in next-generation RFID systems. Another solution is represented by joint localization and mapping systems for automatic indoor navigation (Witrisal et al. 2016).

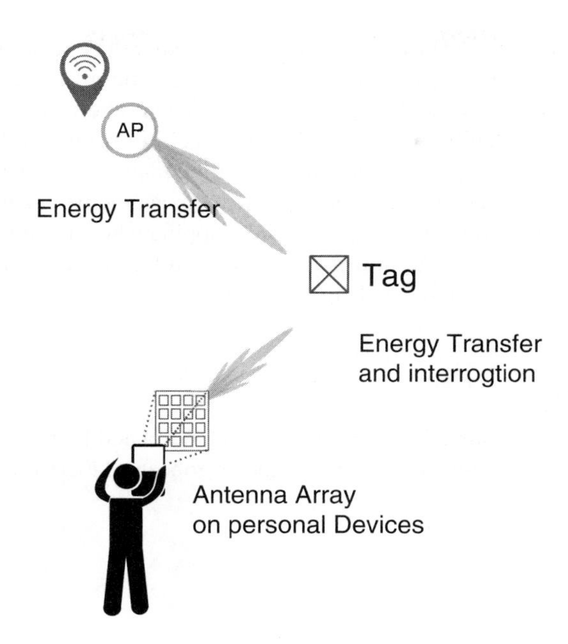

Figure 9.23 Energy transfer mechanism to localize and energize passive tags using mmW/THz massive antenna arrays.

One of the key technologies for future UWB systems is represented by millimeter wave (mmW) and, in the distant future, by the THz band (Pursula et al. 2008). The incoming 5G smartphones will integrate mmW interfaces to boost the communication data rate beyond 1Gbps. Apart from the extremely large bandwidth available at such frequencies, which is beneficial for accurate positioning, mmW technologies also offer other interesting opportunities. First of all, the small signal wavelength (5 mm at 60 GHz) allows hundreds of antenna elements to be packed into a small area, even integrable into smartphones. Such a large number of antenna elements permits a near-pencil beam that is electronically steerable. Thanks to the extremely narrow beam formed, focusing the power flux towards the mobile tag and transferring the energy at several meters will become possible with much higher efficiency than that achievable with today's technology. There will therefore be the possibility of energizing, detecting and localizing tagged items using smartphones at a distance pf several meters, enabling augmented reality applications (see Figure 9.23). No dedicated infrastructure would be required as large antenna arrays already used for communication could be employed as reference nodes with the advantage of permitting both accurate TOA and AOA estimates.

Another possible scenario is that in which access points deployed for indoor communications and equipped with mmW antenna arrays localize the tags present in the surrounding environment and focus the beam to allow the tag to accumulate the energy (e.g., during the night) so that they will be operative whenever a mobile device interrogates them, as shown in Figure 9.23.

In Guidi et al. (2015) the idea of *personal mobile radar* has been suggested, in which narrow beams with electronic-driven steering capabilities could be used for high-accuracy mapping of the environment as a cheap and energy-efficient alternative to laser technology. In fact, the pervasive availability of personal radar capabilities on smartphones will enable high-definition, zero-effort and automatic indoor building mapping by exploiting the possibility of *crowd mapping*. In addition, other new applications would be enabled, such as aids for the visually impaired or pictorial 3D environmental mapping where the third dimension results from integration of the angle/range information collected by the personal radar with the 2D picture coming from a conventional embedded camera.

Acknowledgments

The author would like to thank Nicoló Decarli, Anna Guerra and Francesco Guidi. This work has been performed within the framework of the European H2020 Project XCycle (Grant 635975).

10

Indoor Positioning in WLAN

Francescantonio Della Rosa[1,2], Mauro Pelosi[1] and Jari Nurmi[2]
[1]*Radiomaze Inc, Cupertino, California, USA*
[2]*Tampere University of Technology, Finland*

10.1 Introduction

In the family of the signals of opportunity (SoO) for location-based services (LBSs), wireless local area network (WLAN) signals are the most suitable, being promising but also the most challenging to exploit. Literature proposes terms like WLAN, IEEE 802.11, WiFi and Wi-Fi to define this technology, which is adopted not only for communications but also to estimate the location of mobile devices present in the coverage area (Rainer 2012). The location estimation is provided in several ways, making the end users part of the localization process either voluntarily or involuntarily. Without doubt this is one of the most commonly adopted technologies for positioning or tracking people indoors, seamlessly using 2.4 or 5 GHz bands. Almost every mass market mobile device is equipped with wireless fidelity (WiFi), providing access to geo-located information to everybody and covering almost every building, shopping mall, office, airport, civil and military area with deployed WiFi hotspots. Free WiFi hotspots are becoming a must in public places and the WiFi guest experience is usually linked to additional services for both business and end users. The widespread use of WiFi also makes it a mainstream technology for localization (Liu et al. 2007). LBSs relying on WiFi for positioning or passive tracking, exploiting different methods and techniques adopted not only in experimental activities for academic purposes but also in commercial applications targeting both business to business (B2B) and business to consumers (B2C).

There are two main categories of indoor positioning systems: mobile-based and network-based. In the first the mobile station (MS) gathers data from nearby access points (APs) and processes the location estimation locally. In the second the APs gather the required signals coming from MSs and the location estimation is performed in the network (or in the cloud), making the approach also suitable for passive tracking (Della Rosa et al. 2012). The exploitation of WLAN signals is a tempting approach for researchers and service providers. The WLAN range goes up to 100 m, outstripping that of Bluetooth, Zigbee, radio frequency

Mobile Positioning and Tracking: From Conventional to Cooperative Techniques, Second Edition.
Simone Frattasi and Francescantonio Della Rosa.

identification (RFID) or other dedicated sensors. The use of WLANs for localization does not necessarily need line of sight (LOS) as walls and other environmental impairments provided by static objects can be overcome with mathematical models implemented with off-the-shelf hardware (Rainer 2012).

When the signal strength of nearby hotspots is exploited, a solution for estimating the target location is to use models relating the strength of the received radio signal either to the distance between the MS and the signal emitter or to the location of the MS directly (Della Rosa et al. 2012). These so-called radio frequency identification (RSS)-based positioning methods can be divided into three main categories: cell identifier (cell ID), fingerprinting and path-loss. Although the research literature proposes different approaches to measure the signals, for consumer market applications the RSS observations are considered to be more easily available than angle- or time-based ones as the RSS can be passively listened to from both the WLAN infrastructure or mobile devices, simply by exploiting software development kit (SDK) or native commands for network diagnostics, without adding any extra load to the network.

According to the IEEE 802.11 standard, APs periodically broadcast beacon frames while mobile devices send probe requests containing information for network identification, broadcasting network capabilities, and for other control and management purposes (Della Rosa et al. 2012). In particular, the MS can record information from any beacon it receives while APs can collect data from passing mobile devices. This process is usually performed to determine the AP with the best link quality and to allow MSs to easily connect to already known networks (Della Rosa et al. 2012). Hence, the ease of gathering and exploiting network information paves the way for endless mass market applications.

Indoor positioning needs are so important that the Federal Communications Commission (FCC) NPRM 14-13 has proposed improving the performance of positioning solutions to locate users' phones after emergency calls because "The majority of 911 calls come from wireless phones" located indoors, where conventional satellite-based approaches do not work. Specifically, the new FCC rules propose location accuracy requirements for indoor wireless 911 calls, as well as revisions to the existing E911 rules (FCC-14-13 2014). In addition, 80% of mobile phones are used indoors and emergency calls are made by WiFi-equipped smartphones. This problem has arisen because of the low accuracy standards agreed for outdoor scenarios, which recognize the importance of not only bi-dimensional, but also vertical accuracy, enabling localization possibilities across multiple floors (Divis 2014).

In this chapter, in addition to a description of the conventional approaches and characteristics of WLAN positioning, we describe how human-induced errors should not be ignored when performing experimental activities for indoor positioning using RSS. We show that phenomena like hand grip and body loss have a meaningful effect on conventional positioning methods, highly compromising the positioning accuracy. Although environmental and human-induced perturbations generate systematic errors, if correctly accounted for and cognitively exploited, rather than roughly discarded or mitigated, it might be possible to minimize these and obtain beneficial effects in terms of positioning accuracy (Della Rosa et al. 2011a).

10.2 Potential and Limitations of WLAN

The literature proposes different methods for indoor positioning, where the main concern refers to the accuracy of the proposed solutions compared to previously obtained results. However, more often the claims are not backed by the real potential use and usability for the final stakeholder. In fact it is not yet clear if focusing on accuracy enhancements is a more pressing need compared to solutions targeting a broader use of them.

Although it is common knowledge that indoor positioning based on WLAN (e.g., already deployed WiFi hotspots) has the great advantage of exploiting an already deployed infrastructure, lowering the required capital investment for a basic LBS, the overall burden of deploying, calibrating, delivering and maintaining such a solution on a global scale is still quantitatively unclear. Hence, it is well known that the main benefit of WLAN positioning is the leverage of existing assets exploiting the concept of the unintended use of technology. If a compromise in terms of accuracy requirement is achieved, minimal information about location can be obtained even with one AP, and equipping every mobile device with WiFi makes the potential of LBSs grow exponentially.

However, the inner and simplistic nature of the solution has some limitations while at the same time end users pretend systems and services ideally trespass the physical limits of the available technology. One of the main limitations is given by the inner nature of the RSS itself, as it was not meant to be exploited for positioning but for communication purposes and network diagnostics. The RSS is obtainable almost everywhere, but at the same time it is highly sensitive to the surrounding environment and people presence so the LBSs based on this approach suffer from both short-term and long-term environmental changes (Della Rosa et al. 2011a).

10.3 Empirical Approaches

Empirical approaches for indoor positioning (Figure 10.1) rely on the empirical evidence of hidden behavioral phenomenon characterizing measured signals, data and propagation environment. These approaches have the benefit of gaining massive knowledge of previously unknown effects by means of direct and indirect experimental observations. The collection and analysis of direct observations from experimental activities offer the possibility of empirical evidence that can be quantitatively and qualitatively analyzed. A plethora of practical approaches ise proposed in the literature (e.g., Della Rosa et al. 2012) although a concrete path to a more formal standardization is not yet in place. While literature provides exhaustive theories and models based on assumptions and hypotheses now part of the common knowledge for the research community, empirical observations have the potential to reveal hidden and fresh answers, giving out results only obtainable through experimentation and paving the way to an exponential domino effect of novel considerations and assumptions in the location-based research field (Goodwin 2005). Recently, the effect of human bodies on experimental activities was addressed and the research community has started to describe and quantify the evidence of interactions between humans, devices and environment, which cause unpredictable behavior in measurements as well as jeopardizing the delivery of robust and reliable indoor positioning solutions. With the proliferation of smart devices (smartphones, tablets, smart TVs, smart appliances, etc.) equipped with WiFi, the opportunities to position and track people in any indoor environment become endless and the commercial exploitation of this is gaining major importance from a business perspective, focusing research-oriented empirical approaches more and more on solving real-world problems.

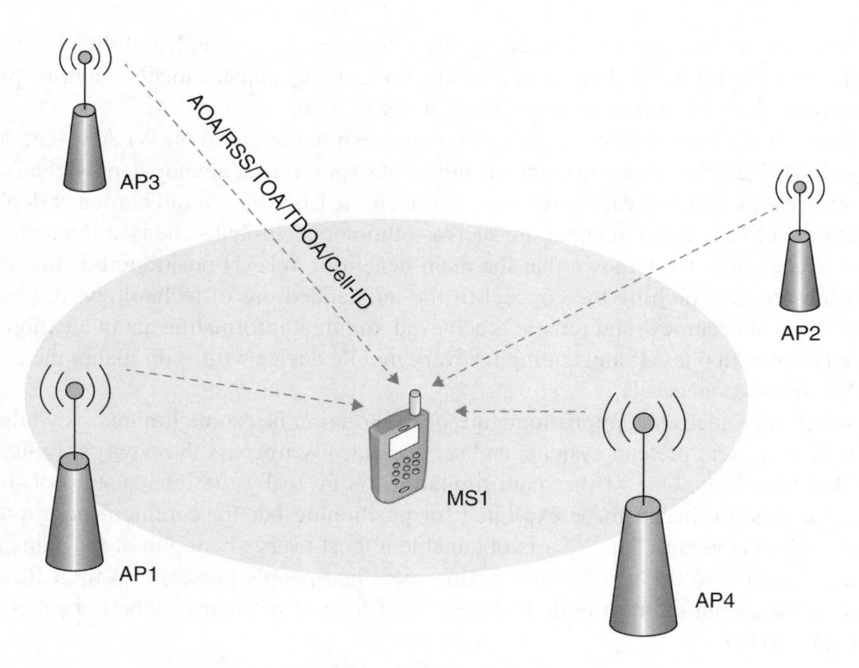

Figure 10.1 Empirical scenario (Della Rosa et al. 2012).

10.3.1 Probe Requests and Beacon Frames

When a WiFi-enabled mobile device (e.g., smartphone, laptop, internet tablet) needs to connect to a nearby WiFi network two solutions can be adopted: (1) scanning and looking for beacon frames broadcast by APs and finding for a known network or (2) periodically broadcasting probe requests (usually broadcasted by smartphones) containing not only information about the medium access control (MAC) address of the device, but also names of known networks, allowing a faster connection to the WLAN service (Keeble 2014). Both beacon frames and probe requests represent a gold mine for LBS providers.

The technical differences between these two sources of information characterize their suitability for different applications (Figure 10.2).

- *Beacon frame*: Every AP periodically sends beacon frames to let nearby clients know about its presence in the covered area. The broadcast frame announces AP information like Basic Service Set Identifier (BSSID), Service Set Identifier (SSID), and the timestamp to nearby network interface cards (NICs), while radio NICs periodically scan the channels analyzing the beacon frames to select the closest AP to connect to. MSs are able to measure the RSS reported in decibel-milliwatts (dBm), but the measured value might vary according to vendors and the hardware characteristics peculiar to the MSs.
- *Probe request*: The probe request is sent by stations to obtain information from other stations. Typically NICs send probe requests to detect APs within range and check if the nearby ones are already in its recorded list of known networks.

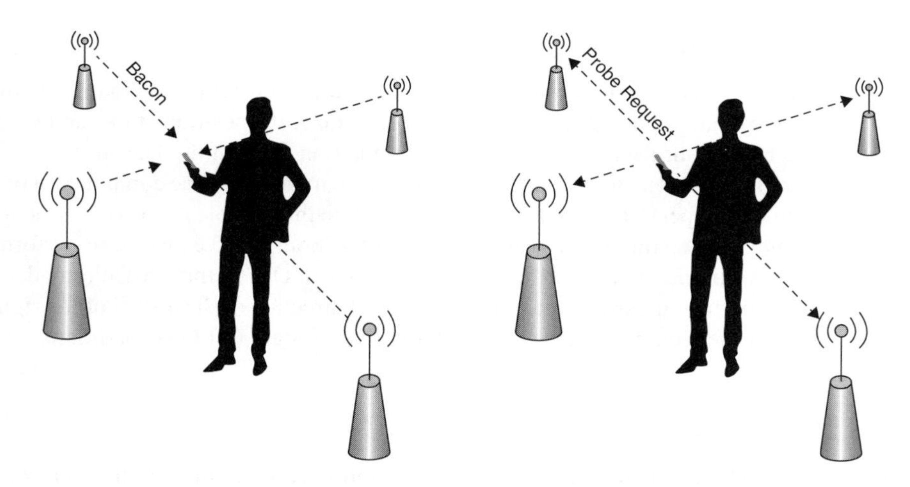

Figure 10.2 Broadcasted frames (Della Rosa et al. 2012).

Thanks to progress of technology, indoor positioning in WLAN has been exploited for a plethora of business applications and services, and most capital investments for indoor positioning solutions that exploit already available WLAN infrastructure mainly optimize internal processing and business analytics or deliver better and novel services to end users. With the explosion of RSS-based applications exploiting beacons or probes, in addition to real time locating systems (RTLSs), a new breed of indoor positioning solutions is growing in importance. It is represented by the passive tracking activity performed on mobiles and users by LBS providers. These solutions can exploit not only the RSS of carried-on MS but also the human body effect on WiFi radiowaves for passive tracking purposes. However, depending on the specific scope of the provider's business, users might not always be aware of the real activity performed on them.

For example, a well-known activity, adopted for retail analytics, is to track and monitor people flow by analyzing the probe requests when passing MSs with WiFi turned on. The method does not require the MS to be connected to the network as the tracking can happen in the AP itself or through dedicated WiFi clients. Although the method is quite powerful, the current randomization of MAC addresses on new smartphones makes it difficult to keep track of or profile individuals, reducing the method to a mere people-counting service.

A different type of tracking focuses exclusively on human bodies and not devices. This method, known as tomographic motion detection, requires a huge number of dedicated WiFi sensors to perform accurate enough tracking in populated areas. Although quite intriguing, a limit of the technique is the low possibility of identifing the tracked body as no unique identifier is associated with the target without additional devices.

10.3.2 Positioning Methods

A plethora of indoor positioning methods have been adopted and tested for WLAN infrastructure and mobile handsets. However, most of them still present several weaknesses and limitations when targeting real applications for commercial exploitation, which are strictly dependent on the adopted technology.

10.3.2.1 Angle of Arrival

The angle of arrival (AOA) in WLANs makes use of array and multi-array antennas to estimate the angle of the line of arrival of the signal, and the estimation of the position of the target MS is placed at the intersection of the lines by adopting triangulation algorithms. The antenna arrays proposed in literature are large in size and this characteristic might force the computation of the position estimation to be performed in the network. Although in principle AOA was adopted for outdoor positioning, interesting results on AOA-based positioning have also been performed indoors and reported in the literature (Della Rosa et al. 2012). Due to the burden provided by this solution when performing experimental activities, the direction of the arrival of the signals is highly sensitive to reflections, fluctuations and non-line-of-sight (NLOS) conditions.

10.3.2.2 Time of Arrival

Time of arrival (TOA) information in WLANs can be retrieved by evaluating the TOA of the signal from MSs to APs, which are the fixed reference points with known coordinates, or vice versa. Synchronization in time of both entities is a must for this method. Once the distance between the MS and at least three APs has been obtained from TOA measurements, position estimation can be performed by using trilateration, that is, through the intersection of circles with radii determined by TOA measurements, with the center at the known AP coordinates. In addition, signal reflections and NLOS conditions highly distort TOA measurements, making the method often not suitable indoors because of the adverse environment conditions.

10.3.2.3 Time Difference of Arrival

The time difference of arrival (TDOA) in WLANs is based on evaluating the difference in the arrival times of signals from two different APs to MSs. TDOA values define the hyperbolas on which the MS location is potentially estimated. The positions of the MSs are then calculated at the intersection of the hyperbolas. This method represents a valid alternative to TOA as the main advantage of this technique, when compared to TOA, is that it does not need prior information of the absolute time of the signal transmission, hence synchronization between transmitters and receivers is not required, simplifying much of the set-up process. However, synchronization among transmitters is still required.

10.3.2.4 Received Signal Strength

RSS-based positioning methods for use indoors can be divided into three main categories: cell ID, fingerprinting and path-loss based. Although theoretical models are available in the literature, their set-up parameters can, in principle, be experimentally determined. In fact, as the method suitable for exploitation in commercial applications, experimental data allows better adaptation of the proposed solution to the surrounding environment. For location-based applications, targeting mass-market RSS is considered to be easier than any other type of measurement as it can be passively listened to from the APs belonging to an already deployed infrastructure (such as a WLAN). Technically APs are built to periodically broadcast beacon

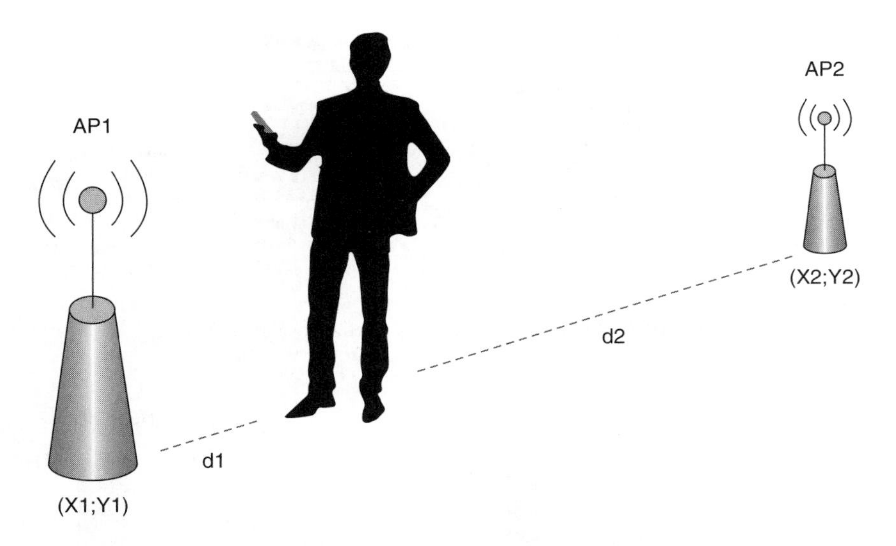

Figure 10.3 Cell-ID (Della Rosa et al. 2012).

frames containing information for network identification, connectivity and quality (e.g., SSID, BSSID, RSS, received signal strength indicator (RSSI), channel, etc.). This process allows a passing MS to determine unique AP network identifiers and signal strengths. In contrast to time-based approaches, synchronization in time is not required (Della Rosa et al. 2012).

Indoor positioning designers usually make choice of one or a combination of the aforementioned solutions in order to optimize and better adapt to different propagation environments. The popularity of RSS-based approaches is so great that it is now common knowledge that a mix of the different RSS-based methods might offer valid solutions to be exploited commercially. However, when real-time needs are in place several aspects might delay the whole process, such as RSS readings, timestamps, server responses, probe requests and responses, reporting intervals and so on (Cisco 2008). RSS and MAC information can be obtained and elaborated from two main sources: the network/server side, if the network infrastructure analyzes the information on the strength of the signal coming from the attached MSs, or client side, if the MS reports and analyzes the RSS from the signal transmitted by the AP.

RSS-based methods are described in this section.

- *Cell-ID* This method (Figure 10.3) is based on prior information on the location of the AP from which a MS receives the strongest RSS value after performing a beacon frame scanning activity. A unique identifier is defined by the MAC address of the AP and stored in a previously created database. This represents the simplest approach to positioning, offering an easy set-up for deployment. However, despite the easiness of the solution, the method is strictly dependent on the MS–AP distance, hence it can be applied in scenarios where only a rough accuracy is sufficient (Della Rosa et al. 2012). Due to the low performances in terms of accuracy (the user position is assumed to have the same coordinates as the AP), the method is only suitable for applications with low accuracy requirements, although it is quite straightforward and has low complexity (Rainer 2012).

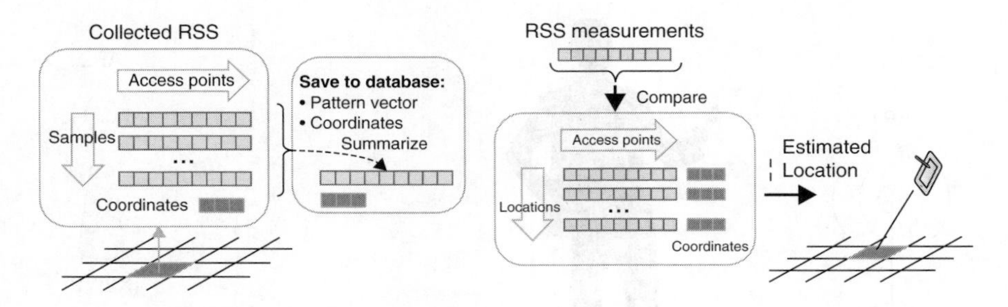

Figure 10.4 Fingerprinting phases (Della Rosa et al. 2012).

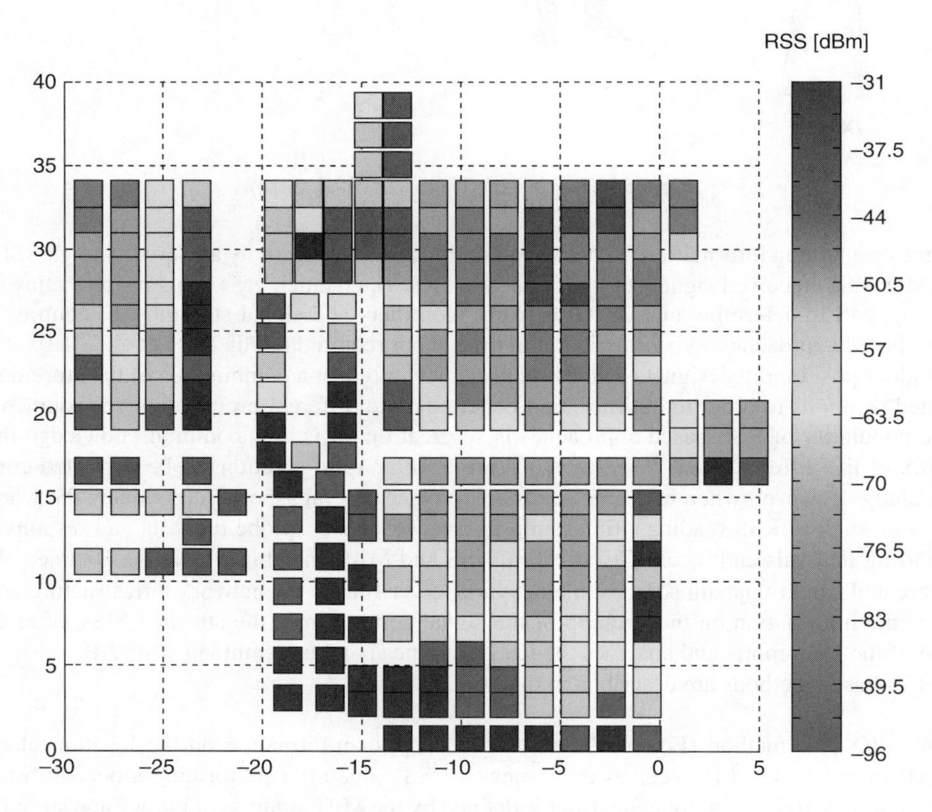

Figure 10.5 Fingerprinting map (Della Rosa et al. 2012).

- *Fingerprinting* Fingerprinting methods (Figures 10.4 and 10.5) are based on models relating RSS measurements coming from nearby APs to the position where the measurements have been performed. The method is straightforward and is based on two phases: off-line and on-line. During the off-line phase the RSS of nearby APs is scanned and stored for each specific location for a defined time. This allows a particular RSS behavior to be assigned to a defined location of the monitored area. Parameters like RSS pattern, average values and variance might help to characterize the fingerprint of a specific location and help the next phase. In this way, at each location, from the off-line collected data, a typical signal pattern is extracted and saved to the fingerprinting database with the coordinates of the location. During the on-line phase, also known as the positioning phase, the current set of RSS measurements collected from the nearby APs in real time are compared to the patterns stored in the database and matched with the coordinate database. The coordinate estimate is obtained from the database entry whose stored signal pattern has the closest match with the measured signal vector. The method is considered to be robust with regard to propagation impairments caused by the surrounding environments (multipath, attenuation caused by walls, etc.) when compared to cell-ID and path-loss-based methods as it makes use of the location-dependent characteristics of the radio signal in the monitored spot. However, the method is not immune to the ever-changing radio signal propagation caused by human bodies and hand-held devices, which greatly decreases the performance of the method (Della Rosa et al. 2011b). A known disadvantage in fingerprinting approaches is the fact that collection of the data for the fingerprinting database is laborious and time-consuming, hence the major drawbacks of the fingerprinting approach are the calibration process in the off-line phase and the maintenance of fingerprinting databases (Rainer 2012). Moreover, the heterogeneity of the WLAN cards embedded into WiFi devices (e.g., laptops, desktop PCs, smartphones, internet tablets) is a big problem for indoor positioning applications based on fingerprinting. Differences of more than 25 dBs between the RSSs of different devices have been reported (Della Rosa et al. 2011b), which make the database created in the off-line phase not suitable for every future device if high accuracy is required because of the continuous re-calibration needs. Even if fingerprinting is more robust, as previously mentioned, the data collection in the calibration phase is still a very laborious and time-consuming procedure, requiring a huge investment in terms of time (and money) when mass-market positioning applications should be implemented for devices with different vendor-related hardware characteristics and in environments with continuous network maintenance (e.g., replacing broken APs). This is because different vendors produce proprietary chipsets with specific radio frequency (RF) accuracy, range of power, sensitivity and scaling, which make the signal values stored into the database highly inaccurate if frequent re-calibrations are ignored (Della Rosa et al. 2011b).

- *Path-loss-based positioning* The path-loss model (Figure 10.6) for positioning has been described in previous sections of this book. However, empirical or semi-empirical models are nowadays adopted when commercial solutions need to be deployed. Path-loss models of WiFi radio signals are typically used to convert RSS measurements into distance estimation between MSs and APs. This operation is the most crucial to perform as any wrong estimation or prediction of the latter could introduce errors to the final position estimation (Della Rosa et al. 2012).

  Specifically, path loss represents the difference between the transmitted signal, measured at face of the transmitting antenna, and the level of the received signal, measured at the face of the receiving antenna. Although path loss might not consider antenna gains and cable losses, when estimated it shows the level of signal attenuation present in the monitored environment where the positioning system should be deployed. Path loss is mainly due to the effects of free space propagation, reflection, diffraction and scattering (Cisco 2008).

  A simplified model for indoor positioning is represented by the following:

$$PL = PL_{1\mathrm{m}} + 10 \log (d^n) + s$$

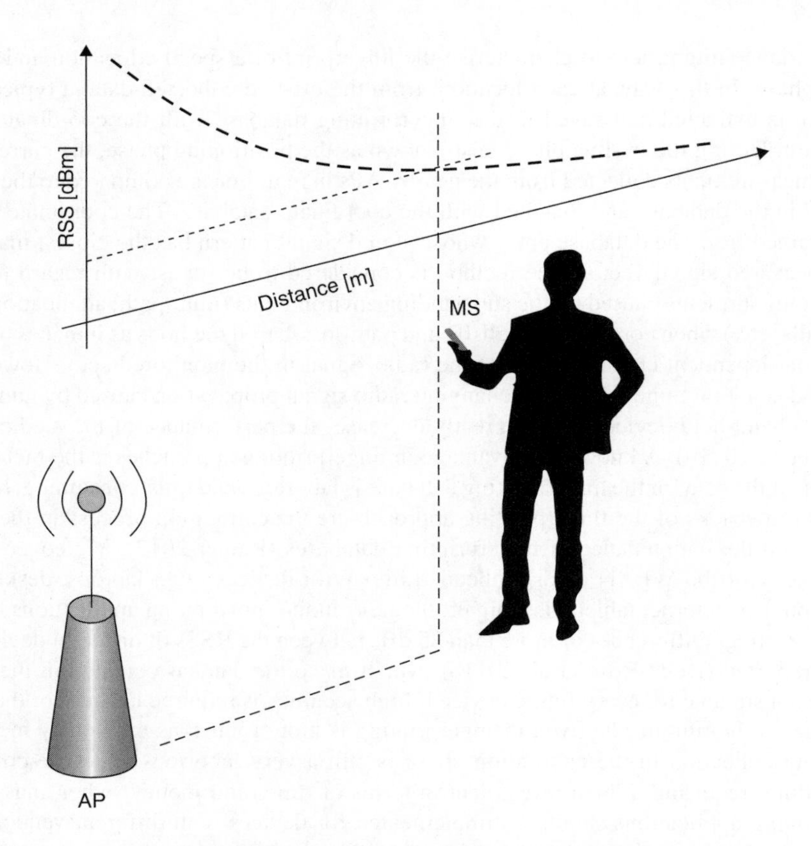

Figure 10.6 Path loss (Della Rosa et al. 2012).

where the increase rate of the exponent n is proportional to the distance from the transmitter (e.g., AP). Path-loss exponent values might vary according to the specific characteristics of the surrounding environment and represent an approximation. Typically values of 2 are adopted for free space, while values greater than 2 are adopted in obstructed environments. Cisco (2008) suggests 3.5 for indoor office environments and values up to 4 in industrial and densely populated commercial environments.

Specifically, PL is the total path loss expected in decibels between the AP and the MS, PL_{1m} is the path loss adopted as reference for the specific frequency considering the distance AP-MS being 1 m, d is the distance between AP and MS, n is the path loss exponent and s is the standard deviation associated with the shadow fading. The AP–MS distance is not the only parameter to take into account when modeling path loss. In fact, the variability in time of the RSS, typically shown in terms of fluctuations, is represented by the standard deviation of the shadow fading(s), sometimes interpreted as noise, which aims to include the burden introduced by antenna orientation, attenuation, multipath and other obstructions. In installations with antenna diversity features, values between 3 and 7 dB are reported (Cisco 2008).

As previously mentioned, the first goal of path-loss-based approaches is to estimate the distance between AP and MS. As consequence it is possible to derive the following:

$$d = 10^{\frac{Tx_{\mathrm{PWR}} - Rx_{\mathrm{PWR}} - Loss_{\mathrm{TX}} - PL_{1\mathrm{m}} + s + Gain_{\mathrm{RX}} - Loss_{\mathrm{RX}}}{10n}}$$

where Rx_{PWR} is the RSS, Tx_{PWR} is the transmitter power output, $Loss_{\mathrm{TX}}$ is the sum of transmitter losses due to connectors and cables, $Gain_{\mathrm{TX}}$ is the transmitter antenna gain, $Loss_{\mathrm{RX}}$ is the sum of receiver losses due to connectors and cables, and $Gain_{\mathrm{RX}}$ is the receiver antenna gain. All of these are positive values expressed in decibels.

Once the distance AP–MS has been found, the distance d becomes the radius of a circle plotted around the target. Doing the same for other sources (APs) makes it possible to provide the right measurements to any trilateration algorithm to finalize the position estimation of the MS.

At this point it is clear how the RSS value scanned by WiFi-enabled devices is a function of the distance between the AP and the MS. The attenuation model describes the dependency in general. However, even though it is commonly adopted in literature, this model does not exhaustively describe the distribution and the overall behavior of the signal during its path in indoor environments, where the propagation conditions are dynamic due to multipath and shadow fading, and many other unpredictable causes coming from external and unknown sources (Rainer 2012).

As mentioned before, when the MS–AP distance is accurately extracted from RSS measurements, common trilateration methods can be used to estimate the position of the MS and the user. RSS measurements from at least three distinct APs are needed. Before reaching this point a valuable empirical approach might be to perform experimental activities to build a custom path-loss model, which is specific for the desired indoor environment. In order to do this, as for the cell-ID method, the MS requires prior information about the MAC addresses and locations of APs, for example from a database. When RSS data are adopted, multipath attenuation caused by walls, structures and even human bodies make the modeling of the signal and the estimation of the AP–MS distance very challenging, leading to counterintuitive conclusions (Della Rosa et al. 2012). A common counterintuitive behavior is when at longer distances the average RSS value measured is higher than the one measured at shorter distances. This is typically caused by a constructive multipath when the experiments are performed in long corridors with walls with reflective material. Because of environmental impairments, the path-loss-based accuracy is typically worse than fingerprinting. On the other hand, methods that use path-loss models to estimate distances are needed for fast deployment of positioning services or when the signal properties of ad hoc WLAN connections between two MSs need to be exploited (e.g., in cooperative schemes), mainly because the dynamic information about moving APs is difficult if not impossible to incorporate in fingerprint databases (Della Rosa et al. 2012).

Building your own empirical path-loss model is a straightforward and not very time-consuming exercise. It can be obtained by placing the MS at several distances, for example from 1 m to 20 m, with respect to the AP broadcasting the signal, measuring the RSS and getting, for example, the average or most frequent value at each step. The step-by-step procedure is shown in Figure 10.7 while Figure 10.8 shows the result obtained. In order to get the desired approximation additional filtering might be needed (Della Rosa et al. 2011a). It is worth noting that while obtaining the measurements described in

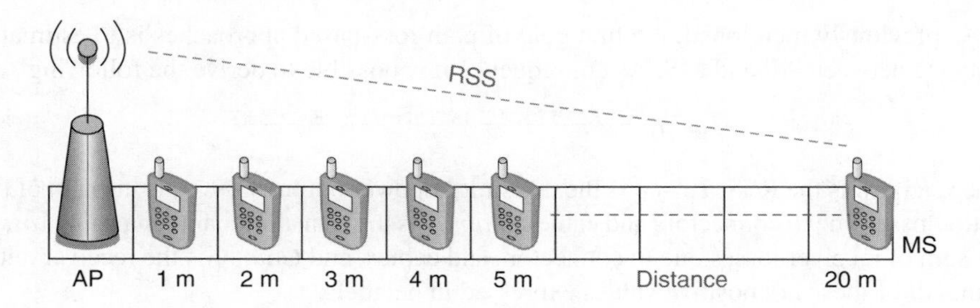

Figure 10.7 Empirical path-loss phases (Della Rosa et al. 2012).

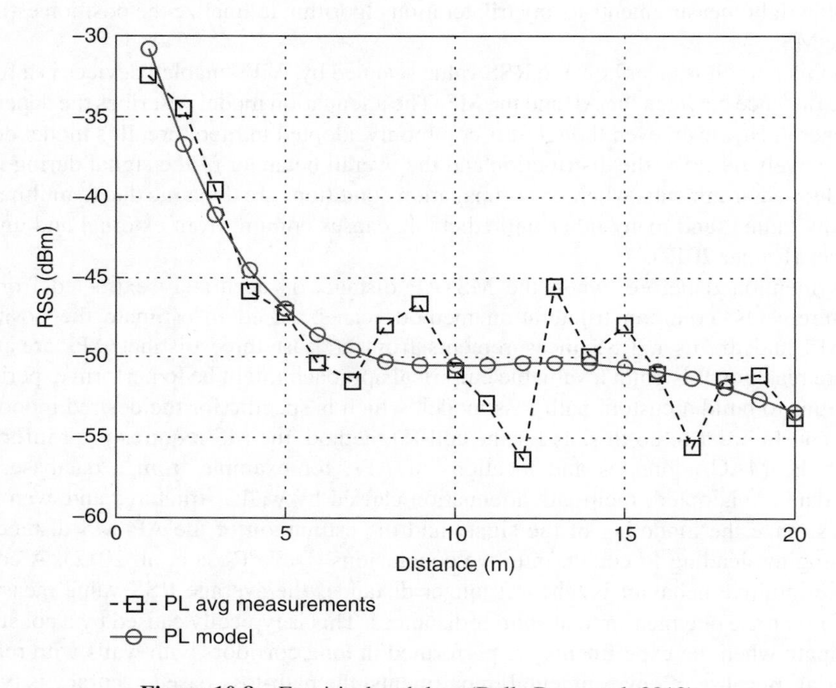

Figure 10.8 Empirical path loss (Della Rosa et al. 2012).

this section, as the distance increases the RSS variations might increase, which makes distance estimation and approximation a more challenging experience. Hence short-range measurements might be considered more reliable and valuable than long-distance ones, as shown in Figure 10.9, due to the different stability of the values. In the end, the obtained empirical path-loss model will clearly match the theoretical expectations of signal decay (Della Rosa et al. 2011a).

10.3.3 Evaluation Criteria for Indoor Positioning Systems Based on WLANs

In order to perform a comparison of the different methods to adopt in a WLAN environment, several parameters can be taken into account. It is well known that the unpredictable

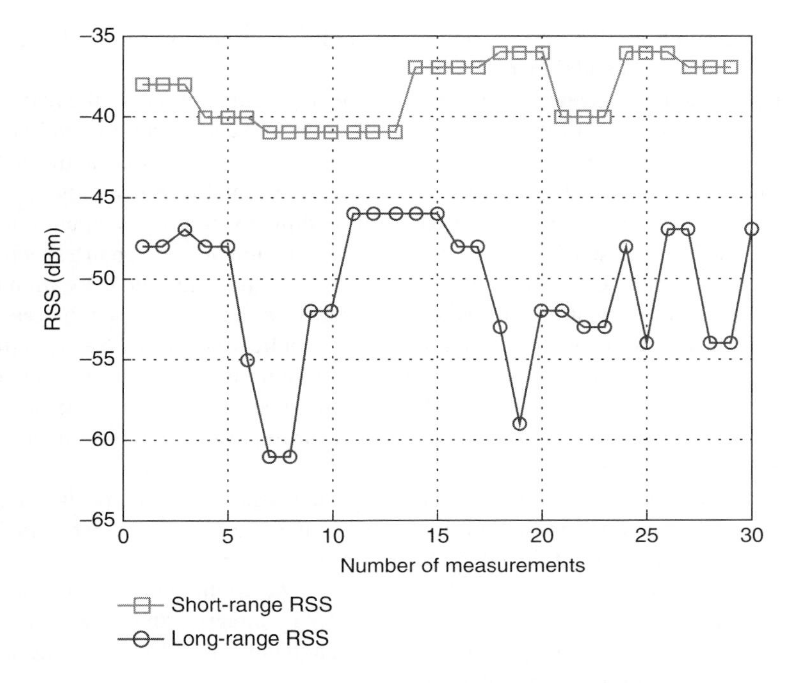

Figure 10.9 Long-range and short-range RSS fluctuations (Della Rosa et al. 2012).

behaviors of the surrounding environment might bias even objective considerations, providing counterintuitive results. Moreover, the large number of criteria that can be found in the literature do not help much in identifying an optimal general solution. The following list identifies the fundamental user requirements in indoor positioning systems based on WLANs. It is possible to distinguish several requirements for positioning systems based on WLANs, for example accuracy, coverage, cost, infrastructure, privacy, update rate, integrity, availability and scalability (Rainer 2012).

- *Accuracy* determines whether or not the solution is suitable for commercial exploitation. It is defined as the degree of closeness of measurements of a quantity to that quantity's true value. When referred to positioning it specifies "the degree of conformance of an estimated or measured position at a given time, to the true value, expressed for the vertical and horizontal components at the 95% confidence level" (Rainer 2012). Accuracy is the key parameter to consider when dealing with indoor positioning or positioning systems in general. However, its combination with additional evaluation parameters will provide a better view of the overall performance of a positioning system.
- *Coverage* specifies the spatial extension of the deployed positioning system, and is divided into local, scalable and global coverage. Indoor positioning based on WLANs is usually intended for small and defined areas (local).
- Even though one of the key concepts of positioning systems for indoor environments based on WLANs is the exploitation of an existing infrastructure, the final deployment *cost* might depend on several factors, for example installation, calibration, maintenance, re-calibration, etc. These factors are time-consuming and might increase the cost of the system, as for

fingerprinting solutions. The use of dedicated hardware might limit the calibration needs but increase the initial capital investment.

- WLAN *infrastructure* is related to the nature of the deployment mode of the network itself. A WLAN is typically in infrastructure mode when an AP acts as an hub and is linked to other APs in the network. However, in indoor positioning infrastructure can mean the group of physical devices deployed in the monitored area describing the positioning system itself.
- *Privacy* concerns are mostly related to systems providing positioning computation from the server side, that is, when sensitive information on the location of a person in a specific indoor environment is centrally estimated, stored and reported to the final user as location information, for example for navigation purposes. Major concerns are raised when the user's device is passively tracked, for example for retail analytics applications or people-flow statistics.
- The *update rate* of the location information is a requirement strictly dependent on the use-case and application. Depending on the particular need, an update of the position estimation and notification might happen regularly or on request. It is worth mentioning that these events might happen at different times.
- *Integrity* describes the probability that an unexpected behavior of the positioning system provides a position estimation that differs from the real one more than a specified amount without notifying the user within a defined period.
- *Availability* represents the percentage of time when the positioning system is expected to be available and adopted to the user, while still considering integrity and accuracy.
- *Scalability* is the capability of a positioning system to handle a growing number of requests from an increasing number of simultaneous users.

Depending on the needs of the final stakeholder or service provider, a valuable positioning system should take into consideration a mix of different parameters and satisfy a plethora of users' needs and requirements.

10.4 Error Sources in RSS Measurements

Indoor positioning systems, based on WLANs, are not immune to errors caused by the ever-changing environment (Figure 10.10). New devices with heterogeneous WiFi cards, different form factors, human bodies walking or shading the LOS, hand grip and so on make it difficult if not impossible to implement and validate objectively not only experimental activities but also commercial exploitations of them. During the last decade, researchers have started to address, isolate and analyze the problems, with every method based on RSS proposed so far highly compromised. At the same time, experimental activities might be too customized and based on assumptions, making the final results not repeatable or comparable with novel ones due to the strict dependence on the surrounding environment. Basically, the intrinsic complexity of the indoor environment (including shadowing, multipath, overlapping channels, objects, and variations in the sensitivity of wireless cards embedded in mobile devices) mixed with human body influence, represent the major sources of errors in measurements, setting great challenges for fast deployment of accurate solutions. In this context, the user's influence on RSS measurements is represented by two main components: body loss and hand grip. Experimental results show how both components are major causes of unpredictable behavior and fluctuations in the measured RSS. In particular, the direction and orientation of the user's body has attracted research interest and has been identified as source

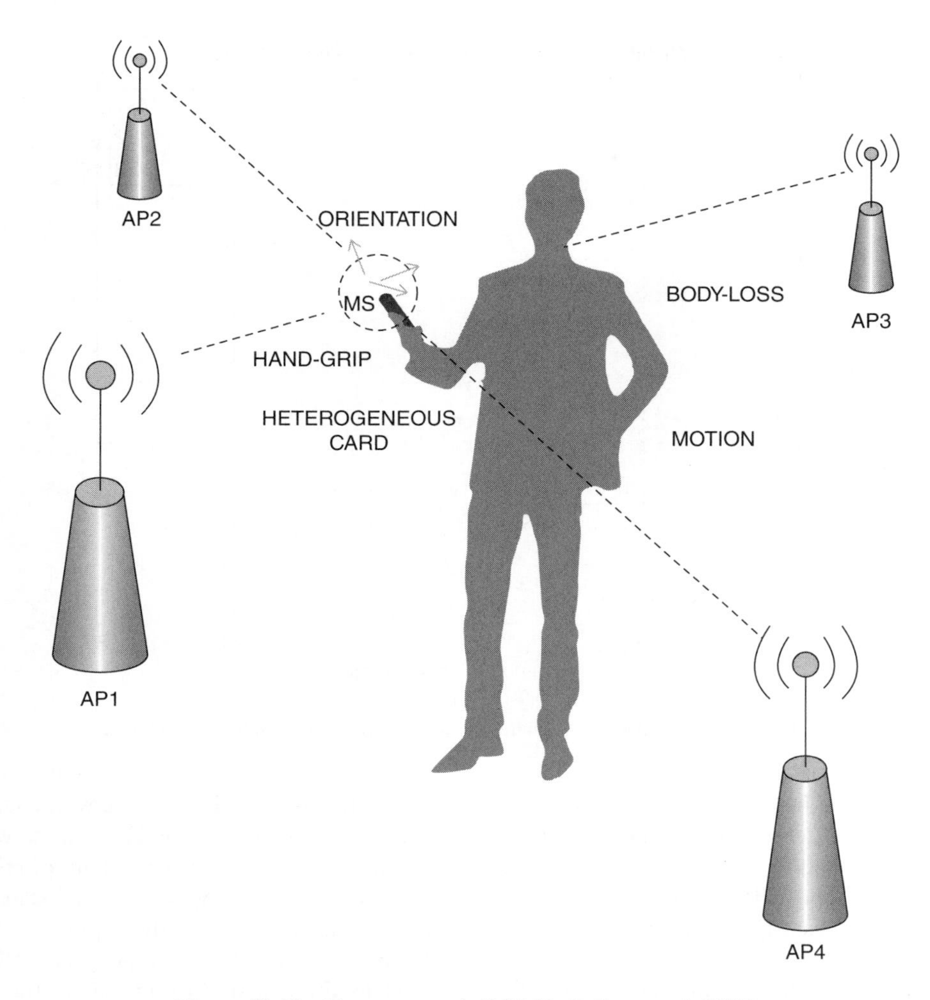

Figure 10.10 Error sources in RSS (Della Rosa et al. 2012).

of errors in location estimation. In fact, only in one direction within a specified range is there a direct propagation path between the WLAN AP and the MS held by the user. In the opposite orientation the user's body obstructs the path, hence causing NLOS with the body itself, which contains almost 70% water, and absorbing part of the 2.4 GHz WLAN radio signal, causing a significant drop in the signal amplitude and consistent errors in RSS, distance and position estimation. However, mobile devices do not stay in the air by themselves. They are held by the users, adding another source of errors in the RSS measurements: the hand-grip.

10.4.1 Heterogeneous WiFi Cards

It is worth mentioning that the measured RSS values vary from device to device and from vendor to vendor (Figure 10.11). This can be seen not only from reading the vendor's specification but also through simple experimental activities.

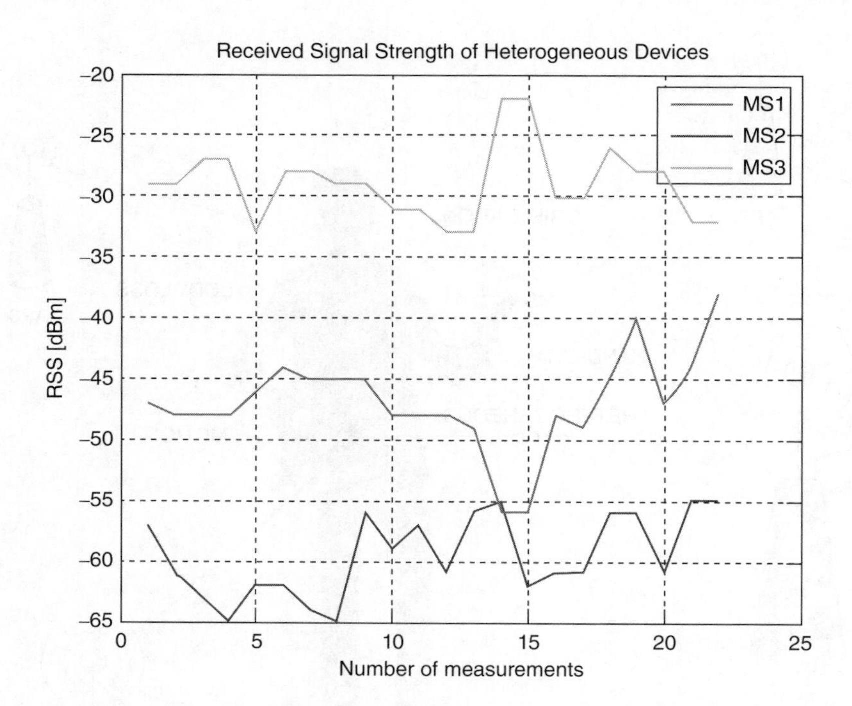

Figure 10.11 RSS comparison for heterogeneous WiFi cards (Della Rosa et al. 2012).

In fact, the heterogeneity of the WLAN cards embedded into WiFi-enabled devices such as APs, WiFi extenders, laptops, desktop PCs, smartphones, tablets, wearables and so on, when exploited for positioning and tracking applications, represents a huge problem for both researchers and LBS providers targeting the mass market. Specifically, for indoor location applications based on RSS, different MSs measure the RSSs coming from the deployed APs differently from each other, reporting differences above 25 dBs for RSSs (as reported in Della Rosa et al. 2011b). This discrepancy is attributed to both hardware and software characteristics of the WLAN cards available on the market and produced by different vendors. It may arise from a lack of rigorous application standardized guidelines, particularly as WLAN cards and APs are not primarily designed for positioning purposes. This is why we can encounter huge variations in characteristics that for positioning applications need to be as stable as possible. At the moment it is not straightforward to provide universally accurate and reliable interpretations of RSS values in indoor scenarios for different target MSs on a large scale. Chipsets are deployed with different RF characteristics, such as sensitivity, range of power and accuracy, causing accuracy problems when implementing indoor positioning applications. The problem is so significant that companies providing services and solutions for positioning purposes need to maintain and update databases of supported clients, which makesit difficult to maintain the positioning service for every approach, including cell ID, fingerprinting and path-loss.

The different sensitivities and granularities reported in research make experimental results vary a lot depending on the selected device, even though the environmental conditions are kept stable in a controlled environment. 802.11 client devices report inconsistent values of RSS, causing inconsistent position estimations and tracking performances. If measurements are performed exclusively on the network side, the problem of maintaining huge databases containing different devices and characteristics might be overcome. In fact this activity might be endless due to the huge number of device vendors on the market. It is neither feasible nor realistic to assume that every WiFi device will be built by a unique vendor, hence a way to overcome the problem or attempt to equalize the heterogeneity is needed. As example Cisco (2008) describes a typical case in whcih a particular positioning system expects to see reported RSS in a defined range, for example from –127 to +127 dBm in 254 increments of 1 dBm each, while other devices located and tracked by the system may only be able to report RSS from –111 to +111 dBm in 74 increments of 3 dBm each. Because the granularity and range of the reported RSS varies a lot, usually the burden to perform such equalization is left to the service provider. A workaround is provided by installing client softwares or apps in hand-held devices to mitigate the problem (Cisco 2008). For example, solutions provided in the literature for handling the heterogeneity of RSS values propose the use of hyperbolic location fingerprinting (Kjaergaardi 2008), making use of the RSSs ratio detected from couples of APs, or simpler techniques involving manual data collection and mapping between the RSSs recorded (Della Rosa et al. 2011b).

10.4.2 Device Orientation

In addition to the aforementioned hardware heterogeneity of devices, an aspect which is not yet well known in the field of indoor positioning based on WLANs using hand-held devices is the orientation of the device itself respect to the APs from where it is measuring the RSS. Although most hand-held device antennas might be intended as omnidirectional, depending on the specific position with respect to the AP, different values of RSS can be measured according to the specific position and orientation. Figures 10.12 and 10.13 show eight values from the set of different orientations available (Della Rosa et al. 2011a). Devices report different RSS values depending on the specific orientation with respect to the AP. More specifically Della Rosa et al. (2011a) have presented the results of an experiment where the device antenna is fully free from the user's hand effect and report standard deviations up to 7.6 dBm. It is easy to understand that by applying an RSS-to-distance conversion model or relying on fingerprinting databases the orientation of the device will greatly affect the final distance and position estimation.

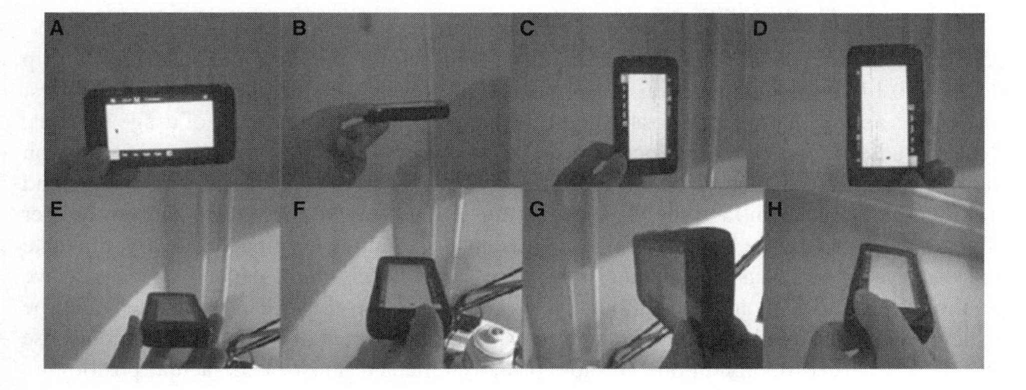

Figure 10.12 Different handheld device orientations (Della Rosa et al. 2012).

	Devices Orientations								
	A	B	C	D	E	F	G	H	std
MSI	–28	–44	–27	–37	–34	–49	–40	–49	7.6
MS2	–42	–43	–50	–41	–44	–45	–45	–51	3.6

Figure 10.13 RSS values for different handheld WiFi orientation (Della Rosa et al. 2012).

10.4.3 Channel in the Presence of the User and Body Loss

Another aspect emerging in research is the effect of the user's body on radio propagation and in particular the effect caused in the near-field. Although the majority of propagation phenomena happen in the space surrounding WiFi devices, allowing the exploitation of conventional propagation models, recent discoveries have highlighted how the reliability of wireless links is affected by the user's body influence. This demands rigorous analysis, especially when exploiting RSS measurements for indoor positioning systems. The user's movements, gestures, hand grip and posture represent a subset of the time-varying conditions that have a direct impact on the RSS measurements and consequently on the final position estimation, whether performed with cell ID, fingerprinting or path-loss-based solutions. Considering the body to be a homogenous dielectric cylinder might be a simplistic attempt to model it electromagnetically in order to study the behavior of absorption, reflection and diffraction phenomena. It is worth mentioning that the direct effect on the 2.4 GHz will be dependent on different factors, for example the dimensions of the body, the dimensions of the body parts, the composition of human tissue, etc. Body parts with dimensions smaller than the wavelength might have negligible effects if counted alone, causing small fluctuations. The experience of having a device rotating around a blocking surface, like a human body, might also produce an emulation of the directional behavior to omni-directional antennas. Additionally, the presence of a body mass between the AP and the device antenna causes a significant drop in the signal, which can be huge if the receiving device is totally over-shadowed by the user's body. Sometimes signal drops can be caused by multipath effects.

10.4.4 The Hand Grip

For years mobile device manufacturers have overlooking the importance of hand grip (Figure 10.14) on communication performance, with usability studies restricted a few test cases. This approach limits the possibility of identifying the full complexity of usage patterns. Recent studies have started to focus not only on operability and ergonomics, but also on the radiating performance of hand-held devices when in contact with the user's body and specifically the hand. In fact, the close proximity of the antenna might have an even greater effect than the rest of the body, making the RSS measured at the terminal level vary, fluctuate and decay in a very sensitive manner which, in some cases, might depend on the relative position of the user's fingers. Although in the literature exceptional performances can be achieved in a controlled environment and in restricted experimental activities, common sense reveals that if real-life conditions are not correctly reflected, most of the actual positioning methods and solutions might even be worse than those in the reference cases. A well-known example of this is the iPhone 4 antenna issue, which has highlight the importance of this issue.

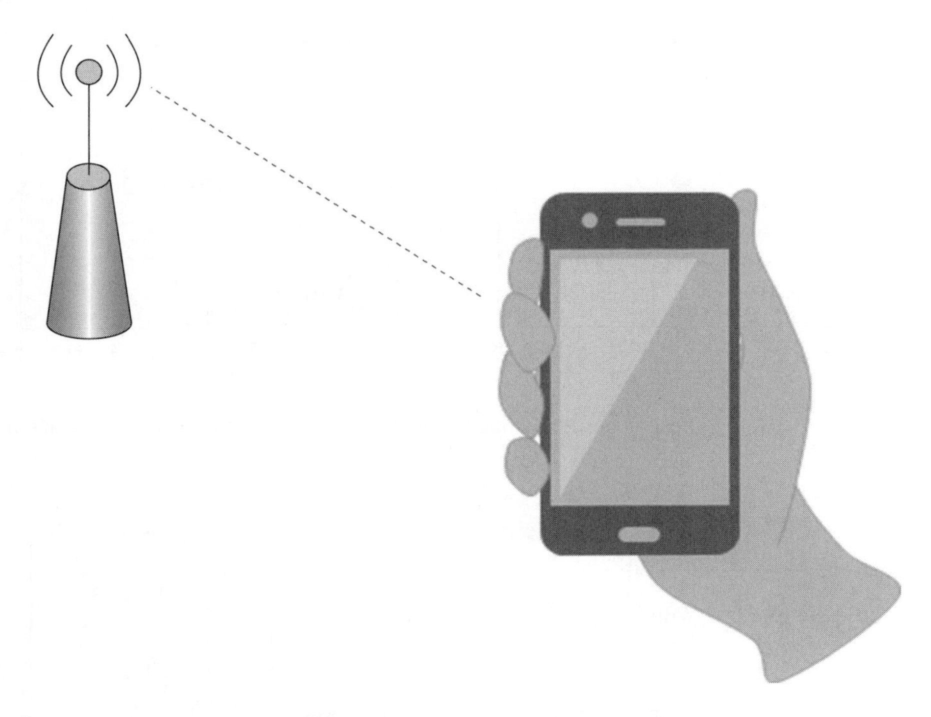

Figure 10.14 Hand grip (Della Rosa et al. 2012).

10.5 Experimental Activities

In this section we present experimental activity results showing the effect of hand grip and the human body generally when performing RSS measurements for the localization of mass-market devices in indoor environments. The experimental activities proposed in this section were performed in the Department of Electronics and Communication Engineering

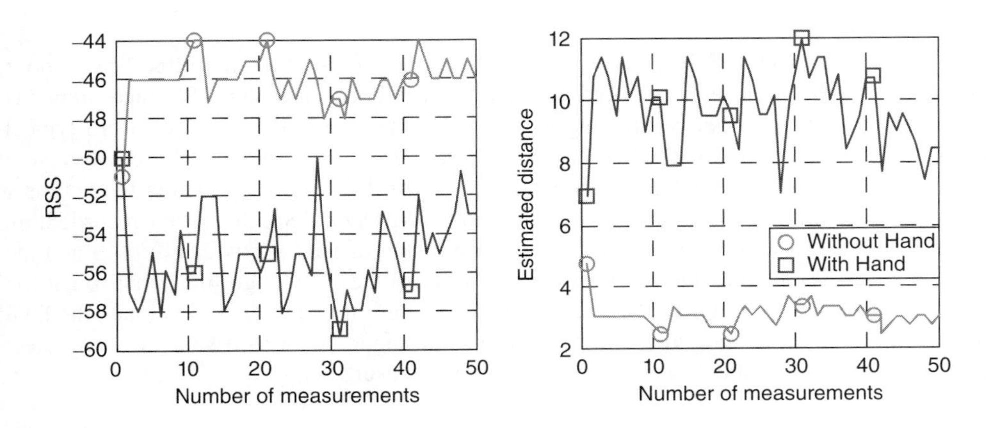

Figure 10.15 Hand-grip effect on RSS measurements and distance estimations (Della Rosa et al. 2012).

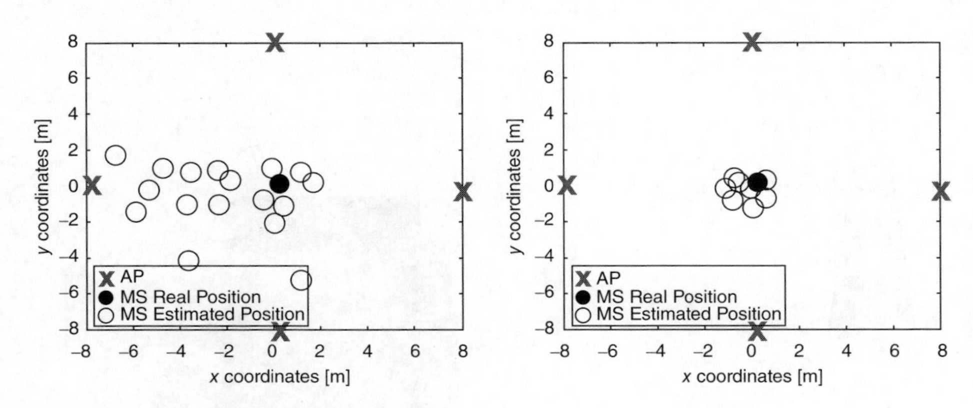

Figure 10.16 Hand-grip effect on position estimation without (left) and with (right) mitigation (Della Rosa et al. 2012).

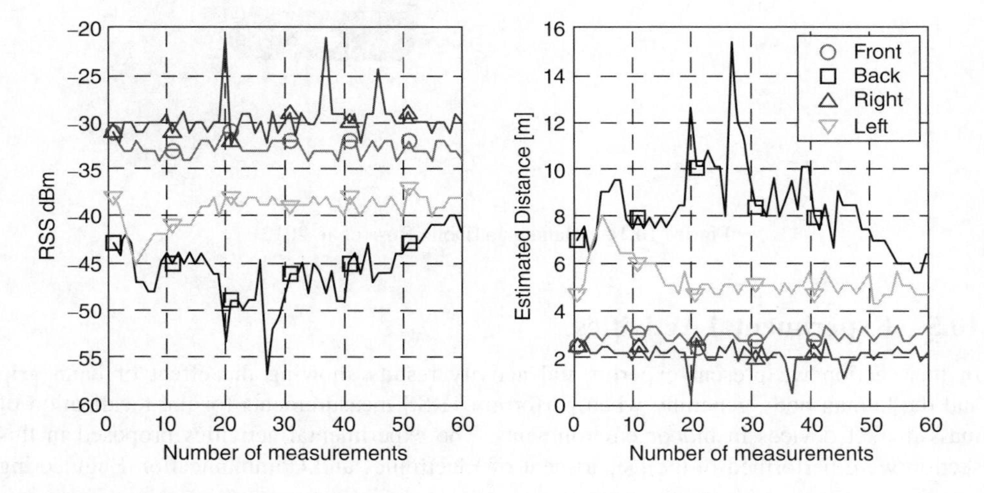

Figure 10.17 Body-loss effect on RSS measurements and distance estimations (Della Rosa et al. 2012).

at Tampere University of Technology. The main goal was to show the effect that indoor environments, heterogeneous devices and the human body have on RSS measurements when implementing indoor positioning systems. The hand-held devices used were typical mass-market smartphones with WiFi capability (Della Rosa et al. 2011a). The devices were placed in a square controlled environment covered with four WiFi APs while the test user was placed in the middle of the area, holding an MS. Once RSSs had been recorded, an empirical path-loss model was applied and the estimated distances between the MS and the four APs were adopted as input for a common least squares (LS) algorithm, first to get the initial estimation to a nonlinear least-squares (NLLS) algorithm and then to estimate the final position. The input was provided without any pre-filtering to better assess the negative effect on the positioning accuracy caused by the erroneous measurements.

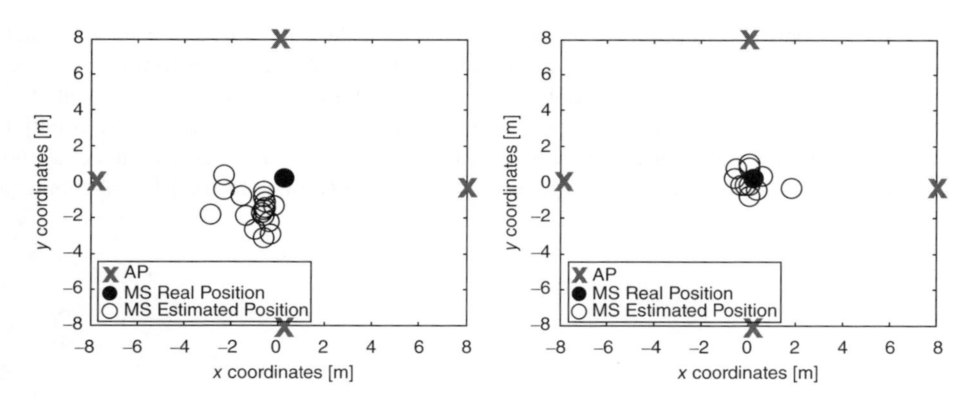

Figure 10.18 Body-loss effect on position estimation without (left) and with (right) mitigation (Della Rosa et al. 2012).

The estimated positions are shown taking into account the different human body effects (Figures 10.15 and 10.17) described in this chapter while the beneficial effect of a priori knowledge of the quantities to mitigate the erroneous measurements is also shown (Figures 10.16 and 10.18). The knowledge of the average loss caused by a human body or hand grip, and the orientation of the device is valuable information that can enhance the performance of a positioning system. The results demonstrate that the human factor cannot be neglected in experimental activities when WiFi-enabled mobile devices are adopted, and that if the effects are mitigated the accuracy of the final position estimation may be enhanced (Della Rosa et al. 2011a).

10.6 Conclusions

This chapter has described indoor positioning solutions based on WLANs, focusing on empirical approaches and problems affecting RSS measurements not only caused by the surrounding environment but also by the user's body and body parts interacting with hand-held devices and related signals. The potential and limitations of conventional methods based on RSS were introduced by reviewing empirical solutions proposed in the literature, asserting the limitations of providing a good level of accuracy and feasibility.

Without doubt the indoor environment is a challenging scenario for positioning. The intrinsic complexity and unpredictability of the signals make it even more interesting from a research point of view as simplistic modeling for wireless scenarios is not able to capture the whole intrinsic variability of indoor environments.

In this chapter the authors described the effect of hardware-related problems on RSS measurements. Particular attention was paid to the effect of the user's body on the far-field part of the wireless link, mainly identifying it as a time-varying blocking object. Moreover, the effects of the near-field perturbations caused by the close environment were presented, showing that the influence of hand grip on mass-market devices is a problem of growing importance in mobile positioning and tracking for commercial applications. This focuses attention on the need for solutions to real-use problems. It does not come as surprise to learn that the user is part of the propagation environment and plays an active role in it. In fact, if the presence of the user is correctly accounted for in WLAN-based positioning activities the beneficial impact in terms of accuracy is huge. All these considerations mean that proper measurement campaigns and experimental activities need to be enforced by situation awareness techniques and through additional sensors information that is able to detect events in both far-field and near-field to let the positioning system recognize the real-life phenomena of signal impairment and take the correct action.

11

Cooperative Multi-tag Localization in RFID Systems: *Exploiting Multiplicity, Diversity and Polarization of Tags*

Tanveer Bhuiyan[1] and Simone Frattasi[2]
[1]*Aalborg University, Denmark*
[2]*Sony Mobile Communications, Sweden*

11.1 Introduction

Radio frequency identification (RFID) is a wireless technology used to identify objects. In particular, RFID is a technology, mountable or attachable to a product, animal or a human being, that enables the transfer of its stored data using radio waves (*RFID Handbook* 2015).

In its simplest architecture, an RFID system comprises RFID tags (also called transponders) and RFID readers (also called interrogators) (see Figure 11.1). RFID readers broadcast queries to tags in their wireless transmission ranges to obtain information stored in the tags such as their unique serial identification number (ID). Readers can be fixed or hand-held and, depending on their power output, may have a variety of read ranges.

RFID tags essentially comprise a chip for storing and processing data, and an antenna for transmitting signals to the reader (see Figure 11.2). RFID tags have limited storage and limited computing capability. There are three types of tags: *passive*, *semi-passive* and *active*.

Active tags transmit their data either in response to a query from the reader or independently. Since they are equipped with a battery, contrarily to passive tags, they are independent of the received power from the reader and are capable of working with signals received from the reader at lower power. Also, they are more effective in "RF-challenged" environments and more easily integrated with sensors such as temperature and humidity sensors. At the same

Mobile Positioning and Tracking: From Conventional to Cooperative Techniques, Second Edition.
Simone Frattasi and Francescantonio Della Rosa.

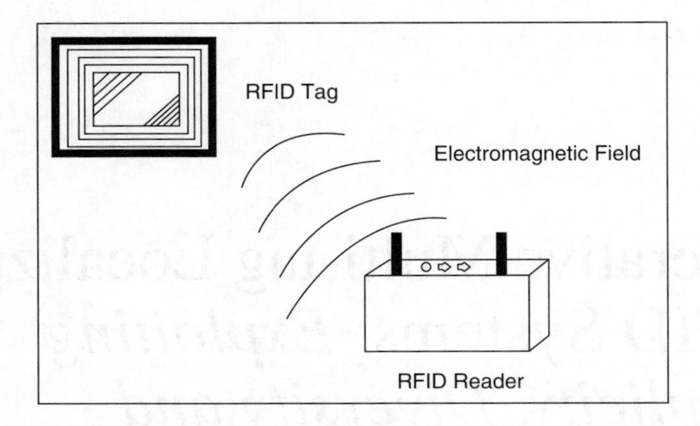

Figure 11.1 Typical RFID system.

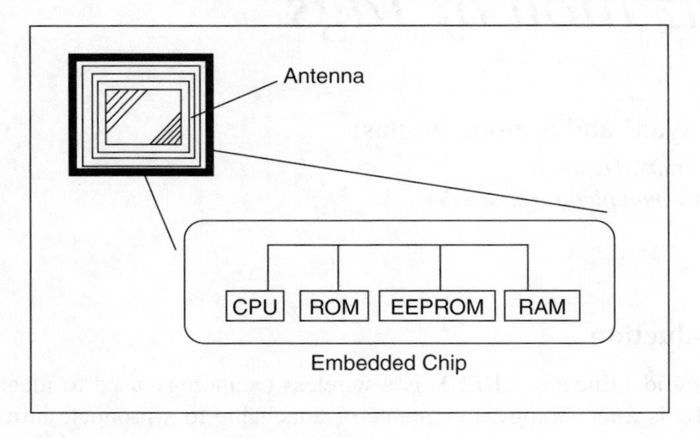

Figure 11.2 RFID tag.

time, due to the battery, they are bigger, more expensive ($5–20) and have a more limited lifetime than the other types of tags.

Semi-passive tags, also known as battery-assisted passive tags, can use their battery to power on in response to a reader's query. These tags are bulkier than passive tags and have a battery life of about 5 years; after that, they act as passive tags. Their read range is in between the read range of active and passive tags (<10 m) and their price is about $1.

Passive tags have no battery, rather the incoming radio frequency (RF) signal from the reader induces a current at the antenna that powers up the CMOS integrated circuit for generating a response to the reader's query. Unlike active tags, they are smaller, cheaper ($0.05–0.20) and have a longer lifetime. Also, they can easily be embedded in stickers and thereby can be used in a plethora of applications. Their main drawback is their shorter read range (3–5 m) with respect to the other types of tags.

Hence, RFID enables contactless identification, even for RFID-tagged objects in motion and/or in non-line-of-sight (NLOS) with respect to the RFID reader. Therefore, RFID finds application, for example, in goods tracking in warehouses, asset tracking, airport luggage management, access control in security systems, etc.

11.2 RFID Positioning Systems

Conventional localization with RFID systems is of a coarse type, as the exact position of the tagged object cannot be retrieved from the detected tag ID but rather whether the object is within the range of a certain reader. To increase the localization accuracy, received signal strength (RSS) measurements can be combined with the information associated with the tag ID. However, the RSS, especially in indoor scenarios, is greatly affected by the propagation environment and may not be stable over time due to shadowing and multipath fading.

RFID positioning systems can entail either *reader localization* or *tag localization*. The former refers to determining the position of a mobile reader with respect to reference tags, whereas the latter refers to determining the position of an object onto which a tag is mounted.

In Pathanawongthum and Chemtanomwong (2010), two methods of reader localization are disclosed. In the first method, the location of the reader is estimated by calculating the average of the locations of the detected tags – consequently, this method may also be referred to as the "centre of gravity" method (Shiraishi et al. 2008). In the second method, the location of the reader is estimated by averaging between the locations of the maximum and the minimum co-ordinates of the detected tags. Naturally, the location accuracy of these methods depends on the density of the reference tags. In Han et al. (2007), mobile reader localization is achieved by placing tags in rectangular and triangular arrangements at a distance of 5 cm from each other. Results show that the triangular arrangement provides higher localization accuracy over the rectangular one. In Ahmad and Mohan (2009), mobile reader localization is achieved by placing 36 passive tags uniformly in a 20 × 20 m room. Results show localization accuracy between 13 and 150 cm. In Hahnel et al. (2004), mobile reader localization is achieved by embedding two RFID antennas in the reader and setting up a grid of reference tags. Results show a location accuracy of less than 2 m.

In Nikitin et al. (2010), tag localization is achieved by using multiple RFID readers and employing the phase difference method.

All the existing RFID localization mechanisms assume that the object contains only a single tag (Vaidya and Das 2008). No results have yet been reported on multi-tag localization and cooperative localization techniques.

11.2.1 Single-tag Localization

According to this approach, the tag is attached to the object or person to be tracked and several fixed readers interrogate the tag. The tag sends a signal encoded with its tag ID to the readers, which extract localization-related information such as RSS, time-of-arrival (TOA) or time difference of arrival (TDOA), whereby the position of the tag can be calculated (e.g., by means of trilateration, with three or four readers for 2D or 3D localization, respectively).

Nevertheless, due to the estimation error intrinsic in the measurements involved, the localization accuracy is low. RSS measurements are the simplest and cheapest in terms of implementation; in fact, from the received signal and an appropriate path-loss model, the range can be easily estimated. However, the accuracy of RSS measurements is low and the estimated range has typically an error of ±50% within the transmission range (Ruixue and Fang 2010). On the other hand, the accuracy of TOA and TDOA measurements is higher (~1–2% error in the estimated range (Leong et al. 2005), but they have as a main drawback the additional burden of synchronization; in fact, TOA needs synchronization at both transmitter and receiver whereas TDOA needs synchronization only at the receiver. Moreover, both TOA and TDOA measurements are highly affected by NLOS (Ruixue and Fang 2010). Note that, since the angle of arrival (AOA) is measured from TDOA, the effect of NLOS will be reflected on the AOA as well.

11.3 Cooperative Multi-tag Localization

Especially in indoor scenarios, where the RSS is highly affected by the propagation environment, whereby its measurements usually display an unstable behavior over time due to shadowing and multipath fading, higher localization accuracy is desirable.

In order to increase the localization accuracy, multiple tags can be deployed on a target (e.g., an object or a person). Consequently, the RSS information obtained from the tags may be combined with the known information about the physical distance between the tags in a manner similar to that described in Chapter 12, which results in a higher estimation accuracy for the target's position.

With respect to a single-tag system (i.e., a system where a single tag is deployed on each target), a multi-tag system (i.e., a system where multiple tags are deployed on each target) has several advantages: multiple/redundant data availability per instance, faster retrieval of statistics per target, higher robustness and quality of service (QoS). Moreover, a multi-tag system has a higher likelihood of line of sight (LOS) as well as a lower likelihood of polarization mismatches between the tags and the reader, which implies a higher number of correct readings.

11.3.1 Multi-tagged Objects and Persons

With the employment of multiple tags, a number of challenges in different application scenarios relevant for localization as well as for identification purposes can be attacked and drastically eased with respect to single-tag systems.

11.3.1.1 Exemplary Application Scenarios

11.3.1.1.1 Package Identification

Let us take the example of an airport where a package is usually labeled with a single RFID tag (see top of Figure 11.3a,b).

If the tag is misread due to NLOS between a reader at a gate and the tag, the airport handling system may lose track of the package or route it incorrectly. This problem can be overcome by the use of multiple tags on the same package, which increases the likelihood of LOS conditions between at least one tag and the reader, and thereby the likelihood of a correct reading and routing of the package (see bottom of Figure 11.3a).

If the tag is misread due to a polarization mismatch between a reader at a gate and the tag, the airport handling system may lose track of the package or route it incorrectly. Again, this problem can be overcome by the use of multiple tags on the same package, in particular placed with different orientations, which lowers the likelihood of a polarization mismatch between the

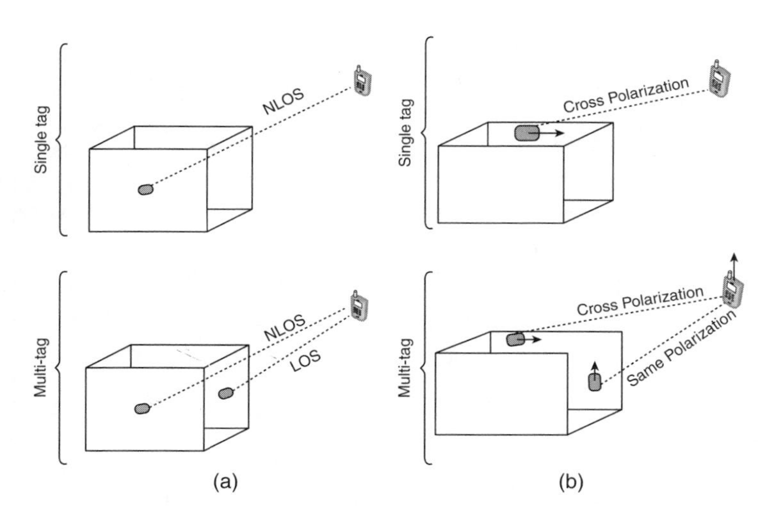

Figure 11.3 (a) LOS/NLOS and (b) polarization match/mismatch.

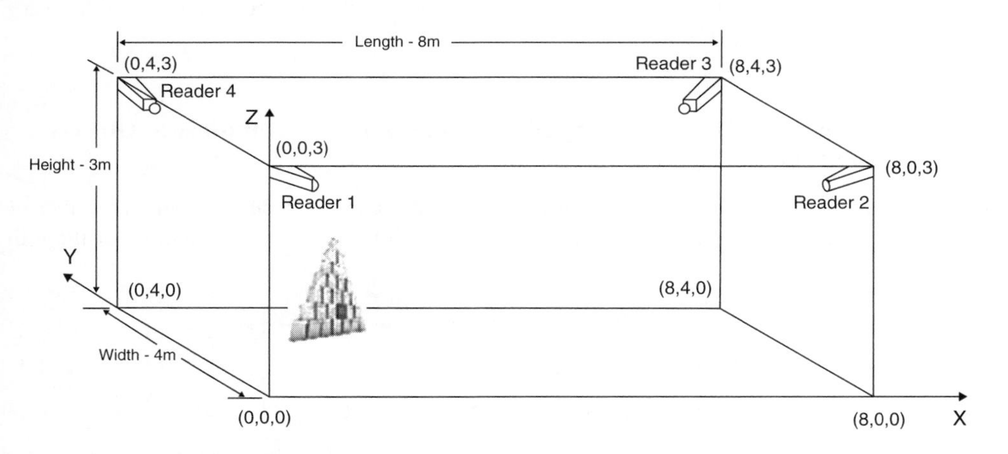

Figure 11.4 Localization of a package in a warehouse.

tags and the reader, and thereby the likelihood of a correct reading and routing of the package (see bottom of Figure 11.3b).

11.3.1.1.2 Package/Person Localization

Let us take the example of a warehouse where a specific package labeled with a single RFID tag is to be localized (see Figure 11.4). Obviously, it would be very time-consuming to check each and every package, even if the latter is within the read range of a reader. At the same time, single-tag localization, for example based on RSS measurements, would be prone to very significant localization errors. Multi-tag localization enables a more accurate localization and retrieval of the package.

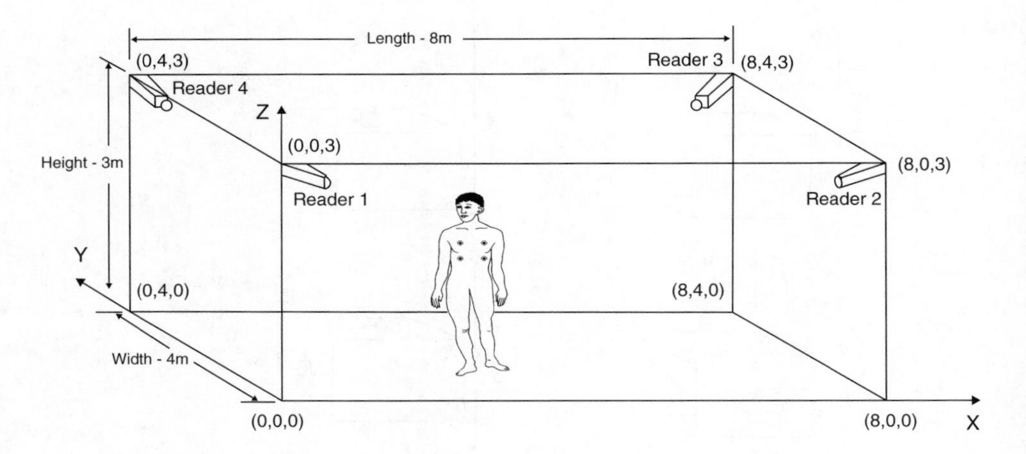

Figure 11.5 Localization of a person in a room.

Similarly, if a person has to be localized in a room, multi-tag localization enables very accurate localization (see Figure 11.5).

11.3.1.2 Cooperative Data Fusion Applied to Localization of Multi-tagged Objects and Persons

Let us assume a number of tags, n_t, in the range of a number of readers, n_r, and a number of RSS measurements per tag, m. The matrix that stores all the RSS measurements for the n_tth tag can be written as:

$$\text{RSS}_{n_t} = \begin{bmatrix} \text{RSS}_1 \\ \cdot \\ \cdot \\ \text{RSS}_{n_r} \end{bmatrix} = \begin{bmatrix} \text{RSS}_{1,1} & \cdots & \text{RSS}_{1,m} \\ \cdot & \cdots & \cdot \\ \cdot & \cdots & \cdot \\ \text{RSS}_{n_{rt},1} & \cdots & \text{RSS}_{n_{rt},m} \end{bmatrix} \tag{11.1}$$

where $\text{RSS}_{n_{rt},m}$ is the mth RSS measurement received at the n_rth reader, and RSS_{n_r} is the mean value of m RSS measurements received at the n_rth reader. Therefore, for n_t number of tags, an n_t number of RSS matrices is obtained. The RSS matrices are then used in the method described below.

STEP 1: From the RSS measurements, the free space path-loss model with a path-loss exponent equal to 3 (Seidel and Rappaport 1992) is used to find the approximate distances of each tag from each reader.

STEP 2: An initial guess is needed for the Coop-WNLLS method (see Chapter 12). For example, the initial guess may be found by applying a lateration technique (David 2015) and the least square method (see Section 5.2).

Let us assume for the 3D case that the distance between a tag located at $X = [x\ y\ z]$ and a reader located at $X_r = [x_i\ y_i\ z_i]$ is d_i, where:

$$d_i^2 = (x_i - x)^2 + (y_i - y)^2 + (z_i - z)^2 \tag{11.2}$$

Subtracting this equation for the ith reader from the equivalent expression from reader 1 gives

$$d_i^2 - d_1^2 = x_i^2 + y_i^2 + z_i^2 - x_1^2 - y_1^2 - z_1^2 \\ - 2x(x_i - x_1) - 2y(y_i - y_1) - 2z(z_i - z_1) \quad (11.3)$$

Equation (11.3) can be rearranged as

$$x(x_i - x_1) + y(y_i - y_1) + z(z_i - z_1) \\ = \frac{1}{2}\left[(d_1^2 - d_i^2) + (x_i^2 + y_i^2 + z_i^2) - (x_1^2 + y_1^2 + z_1^2)\right] \quad (11.4)$$

Equation (11.4) can be written in matrix form as

$$HX = B \quad (11.5)$$

where

$$H = \begin{bmatrix} x_2 - x_1 & y_2 - y_1 & z_2 - z_1 \\ x_3 - x_1 & y_3 - y_1 & z_3 - z_1 \\ . & . & . \\ . & . & . \\ x_n - x_1 & y_n - y_1 & z_n - z_1 \end{bmatrix} \quad (11.6)$$

$$X = \begin{bmatrix} x \\ y \\ z \end{bmatrix} \quad (11.7)$$

$$B = \frac{1}{2}\begin{bmatrix} (d_1^2 - d_2^2) + (x_2^2 + y_2^2 + z_2^2) - (x_1^2 + y_1^2 + z_1^2) \\ (d_1^2 - d_3^2) + (x_3^2 + y_3^2 + z_3^2) - (x_1^2 + y_1^2 + z_1^2) \\ . \\ . \\ (d_1^2 - d_n^2) + (x_n^2 + y_n^2 + z_n^2) - (x_1^2 + y_1^2 + z_1^2) \end{bmatrix} \quad (11.8)$$

The solution for Equation (11.5) is given by

$$X = (H^T H)^{-1} H^T B \quad (11.9)$$

Thus, Equation (11.9) provides the initial guess for the cooperative weighted non-linear least squares (coop-WNLLS) for each tag.

STEP 3: In the final step, the following objective function is to be minimized:

$$\Gamma(X) = \sum_{i=1}^{n_t} \sum_{\substack{j=1 \\ i \neq j}}^{n_t} \gamma_{i,j} \left[f_t(X)\right]^2 + \sum_{i=1}^{n_t} \sum_{j=1}^{n_r} \gamma_{i,j} \left[f_p(X)\right]^2 \quad (11.10)$$

where $f_t(X)$ and$f_p(X)$ are error functions defined as:

$$f_p(X^{(k)}) = \hat{d}^{(i)(j)} - \hat{d}_{k-1}^{(i)(j)} \qquad \begin{cases} i = 1, \ldots\ldots, n_t \\ j = 1, \ldots\ldots, n_r \end{cases} \quad (11.11)$$

and

$$f_t(X^{(k)}) = d^{(i)(j)} - d^{(i)(j)}{}_{k-1} \tag{11.12}$$

where $\hat{d}^{(i)(j)}$ is the distance between the ith tag and the jth reader calculated by the free space path-loss model, $\hat{d}_{k-1}^{(i)(j)}$ is the distance between the ith tag and the jth reader determined at the $(k-1)$th iteration, $d^{(i)(j)}$ is the known distance between the tags and $d^{(i)(j)}{}_{k-1}$ is the distance between the tags determined at the $(k-1)$th iteration.

The distance between the ith tag and the jth reader at the kth iteration is calculated as follows:

$$\hat{d}_k^{(i)(j)} = \sqrt{(x_j - x_{i,k-1})^2 + (y_j - y_{i,k-1})^2 + (z_j - z_{i,k-1})^2} \tag{11.13}$$

where

$$X_{i,k-1} = \begin{bmatrix} x_{i,k-1} \\ y_{i,k-1} \\ z_{i,k-1} \end{bmatrix} \tag{11.14}$$

11.3.2 Localization of Mobile RFID Readers: CoopAOA

Contrarily to *tag localization*, where a number of readers are used to determine the position of a tag, *reader localization* consists of calculating the position of a reader by means of a number of reference tags.

Exemplary applications of localization of mobile RFID readers include, for example, localizing workers carrying hand-held RFID devices in a warehouse, localizing and routing mobile robots equipped with RFID readers, etc.

The accuracy of the reader localization methods found in the literature mainly depends on the density of the reference tags (Pathanawongthum and Chemtanomwong 2010; Han et al. 2007) (see Figure 11.6). In order to reduce the number of reference tags, it might a solution can

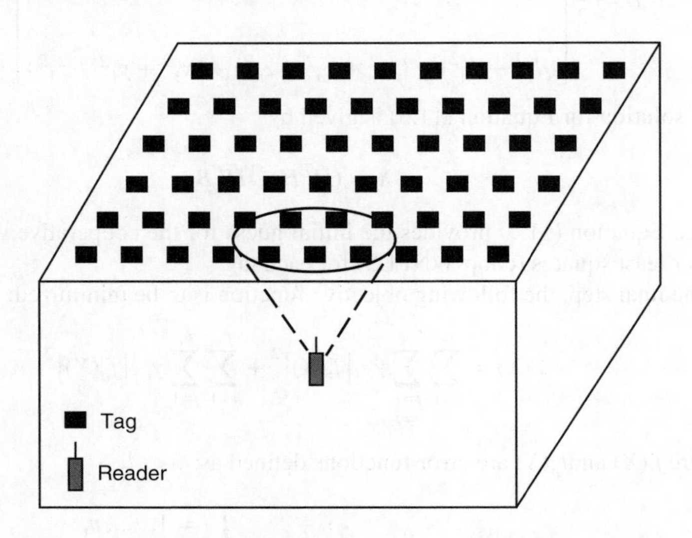

Figure 11.6 Localization of a hand-held RFID reader by means of a high density of reference tags.

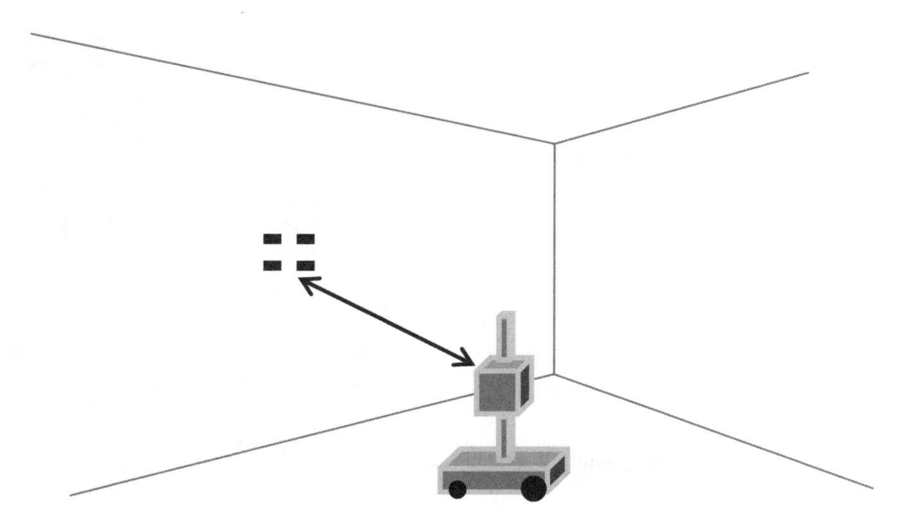

Figure 11.7 Localization of a mobile RFID reader by means of a low density of reference tags.

be envisioned where the mobile RFID reader is equipped with multiple antennas and performs AOA measurements with respect to the reference tags. However, the drawbacks of this solution are that multiple antenna RFID readers are not widespread yet, they are costlier than single antenna readers and they increase the complexity of the positioning system.

To overcome these drawbacks, we propose a new localization technique with a single antenna reader and at least a pair of reference tags (see Figure 11.7), which are separated by a certain correlation distance, to emulate a multiple antenna reader. In particular, a concept similar to virtual/cooperative multiple input–multiple output (MIMO) has been applied to an RFID system, where instead of mobile users being considered as multiple, co-located antennas, fixed, reference tags are considered as multiple, co-located antennas. These perform AOA measurements, thus enabling the system to determine the position of the mobile reader. In the following sections we will refer to this localization technique as Coop-AOA.

Figure 11.8a shows the typical AOA measurement by a two-element antenna array integrated in a reader. Figure 11.8b shows the proposed technique, where the AOA is calculated by a single tag-pair.

In order to use the reference tag-pairs as multiple antennas, the tags should be separated by no more than half a wavelength, λ, as otherwise the channel experienced by each of the tags in a tag-pair will be uncorrelated (Divakaran 2008). For example, if the operating frequency, f, is 900 MHz, the maximum separation distance of the tag-pair must be

$$d_{\max} = \frac{\lambda}{2} = \frac{c}{2 \times f} = 16.67\,\text{cm} \tag{11.15}$$

where c is the speed of light.

The AOA between the reference tag-pair and the reader (i.e., the angle θ in Figure 11.8b) can be determined by calculating the TDOA of the signals received by the reader from the reference tags in the tag-pair. In order to do this, the tags should theoretically be synchronized. However, tags (either passive or active) have no internal clock, therefore it is not possible to synchronize them. Thus, in order to calculate the TDOA at the reader, we propose that

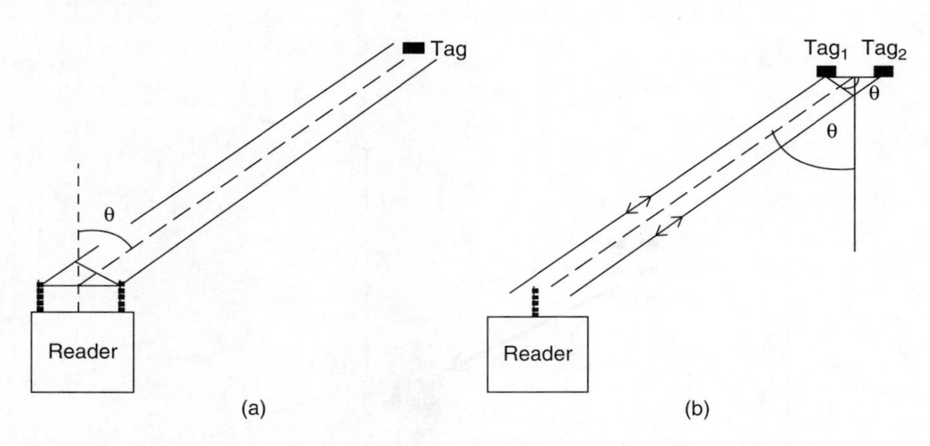

Figure 11.8 (a) AOA by multiple antenna reader. (b) CoopAOA by reference tag-pair.

the readero interrogates the tags and calculates the round trip time (RTT) for each tag of the tag-pair, that is, the time elapse between the transmission of the interrogation signal from the reader and the reception of the response signal from the tag. As a consequence, the issue of tag synchronization is circumvented. The proposed method to calculate the AOA with a single antenna RFID reader and a pair of passive RFID tags is described in detail below.

STEP 1: The reader interrogates the tags by modulating an RF carrier with amplitude shift keying (ASK) modulation (EPC Global 2008). Let us assume that the reader transmits a 32-bit sequence pilot signal, the carrier has a frequency of 900 MHz and the reader TX power is 20 dbm.

Figure 11.9 shows an unmodulated RF carrier, $x(t)$, at 900 MHz. The ASK-modulated signal can then be calculated as

$$y(t) = \begin{cases} x(t) \ldots\ldots\ldots\ldots \textit{if } `1' \\ 0 \ldots\ldots\ldots\ldots\ldots\ldots \textit{if } `0' \end{cases}$$

Figure 11.10 shows the pilot bit sequence (top of Figure 11.10) and the resulting ASK-modulated signal (bottom of Figure 11.10). Since the selected bit rate is 115.2 kbps, the bit duration is 8.67×10^{-6} s.

STEP 2: Being interrogated by the reader, each tag responds by modulating an RF carrier at 900 MHz with binary phase shift keying (BPSK) modulation (EPC Global 2008) (see Figure 11.11).

Figure 11.12 shows an example of two BPSK-modulated signals transmitted by two tags and received by the reader according to the layout of Figure 11.8b. In particular, it is noticeable that the two signals differ in terms of TOA at the reader as well as in terms of initial phase at the tags. Assuming LOS between the tag-pair and the reader, this is because the ASK-modulated signal transmitted by the reader arrives at Tag_2 with a delay with respect to Tag_1, therefore the BPSK-modulated signals transmitted by the tags show different initial phases and arrive at the reader at different times.

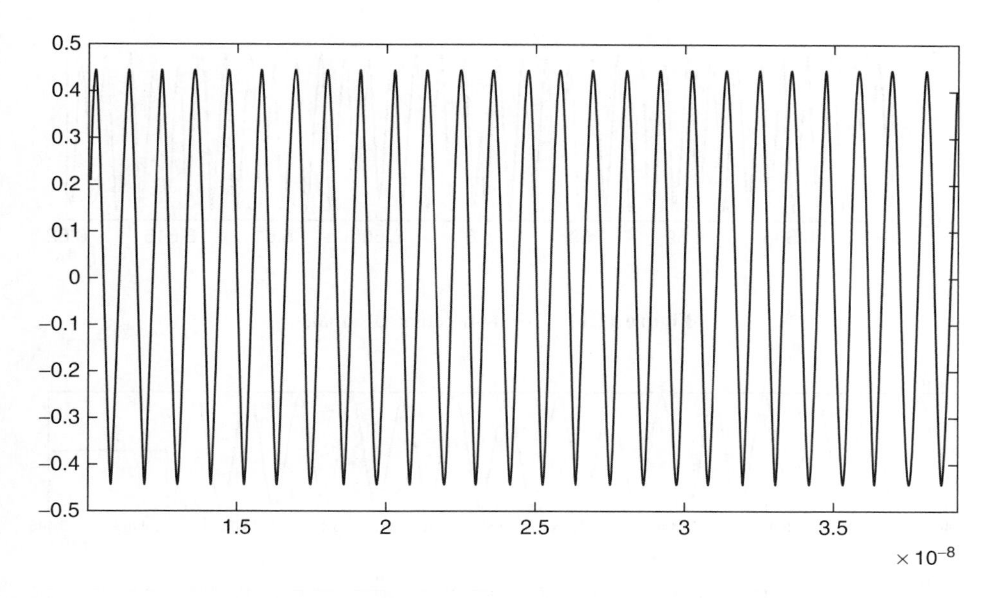

Figure 11.9 Unmodulated RF carrier at 900 MHz.

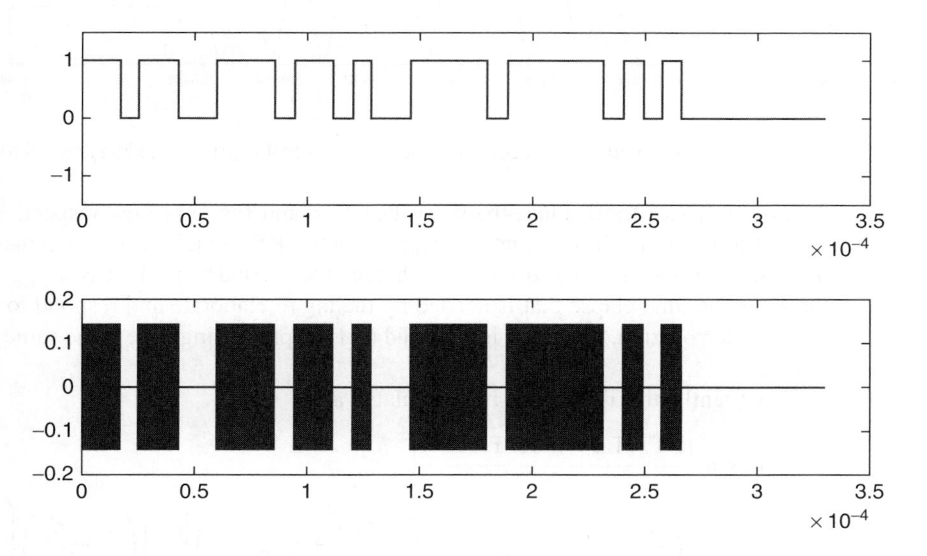

Figure 11.10 Pilot bit sequence (top) and ASK-modulated signal (bottom).

The RTT corresponding to each tag can be formulated as

$$\mathrm{RTT}(\mathrm{Tag}_1) = \frac{2 \times d_1}{c} + t_{\mathrm{processing}} \tag{11.16}$$

$$\mathrm{RTT}(\mathrm{Tag}_2) = \frac{2 \times d_2}{c} + t_{\mathrm{processing}} \tag{11.17}$$

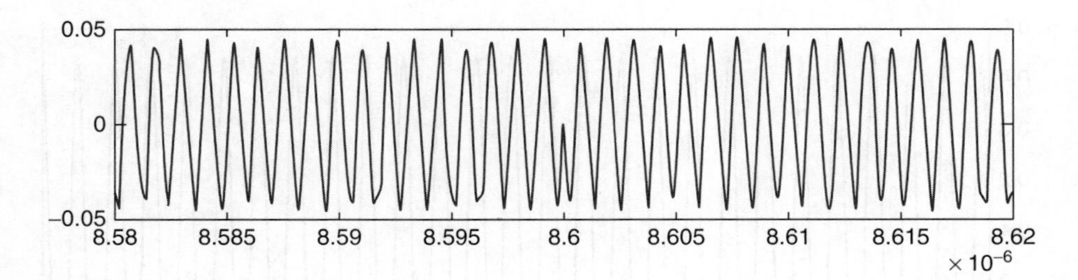

Figure 11.11 BPSK-modulated signal.

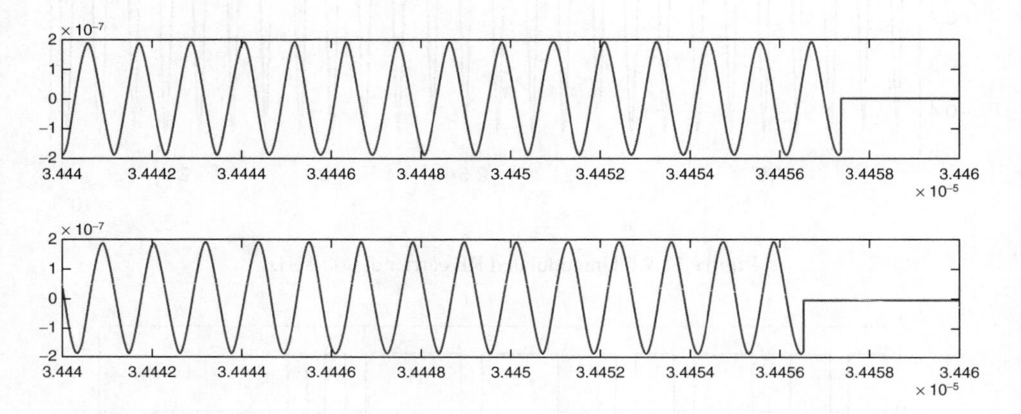

Figure 11.12 BPSK-modulated signal received at the reader from Tag_1 (top) and Tag_2 (bottom).

where d_1 and d_2 are the distances between the reader and Tag_1 and Tag_2, respectively. The first term in the above equations represents the RTT required by the signal to travel between the reader and the tag, whereas the second term, that is, $t_{\text{processing}}$, represents the time elapse that is needed by the tag to elaborate and respond to the reader's interrogation. Herein, it is assumed that the processing time is the same for each tag.

Consequently, the TDOA can be formulated as:

$$\begin{aligned}\text{TDOA} &= \left|\frac{\text{RTT}(\text{Tag}_1) - \text{RTT}(\text{Tag}_2)}{2}\right| \\ &= \left|\left(\frac{2\times d_1}{c} + t_{\text{processing}}\right) - \left(\frac{2\times d_2}{c} + t_{\text{processing}}\right)\right| = \left|\left(\frac{d_1 - d_2}{c}\right)\right|\end{aligned} \tag{11.18}$$

STEP 3: To calculate the TDOA, the cross-correlation of the two BPSK signals is carried out. In particular, the TDOA is the lag of the maximum of the cross-correlation function. Since both signals are BPSK, the cross-correlation of the two signals is in fact the auto-correlation function of one signal shifted by the TDOA (EPC Global 2008).

Let T_{auto} denote the time index corresponding to the maximum of the auto-correlation function and T_{cross} the time index corresponding to the maximum of the

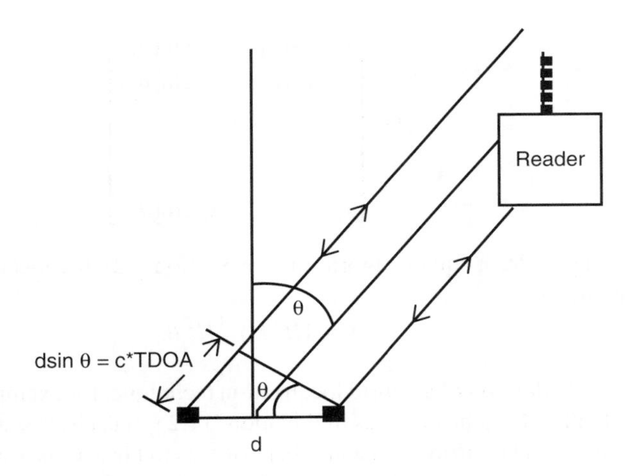

Figure 11.13 AOA determination from TDOA.

cross-correlation function. Then, since we are considering the RTT, the TDOA can be calculated as:

$$\text{TDOA} = \frac{|T_{\text{auto}} - T_{\text{cross}}|}{2} \tag{11.19}$$

STEP 4: The AOA is calculated from the TDOA thanks to a simple trigonometric formula according to the geometry of the system. In particular, referring to Figure 11.13, the AOA corresponds to the angle θ:

$$\theta = \arcsin\left(\frac{c \times \text{TDOA}}{d}\right) \tag{11.20}$$

where d is the known distance between the tags in the tag-pair.

STEP 5: Let us assume that the reader estimates the AOA, θ_i, corresponding to the ith tag-pair. Let (x_i, y_i)denote the midpoint of that tag-pair and (x, y) the location of the reader. Consequently, the following is derivable:

$$(x_i - x)\sin\theta_i = (y_i - y)\cos\theta_i \tag{11.21}$$

If there is an n number of tag-pairs available, then there will be an n number of equations, which can be written in matrix form as:

$$HX = B \tag{11.22}$$

where

$$H = \begin{bmatrix} -\sin(\theta_1) & \cos(\theta_1) \\ -\sin(\theta_2) & \cos(\theta_2) \\ \cdot & \cdot \\ \cdot & \cdot \\ \cdot & \cdot \\ -\sin(\theta_n) & \cos(\theta_n) \end{bmatrix} \tag{11.23}$$

$$B = \begin{bmatrix} y_1 \cos(\theta_1) - x_1 \sin(\theta_1) \\ y_2 \cos(\theta_2) - x_2 \sin(\theta_2) \\ \cdot \\ \cdot \\ \cdot \\ y_n \cos(\theta_n) - x_n \sin(\theta_n) \end{bmatrix} \tag{11.24}$$

By applying the least square algorithm (see Section 5.2), the location of the reader is obtainable by

$$X = (H^T H)^{-1} H^T B \tag{11.25}$$

In the case of adoption of a hybrid localization technique, for example TDOA-AOA, instead of only AOA, the matrices in Equation (11.25) must be cascaded accordingly. Let us consider a tag-array with a number i of tags (a tag-pair is a special configuration of a tag-array comprising only two reference tags). The total number of TDOAs corresponds to taking in turn one tag in the tag-array as a reference and determining the TDOA with respect to each of the remaining tags in the tag-array (e.g., if $i = 3$: TDOA_{12}, TDOA_{13}, TDOA_{23}). Therefore, we can write:

$$d_i - d_1 = c \times \text{TDOA}_{1i} = \Delta_{1i} \tag{11.26}$$

where d_i is the distance between the reader and Tag_i, d_1 is the distance between the reader and Tag_1, and TDOA_{1i} is the TDOA between Tag_1 and Tag_i. Equation (11.26) can be processed and rewritten as:

$$(\Delta_{1i} + d_1)^2 = d_i^2 \tag{11.27}$$

$$\Delta_{1i}{}^2 + 2\Delta_{1i} d_1 + d_1{}^2 = d_i^2 \tag{11.28}$$

$$\Delta_{1i}{}^2 + 2\Delta_{1i} d_1 = d_i^2 - d_1{}^2 \tag{11.29}$$

$$\Delta_{1i}{}^2 + 2\Delta_{1i} d_1 = \left\{(x_i - x)^2 + (y_i - y)^2\right\} - \left\{(x_1 - x)^2 + (y_1 - y)^2\right\} \tag{11.30}$$

$$\Delta_{1i}{}^2 + 2\Delta_{1i} d_1 = (x_i{}^2 + y_i{}^2) - (x_1{}^2 + y_1{}^2) - 2x(x_i - x_1) - 2y(y_i - y_1) \tag{11.31}$$

Equation (11.31) can be written in matrix form as $HX = B$, where:

$$H = \begin{bmatrix} x_2 - x_1 & y_2 - y_1 & \Delta_{12} \\ x_3 - x_1 & y_3 - y_1 & \Delta_{13} \\ \cdot & \cdot & \cdot \\ \cdot & \cdot & \cdot \\ x_n - x_1 & y_n - y_1 & \Delta_{1n} \end{bmatrix} \tag{11.32}$$

$$B = \frac{1}{2} \begin{bmatrix} (x_2{}^2 + y_2{}^2) - (x_1{}^2 + y_1{}^2) - \Delta_{12}{}^2 \\ (x_3{}^2 + y_3{}^2) - (x_1{}^2 + y_1{}^2) - \Delta_{13}{}^2 \\ \cdot \\ \cdot \\ (x_n{}^2 + y_n{}^2) - (x_1{}^2 + y_1{}^2) - \Delta_{1n}{}^2 \end{bmatrix} \tag{11.33}$$

Thus, Equations (11.23) and (11.24) become, respectively:

$$H = \begin{bmatrix} x_2 - x_1 & y_2 - y_1 & \Delta_{12} \\ x_3 - x_1 & y_3 - y_1 & \Delta_{13} \\ \cdot & \cdot & \cdot \\ \cdot & \cdot & \cdot \\ x_n - x_1 & y_n - y_1 & \Delta_{1n} \\ -\sin(\theta_1) & \cos(\theta_1) & 0 \\ -\sin(\theta_1) & \cos(\theta_1) & 0 \\ \cdot & \cdot & \cdot \\ \cdot & \cdot & \cdot \\ \cdot & \cdot & \cdot \\ -\sin(\theta_1) & \cos(\theta_1) & 0 \end{bmatrix} \quad (11.34)$$

$$B = \begin{bmatrix} \frac{1}{2}\left((x_2{}^2 + y_2{}^2) - (x_1{}^2 + y_1{}^2) - \Delta_{12}{}^2\right) \\ \frac{1}{2}\left((x_3{}^2 + y_3{}^2) - (x_1{}^2 + y_1{}^2) - \Delta_{13}{}^2\right) \\ \cdot \\ \cdot \\ \cdot \\ \frac{1}{2}((x_n{}^2 + y_n{}^2) - (x_1{}^2 + y_1{}^2) - \Delta_{1n}{}^2) \\ y_1\cos(\theta_1) - x_1\sin(\theta_1) \\ y_2\cos(\theta_2) - x_2\sin(\theta_2) \\ \cdot \\ \cdot \\ \cdot \\ y_n\cos(\theta_n) - x_n\sin(\theta_n) \end{bmatrix} \quad (11.35)$$

11.3.3 Performance Evaluation

This section deals with the performance evaluation of the proposed algorithm in different simulation scenarios. Both tag and reader localization systems are evaluated.

11.3.3.1 Simulation Scenario

11.3.3.1.1 Tag Localization

As shown in Figure 11.14, an indoor scenario has been considered, specifically an $8 \times 4 \times 3$ m room with a reader mounted at each corner of the ceiling at a height of 3 m. The person who is to be localized inside the room is assumed to be 1.65 m tall and wear one or more tags on his body (e.g., on his chest), with the distance between the tags is about 40 cm.

Figure 11.15 shows the typical arrangements of the tags on a person's body for single and multiple tag configurations.

11.3.3.1.2 Reader Localization

As shown in Figure 11.16, an indoor scenario has been considered, specifically a 10×10 m room with four tag-pairs mounted on each wall and a distance between the tags of 15 cm.

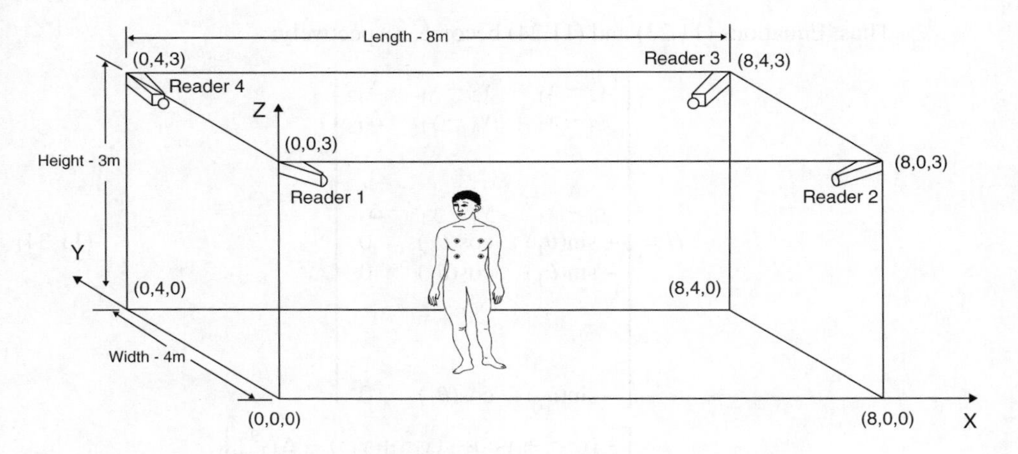

Figure 11.14 Simulation scenario (perspective view).

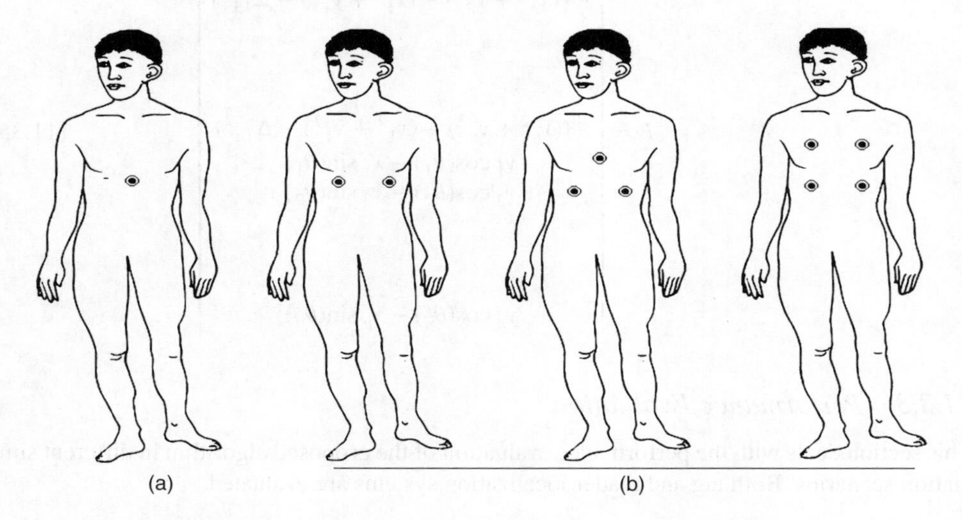

Figure 11.15 Tag configurations with: (a) a single tag; and (b) multiple tags.

The reader that is to be localized inside the room is assumed to be a portable, single-antenna RFID reader.

11.3.3.2 Simulation Parameters for Tags and Readers

Table 11.1 shows the simulation parameters, for both tags and readers, used in our performance evaluation.

11.3.3.3 Channel Model

11.3.3.3.1 Near-field Channel Model

In the near-field channel model, neither multipath fading or shadowing are taken into account, because (1) the attenuation due to multipath fading and shadowing is related to the far-field

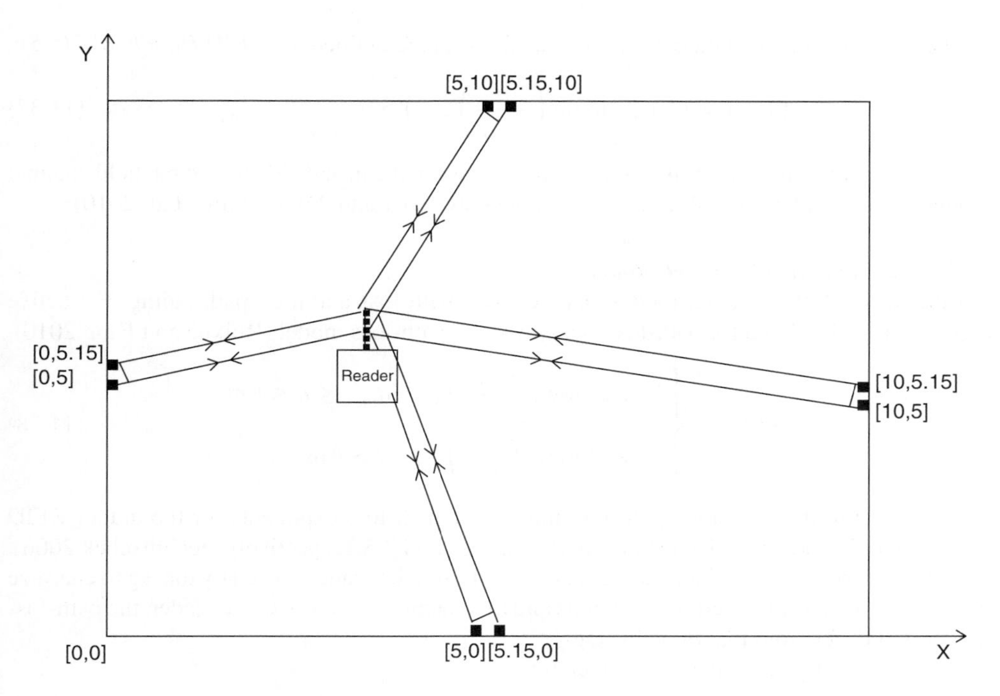

Figure 11.16 Simulation scenario (top view).

Table 11.1 Simulation parameters.

Parameter	Values
Carrier frequency, f_c	900 MHz (UHF)
Reader transmission power, $P_{TX,r}$	10–30 dBm
Active tag transmission power	1 mW
Antenna gain at the reader, G_R	0 dB (omnidirectional)
Antenna gain at the tag, G_T	0 dB (omnidirectional)
Passive tag sensitivity	–20 dBm (Jiang et al. 2006)
Modulation efficiency, η_{eff}	0.25 (Karthaus and Fischer 2003)
Sampling frequency, f_{samp}	5 GHz

model (Han et al. 2007) and (2) the multi-ray channel model does not exhibit relevant differences with respect to the free-space path-loss formula for distances lower than 1–1.5 m (*RFID Handbook* 2015).

The near-field channel model is thus characterized only by path-loss. In Nikitin et al. (2010), it was shown that the path-loss model in the near-field matches with the experimental measures as long as the distance d between the TX and the RX is no lower than 1.5 times λ (Han et al. 2007):

$$\lambda = \frac{c}{f} = \frac{3 \times 10^8\,\text{m/s}}{9 \times 10^8\,\text{Hz}} \approx 0.33\,\text{m} \Rightarrow 1.5 \times \lambda \approx 0.495\,\text{m} \tag{11.36}$$

Therefore, the channel model for the near-field can be defined as (*RFID Handbook* 2015):

$$\mathrm{PL}(d) = 32 + 25\log_{10}\left(\frac{d}{d_{\max}}\right), \quad 1.5 \times \lambda \le d \le d_{\max} \tag{11.37}$$

where $d_{\max}$ represents the break-point distance between the near-field and the far-field channel models. In particular, the values for $d_{\max}$ range between 1 and 2 m (Nikitin et al. 2010).

11.3.3.3.2 Far-field Channel Model

For the far field, the channel model includes both path-loss and multipath fading. For active tags, the model utilized for path-loss is a two-slope path-loss model (Ruixue and Fang 2010):

$$\mathrm{PL}(d) = \begin{cases} 32 + 25\log_{10}\left(\frac{d}{d_{\max}}\right), & d_{\max} \le d \le 8\,\mathrm{m} \\ 32 + 35\log_{10}\left(\frac{d}{d_{\max}}\right), & d > 8\,\mathrm{m} \end{cases} \tag{11.38}$$

According to the two-slope path-loss model, the path-loss exponents for the indoor RFID channel within and beyond a distance of 8 m are 2.5 and 3.5, respectively (Jechlitschek 2006).

In case of passive tags, since the signal received from the reader is used by the tag to energize itself, to perform some internal operations and to transmit a response to the reader, the path-loss model will differ from that of active tags.

The received power at the tag is given by:

$$P_{\mathrm{RX,t}} = P_{\mathrm{TX,r}} G_{\mathrm{T}} G_{\mathrm{R}} \left(\frac{\lambda}{4\pi d}\right)^2 \tag{11.39}$$

where $P_{\mathrm{TX,r}}$ is the transmitted power at the reader, G_{T} is the antenna gain at the transmitter, G_{R} is the antenna gain at the receiver, λ is the signal wavelength and d is the distance between reader and tag. Note that $P_{\mathrm{RX,t}}$ should be greater than the sensitivity of the tag in order to energize the latter.

The received power at the reader is given by

$$P_{\mathrm{RX,r}} = P_{\mathrm{TX,t}} G_{\mathrm{T}} G_{\mathrm{R}} \left(\frac{\lambda}{4\pi d}\right)^2 \tag{11.40}$$

where $P_{\mathrm{TX,t}}$ is the transmitted power at the tag:

$$P_{\mathrm{TX,t}} = \eta_{\mathrm{eff}} \times P_{\mathrm{RX,t}} \tag{11.41}$$

η_{eff} is the modulation efficiency (Divakaran 2008), defined as

$$\eta_{\mathrm{eff}} = \frac{\text{power transmitted by the tag}}{\text{power received by the tag}} \tag{11.42}$$

Therefore, Equation (11.40) can be rewritten as

$$P_{\mathrm{RX,r}} = \eta_{\mathrm{eff}} P_{\mathrm{TX,r}} (G_{\mathrm{T}} G_{\mathrm{R}})^2 \left(\frac{\lambda}{4\pi d}\right)^4 \tag{11.43}$$

In this case the modified path-loss model is (Leong et al. 2005)

$$\mathrm{PL}_{\mathrm{dB}} = 63 - 10\log_{10}\eta + 50\log_{10}d \tag{11.44}$$

Finally, we assume that at least one LOS path is available between the tags and the readers. Hence, the multipath fading can be characterized by a Rician distribution with a K value of 7 (David 2015).

11.3.3.4 Simulation Results

11.3.3.4.1 Tag Localization

Impact of number of tags on localization accuracy Figure 11.17 shows the comparative plots of the cumulative distribution functions (CDFs) of the estimation error for a varying number of active tags. It is noticeable that (1) the configuration with multiple tags clearly outperforms the configuration with a single tag and (2) the localization accuracy increases (though slowly) with the number of tags.

Figure 11.18 shows the comparative plots of the CDFs of the estimation error for a varying number of passive tags. It is noticeable that (1) the configuration with multiple tags clearly outperforms the configuration with a single tag, (2) the localization accuracy increases (though slowly) with the number of tags and (3) the system with multiple active tags outperforms that with multiple passive tags (as expected).

Hence, multi-tag localization shows higher localization accuracy with respect to single-tag localization.

Impact of the polarization tag/reader on the localization accuracy For simplicity, we have considered the following two cases: (1) the reader–tag combination has the same polarization (i.e., both the reader and the tag are in horizontal/vertical polarization) and (2) the reader–tag combination is cross-polarized (i.e., if the reader has a horizontal polarization, the tag has a vertical polarization, and vice versa).

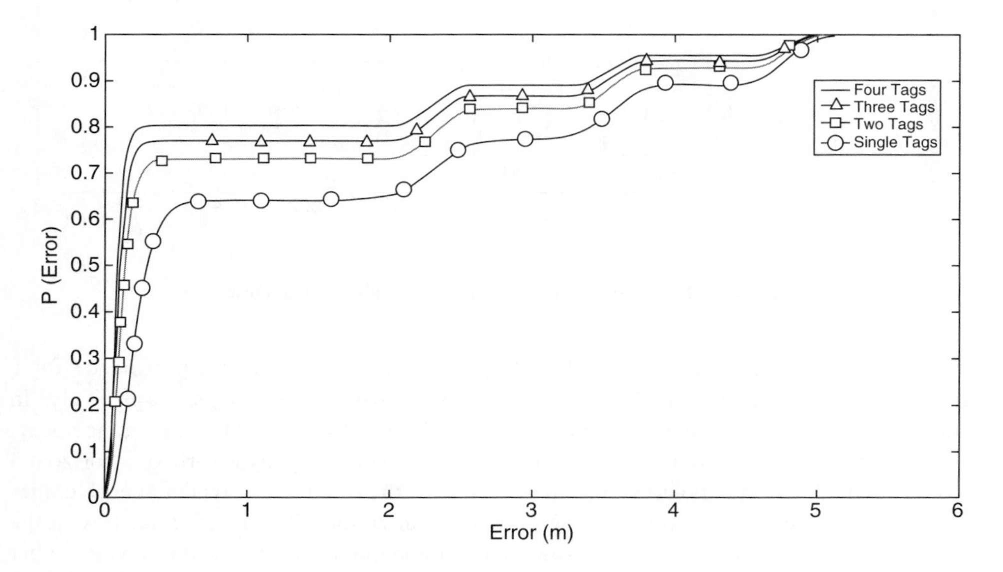

Figure 11.17 Comparison between single and multiple active tags.

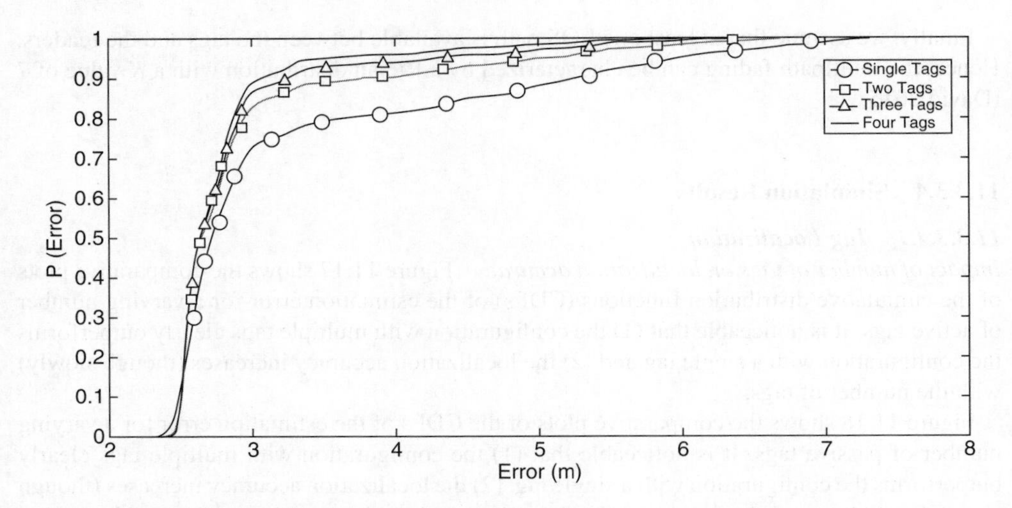

Figure 11.18 Comparison between single and multiple passive tags.

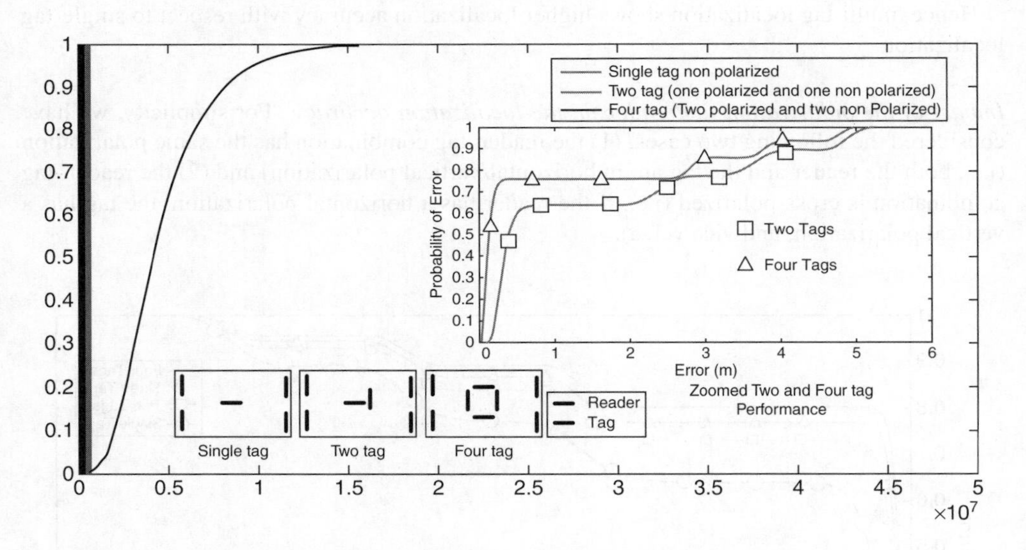

Figure 11.19 Effect of polarization on localization accuracy.

In Figure 11.19 we compare the effect of polarization on the error performance for a single-tag scenario and two multi-tag scenarios with two and four tags, respectively. In particular, the three rectangular boxes inside Figure 11.17 depict the reader–tag arrangements for each scenario. The rectangular box on the left shows a single tag in the form of a horizontal bar (i.e., having horizontal polarization) in the centre of the room with a reader at each corner in the form of a vertical bar (i.e., having vertical polarization). The rectangular box in the middle shows one tag in the form of a horizontal bar and one tag in the form of a vertical bar in the centre of the room with the same reader configuration as in previous scenario. Finally, the rectangular box on the left shows two tags in the form of a horizontal bar and two tags in the form of a vertical bar in the centre of the room with the same reader configuration as in the previous scenarios.

From Figure 11.19 it is noticeable that (1) the configuration with multiple tags clearly outperforms the configuration with a single tag and (2) the localization accuracy increases with the number of tags (see plot inside Figure 11.19). This is because the likelihood of a polarization mismatch between tags and readers is decreased with an increasing number of available tags. In particular, in the case with four tags, two of which have the same polarization as the readers, the weight associated to that tag-pair is high in the data fusion algorithm and is correlated with the known distance between the two tags.

Hence, multi-tag localization shows higher localization accuracy with respect to single-tag localization.

Impact of localization accuracy in height on decision taking Herein, we have carried out a performance evaluation of the error in localization accuracy in the z axis, that is, with respect to the height of the target. In particular, such an analysis is remarkable because decisions can be taken on the base of the estimated height of the target. For example, in case of a patient being monitored, it is of utmost importance to determine whether the patient is standing, sitting or lying on the floor.

Figures 11.20–11.22 show the error performance in height for both single- and multi-tag localization for three vertical settings corresponding to a target lying (50 cm), sitting (1 m) and standing (1.5 m), respectively. The results are also summarized in Table 11.2. Note that we have assumed that correct decisions on the state of the target can be taken with an error in height of $\pm$10 cm from the real height.

From Figures 11.20–11.22 and Table 11.2 it is noticeable that (1) the configuration with multiple tags clearly outperforms the configuration with a single tag and (2) the localization accuracy increases with the number of tags. Hence, multi-tag localization shows higher localization accuracy with respect to single-tag localization.

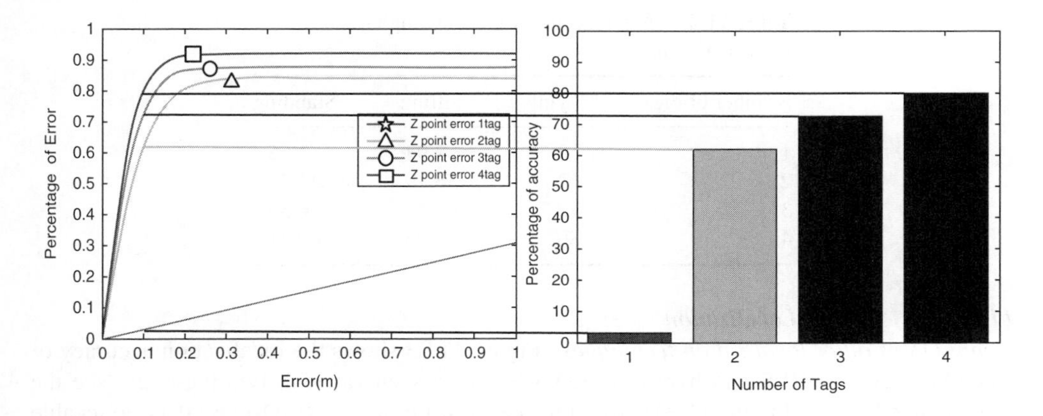

Figure 11.20 CDFs of the localization error in height and corresponding accuracy for a lying target.

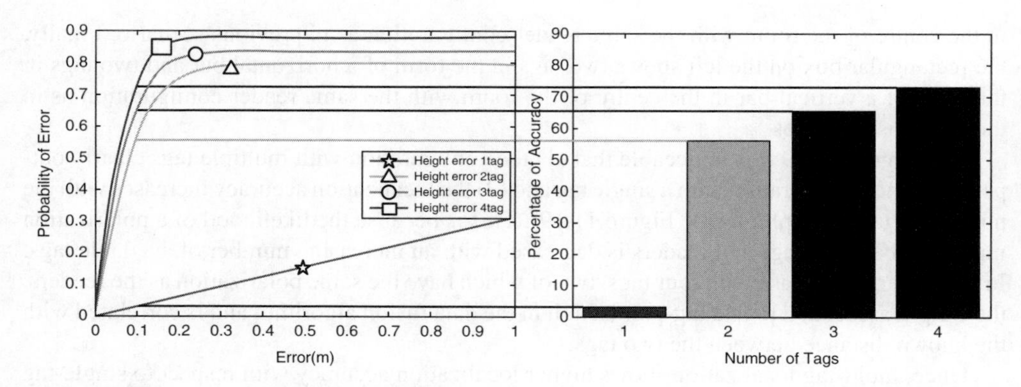

Figure 11.21 CDFs of the localization error in height and corresponding accuracy for a sitting target.

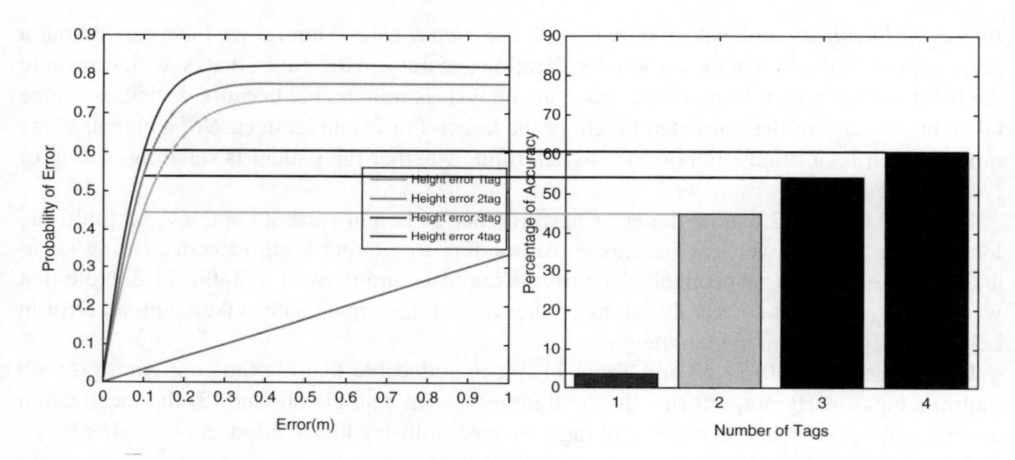

Figure 11.22 CDFs of the localization error in height and corresponding accuracy for a standing target.

Table 11.2 Height accuracy (%) for single- and multi-tag scenarios.

Number of tags	Lying	Sitting	Standing
1	3.23	2.93	3.55
2	62	56.19	44.75
3	72.7	65.37	54.39
4	80	72.95	60.61

11.3.3.4.2 Reader Localization

CoopAOA vs other localization techniques Figure 11.23 shows the localization accuracy of CoopAOA against RSS and hybrid TDOA-RSS for a scenario with two tag-pairs (see the box depicted inside Figure 11.24), wherein the tags are active RFID tags. It is noticeable that CoopAOA clearly outperforms the other two localization techniques, where RSS performs worse.

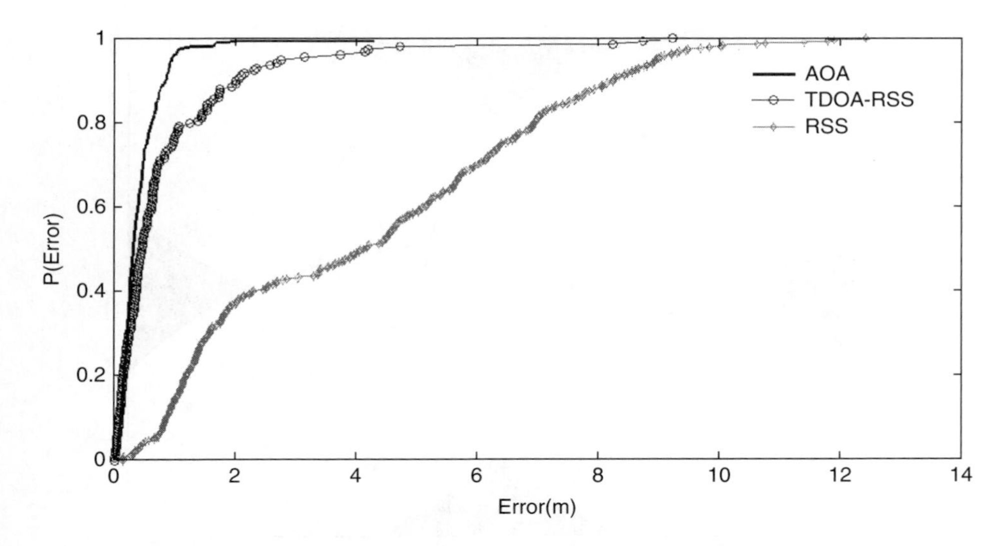

Figure 11.23 CoopAOA vs RSS and TDOA-RSS.

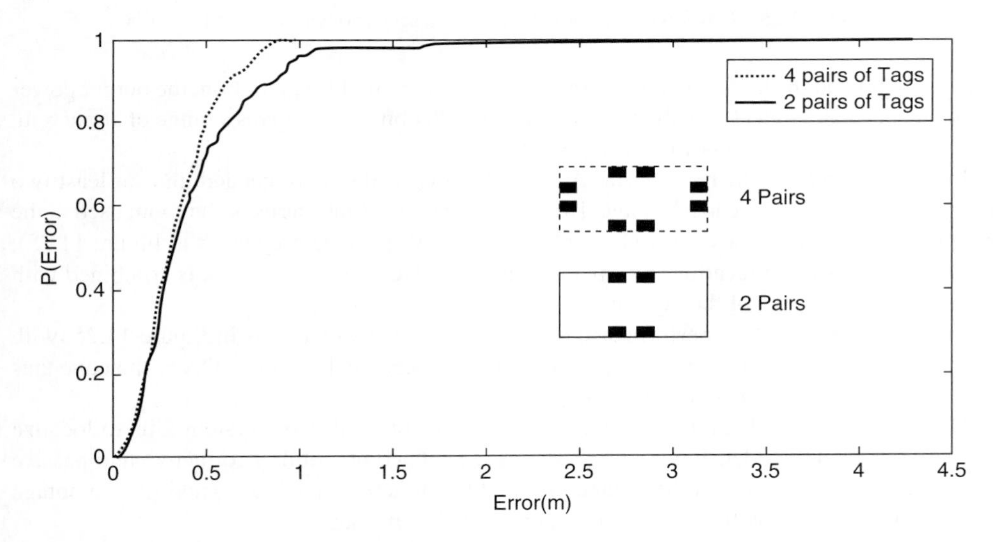

Figure 11.24 Localization accuracy of CoopAOA with two and four tag-pairs.

Impact of number of tag-pairs on the localization accuracy of CoopAOA Figure 11.24 shows the localization accuracy of CoopAOA for a scenario with two and four tag-pairs, respectively (see the boxes depicted inside Figure 11.24). It is noticeable that CoopAOA with four tag-pairs outperforms CoopAOA with two tag-pairs, although marginally.

Impact of active/passive tags on the localization accuracy of CoopAOA In this section, the error performance of CoopAOA with passive tags has been assessed. The main drawback of passive tags is the short range. Since passive tags are energized by the reader power, the reader

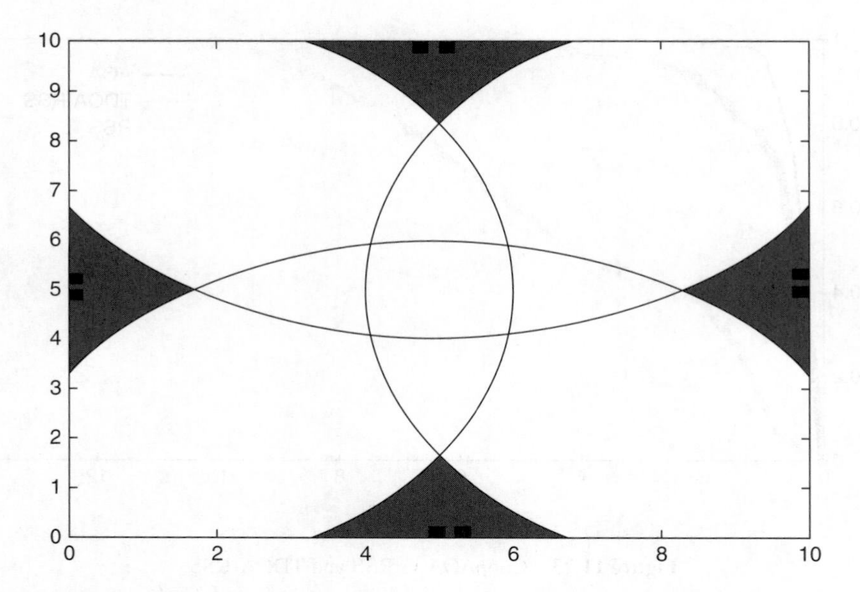

Figure 11.25 Coverage area for the simulation scenario with passive tag-pairs.

power output and the tag's sensitivity are essential factors. In this simulation, the output power of the reader is 30 dbm (1 W), the tag sensitivity is –20 dbm, which gives a range of about 6 m, and the modulation efficiency of the tag is 25%.

In order to localize the reader with AOA, at least two AOAs are needed, thus at least two tag-pairs must be in the reader's range. However, due to the dimensions of the room used in the simulation, there are zones covered only by one tag-pair (see the grey areas in Figure 11.25). As a consequence, no location information can be derived unless the AOA is combined with the RSS from each tag of the tag-pair.

Figure 11.26 shows the performance of CoopAOA for a scenario as in Figure 11.25, with four active and passive tag-pairs, respectively. It is noticeable that CoopAOA with active tags outperforms that with passive tags, as expected.

Herein, "outage probability" is defined as the probability that the system fails to localize the position of the reader. Figure 11.27 shows the outage probability for active and passive tags. It is noticeable that (1) the outage probability for active tags is zero and (2) the outage probability decreases with an increasing number of passive tags.

11.3.3.5 Energy–Cost–Accuracy Analysis

For active tags, the transmission power of the reader is set to 11 dBm (10 mW). Since the sensitivity of active tags is about –97 dBm, the aforementioned power is sufficient to interrogate the tags. If the bit rate is 115.2 Kbps, the duration of the signal transmitted by the reader to the tags (i.e., the pilot signal consisting of 32 bits modulated with ASK) is $32 \times \frac{1}{115.2 \times 10^3} = 0.2778$ ms. Thus, the energy consumed at the reader for this transmission is 1.736×10^{-6} J. If we now assume that two tag-pairs are mounted in the room (one on each wall or corner) and the reader pings the tags every 2 s to encounter for a location update in case of movement (i.e., it

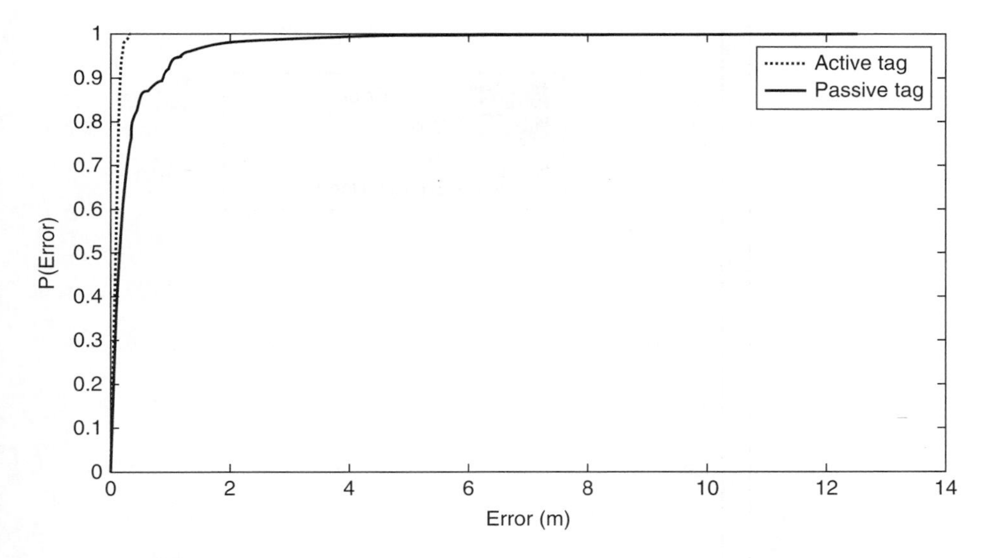

Figure 11.26 Error performance of CoopAOA with active and passive tags.

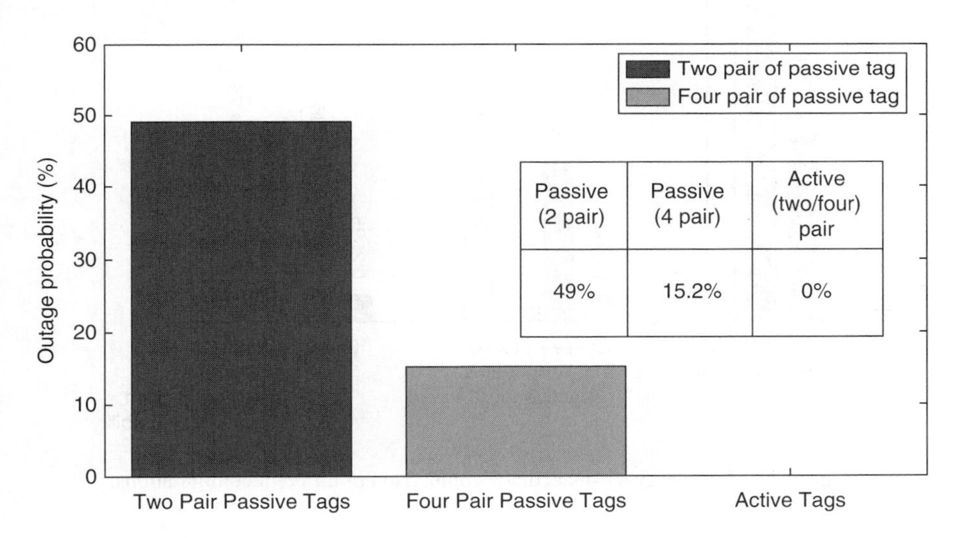

Figure 11.27 Outage probability for CoopAOA with passive and active tags.

interrogates the tags 30 times/min or 1800 times/h), the total energy consumed per hour will be $(1.736 \times 10^{-6}) \times 1800$ J = 0.003125006451 J. Consequently, the total energy consumption for the only transmission from the reader to the tags during an 8-h operation is 0.025 J.

According to the *RFID Handbook* (2015), passive tags require at least –20 dBm (10^{-5}W) for internal operation and backscattering of the signals to the reader. We have therefore set the transmission power of the reader at 30 dBm (1 W), which satisfies the tag sensitivity and provides a read range of 5.5 m. Following the same calculation carried out above in the active tags case, the energy consumed at the reader for the transmission of the pilot signal is

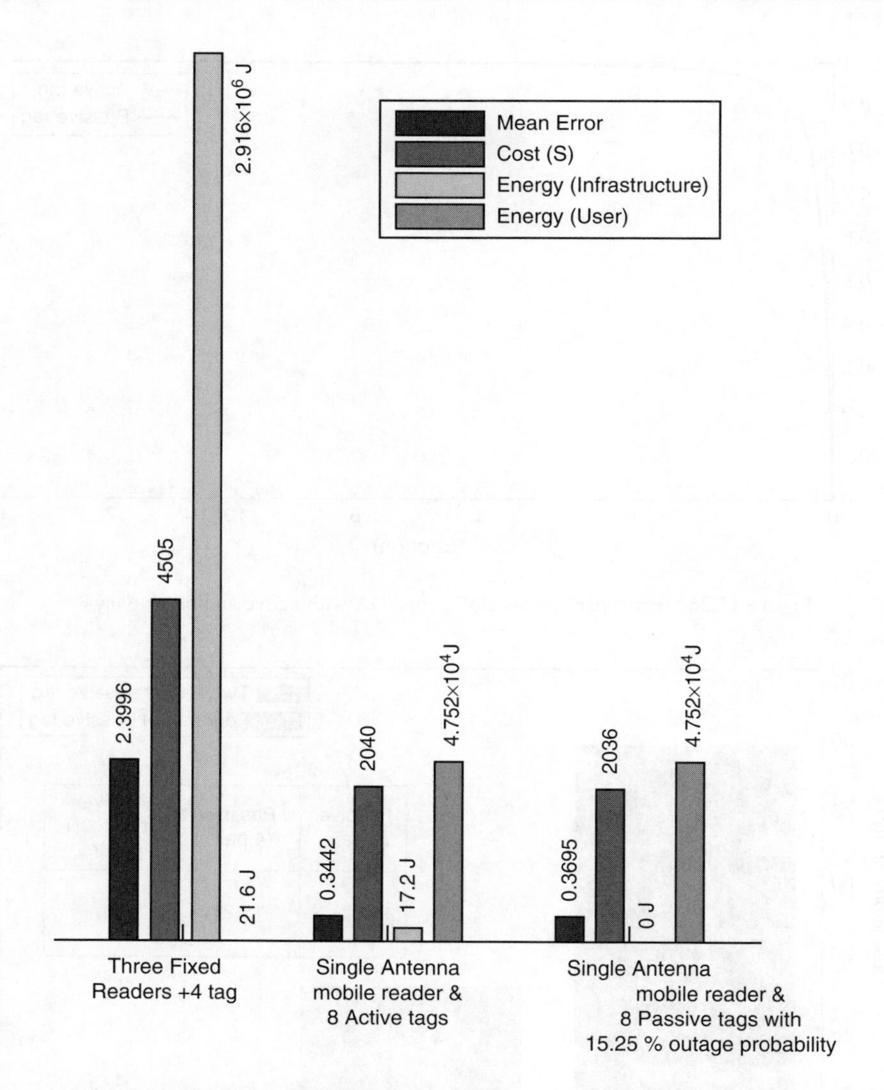

Figure 11.28 Energy–cost–accuracy comparison of different configurations.

$1.041668817 \times 10^{-4}$ J. Consequently, the total energy consumption in case of a mobile reader and two passive tag-pairs during an 8-h operation is $(1.041668817 \times 10^{-4}) \times 1800 \times 8$ J = 1.500003096738612 J.

Although we did not consider the energy consumption spent in performing the signal reception and the cross-correlation, the energy consumption at the reader side is very small with respect to the total energy stored in its battery. In particular, a standard battery would be consumed for a continuous operation in 7.99 h in the case of active tags and in 7.84 h in the case of passive tags.

Figure 11.28 shows the energy–cost–accuracy of different configurations assuming the costs in Table 11.3.

Table 11.3 Costs of readers and tags.

Item	Cost ($)
Fixed reader	1500
Mobile reader	2000
Active tag	5
Passive tag	0.50

From Figure 11.28 it is noticeable that (1) multi-tag localization exhibits a higher energy consumption on the infrastructure side (i.e., three readers) than on the user side (i.e., four tags) due to the number of readers involved in the interrogation of the tags, (2) reader localization by CoopAOA exhibits a higher energy consumption on the user side (i.e., 1 reader) than the infrastructure side (i.e., four pairs of active/passive tags) due to the reader involved in the interrogation of the tags and (3) reader localization by CoopAOA exhibits a higher energy consumption on the infrastructure side with active tags than with passive tags, as expected.

11.3.4 Experimental Activity for Tag Localization

11.3.4.1 Scenario

The experiment (including both the calibration and the localization) was carried out in the NOVI building at Aalborg University. Figure 11.29 shows a scenario with three readers located at three corners of an 8 × 5 m open space within the building. For practical reasons, the three readers were mounted on three chairs (see Figure 11.30). Note that, since we used three readers, 2D localization was performed.

The grid in Figure 11.29 represents the positions of the four active tags used during each trial. For practical reasons, the tags were mounted on a board, where the distance between the tags was 30 cm (see Figure 11.31).

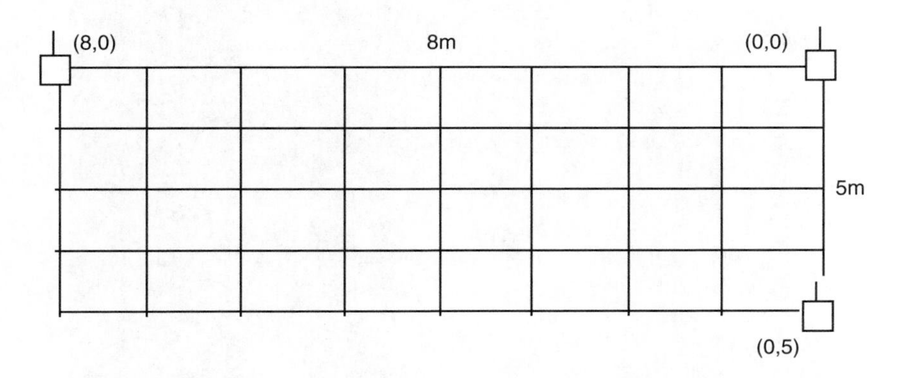

Figure 11.29 Scenario of the experimental setup.

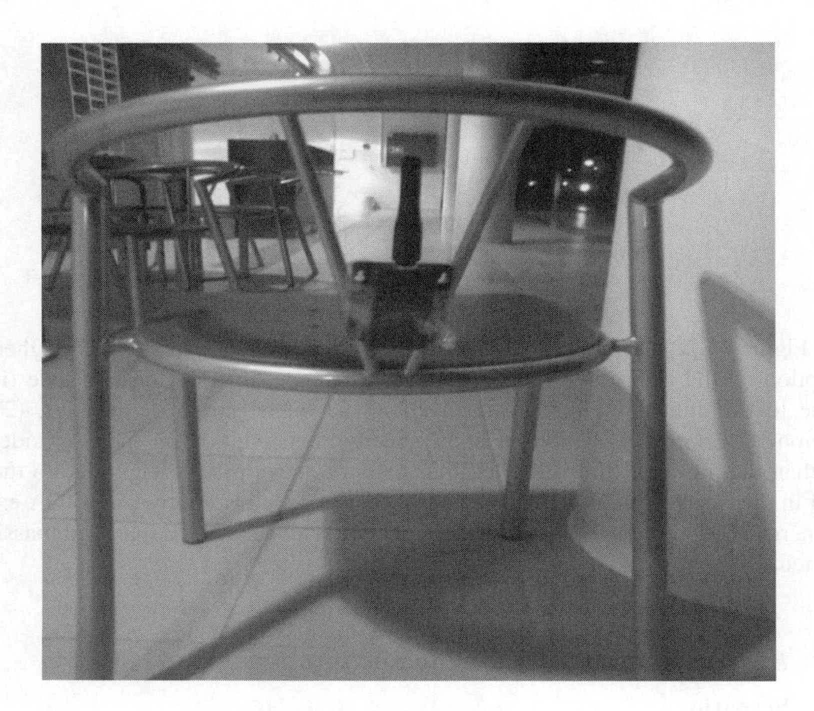

Figure 11.30 A reader mounted on a chair.

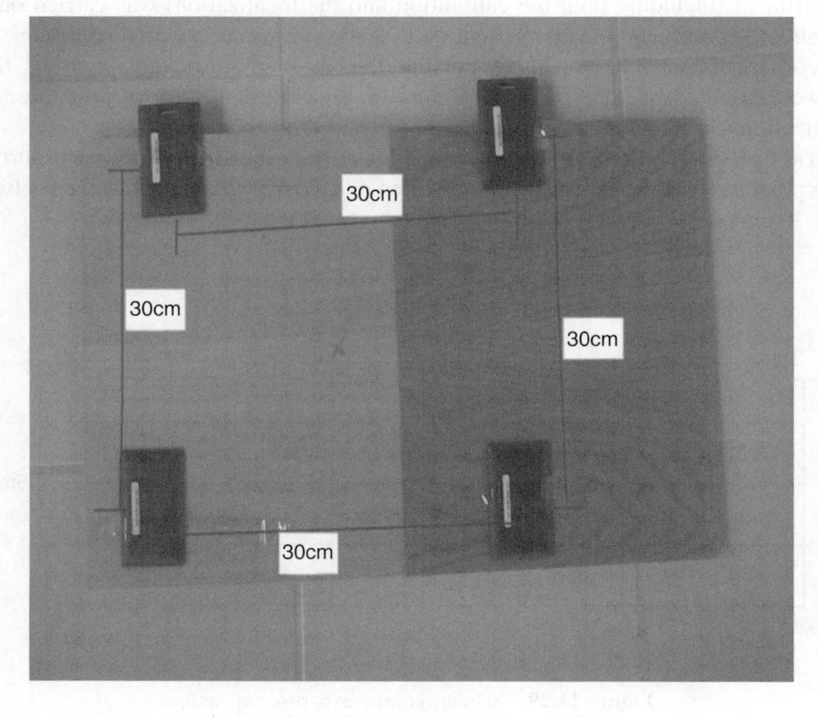

Figure 11.31 Arrangement of the four tags.

Table 11.4 Reader (Wavetrend L-RX201).

Parameter	Value
Frequency	433 MHz
Modulation	ASK
Sensitivity	–103 dBm
Supply voltage	6–18 V DC
Maximum current consumption	±72–90 mA
Size	84 × 40 × 19 mm
Weight	45 g

Table 11.5 Active tag (Wavetrend).

Parameter	Value
Frequency	433.92 MHz
Power output	38 uW, –14 dBm, 93 dBuV
Modulation	ASK
Weight	<15 g (L-TG100 excluding antenna)
Battery	Lithium

11.3.4.2 Equipment

The technical specifications of the reader and tags used in the experiments are given in Tables 11.4 and 11.5.

11.3.4.3 Channel Calibration

Since we ran experiments based on RSS measurements, the channel modeling is very important. We developed the channel model by gathering RSS measurements, in particular received signal strength indicator (RSSI) values ranging from 0 to 255, at a distance ranging from 0 to 20 m from each reader. The resulting data were smoothed by means of a moving averaging filter. The latter smooths data by replacing each data point with the average of the neighboring data points defined within a certain span. Let y denote the unsmoothed data, then the smoothed data after applying the moving average filter can be derived as follows:

$$y_s(i) = \frac{1}{2N+1}(y(i+N) + y(i+N-1) + \ldots + y(i-N)) \tag{11.45}$$

where $y_s(i)$ is the smoothed value for the ith data point, N is the number of neighboring data points on either side of $y_s(i)$, and $2N + 1$ is the span. Finally, we applied different curve fittings (see Figures 11.32–11.34).

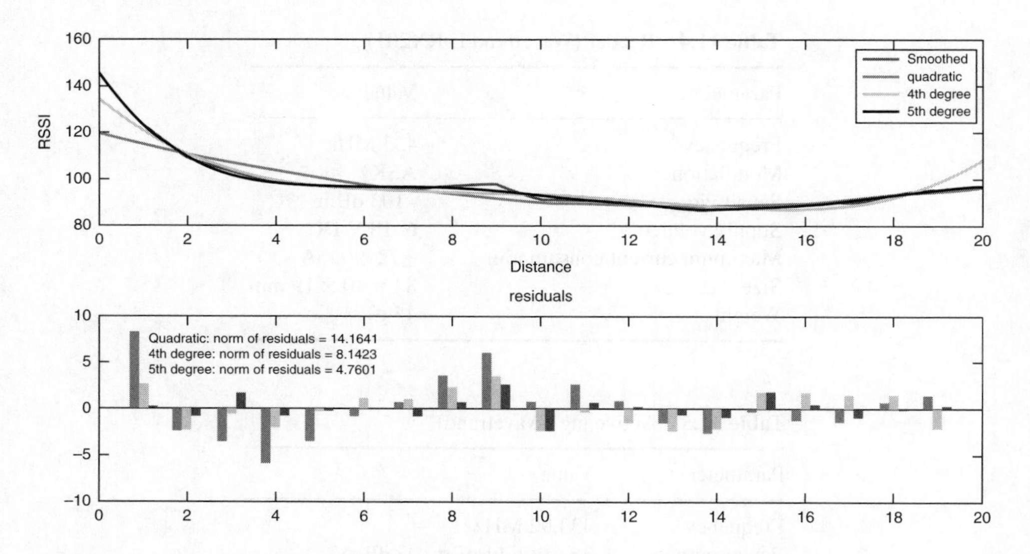

Figure 11.32 Different curve fittings for reader 1.

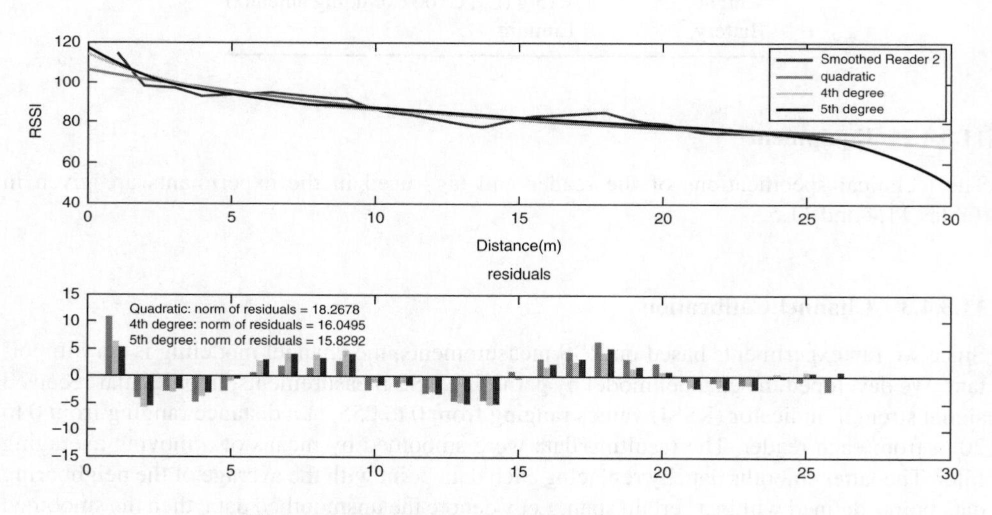

Figure 11.33 Different curve fittings for reader 2.

From the above figures, it transpires that the fifth-degree polynomial is the best fitting as it provides the fewest residuals with respect to other curve fittings. Hence, we used the fifth-degree polynomial to characterize the channel model according to the following equations:

$$\begin{aligned}\text{Reader 1}: y = &-0.0005431 \times x5 + 0.03153 \times x4 - 0.6709 \times x3 \\ &+ 6.473 \times x2 - 28.89 \times x + 145\end{aligned} \quad (11.46)$$

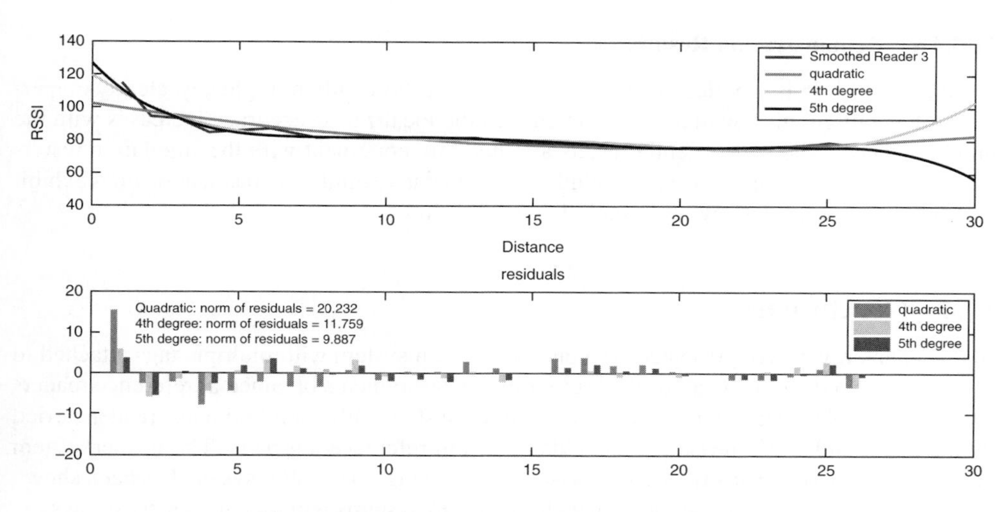

Figure 11.34 Different curve fittings for reader 3.

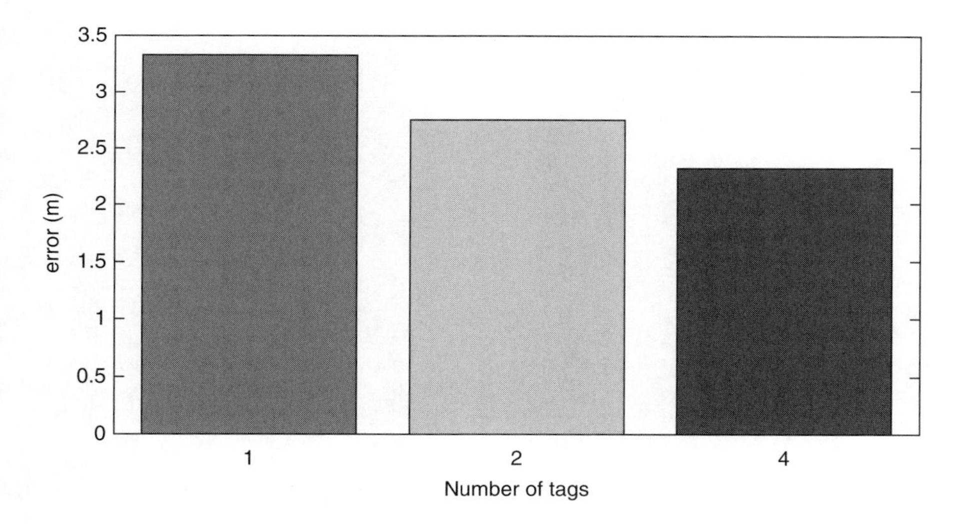

Figure 11.35 Localization accuracy with a different number of tags.

$$\text{Reader 2}: y = -5.353e-005 \times x5 + 0.\ 0.003827 \times x4 - 0.1045 \times x3 + 1.37 \times x2 - 9.644 \times x + 118.8 \tag{11.47}$$

$$\text{Reader 3}: y = -0.0001028 \times x5 + 0.007957 \times x4 - 0.2306 \times x3 + 3.077 \times x2 - 19.06 \times x + 126.6 \tag{11.48}$$

where y represents the RSSI value at the x distance.

11.3.4.4 Demonstration Results

From Figure 11.35 it is noticeable that (1) the configuration with multiple tags clearly outperforms the configuration with a single tag and (2) the localization accuracy increases with the number of tags. The experimental results are therefore congruent with the simulation results (see Figure 11.18). Hence, we can conclude that cooperative multi-tag localization does exhibit higher localization accuracy over single-tag localization.

11.4 Conclusions

In this chapter, we have presented (i) a tag localization system with multiple tags attached to the target and a data fusion algorithm combining measurements obtained at reference readers from the multiple tags; and (ii) a reader localization system with a single antenna reader carried by the target and AOA measurements obtained from reference tag-pairs. The former system has been shown to outperform systems based on single tags. The latter system has been shown to provide a low-cost and high-accuracy alternative to systems with a high density of tags.

12

Cooperative Mobile Positioning

Simone Frattasi[1], Joaõ Figueiras[2] and Francescantonio Della Rosa[3,4]
[1]*Sony Mobile Communications, Lund, Sweden*
[2]*Aalborg University, Aalborg, Denmark*
[3]*Radiomaze Inc, Cupertino, California, USA*
[4]*Tampere University of Technology, Finland*

12.1 Introduction

As described in previous chapters, the underlying technologies used in mobile positioning systems are either *satellite-based* or *terrestrially based*, and differ from each other in terms of accuracy, coverage, cost, energy consumption, application environment (outdoor or indoor) and system impact. Satellite-based technologies are employed mainly for outdoor applications and are mainly represented by the GPS. Even though the latter is the most popular solution on the market because of its navigation applications, the introduction of mobile handsets with built-in receivers for the third-generation (3G) system led to increased cost, size and battery consumption, and took a long time to reach full market penetration (Sayed et al. 2005a). Furthermore, in difficult environments, such as outdoor urban canyons and indoor environments, which are of greatest importance to service providers, it is difficult, if not impossible, to obtain any sort of location information. This is due to the unfeasibility of having a clear view of at least four satellites, or to signal blocking and multipath effects. Current terrestrially based (GPS-free) technologies have the same drawbacks in multipath environments and in NLOS conditions, in which none of the costly accurate environmental information is available. Hence, investigations have started in connection with the fourth-generation (4G) system to define a solution that overcomes the drawbacks of the GPS and the current terrestrially based technologies, and provides location information with a high level of accuracy *anywhere and anytime*. As described in Frattasi (2007), wireless cooperation gives designers the potential to achieve enhancements in terms of coverage, spectrum usage and energy consumption. In this chapter, we will discuss the impact of wireless cooperation on location estimation accuracy.

Cooperative localization is an approach that has been used in robot networks and wireless sensor networks (WSNs) in order to enhance the location estimation accuracy and thus enlarge

Mobile Positioning and Tracking: From Conventional to Cooperative Techniques, Second Edition.
Simone Frattasi and Francescantonio Della Rosa.

the range of applications of such networks. In practice, in addition to using measurements from known to unknown devices, measurements between unknown devices are also exploited, which permits a more robust and precise localization to be achieved. In particular, WSNs employ clustering as a technique to reduce the volume of data to be transmitted to the remote station by aggregating location measurements from co-located sensors in a master device, which then takes care of the long-range transmissions. Recently, the cooperative localization approach has been ported to wireless mobile networks, where the actors are no longer robots or sensors but users, with their mobility patterns and social behaviors. In this context, we have coined the expressions *cooperative mobile positioning* and *cooperative augmentation system* to refer to cooperative localization applied to wireless mobile networks and to localization systems employing such an approach, respectively.

Cooperative mobile positioning is becoming increasingly important as a promising new branch of wireless location in which several research directions are being explored (e.g., positioning, tracking, data fusion and clustering). Its identifying concept is the exploitation of likely reliable mobile-to-mobile measurements to increase the location estimation accuracy of a satellite-based or terrestrially based system, which would have been provided otherwise only by fixed-to-mobile measurements. In this chapter, we use the framework of cooperative mobile positioning framework and introduce the COMET as an innovative solution for position determination in fourth generation (4G) wireless networks. COMET is supported by the "ad coop" architecture specified in Frattasi (2007). Specifically, peer-to-peer (P2P) communications are exploited in a mesh fashion within cellularly established clusters for cooperation-aided localization purposes. On receiving a location information request from a mobile station (MS), the home base station (BS) forms a cluster in the surroundings of that MS, where the latter can be chosen as the cluster head (CH) and its neighbors as cluster members (CMs). While the available BSs perform time/time-difference-based and angle-based measurements for each BS–CM link, each CM performs range-based measurements for each CM–CM link. Finally, the positions of all the CMs are obtained by novel data fusion methods, which appropriately combine the available long- and short-range location information. The numerical results presented in Section 12.4 show that thanks to the cooperation established within a group of mobiles, our proposal has the potential to enhance the location estimation accuracy with respect to conventional terrestrial positioning systems in stand-alone cellular networks.

The rest of the chapter is organized as follows. Section 12.2 presents a brief survey of cooperative localization in robot networks and WSNs, and introduces the cooperative mobile positioning approach and its overall benefits. Section 12.3 illustrates the data fusion and filtering techniques of Chapter 5, which have been adapted to cope with the cooperative framework. Section 12.4 describes COMET, an example of a cooperative augmentation system (CAS) in a cellular network, with its system architecture and its data fusion methods; in particular, Section 12.4.3 discusses its performance, which has been evaluated via computer simulations. Finally, our concluding remarks are made in Section 12.6.

12.2 Cooperative Localization

12.2.1 Robot Networks

Robotics is a field that has been around for decades and ranges from the simplest domestic appliances to the more sophisticated robots used in manufacturing plants and warehouses, and the ones used in the realm of biomedical engineering, for example for teleguided microsurgery. In particular, when we move from fixed to mobile robots, an avenue of other applications pops up (MobileRobots 2009; Parker 1996): delivery, guidance, mapping, patrolling, environmental monitoring and inspection, search-and-rescue missions, archaeological investigations, planetary explorations, etc. Even though a multiskilled single robot may accomplish several of these tasks, there are many applications in which, owing to the inherent characteristics of the mission/environment or a shortage of time, it is faster and cheaper to deploy a team of robots. Consider, for instance, the case of patrolling a museum: security personnel could remotely control from a central office a squad of mobile robots deployed in different zones of the museum by changing their directions, shifting their camera angles, and deciding what type of measurements to take and what mechanical movements to perform (Butler 2003). Moreover, we are able to increase the reliability and robustness of the system by combining several robots (Lima 2007). Based on the mission to be accomplished, we can also decide whether to deploy homogeneous or heterogeneous robots, that is, robots with the same or different type of features (e.g., measuring equipment).

Regardless of their specific characteristics and tasks, being able to localize a team of mobile robots is one of the enabling features for the aforementioned applications. In particular, we emphasize that while in the case of a single robot we are interested in its *absolute* position, in the case of a team of robots we are more interested in the *relative* position of each robot with respect to its team mates. This is mainly because multi-robot teams usually perform actions that require a high level of coordination. As mentioned in Howard et al. (2003), "Consider, for example, a team of robots executing a formation behavior: these robots need not to know their latitude and longitude, but *must* know the relative pose of their neighbors." In this context, by exploiting the fact that mobile robots often need to communicate in order to coordinate their efforts, an approach called *cooperative localization* has been developed. In practice, each robot performs positioning measurements (e.g., radio- or image-based) with respect to its team mates via its in-built sensors (see Figure 12.1), and by exchanging these data over the air, more accurate location information for the whole team can be achieved (Mourikis and Roumeliotis 2006b). For a survey of related work on cooperative localization in robot networks, the reader is referred to Mourikis and Roumeliotis (2006a).

12.2.2 Wireless Sensor Networks

The interest in WSNs has recently grown at a tremendous pace. Like robot networks, WSNs facilitate a plethora of applications (Alhmiedat and Yang 2007): healthcare, environmental and species monitoring, military surveillance, search and rescue, tracking soldiers and cars, etc. As mentioned in Patwari et al. (2005), "Smart structures will actively respond to earthquakes and make buildings safer; precision agriculture will reduce costs and environmental impact by watering and fertilizing only where necessary and will improve quality by monitoring storage conditions after harvesting; condition-based maintenance will direct equipment servicing exactly when and where it is needed based on data from wireless sensors; traffic monitoring systems will better control stoplights and inform motorists of alternate routes in the case of traffic jams; and environmental monitoring networks will sense air, water, and soil quality and identify the source of pollutants in real time." In general, many of these applications are enabled by the use of a large number of sensor nodes, and therefore it is highly desirable that such a network is built up in a cheap, low-energy-consuming and, as much as

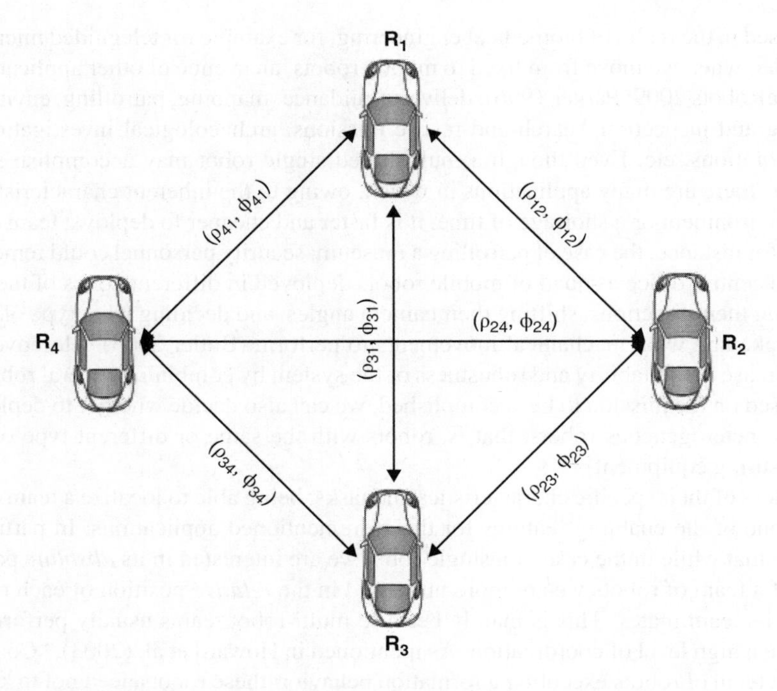

Figure 12.1 Cooperative localization in a network of mobile robots.

possible, self-organizing way. As a consequence, a sensor node – which is constituted of a battery, a sensor, a processor, a transceiver and a memory – usually has the following characteristics (Alhmiedat and Yang 2007): (1) small physical size, (2) low power consumption, (3) limited processing power, (4) limited communication capabilities (only short-range) and (5) limited memory storage.

As in the case of robot networks, localization is a key feature of WSNs. Indeed, "a sensor's location must be known for its data to be meaningful" (Patwari et al. 2005). Specifically, we can classify the localization techniques envisioned for WSNs as shown in Figure 12.2. First we distinguish between single-hop and multi-hop techniques. While *single-hop* refers to scenarios in which the unknown-location node is a one-hop neighbor of a sufficient number of known-location nodes (also called anchor nodes), *multi-hop* refers to scenarios in which the unknown-location node is $\kappa > 1$ hops from the anchor nodes. Within the realm of multi-hop techniques, we can divide these techniques further into range-free and range-based techniques. Unlike the former techniques, which assume that each node exploits some network-related information (e.g., connectivity information) to retrieve its own position, the latter techniques rely on distance-based measurements between nonanchor nodes. Finally, we can differentiate range-based techniques into centralized and distributed approaches, depending on whether the information obtained by each unknown-location node is sent to a central processing node or processed independently. For a survey of work on localization in WSNs, the reader is referred to Mao et al. (2007).

In practice, single-hop techniques embrace the localization techniques described in Chapter 5. The multi-hop case, however, monopolizes most of the attention, as it is the most

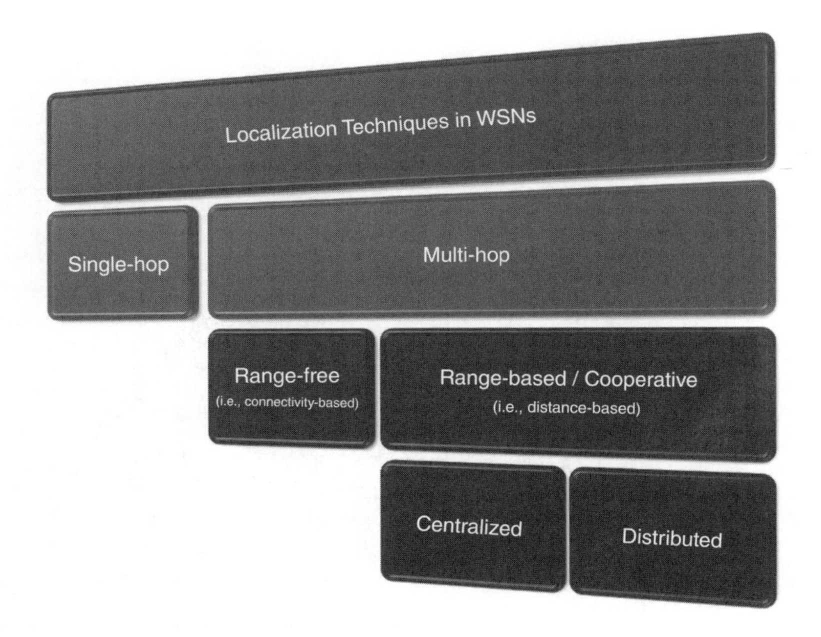

Figure 12.2 Taxonomy of localization techniques for WSNs.

common case in WSNs. Indeed, as mentioned before, the latter are usually vast networks and have to be kept low cost and low power consuming, and therefore only very few nodes can be considered as anchor nodes (e.g., GPS-equipped nodes), and the probability for an unknown-location node to be close to them will consequently be very low. In this context, similarly to the case of relative positioning in robot networks, we can use the cooperative localization approach, where, in addition to measurements between unknown-location nodes and anchor nodes, each unknown-location node performs local measurements with respect to its one-hop neighbors (Figure 12.3). In particular, if all this information is conveyed to a location server, the latter can calculate the estimated "absolute" positions of all the nodes in the network at once, and the accuracy and robustness of the localization system will increase (Patwari et al. 2005).

12.2.2.1 Clustering

As described above, WSNs are characterized by a high node density and the possibility of data aggregation. In particular, since the measurements performed by neighboring sensor nodes are correlated, local aggregation can be done in order to reduce the amount of data to be sent to the remote station (Mhatre and Rosenberg 2005). As a consequence, the classical operation of clustering or grouping of the sensors is performed (see Section 6.4.4.1). Practically, the space covered by the network is divided into several logical spots, inside which CHs, CMs and cluster gateways (CGs) are defined (Figure 12.4). In particular, the CH is the controller and aggregator of the group, the CMs are the sensor nodes belonging to the cluster and obeying the CH, and the CGs are the interfaces between different clusters. Besides its original purpose

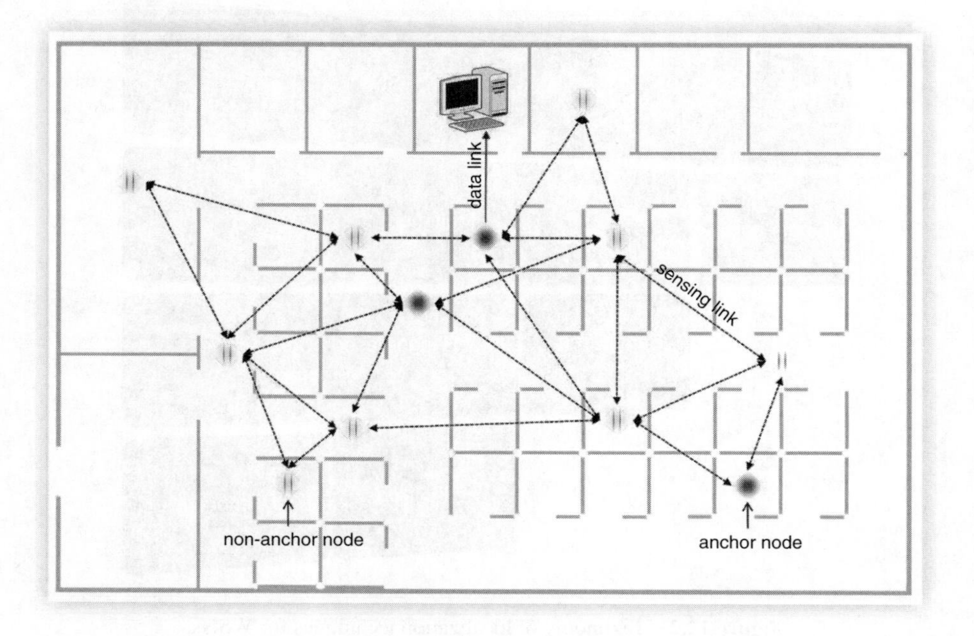

Figure 12.3 Cooperative localization in WSNs.

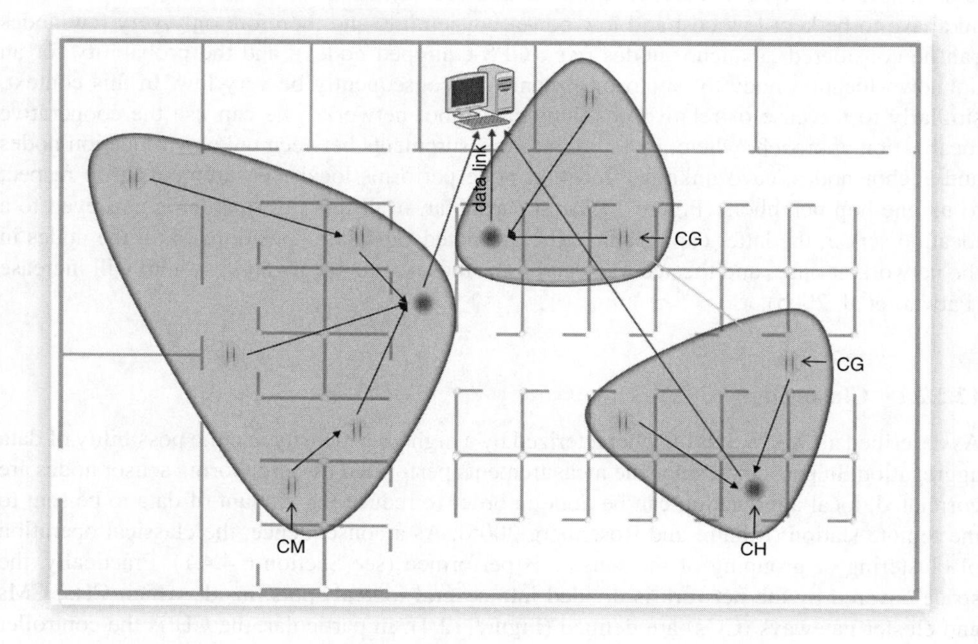

Figure 12.4 Clustering in WSNs.

of lightening the process of data aggregation over a vast network, clustering also brings the following advantages (Heinzelman et al. 2002; Mhatre and Rosenberg 2004): scalability, bandwidth reuse, increased system capacity, smarter resource allocation, improved power control, maximized network lifetime and fault tolerance. For a survey of work on clustering in WSNs, the reader is referred to Wei and Chan (2005).

12.2.3 Wireless Mobile Networks

As described in the previous sections, cooperative localization is an approach that was originally used in robot networks (for relative positioning) and later widely adopted in WSNs (also in this case for absolute positioning, provided there is a sufficient number of anchor nodes). Such an approach has been ported recently to wireless mobile networks, that is, networks composed of MSs immersed in a wireless or cellular layout, where the goal is to improve the accuracy of standard localization systems (satellite-based or terrestrially based) in environments where the latter alone do not suffice. In this sense, we refer to a localization system which employs the cooperative localization approach as a CAS. Practically, using the formalism of WSNs, besides having measurements between anchor nodes (i.e., BSs, APs or satellites) and unknown-location nodes (i.e., MSs) at our disposal, we rely also on measurements between unknown-location nodes. Specifically, while satellite-based and terrestrially based localization systems on their own provide high accuracy in rural and suburban scenarios, CASs are best exploitable in (1) indoor environments, where both global navigation satellite systems (GNSSs) and cellular-based localization systems provide poor accuracy, and WLAN-based localization systems have to deal with challenging propagation conditions, and (2) outdoor environments in the presence of urban canyons, where both GNSSs and cellular-based localization systems provide poor accuracy, and a dense wireless local area network (WLAN) might not be available.

As shown in Figure 12.5, the presence of multipath effects and shadowing between the majority of MSs and the available satellites or BSs might compromise the accuracy of the localization system. In this context, the employment of the cooperative localization approach has two main benefits:

1. Owing to the short distances between users, the P2P measurements between MSs are likely to be very accurate and therefore even if all MSs are in NLOS conditions with respect to all available reference points, the accuracy of the localization system can be enhanced (Frattasi 2007; Frattasi and Figueiras 2007).
2. Owing to the spatial diversity of the MSs, if at least one of them is in LOS with respect to one reference point (e.g., MS_1 with respect to S_1 or MS_2 with respect to BS_1), we can circumvent co-located shadowing effects. Using the terminology of WSNs, the MSs in LOS can be treated as anchor nodes and the accuracy of the localization system can be reduced and equalized to the accuracy of one of those MSs over the region considered (Frattasi and Monti 2007b).

Such a remarkable discovery opens up previously unimaginable scenarios, which will lead to new business models incorporating incentives to enable cooperation. For instance, we could imagine a situation where users will actually possess a virtual GPS on their mobiles, as they would be able to rely on the equipment of their neighbors. Therefore, even less sophisticated

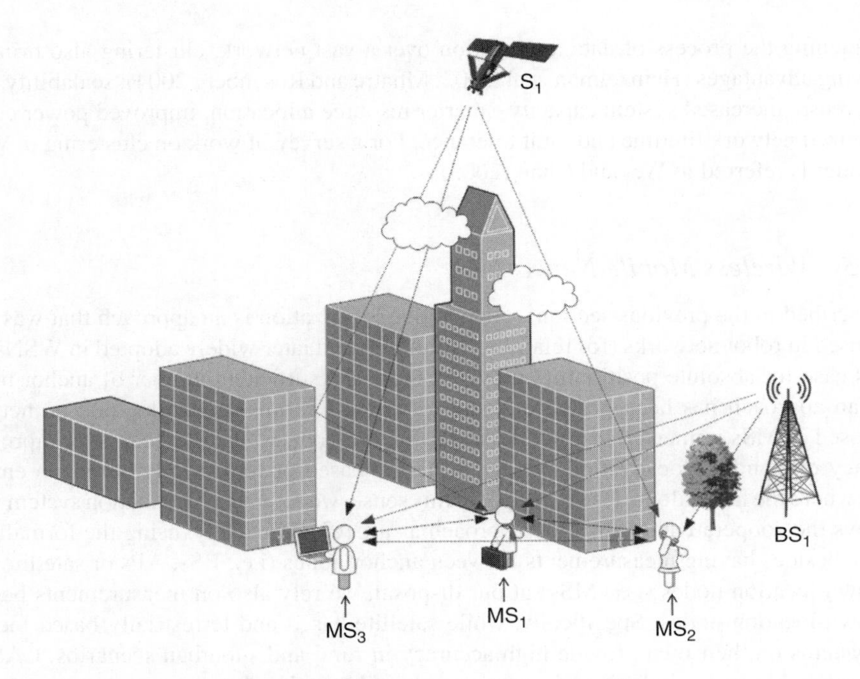

Figure 12.5 Typical "urban canyon" scenario.

and less costly devices will be able to achieve the same accuracy level as if they had a GPS receiver embedded. On the other hand, the latter will help the former to obtain a continuous and accurate positioning service in areas where the GPS might not suffice (e.g., in outdoor urban canyons). Hence, there is in any case a benefit for all MSs (GPS-equipped and not), which might represent a sufficient incentive for utilizing the service. In addition to the previous two benefits, it has also to be considered that, as in the case of WSNs, the accuracy of a CAS increases with the density of cooperating MSs (Frattasi 2007). Therefore, since the aforementioned scenarios are the most obvious environments in which there is a high user density, one can immediately appreciate the great potential of this approach and the breakthrough that it represents which we refer to as *cooperative mobile positioning*: cooperative localization applied to wireless mobile networks.

Cooperative mobile positioning is an approach that applies naturally to heterogeneous settings where MSs embed technologies suited for both long- and short-range communications. It is possible to envisage several combinations of systems and technologies that could occur. In a macroscale indoor scenario, for example, a WLAN could be used for performing location measurements on both the AP–MS links and the MS–MS links, for instance this could be implemented by switching the MS's Wi-Fi from infrastructure to ad hoc mode and back again (Rosa et al. 2007b). In a microscale indoor scenario, it would be better to use a WPAN (e.g., a Bluetooth-, ZigBee- or UWB-based system) to perform the short-range measurements. Finally, in an outdoor scenario, a cellular system (e.g., GSM, 3GPP-LTE or WiMAX), a GNSS (e.g., GPS) and a WLAN system could be used for performing location measurements on the BS–MS links, the satellite–MS links and the MS–MS links, respectively.

As a consequence of all of these considerations, it clearly emerges that *cooperative mobile positioning could represent a global (indoor/outdoor) augmentation solution for the mobile localization problem.*

12.3 Cooperative Data Fusion and Filtering Techniques

In this section, we extend the algorithms shown in Chapter 5 to the cooperative framework. We also describe some specific examples of their application, which will be extensively used in the next section.

12.3.1 Coop-WNLLS: Cooperative Weighted Nonlinear Least Squares

Given a vector function $f : \mathbb{R}^q \mapsto \mathbb{R}$, a cooperative weighted nonlinear least squares (coop-WNLLS) problem is of the form

Find

$$\hat{X} = \arg\min_X \mathcal{J}(X),$$

where

$$\mathcal{J}(X) = \sum_{i=1}^{l} \sum_{j=1}^{m} \gamma_{i,j}[f_{i,j}(X)]^2, \tag{12.1}$$

with $l + m \geq q$ and $\gamma_{i,j} \geq 0$. For practical solutions to this minimization problem, the reader is referred to Madsen et al. (2004).

12.3.1.1 Example of Application[1]

Let us consider a co-located number n_{ms} of MSs ($\text{MS}_1, \ldots, \text{MS}_{n_{\text{ms}}}$) in the range of a geographically distributed number n_{bs} of cellular BSs ($\text{BS}_1, \ldots, \text{BS}_{n_{\text{bs}}}$). Moreover, let us assume that while each MS performs either m_{bs} TOA or TDOA measurements with respect to each available BS, only the home BS, BS_1, performs m_{bs} AOA measurements with respect to each MS. In addition, let us suppose that each MS is in the range of the other MSs and performs m_{ms} RSS measurements with respect to each of them. Hence, after the estimation process, at the location server we have the following available sets of data for each MS: (1) (**TOA**, **AOA**, **RSS**) or (2) (**TDOA**, **AOA**, **RSS**), with entries defined by Equations (5.45)–(5.47) and

$$\mathbf{RSS} = \begin{bmatrix} \mathbf{RSS}_1 \\ \vdots \\ \mathbf{RSS}_{n_{\text{ms}}-1} \end{bmatrix} = \begin{bmatrix} RSS_{1,1} \cdots RSS_{1,m_{\text{ms}}} \\ \vdots \\ RSS_{n_{\text{ms}}-1,1} \cdots RSS_{n_{\text{ms}}-1,m_{\text{ms}}} \end{bmatrix}. \tag{12.2}$$

Let us write this problem in a coop-WNLLS form.

[1] S. Frattasi, M. Monti, "Ad-Coop positioning system (ACPS): Positioning for cooperative users in hybrid cellular ad-hoc networks", *European Transactions on Telecommunications Journal*, Wiley, vol. 19, no. 8, pp. 923–924, May, 2007. Reproduced in part by permission of © 2007 John Wiley & Sons Ltd.

STEP 0: In addition to the definitions presented in Section 5.2.3.1, which can be simply generalized by replacing the index m with m_{bs}, we start by introducing the following definitions (Frattasi and Monti 2007b):

$$\hat{X} = \begin{bmatrix} \hat{X}^{(1)} \\ \vdots \\ \hat{X}^{(n_{\mathrm{ms}})} \end{bmatrix} = \begin{bmatrix} \hat{x}^{(1)} & \hat{y}^{(1)} \\ \vdots & \vdots \\ \hat{x}^{(n_{\mathrm{ms}})} & \hat{y}^{(n_{\mathrm{ms}})} \end{bmatrix}, \tag{12.3}$$

where $\hat{X}$ represents the estimated location coordinates of $\mathrm{MS}_1, \ldots, \mathrm{MS}_{n_{\mathrm{ms}}}$ in a Cartesian coordinate system:

$$\hat{d}^{(i)(j)} = 10^{(\alpha - p^{(i)(j)})/10\beta}, \quad \begin{cases} i = 1, \ldots, n_{\mathrm{ms}}, \\ j = 1, \ldots, n_{\mathrm{ms}}, \\ j \neq i, \end{cases} \tag{12.4}$$

where $\hat{d}^{(i)(j)}$ is the estimated distance between MS_i and MS_j derived from $p^{(i)(j)}$, the RSS of the MS_j signal at MS_i, such that the parameters α and β are given by Equation (4.1), and

$$\hat{d}_k^{(i)(j)} = \sqrt{\{\hat{X}_{k-1}^{(i)} - \hat{X}_{k-1}^{(j)}\}^{\mathrm{T}}\{\hat{X}_{k-1}^{(i)} - \hat{X}_{k-1}^{(j)}\}}, \tag{12.5}$$

where $\hat{d}_k^{(i)(j)}$ has a similar meaning to that in Equation (12.4), but with the difference that here, instead of being calculated directly from the RSS measurements, it is derived from $\hat{X}_{k-1}^{(i)} = [\hat{x}_{k-1}^{(i)}, \hat{y}_{k-1}^{(i)}]^{\mathrm{T}}$ and $\hat{X}_{k-1}^{(j)} = [\hat{x}_{k-1}^{(j)}, \hat{y}_{k-1}^{(j)}]^{\mathrm{T}}$, the estimated location coordinates of MS_i and MS_j at iteration $k-1$ of the minimization routine.

STEP 1: The coop-WNLLS method requires initial guesses

$$\hat{X}_0 = [\hat{X}_0^{(1)}, \ldots, \hat{X}_0^{(n_{\mathrm{ms}})}]^{\mathrm{T}}$$

at the beginning of its minimization routine. These can be chosen, for example, as the outputs of an ordinary least-squares algorithm (see Section 5.2).

STEP 2: As well as Step 2 of Section 5.2.3.1, we may apply the median operator to each of the rows in Equations (5.45)–(5.47) and (12.2), and, by using the formalism introduced previously, we obtain Equations (5.55)–(5.57) and

$$\mathbf{RSS} = \begin{bmatrix} \widetilde{\mathbf{RSS}_1} \\ \vdots \\ \widetilde{\mathbf{RSS}_{n_{\mathrm{ms}}}} \end{bmatrix} = \begin{bmatrix} p^{(1)(2)} \\ \vdots \\ p^{(1)(n_{\mathrm{ms}})} \end{bmatrix}. \tag{12.6}$$

STEP 3: Finally, we can write the objective function $\mathcal{J}(X)$ in Equation (12.1) for each of the two envisioned cases, that is, (**TOA**, **AOA**, **RSS**) and (**TDOA**, **AOA**, **RSS**), as (Frattasi 2007)

$$\mathcal{J}(\hat{X}) = \sum_{i=1}^{n_{\text{ms}}} \sum_{j=1}^{n_{\text{bs}}} \gamma_t^{(i)[j]} \{f_t(\hat{X}^{(i)})\}^2 + \sum_{i=1}^{n_{\text{ms}}} \gamma_\theta^{(i)[1]} \{f_\theta(\hat{X}^{(i)})\}^2 + \sum_{i=1}^{n_{\text{ms}}} \sum_{j=1, j\neq i}^{n_{\text{ms}}} \gamma_p^{(i)(j)} \{f_p(\hat{X}^{(i)})\}^2 \tag{12.7}$$

and

$$\mathcal{J}(\hat{X}) = \sum_{i=1}^{n_{\text{ms}}} \sum_{j=1}^{n_{\text{bs}}} \gamma_{\text{t}}^{(i)[j]} \{f_{\text{t}}(\hat{X}^{(i)})\}^2 + \sum_{i=1}^{n_{\text{ms}}} \gamma_\theta^{(i)[1]} \{f_\theta(\hat{X}^{(i)})\}^2 + \sum_{i=1}^{n_{\text{ms}}} \sum_{j=1, j\neq i}^{n_{\text{ms}}} \gamma_p^{(i)(j)} \{f_p(\hat{X}^{(i)})\}^2, \tag{12.8}$$

where the functions f_t, f_{t} and f_θ are defined as in Equations (5.60)–(5.62) and f_p is defined as

$$f_p(\hat{X}^{(i)}) = \hat{d}^{(i)(j)} - \hat{d}_k^{(i)(j)}, \qquad \begin{cases} i = 1, \ldots, n_{\text{ms}}, \\ j = 1, \ldots, n_{\text{ms}}, \\ j \neq i. \end{cases} \tag{12.9}$$

The weights γ_t, γ_{t}, γ_θ and γ_p should be appropriately selected to reflect the reliability of the available TOA, TDOA, AOA and RSS measurements. In particular, γ_t, γ_{t} and γ_θ can be chosen as in Equations (5.63)–(5.65), and γ_p as (Frattasi 2007)

$$\gamma_p^{(i)(j)} = \frac{1}{\sigma^2_{\mathbf{RSS}_j}}, \qquad j = 1, \ldots, n_{\text{ms}}, \tag{12.10}$$

where $\sigma^2_{\mathbf{RSS}_j}$ is the variance directly derived from the RSS measurements available at MS_i.

NB The coop-WNLLS method extracts the set of location estimates which maximizes the probability that the polygon created by the conjunction of the initial position guesses has, in total, a combination of edges to the distances defined by the available sets of data. Since MS–MS links are likely to have a better channel quality than BS–MS links, the weights γ_p will have a higher value in the objective function than γ_t, γ_τ and γ_α, and thus they will tend to geometrically constrain the final solution, diminishing its location error. Therefore, it can be expected that a higher location estimation accuracy can be achieved in both LOS and NLOS conditions compared with the noncooperative solution of Section 5.2.3.1. In particular, as mentioned in Section 12.2.3, owing to the spatial diversity of the available BSs experienced by nearby MSs, there can also be the possibility that some BS–MS links are in LOS. As a consequence, the coop-WNLLS method will treat the corresponding MSs as anchors for the others, and the remarkable result is that the location estimation accuracy can be improved and equalized over the region considered for the value for the MSs in LOS (Frattasi and Monti 2007b).

However, in the case of the coop-WNLLS method when TDOA measurements are available, to declass an NLOS localization problem into an LOS localization problem is a harder matter. This is because each TDOA measurement is a function of two BSs, the home BS and a neighboring one, which is likely to be in NLOS (Cong and Zhuang 2004). Therefore, the accuracy of the AOA measurement in that case is decisive (Frattasi 2007).

12.3.2 Coop-EKF: Cooperative Extended Kalman Filter

In this section we apply the EKF to the cooperative framework, taking into consideration the example described in the previous section. We assume further that each MS is static, so that we can lighten the mathematical treatment of the problem (Frattasi and Figueiras 2007).

12.3.2.1 Example of Application

STEP 0: In order to model the filter, it is required to first define the hidden and observable vectors, as shown in Section 5.3.1. Since it is necessary to ensure cooperation among all the devices, the state space to be estimated has to include their coordinates

$$\hat{X}_k = [[\hat{X}_k^{(1)}]^{\mathrm{T}} \cdots [\hat{X}_k^{(n_{\mathrm{ms}})}]^{\mathrm{T}}]^{\mathrm{T}}, \tag{12.11}$$

where $\hat{X}_k^{(i)} = [\hat{x}_k^{(i)} \quad \hat{y}_k^{(i)}]^{\mathrm{T}}$, the superscript identifies the MS and the index k refers to a generic iteration of the estimation routine.

STEP 1: For the TOA, TDOA and RSS measurements, it is possible to define a vector of measurements for each MS:[2]

$$T_k^{(i)} = [t_k^{(i)[1]} \cdots t_k^{(i)[n_{\mathrm{bs}}]}]^{\mathrm{T}}, \tag{12.12}$$

$$\mathrm{T}_k^{(i)} = [\mathrm{t}_k^{(i)[2]} \cdots \mathrm{t}_k^{(i)[n_{\mathrm{bs}}]}]^{\mathrm{T}} \tag{12.13}$$

and

$$P_k^{(i)} = [p_k^{(i)(1)} \quad \cdots \quad p_k^{(i)(i-1)} \quad p_k^{(i)(i+1)} \quad \cdots \quad p_k^{(i)(n_{\mathrm{ms}})}]^{\mathrm{T}}, \tag{12.14}$$

where $t_k^{(i)[j]}$ is a TOA measurement between MS_i and BS_j, $\mathrm{t}_k^{(i)[j]}$ is a TDOA measurement obtained by BS_j and BS_1 with respect to MS_i, and $p_k^{(i)(j)}$ is a cooperative power measurements between MS_i and MS_j. The full set of measurements is then obtained by integrating all the measurements of the same kind into the same vector:

$$T_k = [[T_k^{(1)}]^{\mathrm{T}} \cdots [T_k^{(n_{\mathrm{ms}})}]^{\mathrm{T}}]^{\mathrm{T}}, \tag{12.15}$$

$$\mathrm{T}_k = [[\mathrm{T}_k^{(1)}]^{\mathrm{T}} \cdots [\mathrm{T}_k^{(n_{\mathrm{ms}})}]^{\mathrm{T}}]^{\mathrm{T}}, \tag{12.16}$$

$$\Theta_k = [\theta_k^{(1)[1]} \quad \cdots \quad \theta_k^{(n_{\mathrm{ms}})[1]}]^{\mathrm{T}} \tag{12.17}$$

and

$$P_k = [[P_k^{(1)}]^{\mathrm{T}} \cdots [P_k^{(n_{\mathrm{ms}})}]^{\mathrm{T}}]^{\mathrm{T}}, \tag{12.18}$$

[2] Since for each MS there is only a single BS that performs measurements of AOA, there is no need to specify a vector for AOA measurements.

for TOA, TDOA, AOA and RSS, respectively. In Equation (12.17), $\theta_k^{(i)[1]}$ represents the AOA measurement obtained by the home BS, BS_1, and MS_i. Finally, the full measurements vector is defined according to the type of measurements available:

$$Z_k = [[T_k]^T \quad [\Theta_k]^T \quad [P_k]^T]^T \tag{12.19}$$

and

$$Z_k = [[\mathrm{T}_k]^T \quad [\Theta_k]^T \quad [P_k]^T]^T, \tag{12.20}$$

where (**TOA**, **AOA**) measurements and (**TDOA**, **AOA**) measurements, respectively, are available.

STEP 2: After defining the state space and the measurement space, it is necessary to specify the motion model and the observation model (see Section 5.3.1). If we assume that the MSs are static and that there is no external excitation, the new predicted state $\hat{X}_{k|k-1}$ is equal to the previous state:

$$\hat{X}_{k|k-1} = \mathcal{F}(\hat{X}_{k-1|k-1}, 0, 0) = \hat{X}_{k-1|k-1}. \tag{12.21}$$

Concerning the observation model, it is simply necessary to obtain the vector $\hat{Z}_k$ based on the models defined by Equations (5.54), (5.50), (5.51) and (12.4). In particular, we can rewrite those expressions as

$$\hat{t}^{(i)[j]} = \mathcal{H}_t(\hat{d}^{(i)[j]}) = \frac{1}{c}\hat{d}^{(i)[j]} + t_{\mathrm{tx}}, \tag{12.22}$$

$$\hat{\mathrm{t}}^{(i)[j]} = \mathcal{H}_{\mathrm{t}}(\hat{d}^{(i)[j]} - \hat{d}^{(i)[1]}) = \frac{1}{c}\{\hat{d}^{(i)[j]} - \hat{d}^{(i)[1]}\}, \tag{12.23}$$

$$\hat{\theta}^{(i)} = \mathcal{H}_\theta(\hat{X}^{(i)}) = \mathrm{atan}\frac{\hat{y}^{(i)} - y^{[1]}}{\hat{x}^{(i)} - x^{[1]}}, \tag{12.24}$$

$$\hat{p}^{(i)(j)} = \mathcal{H}_p(\hat{d}^{(i)(j)}) = \alpha - 10\beta\log(\hat{d}^{(i)(j)}). \tag{12.25}$$

Consequently,

$$\hat{Z}_k = \mathcal{H}_\star(\hat{X}_{k|k-1}), \tag{12.26}$$

where the entries of $\hat{Z}_k$ are calculated using $\mathcal{H}_\star(\bullet)$, which is an entrywise function that denotes the appropriate function to be used for each entry of $\hat{Z}_k$ (i.e., $\mathcal{H}_t(\bullet)$, $\mathcal{H}_\tau(\bullet)$, $\mathcal{H}_\alpha(\bullet)$ and $\mathcal{H}_p(\bullet)$).

STEP 3: Another important step in the design of the cooperative extended Kalman filter (coop-EKF) is the determination of the process and measurement noise covariance matrices. If we assume that the MSs are static, no process noise exists, and consequently Q should be equal to the zero matrix. However, in order to allow a faster convergence of the filter, a process noise adaptation method such as annealing (Merwe and Rebel 2006) can be employed, where the adaptation parameters of the method are adjusted by tuning. Concerning the measurement noise, we consider R as a diagonal matrix (i.e., there is no correlation between measurements):

$$R = \begin{bmatrix} \sigma_{\mathrm{TOA}}^2 I & 0 & 0 \\ 0 & \sigma_{\mathrm{AOA}}^2 I & 0 \\ 0 & 0 & \sigma_{\mathrm{RSS}}^2 I \end{bmatrix} \tag{12.27}$$

and

$$R = \begin{bmatrix} \sigma^2_{\mathrm{TDOA}} I & 0 & 0 \\ 0 & \sigma^2_{\mathrm{AOA}} I & 0 \\ 0 & 0 & \sigma^2_{\mathrm{RSS}} I \end{bmatrix}, \tag{12.28}$$

for (**TOA**, **AOA**) measurements and (**TDOA**, **AOA**) measurements, respectively, and I represents the identity matrix. The variances σ^2_{TOA}, σ^2_{TDOA}, σ^2_{AOA} and σ^2_{RSS} can be derived from Monte Carlo simulations. Practically, two different approaches can be used in order to set these values (Figueiras 2008):

- The position of the MS is set as constant, and the standard deviations are determined after running Monte Carlo simulations of the wireless channel conditions. This approach represents the case when a priori information concerning the noise statistics is available.
- The position of the MS is changed for every new run of the Monte Carlo simulations of the wireless channel conditions. Thus, in this case, the noise statistics are generalized and taken into account independently of the position. This approach represents the case when no a priori information concerning the noise statistics is available.

STEP 4: In order to completely model the filter, we still need to define several of the matrices in Equations (5.92) and (5.93). The matrices A_k and H_k represent the Jacobians of Equations (12.21) and (12.26), respectively, with respect to X_k. If we assume that the MSs are static, A_k is equal to the identity matrix I.

12.4 COMET: A Cooperative Mobile Positioning System

In this section we introduce an example of a CAS, named COMET, specifically applied to cellular networks in outdoor environments (Rosa et al. 2007a). In particular, we will mainly describe two data fusion methods, which use as kernel techniques the ones illustrated in the previous section and in Chapter 5. Finally, we will reveal the potential of such a system by means of simulation results.

12.4.1 System Architecture[3]

Figure 12.6 shows the hybrid cellular ad hoc system architecture taken into consideration for COMET. Upon receiving a location information request from an MS (MS_1), the home BS (BS_1) forms a cluster in the surroundings of that MS (CM_1, CM_2 and CM_3). In particular, we simply assume that the CH is the MS that sent the location request to the home BS and that the CMs are a selection of the neighboring MSs, which have been sorted in descending order according to their SNR levels on the BS–MS links. Since we will suppose in Section 12.4.3 that all MSs are in LOS with each other, we do not consider the channel conditions on the MS–MS links when performing the clustering. In general, a clustering method can also

[3] S. Frattasi, M. Monti, "Ad-Coop positioning system (ACPS): Positioning for cooperative users in hybrid cellular ad-hoc networks", *European Transactions on Telecommunications Journal*, Wiley, vol. 19, no. 8, pp. 923–924, May, 2007. Reproduced in part by permission of © 2007 John Wiley & Sons Ltd.

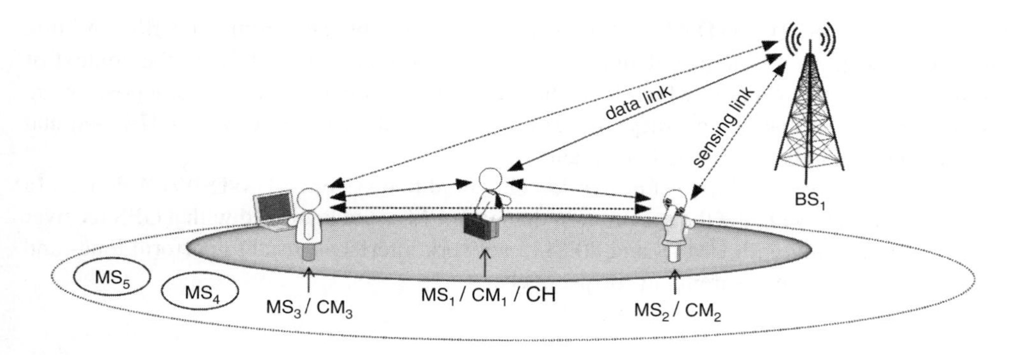

Figure 12.6 COMET: system architecture.

take into consideration parameters such as speed, remaining battery power and homogeneity/heterogeneity of the candidate terminals.

When the cluster has been formed, three types of measurements are carried out for location purposes.

1A. TOA measurements: Each CM can measure the time of arrival of the pilot signals from the available BSs by cross-correlating each of them with an internally generated pilot signal. This type of measurement can only be used if each CM has a clock accurately synchronized with the available BSs.

1B. TDOA measurements: Each CM can measure the time difference between the arrival of a pilot signal from the home BS and a neighboring BS by cross-correlating them (Knapp and Carter 1976). The pilot signal from each of the neighboring BSs can only be used if its SINR at the receiving CM is above a certain threshold.

2. AOA measurements: With an adaptive antenna array, each of the available BSs steers its antenna spot beam to track the dedicated backward-link pilot signal from each CM and thus obtains its arriving azimuth angle (with respect to a specified reference direction). If we consider a cellular system based on CDMA, in order to avoid signal degradation due to the near–far effect, this type of measurement can only be performed by the home BS (Cong and Zhuang 2002).
3. RSS measurements: Each CM can measure the received power of the signals coming from neighboring CMs. This may be done during normal data communications without any additional bandwidth or energy requirements (Frattasi and Monti 2007a). For details on each type of measurement, the reader is referred to Chapter 4.

The determination of the location of each CM, which is performed by one of the data fusion methods that will be described in the next section, can be either *CH-based* or *CH-assisted/network-based* (the choice depends mainly on the computational power at the CH). In the first case, while each available BS collects on the backward link the AOA measurements for each BS–CM link, the CH relays back to the home BS all the TOA/TDOA and RSS measurements obtained from each BS–CM link and CM–CM link, respectively, and the calculations are performed by a specific server in the network. In the second case, while the home BS transmits all the AOA measurements to the CH via the forward link, the

CH collects all the TOA/TDOA and RSS measurements obtained from each BS–CM link and CM–CM link, and the calculations are performed directly at the CH. In the context of the main aim of this section, which is to show the impact of cooperative mobile positioning on the location estimation accuracy of a cellular system, the choice between CH-based and CH-assisted methods is not strictly relevant.

In this section we consider as our hybrid cellular ad hoc system a UMTS/WLAN 802.11a system,[4] in which we make the assumption that each CM is not equipped with a GPS receiver but is equipped only with UMTS and 802.11a network interfaces, used to perform long- and short-range location measurements, respectively.

12.4.2 Data Fusion Methods

In this section we describe two data fusion methods developed for COMET. While the first employs both long- and short-range location measurements together to provide absolute location coordinates for all the CMs at once, the second utilizes the measurements in parallel and may supply relative as well as absolute location information for all the CMs at once.

12.4.2.1 1L-DF: One-Level Data Fusion[5]

Figure 12.7 shows a flow-chart of 1L-DF, which may be described in detail as follows. For each MS/CM, a first location estimate is obtained from a conventional location algorithm taken from the literature. As a reference, we have considered the hybrid TOA/AOA positioning (HTAP) algorithm (Deng and Fan 2000) and the hybrid lines of position (HLOP) algorithm (Venkatraman and Caffery 2004) for use when TOA and AOA measurements are available, and the hybrid TDOA/AOA positioning (HTDOA) algorithm (Chen and Feng 2005a) for use when TDOA and AOA measurements are available (see Section 4.4.5). Considering that we usually have at our disposal more than a single sample for each type of measurement, the quantities used in this step are the median values calculated over each set of measurements (for the reference statistical distribution of the error, see Section 12.4.3.1). Note that the median value is preferred to the mean value, owing to its robustness against possibly biased measurements (Frattasi and Monti 2007b). If cooperation is *off*, in order to determine the final position estimate for each MS, each initial position guess obtained in the first step of the algorithm and each set of TOA/TDOA and AOA estimates relative to each MS are used independently in an unconstrained WNLLS minimization procedure or in an EKF (see Chapter 5). If cooperation is *on*, in order to determine the final position estimates of all the CMs at some time, all initial position guesses obtained in the first step of the algorithm and all sets of TOA/TDOA, AOA and RSS estimates relative to all the CMs are used simultaneously in an unconstrained coop-WNLLS minimization procedure or in a coop-EKF.

[4] For related work on cooperative mobile positioning in a WiMAX/WLAN 802.11a system, the reader is referred to Rosa (2007).

[5] S. Frattasi, M. Monti, "Cooperative mobile positioning in 4G wireless networks", *Cognitive Wireless Networks: Concepts, Methodologies and Visions Inspiring the Age of Enlightenment of Wireless Communications*, Springer, pp. 213–233, September, 2007. Reproduced in part with kind permission of © 2007 Springer Science and Business Media.

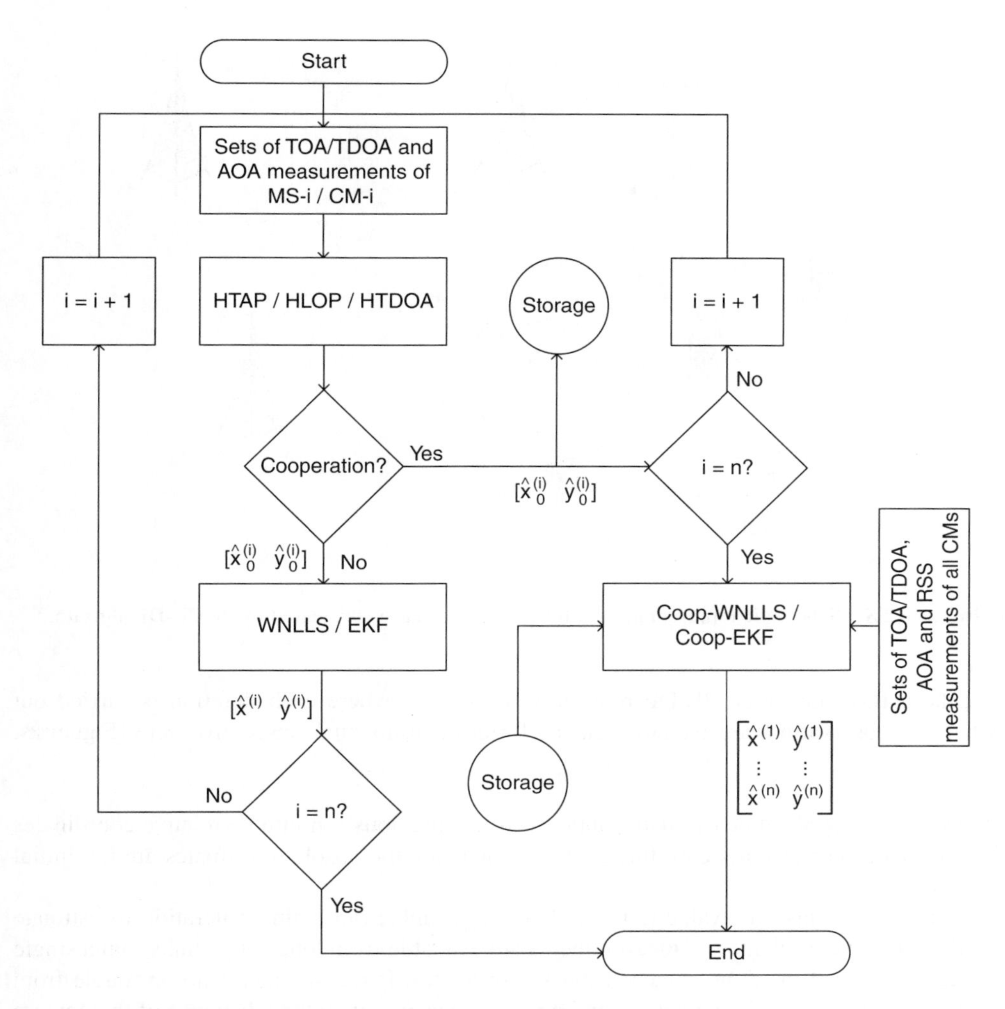

Figure 12.7 Operational representation of one-level data fusion (1L-DF).

12.4.2.2 2L-DF: Two-Level Data Fusion[6]

In this section we describe two-level data fusion (2L-DF), which consists of decoupling relative and absolute localization by using independently the measurements retrieved from the short- and long-range links, respectively. As we can derive from the arrows in Figure 12.8, while relative localization is performed among the cooperating devices by estimating distances between pairs of CMs, absolute localization is performed by estimating the coordinates and orientation of a cluster within the cellular network.

[6] © 2008 IEEE. Reprinted, with permission, from Figueiras, J, Frattasi, S and Schwefel, HP, "Decoupling Estimators in Mobile Cooperative Positioning for Heterogeneous Networks" in *Proceeding of the IEEE 68th Vehicular Technology Conference*.

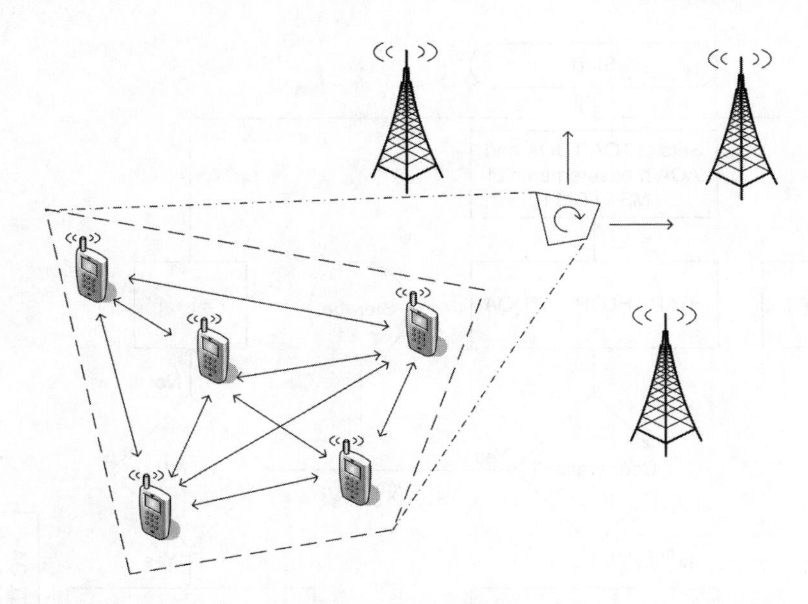

Figure 12.8 Schematic representation of the general scenario considered by the 2L-DF algorithm.

Like a Bayesian filter, 2L-DF runs in a cyclic way, where each iteration is carried out whenever observations are available, and performs the following consecutive steps (Figueiras, 2008):

1. Decouple the absolute estimates obtained in the previous run into (i) relative coordinates and (ii) center-of-mass coordinates. In the first run, the absolute estimates are the initial guesses.
2. If measurements are available from short-range links, run a single iteration to estimate the relative coordinates. If measurements are available from long-range links, run a single iteration to estimate the center-of-mass coordinates. If measurements are available from just one domain, only the corresponding substep is executed; the advantage of this feature is that it allows short- and long-range subsystems to operate with different measurement rates.
3. Couple the newly obtained relative and center-of-mass coordinates to retrieve the current absolute estimates.

Figure 12.9 shows a flowchart of 2L-DF. In particular, it is important to notice that, if there is no request for positioning during a certain execution cycle, the coupling is not performed; this lowers the amount of computation required. Moreover, the link between the relative and absolute positioning blocks is necessary to define distances between CMs and BSs.

12.4.2.2.1 Decoupling and Coupling

The decoupling of the absolute coordinates into relative coordinates, group position and orientation is carried out by applying a transformation of coordinates. The latter is necessary to

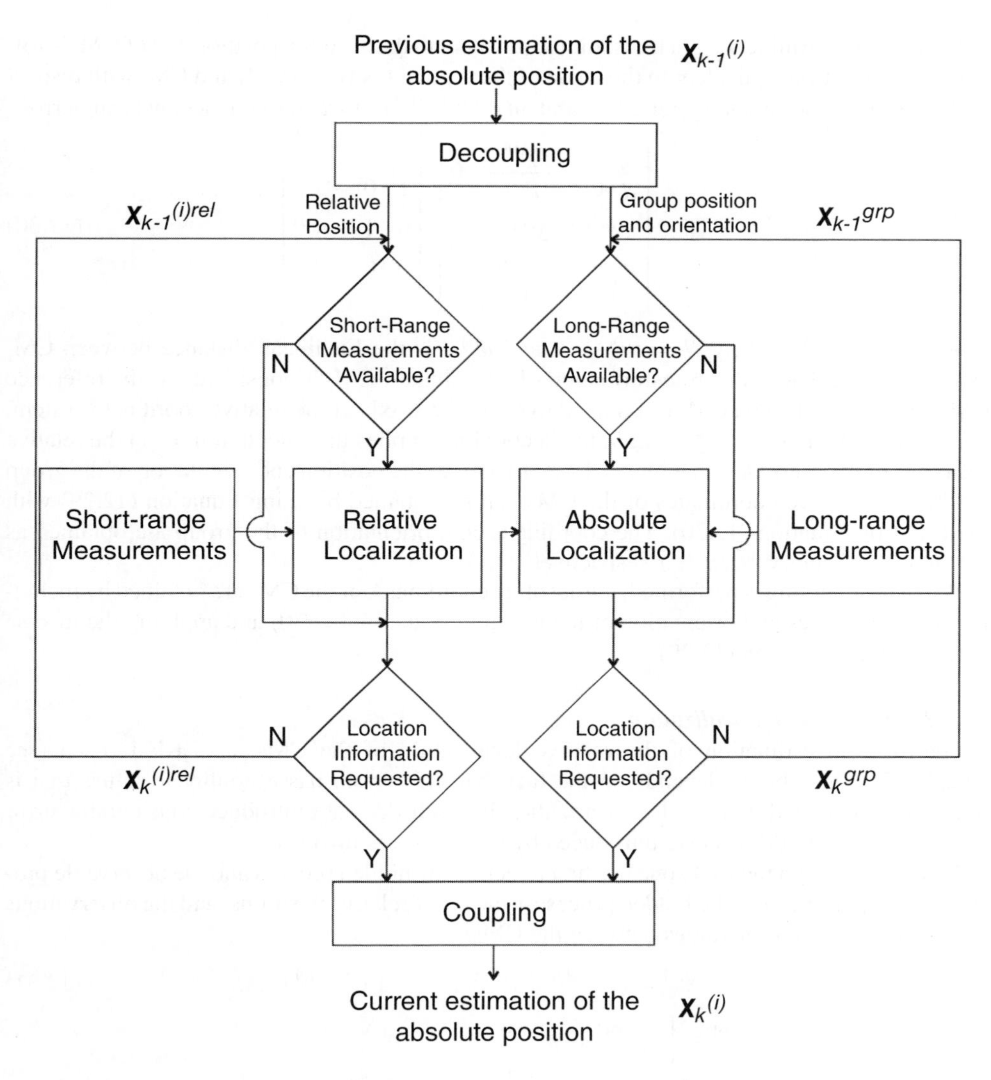

Figure 12.9 Operational representation of 2L-DF.

obtain the current transformation matrix (CTM), which corresponds to a translation followed by a rotation of the axis. Let us assume that the absolute coordinates of CM_i are defined by $X^{(i)}$ and its relative coordinates by $X^{(i)\mathrm{rel}}$. Moreover, let us suppose that $\mathbf{T}_{\mathrm{ctm}}$ is the CTM for the transformation of coordinates. Then, the following relation can be written:

$$\begin{bmatrix} X^{(i)\mathrm{rel}} \\ 1 \end{bmatrix} = \mathbf{T}_{\mathrm{ctm}} \begin{bmatrix} X^{(i)} \\ 1 \end{bmatrix}, \tag{12.29}$$

where the last component of the vector on the right-hand-side of Equation (12.29) can be added to allow transformations independent of $X^{(i)}$.

In order to determine $\mathbf{T}_{\text{ctm}}$, a translation equivalent to the absolute position $X^{(i)}$ of CM_i is followed by a rotation equivalent to the angle of the segment between CM_j and CM_i with respect to the absolute coordinate system, $\theta = \arctan(y^{(i)(j)}/x^{(i)(j)})$. As a consequence, we can derive

$$\mathbf{T}_{\text{ctm}} = \begin{bmatrix} \frac{x^{(i)(j)}}{d^{(i)(j)}} & -\frac{y^{(i)(j)}}{d^{(i)(j)}} & 0 \\ \frac{y^{(i)(j)}}{d^{(i)(j)}} & \frac{x^{(i)(j)}}{d^{(i)(j)}} & 0 \\ 0 & 0 & 1 \end{bmatrix} \begin{bmatrix} 1 & 0 & -x^{(i)} \\ 0 & 1 & -y^{(i)} \\ 0 & 0 & 1 \end{bmatrix}, \tag{12.30}$$

where $x^{(i)(j)} = x^{(j)} - x^{(i)}$, $y^{(i)(j)} = y^{(j)} - y^{(i)}$ and $d^{(i)(j)}$ is the Euclidean distance between CM_i and CM_j. Note that in this transformation of coordinates, CM_i is considered as the reference CM, that is, the CH, and CM_j is assumed to be on the x-axis of the relative coordinate system.

As is noticeable from Figure 12.9, the decoupling implies the calculation of (1) the relative locations of the users ($X^{(i)\text{rel}}$, with $i = 1, \ldots, n$) and (2) the position and orientation of the group (X^{grp}). The relative coordinates of the CMs can be obtained by using Equation (12.29) with the CTM of Equation (12.30). The coordinates and orientation of the group are obtained as $X^{(i)}$ and $\theta = \arctan(y^{(i)(j)}/x^{(i)(j)})$, respectively.

When the coupling is performed, the absolute coordinates of the CMs are obtained by including the coordinates and orientation of the group in Equation (12.30) and applying the inverse calculation of Equation (12.29).

12.4.2.2.2 *Relative Localization*

To perform an estimation of the relative localization of the CMs, an EKF is used (see Chapter 5). This choice depends on the fact that the system has a nonlinear nature that is not very accentuated, that is, the linearization by itself does not introduce considerable error when compared with the noise introduced by the processes involved.

In order to design the EKF, one has first to define the hidden process and the observable process. In the present case, the hidden process is the set of relative positions, and the observations are the measured power values between the CMs:

$$X^{\text{rel}} = [x^{(2)\text{rel}} \quad \{X^{(3)\text{rel}}\}^{\text{T}} \cdots \{X^{(n_{\text{ms}})\text{rel}}\}^{\text{T}}]^{\text{T}}, \tag{12.31}$$

$$Z^{\text{rel}} = \Big[\begin{array}{llll} p^{(1)(2)}, & p^{(1)(3)}, & \cdots & p^{(1)(n_{\text{ms}})}, \\ & p^{(2)(3)}, & \cdots & p^{(2)(n_{\text{ms}})}, \\ & & \ddots & \vdots \\ & & & p^{(n_{\text{ms}}-1)(n_{\text{ms}})} \end{array}\Big]^{\text{T}}. \tag{12.32}$$

Note that Equation (12.31) does not include the coordinates of CM_1 and the y coordinate of CM_2, owing to the fact that they are equal to zero in the relative coordinate system; in order to reduce the state space, those zeros were excluded. It is also important to notice that Z^{rel} is a column vector, as we assume that short-range links are symmetric, that is, for each pair of CMs i and j in Equation (12.32) there is a single observation $p^{(i)(j)}$, with $i < j$. If the links were not symmetric, Equation (12.32) would need to include $p^{(i)(j)}$ with $i \neq j$.

The second step in the design of the filter is the definition of the motion model and the observation model. Since we will assume static devices in the next section, the predicted state is simply equal to the previous estimation, while the observation model is defined by the path-oss equation:

$$\hat{X}^{\mathrm{rel}}_{k|k-1} = \hat{X}^{\mathrm{rel}}_{k-1}, \tag{12.33}$$

$$\hat{p}^{(i)(j)\mathrm{rel}}_{k} = \alpha - 10\beta \log(\hat{d}^{(i)(j)}), \tag{12.34}$$

where the subscript $k|k-1$ refers to a prediction at time k given observations up to time $k-1$. In Equation (12.34), α and β correspond to the parameters in Equation (4.1).

The final design issue is the definition of the process noise Q and measurement noise R. It is assumed that the noise components are uncorrelated, that is, the covariance error matrix Q and the observation covariance error matrix R are diagonal. Concerning the measurement noise, the standard deviation of the measurements σ_p^2 is used for the entries of R:

$$R = \sigma_p^2 I, \tag{12.35}$$

where the matrix I is the identity matrix with the same dimension as the observation vector in Equation (12.32). Although σ_p^2 is defined based on the observations, its value could instead be obtained from the literature, computer simulations or a precalibration phase. Concerning the process noise Q, we can assume it to be zero, since we suppose that we have static devices. However, this approach could imply a need for a large number of measurements to allow us to ignore the additional errors introduced into the position estimation by the slow convergence of the algorithm. For this reason, a process noise adaptation method (e.g., annealing) is used to allow a faster convergence of the filter, where the adaptation parameters of that method are adjusted by tuning.

12.4.2.2.3 Absolute Localization

In contrast to relative localization, which is performed by means of power measurements, absolute localization shows a strongly nonlinear behavior owing to the observation model for angle-of-arrival measurements. Since the latter are related to the position coordinates by an arctan function, the linearization turns out to introduce additional errors into the process estimation. For this reason a UKF is used, as it is expected to perform better than an EKF (see Section 5.3.1).

In order to design the UKF, one has first to define the hidden process and the observable process. In the present case, the hidden process is the set of absolute positions, and the observations are the measured time and angle values between the CMs and the BSs:

$$X^{\mathrm{grp}} = \left[x^{(1)}\ \ y^{(1)}\ \ \arctan(\tfrac{y^{(1)(2)}}{x^{(1)(2)}})\right]^{\mathrm{T}}, \tag{12.36}$$

$$Z^{\mathrm{grp}} = \left[\{T^{(1)}\}^{\mathrm{T}}\quad \{T^{(n_{\mathrm{ms}})}\}^{\mathrm{T}}\quad \theta^{(1)[1]} \ \ldots\ \theta^{(n_{\mathrm{ms}})[1]}\right]^{\mathrm{T}}, \tag{12.37}$$

$$T^{(i)} = [t^{(i)[1]}\ \ldots\ t^{(i)[n_{\mathrm{bs}}]}]^{\mathrm{T}}, \tag{12.38}$$

where n_{bs} is the number of BSs and n_{ms} is the number of MSs.

The second step in the design of the filter is the definition of the motion model and the observation model. Since we will assume static devices in the next section, the predicted state is simply equal to the previous estimation, while the observation model is given by three different expressions, one for the angle measurements and two for the time measurements, for the TOA and TDOA:

$$\hat{X}^{\text{grp}}_{k|k-1} = \hat{X}^{\text{grp}}_{k-1}, \tag{12.39}$$

$$\theta^{(i)[1]} = \arctan\left(\frac{\hat{y}^{(i)} - y^{[1]}}{\hat{x}^{(i)} - x^{[1]}}\right), \tag{12.40}$$

$$t^{(i)[1]} = cd^{(i)[1]}, \tag{12.41}$$

$$\text{t}^{(i)[j]} = c(d^{(i)[j]} - d^{(i)[1]}), \tag{12.42}$$

where c is the speed of light and $d^{(i)[j]}$ is the distance between CM_i and BS_j. It is important to notice that the position of CM_i is obtained from X^{grp} and the relative distances between the CMs, assuming that those distances are fixed parameters. This step corresponds to the link between the relative and absolute localization blocks in Figure 12.9.

The final design issue is the definition of the process noise Q and measurement noise R. As in Section 12.4.2.2.2, it is assumed that the noise components are uncorrelated, that is, the covariance error matrix Q and the observation covariance error matrix R are diagonal. Concerning the measurement noise, the standard deviation of the time measurements σ_t^2 and angle measurements σ_θ^2 are used in the entries of R:

$$R = \left[\begin{array}{c|c} \sigma_t^2 I & 0 \\ \hline 0 & \sigma_\theta^2 I \end{array}\right], \tag{12.43}$$

where the matrix I is the identity matrix with the same dimension as the observation vector in Equation (12.39). Although σ_t^2 and σ_θ^2 are defined based on the observations, their values could instead be obtained from the literature, computer simulations or a precalibration phase. Concerning the process noise, Q is modeled as described for the case of relative localization.

12.4.2.2.4 Properties of 2L-DF

The decoupling and coupling of the position estimators in 2L-DF has the following advantages:

- **Different observation rates.** Positioning solutions generally operate with different measurement rates depending on the technology in use. Since it is common that long- and short-range communications are performed by different technologies, their integration in terms of positioning data requires attention. The problem is addressed by the decoupling of the position estimators, which allows the two domains of data sources to operate independently (Figure 12.9).
- **Computational effort and distributed computing.** Given the structure of 2L-DF, the computational effort is at most as large as that of a data fusion where all measurements are treated equally. Additionally, owing to the decoupling of the estimators, distributed computation is possible to a certain extent. For instance, relative localization could be carried out by the CMs, while absolute localization could be carried out by the cellular network. Then, depending on where the information is requested, the necessary exchange of data could be performed.
- **Group mobility.** The decoupling of short- and long-range measurements permits one to use group mobility models (see Chapter 6) instead of individual mobility models. In this way,

the CMs can be viewed, on the one hand, as a group with correlated mobility patterns, and, on the other hand, as individuals characterized by some uncorrelated mobility within the group. A consequence of such an approach is that management of groups is required when devices move out of range.

12.4.3 Performance Evaluation

12.4.3.1 Simulation Models[7]

Below, we list the models used to obtain the simulation results illustrated in the next section. The statistical models for the TOA and AOA estimation errors were taken from Deng and Fan (2000), where the differentiation between LOS and NLOS is based on results concerning the LOS probability from Wang et al. (2004), from which the channel model used for the estimation of the RSS was also extracted. In particular, in Wang et al. (2004) a statistical channel model was derived from a ray-tracing simulation of the city of Bristol for both BS–MS and MS–MS links at frequencies of 2.1 GHz (UMTS) and 5.2 GHz (WLAN 802.11a), respectively.

12.4.3.1.1 Statistical Models for Time and Angle-of-Arrival Estimation Errors[8]

- **Line-of-sight.** The estimation errors are small and are primarily due to equipment measurement errors. Traditionally, they have been assumed to be normally distributed with zero mean and small standard deviation: $c\sigma_T = 30$ m and $\sigma_\Theta = 1$ deg for TOAs and AOAs, respectively (Deng and Fan 2000).[9] Note that we suppose that the standard deviations for TOA measurement errors associated with different BSs are identical.
- **Non-line-of-sight.** The estimation errors are large and are primarily due to reflection or diffraction of the signal between a BS and a CM. Consequently, the following exponential probability density function can be used to model the excess delay (Greenstein et al. 1997):

$$\mathrm{P}(\tau) = \frac{1}{\tau_{\mathrm{rms}}} \exp \; -\tau/\tau_{\mathrm{rms}}, \tag{12.44}$$

where τ_{rms} is the RMS delay spread:

$$\tau_{\mathrm{rms}} = \widetilde{\tau_{\mathrm{rms}}}\{\hat{d}^{(i)[j]}\}^{\varepsilon}\chi, \tag{12.45}$$

where $\widetilde{\tau_{\mathrm{rms}}}$ is the median value of the RMS delay spread in μs at 1 km, $\hat{d}^{(i)[j]}$ is the distance between BS_j and CM_i in km, ε is an exponent with a value between 0.5 and 1, and χ represents the log-normal shadow fading, so that $X = 10\log\chi$ is a Gaussian random variable with zero mean $\mu_\chi = 0$ and a standard deviation σ_χ that lies between 2 and 6 dB

[7] S. Frattasi, M. Monti, "Cooperative mobile positioning in 4G wireless networks", *Cognitive Wireless Networks: Concepts, Methodologies and Visions Inspiring the Age of Enlightenment of Wireless Communications*, Springer, pp. 213–233, September, 2007. Reproduced in part with kind permission of © 2007 Springer Science and Business Media.

[8] S. Frattasi, M. Monti, "Cooperative mobile positioning in 4G wireless networks", *Cognitive Wireless Networks: Concepts, Methodologies and Visions Inspiring the Age of Enlightenment of Wireless Communications*, Springer, pp. 213–233, September, 2007. Reproduced in part with kind permission of © 2007 Springer Science and Business Media.

[9] In practice, the standard deviation of the measurement errors depends on the chip rate, the propagation environment and the home BS antenna parameters. It can be estimated based on the SNR and its typical values for various propagation conditions (Cong and Zhuang 2004).

Table 12.1 Parameter settings for different environmental types (Greenstein et al. 1997). (S. Frattasi, M. Monti, "Cooperative mobile positioning in 4G wireless networks", *Cognitive Wireless Networks: Concepts, Methodologies and Visions Inspiring the Age of Enlightenment of Wireless Communications*, Springer, pp. 213–233, September, 2007. Reproduced in part by kind permission of © Springer Science and Business Media.)

Channel type	$\widetilde{\tau_{\rm rms}}$ (μs)	ε	σ_χ (dB)
A Urban macrocells	0.4–1.0	0.5	1.9–3.6
B Urban microcells	0.4	0.5	2.3
C Suburban areas	0.3	0.5	2.0–4.7
D Rural areas	0.1	0.5	4.0–5.3
E Mountainous areas	≥ 0.5	1.0	2.4–3.2

(Table 12.1). Specifically, the autocorrelation of the shadow fading among the CMs is modeled by (Senarath 2006)

$$\rho(\Delta x) = e^{-(|\Delta x|/d_{\rm cor})\ln 2}, \tag{12.46}$$

where ρ is the correlation coefficient, Δx is the distance between the CH and a given CM, and $d_{\rm cor} = 20$ m is the decorrelation distance. For example, if χ_1 is the log-normal component at position P_1, the component at position P_2, χ_2, which is Δx away from P_1, is normally distributed with mean $\mu_2 = \rho(\Delta x)\chi_1$ dB and standard deviation $\sigma_2 = \sigma_\chi\sqrt{1-[\rho(\Delta x)]^2}$ dB.

The estimation error in the angle measurement is considered to be Gaussian distributed with zero mean $\mu_\Theta = 0$ and a standard deviation given by (Deng and Fan 2000)

$$\sigma_\theta = \frac{c\tau}{d^{(i)[j]}}, \tag{12.47}$$

where τ is the excess delay determined by use of Equation (12.44).

12.4.3.1.2 Statistical Channel Model for Received-Signal-Strength Estimation[10]

- **Line-of-sight probability.** Figure 12.10 shows the LOS probability as a function of the separation distance between the transmitter (TX) and receiver (RX), for both BS–CM and CM–CM links. The following features can be observed.
  1. When the separation distance between TX and RX lies in the region of 10–20 m, which broadly corresponds to the width of a street, the LOS probability is 1 for this particular database and set of locations (Wang et al. 2004).
  2. When the separation distance between TX and RX increases further, the LOS probability is higher for BS–CM links than for CM–CM links. This is because when the height

[10] S. Frattasi, M. Monti, "Ad-Coop positioning system (ACPS): Positioning for cooperative users in hybrid cellular ad-hoc networks", *European Transactions on Telecommunications Journal*, Wiley, vol. 19, no. 8, pp. 923–924, May, 2007. Reproduced in part by permission of © 2007 John Wiley & Sons Ltd.

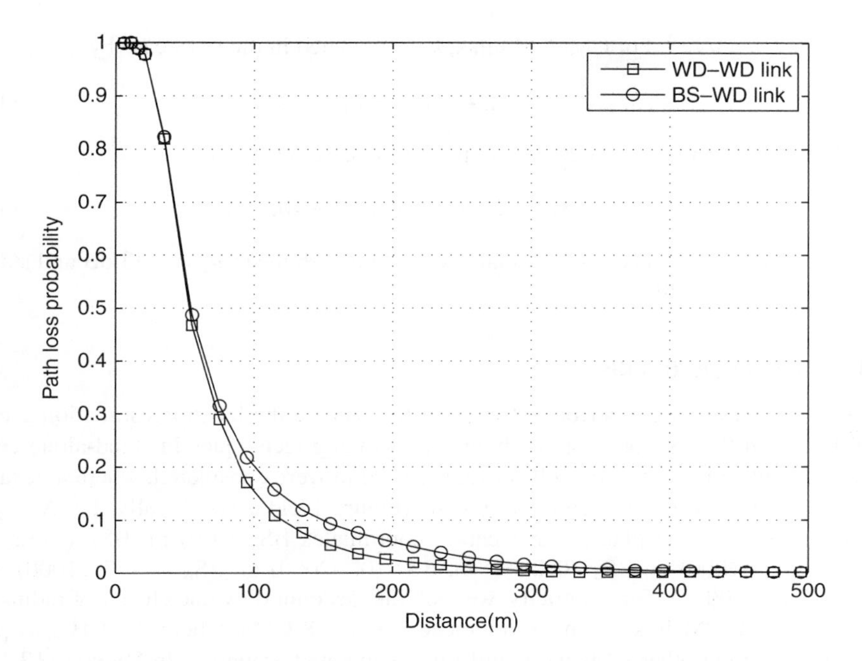

Figure 12.10 LOS probability vs separation distance between TX and RX (Frattasi and Monti 2007a).

of TX is reduced from 15 m for BS–CM links to 1.5 m for CM–CM links, the signal propagation paths are affected not only by the surrounding buildings in the vicinity of TX but also by the terrain. As a consequence, CM–CM links can also be in NLOS even when no buildings lie in the propagation path.

- **Path loss.** The following slope/intercept statistical model is used for the mean outdoor path loss:

$$L = L_0 + 20\log_{10}(f) + 10\beta\log_{10}(d), \tag{12.48}$$

where d is the separation distance between TX and RX (m), f is the operating frequency (MHz) and L_0 is the path loss (dB) at a short reference distance $d_0 = 1$ m. For CM–CM links, L is modeled in the LOS case with $L_0 = -27.6$ dB and $\beta = 2$ (free-space path loss), and in the NLOS case with $L_0 = -51.22$ dB and $\beta = 5.82$.

- **Shadowing.** The shadowing process χ is characterized by a log-normal distribution, that is, a normal distribution in dB, with a distance-dependent standard deviation given by

$$\sigma_\chi = s[1 - e^{-(d-d_0)/D_s}], \tag{12.49}$$

where s is the maximum standard deviation (dB) and D_s is the growth distance factor (m). For CM–CM links, σ_χ is modeled in the LOS case with $s = 2$ dB, $D_s = 36$ m and $d_0 = 0$ m, and in the NLOS case with $s = 23.4$ dB, $D_s = 36$ m and $d_0 = 10$ m.

- **Fast fading.** The most popular models for fast fading in LOS and NLOS are, respectively, the Rice and Rayleigh models, where the ratio between the expected power of the dominant path ρ^2 and the power of the Rayleigh components $2\sigma^2$ is often expressed by the *Ricean*

K-factor $\kappa_0 = \rho^2/2\sigma^2$. For CM–CM links, κ_0 is modeled in the LOS case by

$$\kappa_0 = -n_k d + L_k, \tag{12.50}$$

with $L_k = 23$ dB and $n_k = 0.029$, and in the NLOS case by

$$\kappa_0 = \log \text{Norm}(\mu_k, \sigma_k) - 10, \tag{12.51}$$

where $\log \text{Norm}(\mu_k, \sigma_k)$ is a log-normal distribution with mean $\mu_k = 2.43$ dB and standard deviation $\sigma_k = 0.45$ dB.

12.4.3.2 Simulation Results

Computer simulations were performed in order to compare the location estimation accuracy of COMET with that of conventional hybrid positioning techniques in stand-alone cellular networks. Simulated urban microcells of radius 1000 m were considered, where a hexagonal test cell is surrounded by six neighboring cells (Figure 12.11). Specifically, $1 \leq N \leq 7$ BSs were positioned in a 2D plane with location coordinates BS_1 $(0, 0)$ m, BS_2 $(1732, 1000)$ m, BS_3 $(1732, -1000)$ m, BS_4 $(0, 2000)$ m, BS_5 $(0, -2000)$ m, BS_6 $(-1732, 1000)$ m and BS_7 $(-1732, -1000)$ m. For simplicity, we took into account only one cluster of radius 25 m ($P_{\text{los}} = 1$ for CM–CM links), which embraced $1 \leq n \leq 8$ CMs, where the CH was placed in the center and the other CMs were uniformly generated around it. In Sections 12.4.3.2.1 and 12.4.3.2.2, only n MSs are generated around the CH and considered as CMs. In Section 12.4.3.2.3, we show the effect of clustering by generating up to 25 MSs and selecting n of those according to their SNR levels with respect to the available BSs (see earlier in

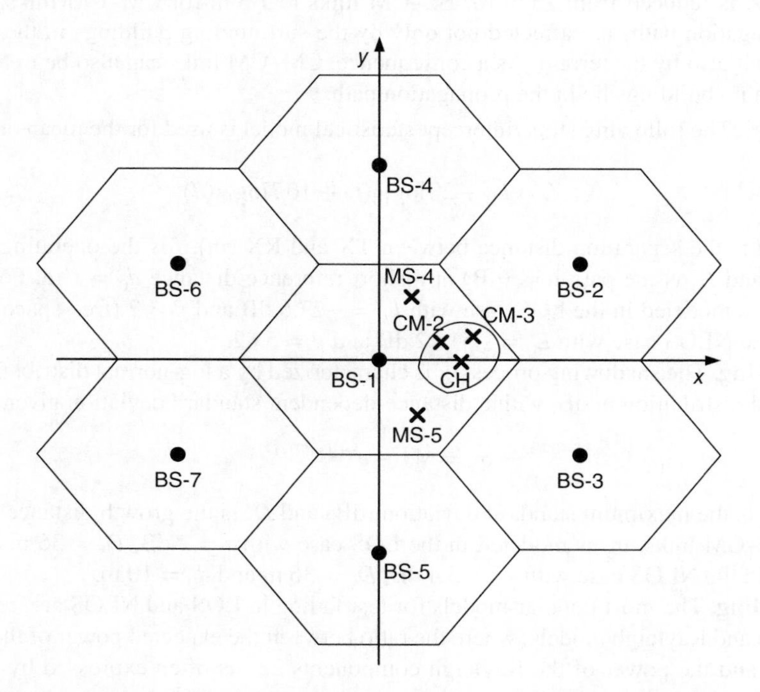

Figure 12.11 The seven-cell and one-cluster system layout considered (Frattasi et al. 2006).

Table 12.2 Simulation parameters (S. Frattasi, M. Monti, "Cooperative mobile positioning in 4G wireless networks", *Cognitive Wireless Networks: Concepts, Methodologies and Visions Inspiring the Age of Enlightenment of Wireless Communications*, Springer, pp. 213–233, September, 2007. Reproduced in part by kind permission of © Springer Science and Business Media.)

Parameter	Value
Cell radius	1000 m
Cluster radius	25 m
Number of BSs	$N = 1–7$
Number of CMs	$n = 1–8$
Measurements per BS	$I = 100$
Measurements per CM	$i = 100$
Channel type (see Table 12.1)	B

Section 12.4). Finally, no interference was considered, and 1L-DF with coop-WNLLS was used as a data fusion method, but similar results can be obtained with 1L-DF and the coop-EKF or with 2L-DF (Figueiras 2008). The values of some of the parameters used in the simulations are summarized in Table 12.2.

The performance of COMET was presented in the form of cumulative distribution functions (CDFs) of the cooperative root mean square error (CRMSE) (Frattasi and Monti 2006):

$$\text{CRMSE} = E[\mathbf{RMSE}], \tag{12.52}$$

where $E[\bullet]$ is the operator "mean" and $\mathbf{RMSE} = \left[RMSE_1 \cdots RMSE_n\right]^{\mathrm{T}}$ is the RMSEs for all the CMs, which are defined as

$$RMSE_i = \sqrt{\{\hat{X}^{(i)} - X_{\mathrm{r}}^{(i)}\}^{\mathrm{T}}\{\hat{X}^{(i)} - X_{\mathrm{r}}^{(i)}\}}, \tag{12.53}$$

where $X_{\mathrm{r}}^{(i)}$ is the vector representing the real location coordinates of CM_i. If cooperation is *off*, the CRMSE becomes equal to the RMSE, which is calculated by considering only one MS, the CH. From Equation (12.52) we can observe further that if cooperation is *on*, the accuracy experienced by a single MS has actually been replaced with the accuracy experienced by all the CMs.

12.4.3.2.1 Dependence of Performance on the Number of Cluster Members[11]

- **HTAP** Figure 12.12 shows a comparison, in terms of location estimation accuracy, between the results obtained with 1L-DF with and without cooperation for a variable number of CMs, when the CH is placed at (500, 0) m ($P_{\text{los}} = 0$ for BS–CM links) and the initial guesses are calculated by HTAP. The following can be observed.

[11] S. Frattasi, M. Monti, "Ad-Coop positioning system (ACPS): Positioning for cooperative users in hybrid cellular ad-hoc networks", *European Transactions on Telecommunications Journal*, Wiley, vol. 19, no. 8, pp. 923–924, May, 2007 and S. Frattasi, M. Monti, "Cooperative mobile positioning in 4G wireless networks", *Cognitive Wireless Networks: Concepts, Methodologies and Visions Inspiring the Age of Enlightement of Wireless Communications*, Springer, pp. 213–233, September, 2007. Reproduced in part by permission of © 2007 John Wiley & Sons Ltd. © 2007 Springer Science and Business Media.

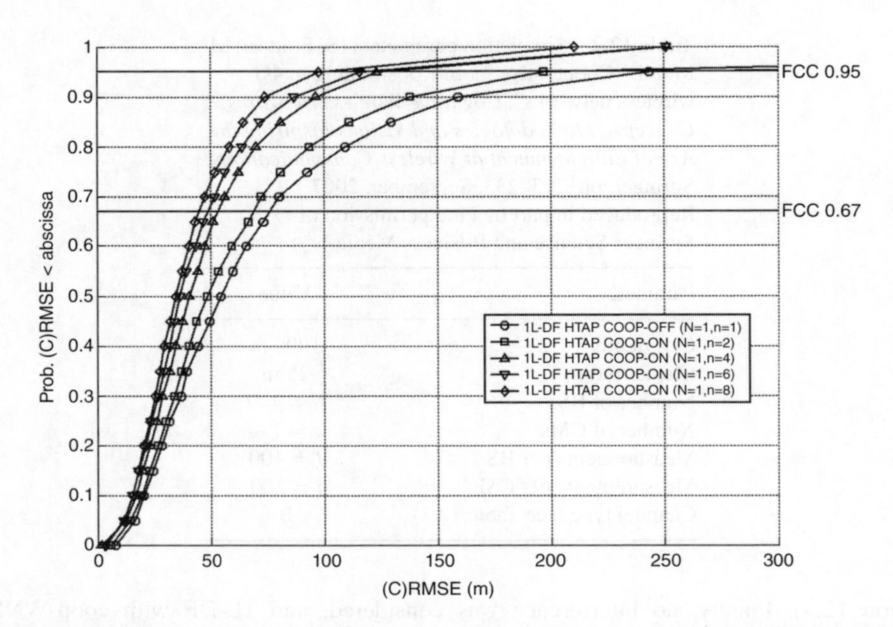

Figure 12.12 (C)RMSE vs number of CMs (HTAP case). (S. Frattasi, M. Monti, "Cooperative Mobile positioning in 4G wireless networks", *Cognitive Wireless Networks: Concepts, Methodologies and Visions, Inspiring the Age of Enlightenment of Wireless Communications*, Springer, pp. 213–233, September, 2007. Reproduced in part by kind permission of © Springer Science and Business Media.)

1. Regardless of the number of CMs involved, the case with cooperation on always outperforms the case with cooperation off. In particular, for a setup with $n = 8$, the location estimation accuracy is increased on average by about 48% (Table 12.3).
2. The CRMSE drops considerably when the number of CMs is increased. This is because the location estimates obtained in the first step of the data fusion method must represent a polygonal configuration in space that respects the geometrical constraints imposed by the knowledge about the relative distances between the CMs. As a consequence, the more cooperative users join the cluster, the more constraints the selected solution has to respect and the more accurate the final location estimation is. For example, on going from $n = 2$ to $n = 8$, the location estimation accuracy is increased on average by about 40% (Table 12.3).
3. For the specific simulation models considered here, both cases (cooperation on and cooperation off) always meet the FCC requirements.

- **HLOP** Figure 12.13 shows a comparison, in terms of location estimation accuracy, between the results obtained with 1L-DF with and without cooperation for a variable number of CMs, when the CH is placed at (500, 0) m ($P_{\text{los}} = 0$ for BS–CM links) and the initial guesses are calculated by HLOP. The following can be observed.

  1. Regardless of the number of CMs involved, the case with cooperation on always outperforms the case with cooperation off. In particular, for a setup with $n = 8$, the location estimation accuracy is increased on average by about 61% (Table 12.3).

Table 12.3 (C)RMSE statistics for a variable number of CMs and BSs.

Data fusion algorithm	Cooperation	N	$n = 1$	$n = 2$	$n = 4$	$n = 6$	$n = 8$	Metrics
HTAP-WNLLS	Off	1	80.66, 91.85 76.07, 244.4	–, – –, –	–, – –, –	–, – –, –	–, – –, –	
	On		–, – –, –	69.78, 71.45 68.79, 207.1	50.53, 37.22 52.35, 125.0	46.51, 35.65 49.30, 116.1	41.79, 28.50 44.60, 98.51	
HLOP-WNLLS	Off	3	74.18, 52.54 84.94, 167.8	–, – –, –	–, – –, –	–, – –, –	–, – –, –	
	On		–, – –, –	57.67, 46.04 66.49, 144.2	36.03, 29.18 42.79, 93.89	32.64, 25.37 39.22, 80.48	29.16, 23.37 33.74, 77.56	
HLOP-WNLLS	Off	5	38.01, 24.36 42.91, 83.48	–, – –, –	–, – –, –	–, – –, –	–, – –, –	$\begin{bmatrix} \mu & , & \sigma \\ 67\% & , & 95\% \end{bmatrix}$ (m, m)
	On		–, – –, –	30.59, 21.44 33.13, 71.53	20.98, 14.30 23.70, 48.92	18.77, 12.83 21.30, 43.87	17.99, 12.18 19.85, 41.51	
HTDOA-WNLLS	Off	3	81.37, 51.02 97.72, 151.08	–, – –, –	–, – –, –	–, – –, –	–, – –, –	
	On		–, – –, –	67.79, 44.64 83.72, 138.34	42.09, 29.12 50.77, 94.24	37.67, 25.54 45.87, 80.58	31.31, 20.76 35.94, 70.45	
HTDOA-WNLLS	Off	5	55.42, 51.78 58.42, 124.81	–, – –, –	–, – –, –	–, – –, –	–, – –, –	
	On		–, – –, –	44.91, 38.85 48.81, 107.08	32.34, 27.05 35.67, 77.32	29.28, 22.51 32.12, 66.24	27.29, 20.59 30.80, 59.95	

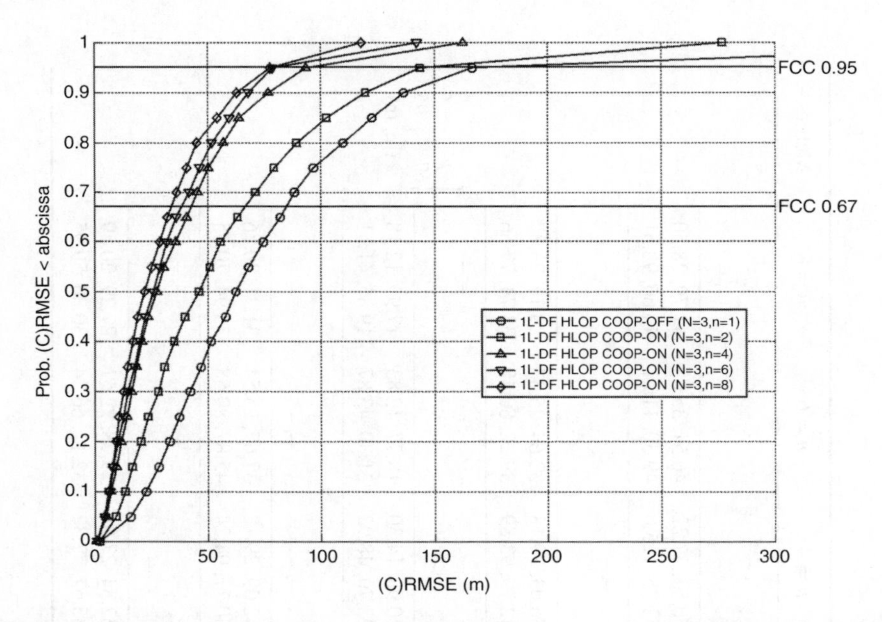

Figure 12.13 (C)RMSE vs. number of CMs (HLOP case). (S. Frattasi, M. Monti, "Cooperative Mobile positioning in 4G wireless networks", *Cognitive Wireless Networks: Concepts, Methodologies and Visions, Inspiring the Age of Enlightenment of Wireless Communications*, Springer, pp. 213–233, September, 2007. Reproduced in part by kind permission of © Springer Science and Business Media.)

2. The CRMSE drops considerably when the number of CMs is increased. For example, on going from $n = 2$ to $n = 8$, the location estimation accuracy is increased on average by about 49% (Table 12.3).
3. 1L-DF with HLOP always performs better than 1L-DF with HTAP and HTDOA (see the next section). For example, for a setup with $n = 8$, the location estimation accuracy is increased on average by about 30% and 7% with respect to 1L-DF with HTAP and with HTDOA, respectively (Table 12.3).
4. For the specific simulation models considered here, both cases (cooperation on and cooperation off) always meet the FCC requirements.

- **HTDOA** Figure 12.14 shows a comparison, in terms of location estimation accuracy, between the results obtained with 1L-DF with and without cooperation for a variable number of CMs, when the CH is placed at (500, 0) m ($P_{\text{los}} = 0$ for BS–CM links) and the initial guesses are calculated by HTDOA. The following can be observed.
  1. Regardless of the number of CMs involved, the case with cooperation on always outperforms the case with cooperation off. In particular, for a setup with $n = 8$, the location estimation accuracy is increased on average by about 61% (Table 12.3).
  2. The CRMSE drops considerably when the number of CMs is increased. For example, on going from $n = 2$ to $n = 8$, the location estimation accuracy is increased on average by about 54% (Table 12.3).
  3. For the specific simulation models considered here, both cases (cooperation on and cooperation off) always meet the FCC requirements.

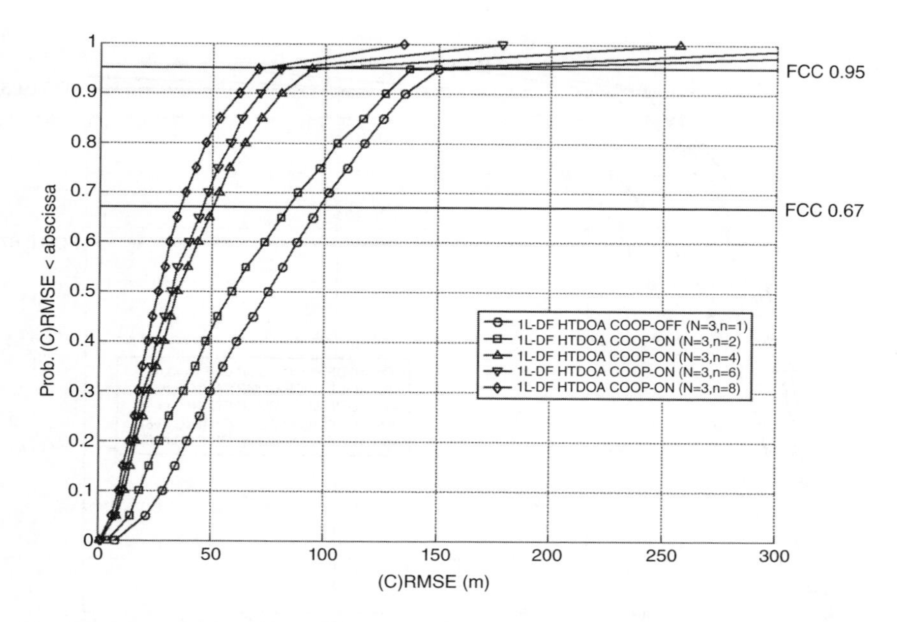

Figure 12.14 (C)RMSE vs. number of CMs (HTDOA case).

12.4.3.2.2 Dependence of Performance on the Number of Base Stations[12]

- **HLOP** Figure 12.15 shows a comparison, in terms of location estimation accuracy, between the results obtained with 1L-DF with and without cooperation for a variable number of BSs, when the CH is placed at (500, 0) m ($P_{\text{los}} = 0$ for BS–CM links) and the initial guesses are calculated by HLOP. The following can be observed.
  1. The CRMSE drops considerably when the numbers of BSs and CMs are increased. For example, on going from $[N = 3, n = 8]$ to $[N = 5, n = 8]$, the location estimation accuracy is increased on average by about 38% (Table 12.3).
  2. For the specific simulation models considered here, COMET achieves an accuracy very much comparable to that of the GPS for the combination $[N = 5, n = 8]$.
- **HTDOA** Figure 12.16 shows a comparison, in terms of location estimation accuracy, between the results obtained with 1L-DF and without cooperation for a variable number of BSs, when the CH is placed at (500, 0) m ($P_{\text{los}} = 0$ for BS–CM links) and the initial guesses are calculated by HTDOA. The following can be observed.
  1. The CRMSE drops considerably when the numbers of BSs and CMs are increased. For example, on going from $[N = 3, n = 8]$ to $[N = 5, n = 8]$, the location estimation accuracy is increased on average by about 13% (Table 12.3).

[12] S. Frattasi, M. Monti, "Cooperative mobile positioning in 4G wireless networks", *Cognitive Wireless Networks: Concepts, Methodologies and Visions Inspiring the Age of Enlightenment of Wireless Communications*, Springer, pp. 213–233, September, 2007. Reproduced in part with kind permission of © 2007 Springer Science and Business Media.

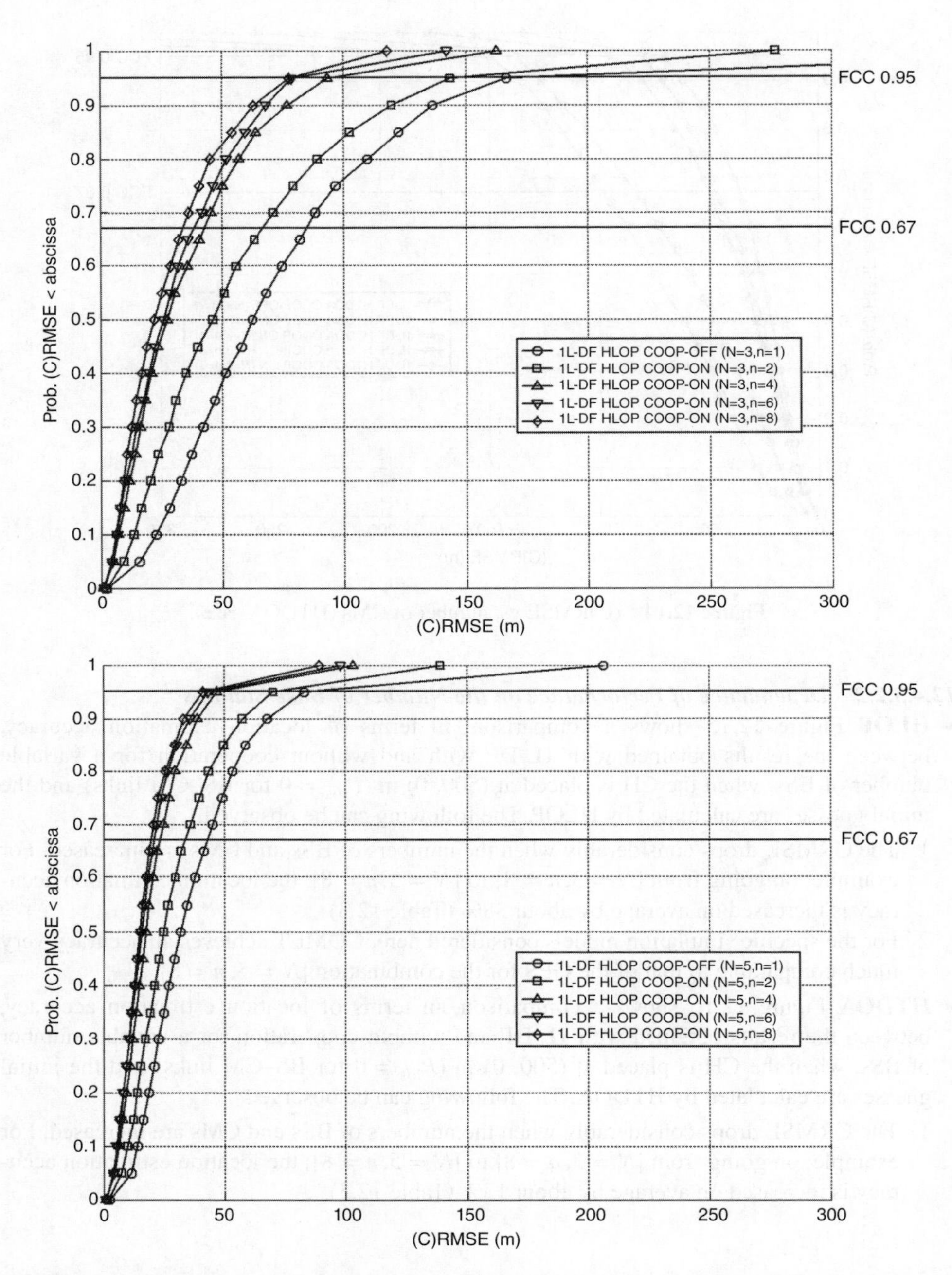

Figure 12.15 (C)RMSE vs number of BSs (HLOP case). (S. Frattasi, M. Monti, "Ad-Coop positioning system (ACPS): Positioning for cooperative users in hybrid cellular ad-hoc networks", *European Transactions on Telecommunications Journal*, Wiley, vol. 19, no. 8, pp. 923–924, May, 2007. Reproduced in part by permission of © 2007 John Wiley & Sons Ltd.)

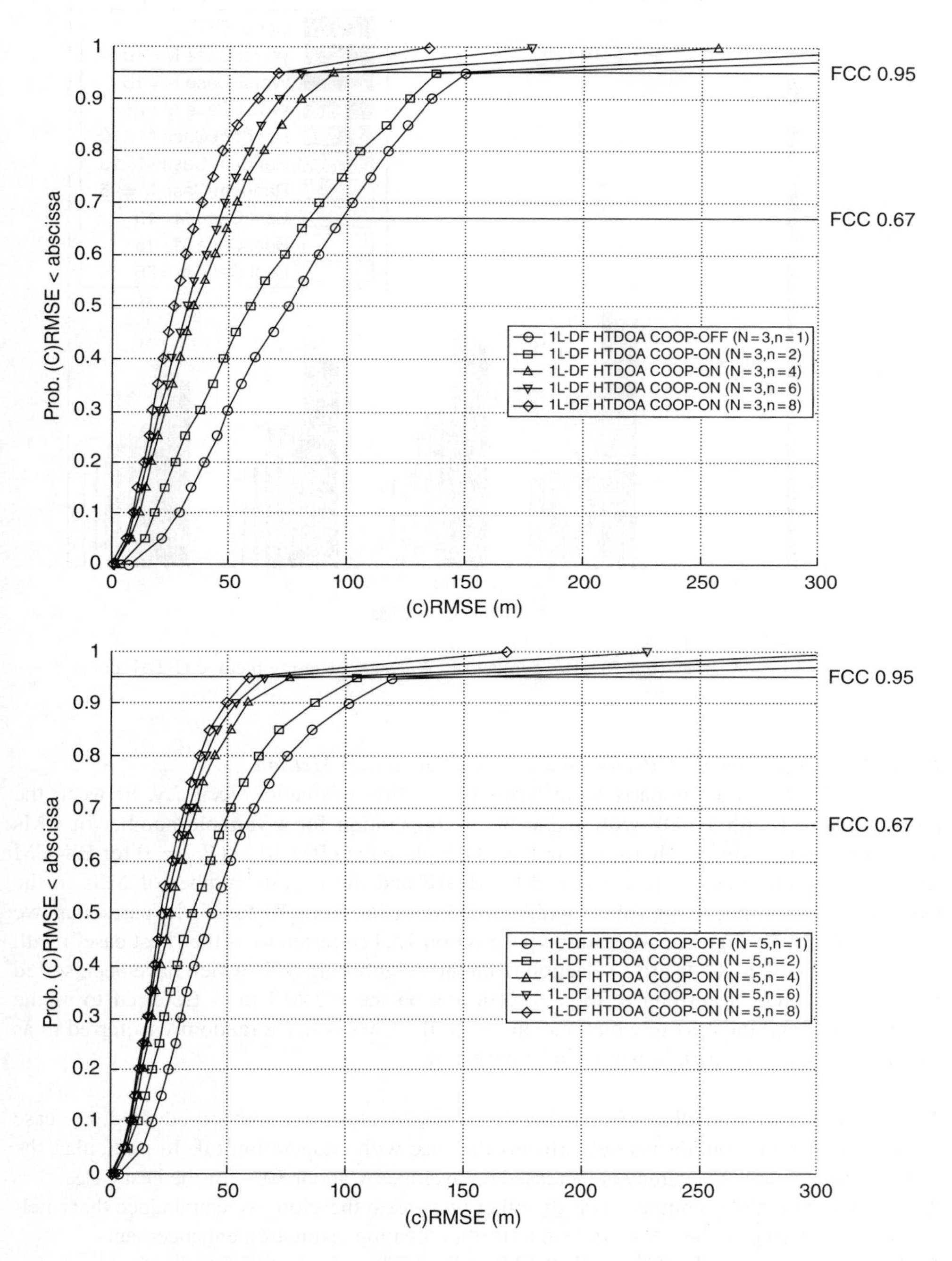

Figure 12.16 (C)RMSE vs number of BSs (HTDOA case).

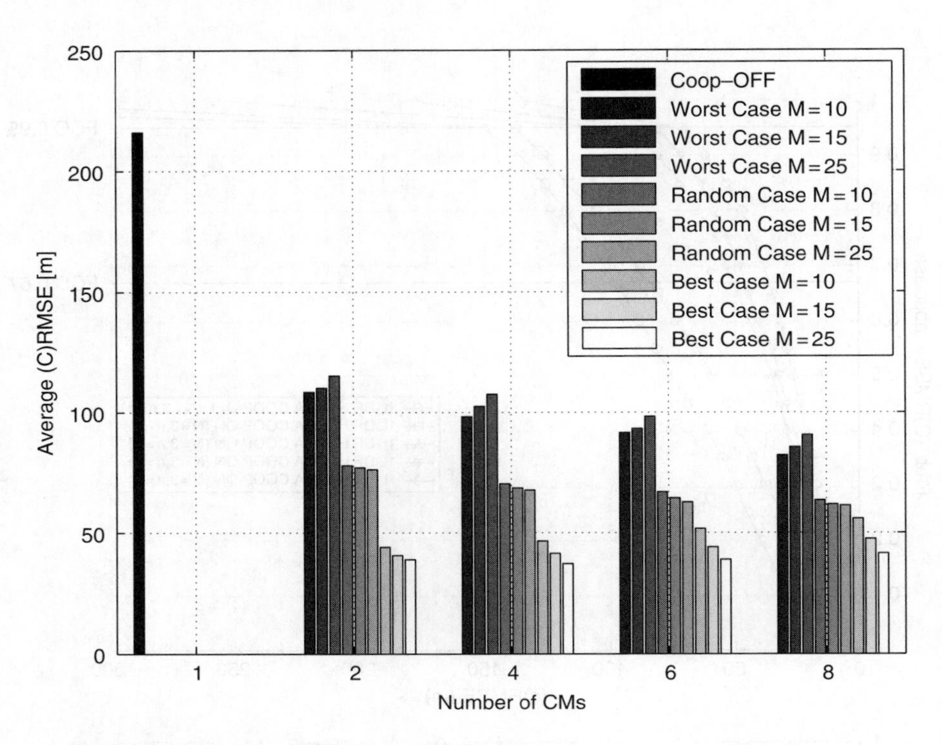

Figure 12.17 Average (C)RMSE vs number of CMs and clustering method (HTAP case).

12.4.3.2.3 Dependence of Performance on the Clustering Method

Figure 12.17 shows a comparison, in terms of location estimation accuracy, between the results obtained with 1L-DF with and without cooperation for a variable number of CMs and different clustering methods, where the CH is placed at (500, 0) m ($P_{\text{los}} = 0$ for BS–CM links), the initial guesses are calculated by HTAP and the overall number of MSs in the surroundings of the CH (i.e., the candidate CMs) varies from 10 to 25. In particular, we compare the clustering method described in Section 12.4 (referred to as the "best case") with the case in which the CMs are a selection from the neighboring MSs which have been sorted in ascending order according to their SNR levels on the BS–MS links (referred to as the "worst case") and the case in which the choice of the CMs is made randomly (referred to as the "random case"). The following can be observed.

1. Regardless of the number of CMs involved and the clustering method adopted, the case with cooperation on always outperforms the case with cooperation off. In particular, the location estimation accuracy is increased on average by about 80% for the best case.
2. The best case always outperforms the other two cases, therefore we can deduce that intelligent clustering of the CMs can lead to further location estimation enhancements.
3. The higher the number of candidate MSs, the higher the location estimation accuracy achieved for the best case. This is because increasing the number of MSs in the surroundings of the CH offers a larger choice of candidate CMs, that is, a larger range of SNR

levels available. As expected, the reverse is valid for the worst case, while the results are practically invariant for the random case.

4. In contrast to the other two cases, the location estimation accuracy decreases slightly with the number of CMs for the best case. This is because the candidate CMs are sorted in decreasing order of SNR levels. Hence, there is a delicate balance when a new CM is chosen to be part of the cluster as the accuracy would tend to decrease owing to the decreasing SNR, although, at the same time, as explained in previous sections, the newly available measurements will create more constraints on the geometry of the problem. Nevertheless, in addition to this minor trend, we can deduce that a few CMs can be enough to achieve the same accuracy as that obtained with many. This is a remarkable result, which implies that we can sensibly reduce the complexity of the localization system.

12.5 Experimental Activity in a Cooperative WLAN Scenario

In this section we introduce an example of a typical experimental activity performed by adopting multiple MSs, specifically applied to WLAN in indoor environments (Figure 12.18). In particular, we will describe the results obtained by performing a cooperative data fusion method based on NLLS, which uses as kernel techniques the ones illustrated in the previous chapters, revealing the potential of such a system by means of experimental results. Our

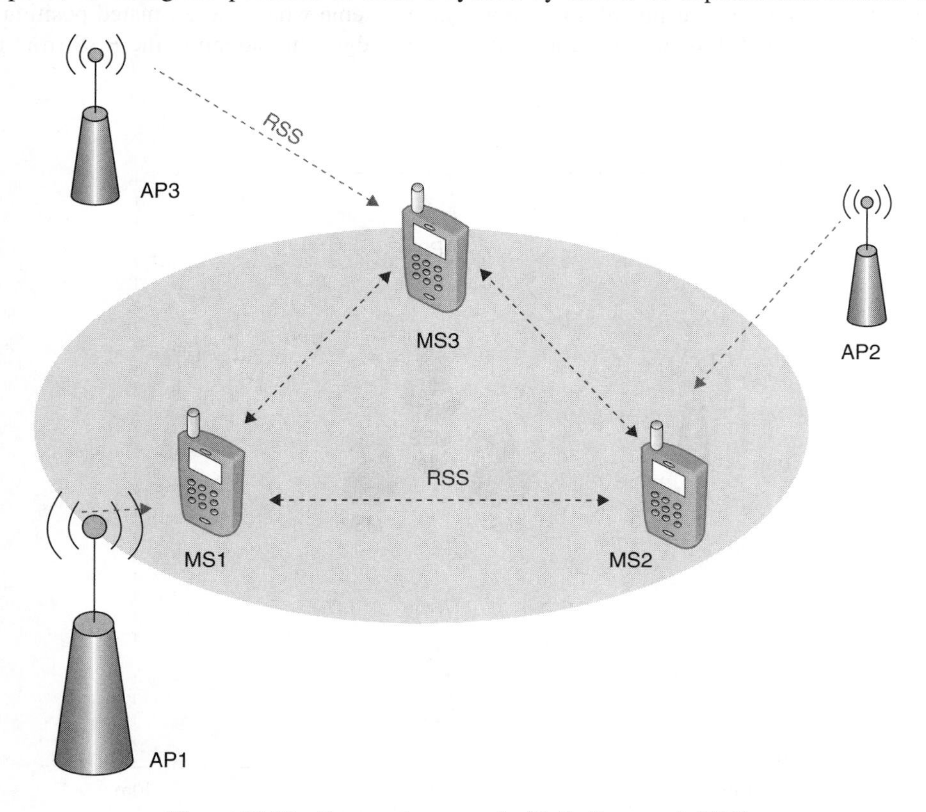

Figure 12.18 Cooperative scenario (Della Rosa et al. 2012).

contribution shows an experimental investigation with analysis of the errors introduced in the distance estimation and exploitation of the knowledge of environmental perturbations with a NLLS algorithm, highlighting the limits and potential of cooperative schemes in real-life scenarios.

12.5.1 *Scenario*

The experimental activity proposed in this section was performed at Tampere University of Technology, Department of Electronics and Communication Engineering. The experimental space is an open area with dimensions of 20 × 20 m. Three conventional mass-market devices (specifically android smartphones) with WiFi capability were placed in the experimental area, as shown in Figure 12.19. A total of four APs 802.11n were deployed according to the configuration shown in Figure 12.19, with the following relative coordinates expressed in meters from the point (0, 0): AP1 (-8, 0), AP2(0, 8), AP3(8, 0), AP4(0, –8), MS1(–1, 0), MS2(0, 1) and MS3(1, 0), with MS1, MS2 and MS3 facing, respectively, AP1, AP3 and AP4. The chosen configuration allows the effect due to both the environment and the user's body to be understood.

12.5.2 *Results*

Figure 12.20 shows the results of a noncooperative scheme where the estimated position of the MSs is obtained through a noncooperative NLLS algorithm adopting the RSS from the

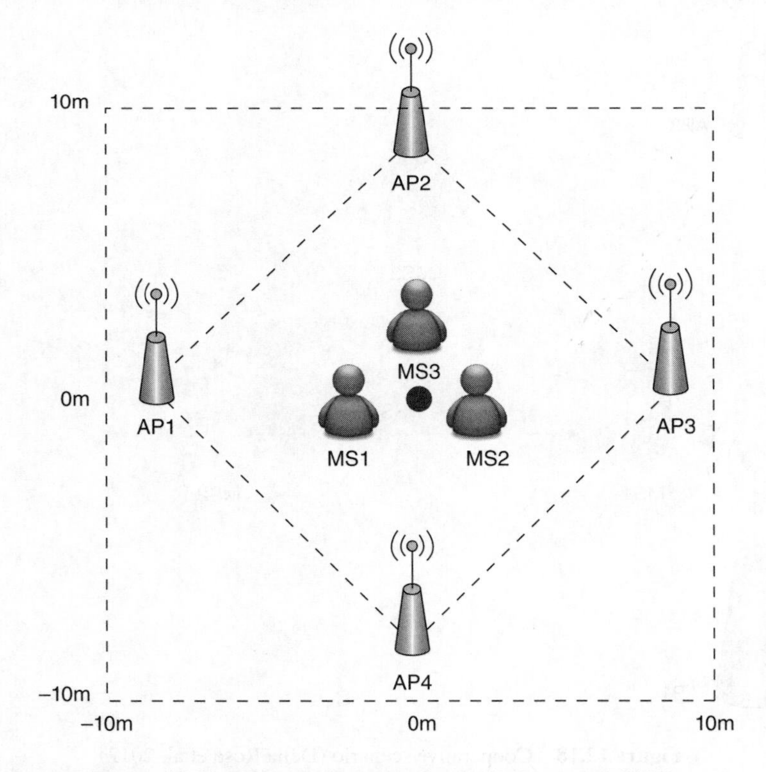

Figure 12.19 Cooperative scenario (Della Rosa et al. 2013).

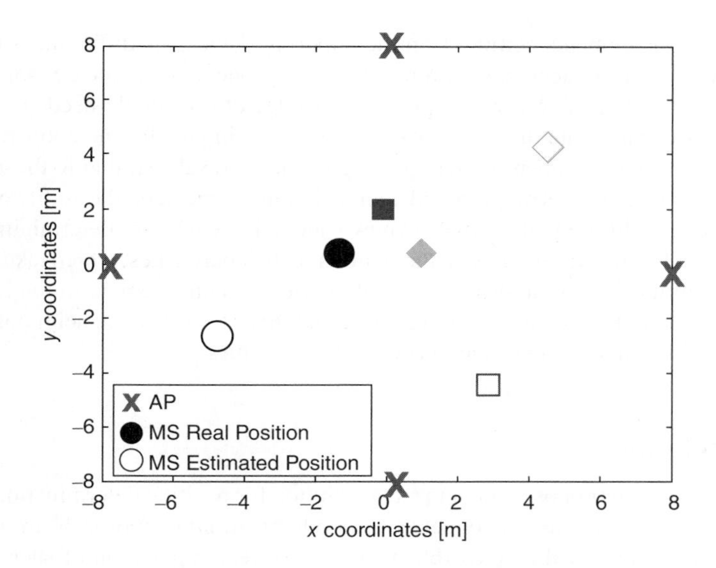

Figure 12.20 Noncooperative case estimated positions (Della Rosa et al. 2013).

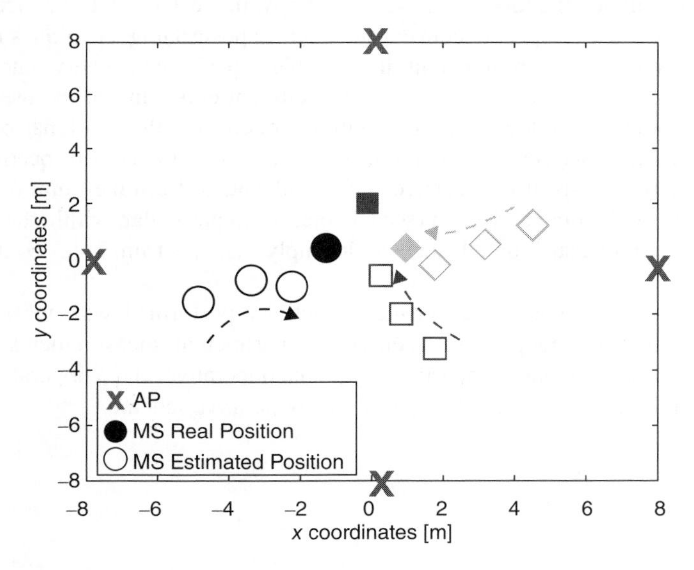

Figure 12.21 Cooperative case estimated positions obtained by exploiting nearby RSS measurements from neighboring devices (Della Rosa et al. 2013).

AP-MSs links. Due to the fluctuations in the RSS caused by the surrounding environment and the human body, such as body loss, hand grip and device-related characteristics, there are large discrepancies between the real and the estimated position of the target MSs. The results obtained suggest that it would be advisable in such conditions to exploit the compensation effects provided by the peer-to-peer (P2P) links which are not possible to exploit in noncooperative schemes. The case of Figure 12.21, where we add the beneficial effect of the most likely reliable measurements coming from neighboring devices, is different. Although there is a clear enhancement in accuracy with respect to the noncooperative case, some additional enhancement might be achieved with a priori knowledge of potentially needed compensation parameters to better mitigate the body-loss and hand-grip impact for the cooperative case. In extreme cases, a lack of compensation might degenerate in results similar to those obtained in the noncooperative case. This range-based approach might depend on the ability of the applied signal processing to differentiate between known and unknown biases determining the quality of the P2P links. The results emphasize that more holistic approaches, which take into account the real-world limitations of algorithms, might provide an enhancement in the level of accuracy of such positioning systems. Moreover, the reliability of short-range neighboring P2P link information plays a fundamental role in cooperative schemes.

12.6 Conclusions

In this chapter we have proposed an innovative solution for position determination in 4G wireless networks by introducing the cooperative mobile positioning approach. In particular, we have described a system, called COMET, which is a direct application of such an approach to cellular networks. The numerical results presented in the chapter have demonstrated that, regardless of the number of cooperating users or BSs available, COMET enhances the location estimation accuracy with respect to conventional hybrid positioning techniques in stand-alone cellular networks and, to a certain extent, it can achieve performance very much comparable to the GPS. Moreover, we have shown that an intelligent clustering of the users can further improve the accuracy of the localization system. In conclusion, this work has proved that the emerging paradigm of wireless cooperation has a beneficial impact on location estimation accuracy. Indirectly, this is also translated into an advantage from the point of view of handover, routing, beamforming and so on (see Chapter 3). In particular, within the framework of cognitive radio systems, such an advance could imply that spectrum holes could be detected more precisely.

Additionally, we presented an experimental activity performed with mass-market WiFi devices in a controlled cooperative scenario. Experimental measurements and position estimations were carried out comparing both noncooperative and cooperative achievable accuracy, highlighting the beneficial effect of the cooperative scheme.

References

3GPP Technical Specification Group (2005) *GSM/EDGE Radio Access Network: Radio Subsystem Link Control (3GPP Standard TS 05.08 V8.23.0)*, 3GPP.

3GPP Technical Specification Group (2006) *Technical Specification Group Radio Access Network. Physical layer aspects for evolved Universal Terrestrial Radio Access (UTRA): Release 7 (Draft Standard 3GPP TR 25.814 V7.1.0)*, 3GPP.

3GPP Technical Specification Group (2008a) *Radio Access Network: Stage 2 Functional Specification of User Equipment (UE) Positioning in UTRAN (3GPP Standard TS 25.305)*, 3GPP.

3GPP Technical Specification Group (2008b) *Services and System Aspects: Functional Stage 2 Description of Location Services (LCS) (3GPP Standard TS 23.271 V8.0.0)*, 3GPP.

3GPP Technical Specification Group (2008c) *Services and System Aspects: Location Services (LCS): Service Description: Stage 1 (3GPP Standard TS 22.071 V8.1.0)*, 3GPP.

3GPP Technical Specification Group (2008d) *Services and System Aspects: Universal Geographical Area Description (GAD) (3GPP Standard TS 23.032 V8.0.0)*, 3GPP.

3GPP Technical Specification Group (2009) *GSM/EDGE Radio Access Network: Radio Subsystem Synchronization (3GPP Standard TS 45.010 V8.4.0)*, 3GPP.

3GPP TS 36.355. *Evolved Universal Terrestrial Radio Access (E-UTRA); LTE Positioning Protocol (LPP)*, 3GPP.

3GPP TS 36.455. *Evolved Universal Terrestrial Radio Access (E-UTRA); LTE Positioning Protocol A (LPPa)*, 3GPP.

3GPP TS 36.211. *Evolved Universal Terrestrial Radio Access (E-UTRA); Physical Channel and Modulation*, 3GPP.

3GPP TS 36.214. *Evolved Universal Terrestrial Radio Access (E-UTRA); Physical Layer Measurements*, 3GPP.

3GPP TS 37.857. *Study on indoor positioning enhancements for UTRA and LTE (Release 13)*, 3GPP.

3GPP TSG SA WG1 SP-150818. *New WID on Study on SMARTER Critical Communications*, 3GPP.

3GPP (2015) *TS 22.071 V14.1.0, Location services (LCS); Service Description; Stage 1.*

AAA (2009) American Automobile Association, http://www.aaa.com [10 March 2010].

Ad2Hand (2009) http://www.ad2hand.com/ [10 March 2010].

Ahmad MY and Mohan AS (2009) *RFID reader localization using passive RFID tags. Proceedings of the Asia Pacific Microwave Conference (APMC)*, pp. 606–609.

Alamouti SM (1998) A simple transmit diversity technique for wireless communications. *IEEE Journal on Selected Areas in Communications*, **16**(8), 1451–1458.

Mobile Positioning and Tracking: From Conventional to Cooperative Techniques, Second Edition.
Simone Frattasi and Francescantonio Della Rosa.

Alhmiedat T and Yang S (2007) A survey: Localization and tracking mobile targets through wireless sensor networks. *Proceedings of the 8th Annual Postgraduate Symposium on the Convergence of Telecommunications, Networking and Broadcasting*, Liverpool, UK.

Allen B, Dohler M, Okon E, Malik W, Brown A and Edwards D (2007) *Ultra-wideband Antennas and Propagation for Communications, Radar, and Imaging*, John Wiley & Sons, Ltd, Chichester.

Aloi D and Korniyenko O (2007) Comparative performance analysis of a Kalman filter and a modified double exponential filter for GPS-only position estimation of automotive platforms in an urban-canyon environment. *IEEE Transactions on Vehicular Technology*, **56**(5), part 2, 2880–2892.

Anon (1991) European Cooperation in the Field of Scientific and Technical Research EURO-COST 231, Urban Transmission Loss Models for Mobile Radio in the 900 and 1800 MHz Bands, Revision 2, The Hague, September 1991.

Anon (2007) IST-4-027756 WINNER II, D1.1.2 V1.2, *Winner II channel models*.

Antonini M, Ruggieri M, Prasad R, Guida U and Corini G (2004) Vehicular remote tolling services using EGNOS. *Proceedings of the ION/IEEE Position Location and Navigation Symposium (PLANS 2004)*, Monterey, CA, pp. 375–379.

Arnitz D, Witrisal K and Muehlmann U (2009) Multifrequency continuous-wave radar approach to ranging in passive UHF RFID. *IEEE Transactions in Microwave Theory Technology*, **57**(5), 1398–1405.

Arnitz D, Muehlmann U and Witrisal K (2010) UWB ranging in passive UHF RFID: Proof of concept. *IEEE Electronics Letters*, **46**(20), 1401–1402.

Arnitz D, Muehlmann U and Witrisal K (2012) Characterization and modeling of UHF RFID channels for ranging and localization. *IEEE Transactions in Antennas and Propagation*, **60**(5), 2491–2501.

Arulampalam M, Maskell S, Gordon N, Clapp T, Sci D, Organ T and Adelaide S (2002) A tutorial on particle filters for online nonlinear/non-Gaussian Bayesian tracking. *IEEE Transactions on Signal Processing*, **50**(2), 174–188.

Baert AE and Seme D (2004) Voronoi mobile cellular networks: Topological properties. *Proceedings of the International Symposium on Parallel and Distributed Computing in Association with HeteroPar'04*, Cork, Ireland, pp. 29–35.

Baghaei-Nejad M, Mendoza D, Zou Z, Radiom S, Gielen G, Zheng L-R and Tenhunen H (2009) A remote-powered RFID tag with 10Mb/s UWB uplink and 18.5dBm sensitivity UHF downlink in 0.18 CMOS. *Proceedings of the IEEE International Solid-State Circuits Conference (ISSCC)*, pp. 198–199.

Bai F, Sadagopan N and Helmy A (2003) IMPORTANT: A framework to systematically analyze the impact of mobility on performance of routing protocols for adhoc networks. *Twenty-Second Annual Joint Conference of the IEEE Computer and Communications Societies INFOCOM 2003*, San Francisco, CA, Vol. 2, pp. 825–835.

Bandara U, Hasegawa M, Inoue M, Morikawa H and Aoyama T (2004) Design and implementation of a Bluetooth signal strength based location sensing system. *Proceedings of the Radio and Wireless Conference*, Atlanta, GA, pp. 319–322.

Bar-Shalom Y, Li X and Kirubarajan T (2001) *Estimation with Applications to Tracking and Navigation*, John Wiley & Sons, New York.

Bassiouni MA, Chiu MH, Loper M, Garnsey M and Williams J (1997) Performance and reliability analysis of relevance filtering for scalable interactive distributed simulation. *ACM Transactions on Modeling and Computer Simulation (TOMACS)*, **7**(3), 293–331.

Bellovin S (2008) Internet Privacy: Big Brother and Little Brother. *Proceedings of NICT Workshop on 18 May 2008*, Tokyo, Japan.

Bluetooth Special Interest Group (2003) Bluetooth core specification v1.2.

Bøgsted M, Olsen RL and Schwefel HP (2010) Probabilistic models for access strategies to dynamic information. *Journal of Performance Evaluation*, **67**(1), 43–60.

Bolcskei H, Gesbert D and Paulraj AJ (2002) On the capacity of OFDM-based spatial multiplexing systems. *IEEE Transactions on Communications*, **50**(2), 225–234.

Borman S (2004) The Expectation Maximization Algorithm: A Short Tutorial. Lecture Notes, The Natural Language Group, University of South California.

Bowditch N (2002) *American Practical Navigator: An Epitome of Navigation*, Paradise Cay Publications, pp. 99–104.

Box G and Jenkins G (1976) *Time Series Analysis, Forecasting and Control*, Holden-Day Series in Time Series Analysis, Holden-Day, San Francisco, CA.

Brenner P (1997) A technical tutorial on the IEEE 802.11 protocol, http://sss-mag.com/pdf/802_11tut .pdf (10 March 2010).

Brimicombe A (2002) GIS: Where are the frontiers now? *Proceedings of the 1st International Conference on GIS (GIS 2002)*, Manama, Bahrain, pp. 33–45.

Butler J (2003) Mobile robots as gateways into wireless sensor networks, http://www.linuxdevices.com/ articles/AT2705574735.htm (10 March 2010).

Caffery JJ (1999) *Wireless Location in CDMA Cellular Radio Systems*, Kluwer, Norwell, MA, Chapter 7.

Caffery JJ and Stüber GL (1998b) Overview of radiolocation in CDMA cellular systems. *IEEE Communications Magazine*, **36**(4), 38–45.

Caffery JJ and Stüber GL (1998a) Subscriber location in CDMA cellular networks. *IEEE Transactions on Vehicular Technology*, **47**(2), 406–416.

Capkun S, Hamdi M and Hubaux JP (2002) GPS-free positioning in mobile ad-hoc networks. *Cluster Computing*, **5**(2), 157–167.

Casadei V, Nanna N and Dardari D (2011) Experimental study in breath detection and human target ranging in the presence of obstacles using ultra-wideband signals. *International Journal of Ultra Wideband Communications and Systems – Special Issue on Applications of Ultra Wideband Technology in Healthcare*, **2**(2), 116–123. http://inderscience.metapress.com/content/A80833LN716110N8.

Casas R, Marco A, Guerrero JJ and Falco J (2006) Robust estimator for non-line-of-sight error mitigation in indoor localization. *EURASIP Journal of Applied Signal Processing*, **2006**(1), 1–8.

Cassioli D, Win MZ and Molisch AF (2002) The ultra-wide bandwidth indoor channel: from statistical model to simulations. *IEEE Journal of Selected Areas of Communication*, **20**(6), 1247–1257.

CCIR (1989) *Spectrum Conserving Terrestrial Frequency Assignments for Given Frequency Distance Separations*, CCIR.

CELLO (2003) Cellular network optimization based on mobile location, http://www.telecom.ntua.gr/ cello (17 August 2009).

Celusion (2009) FDMA, TDMA, CDMA & GSM – What is the future? http://www.celusion.com/ whitepapers/fdma_tdma_cdma.htm (10 March 2010).

Chan YT and Ho KC (1994) A simple and efficient estimator for hyperbolic location. *IEEE Transactions on Signal Processing*, **42**(8), 1905–1915.

Chan YT, Hang HYC and Ching PC (2006a) Exact and approximate maximum likelihood localization algorithms. *IEEE Transactions on Vehicular Technology*, **55**(1), 10–16.

Chan YT, Tsui WY, So HC and Ching PC (2006b) Time of arrival based localization under NLOS conditions. *IEEE Transactions on Vehicular Technology*, **55**(1), 17–24.

Chaudhari N (2007) *Location-aided operation of future mobile wireless networks*, White Paper, Wipro Technologies.

Chawla V and Ha D-S (2007) An overview of passive RFID. *IEEE Communications Magazine*, **45**(9), 11–17.

Chen PC (1999) A non-line-of-sight error mitigation algorithm in location estimation. *Proceedings of the IEEE International Conference on Wireless Communications and Networking (WCNC)*, New Orleans, LA, Vol. 1, pp. 316–320.

Chen C and Feng K (2005a) Hybrid location estimation and tracking system for mobile devices. *Proceedings of the 61st IEEE Vehicular Technology Conference (VTC-Spring)*, Stockholm, Sweden, Vol. 4, pp. 2648–2652.

Chen CL and Feng KT (2005b) An efficient geometry-constrained location estimation algorithm for NLOS environments. *Proceedings of the IEEE International Conference on Wireless Networks, Communication, Mobile Computing*, Hawaii, pp. 244–249.

Cheng C and Sahai A (2004) Estimation bounds for localization. *Proceedings of the IEEE International Conference on Sensors and Ad-Hoc Communications and Networks (SECON)*, Santa Clara, CA, pp. 415–424.

Cheng YC, Chawathe Y and Krumm J (2005) *Accuracy characterization for metropolitan-scale Wi-Fi localization*. Technical Report IRS-TR-05-003, Intel Research.

Chernick MR (1999) *Bootstrap Methods: A Practitioner's Guide*, John Wiley & Sons, New York.

Chiang CC (1998) Wireless network multicasting. PhD thesis, University of California, Los Angeles.

Chiu MH and Bassiouni M (2000) Predictive schemes for handoff prioritization in cellular networks based on mobile positioning. *IEEE Journal on Selected Areas in Communications*, **18**(3), 510–522.

Choi BJ and Shen X (2010) Distributed Clock Synchronization in Delay Tolerant Networks. *Proceedings of the IEEE International Conference on Communications (ICC)*, CApe Town, South Africa.

Choi S and Shin G (1998) Predictive and adaptive bandwidth reservation for hand-offs in QoS-sensitive cellular networks. *ACM SIGCOMM*, Vancouver, Canada, Vol. 28, issue 4, pp. 155–166.

Chon HD, Jun S, Jung H and An SW (2004) Using RFID for accurate positioning. *Journal of Global Positioning Systems*, **3**(1–2), 32–39.

Chor B and Gilboa M (1997) *Computationally Private Information Retrieval*. *ACM Symposium on Theory of Computing*, 303–313.

Cisco, (2008) *Wi-Fi Location-Based Services 4.1 Design Guide*. Americas Headquarters, Cisco Systems, Inc., May 20.

Cohn S, Sivakumaran N and Todling R (1994) A fixed-lag Kalman smoother for retrospective data assimilation. *Monthly Weather Review*, **122**(12), 2838–2867.

Coleman TF and Li Y (1996) An interior trust region approach for nonlinear minimization subject to bounds. *SIAM Journal on Optimization*, **6**, 418–445.

Commission of the European Communities (2007) Commission Decision of 21 February 2007 on allowing the use of the radio spectrum for equipment using ultra-wideband technology in a harmonised manner in the Community. *Official Journal of the European Union*, 2007/131/EC, 23 February, http://eur-lex.europa.eu/LexUriServ/site/en/oj/2007/l_055/l_05520070223en00330036.pdf.

Cong L and Zhuang W (2001) Non-line-of-sight error mitigation in TDOA mobile location. *Proceedings of the IEEE Global Telecommunication Conference (GLOBECOM)*, San Antonio, TX, Vol. 1, pp. 680–684.

Cong L and Zhuang W (2002) Hybrid TDOA/AOA mobile user location for wideband CDMA cellular systems. *IEEE Transactions on Wireless Communications*, **1**, 439–447.

Cong L and Zhuang W (2004) Non-line-of-sight error mitigation in mobile location. *Proceedings of the 23rd IEEE Annual Joint Conference of the Computer and Communications Societies (INFOCOM)*, Hong Kong, pp. 650–659.

Cong L and Zhuang W (2005) Nonline-of-sight error mitigation in mobile location. *IEEE Transactions on Wireless Communications*, **4**(2), 560–573.

Conti A, Dardari D, Guerra M, Mucchi L and Win M Z (2014) Experimental characterization of diversity navigation. *IEEE Systems Journal*, **8**(1), 115–124.

D'Amico A, Mengali U and Taponecco L (2008) Energy-based TOA estimation. *IEEE Transactions in Wireless Communications*, **7**(3), 838–847.

Dardari D (2004) Pseudo-random active UWB reflectors for accurate ranging. *IEEE Communications Letters*, **8**(10), 608–610.

Dardari D and Win MZ (2009) Ziv-Zakai bound of time-of-arrival estimation with statistical channel knowledge at the receiver. *Proceedings of the IEEE International Conference on Ultra-Wideband, ICUWB 2009*, Vancouver, BC, Canada, pp. 624–629.

Dardari D, Chong C-C and Win MZ (2008) Threshold-based time-of-arrival estimators in UWB dense multipath channels. *IEEE Transactions in Communication*, **56**(8), 1366–1378.

Dardari D, Conti A, Lien J and Win M (2008) The effect of cooperation on localization systems using UWB experimental data. *EURASIP Journal of Advances in Signal Processing*, **2008** (Article ID 513873).

Dardari D, Conti A, Ferner U, Giorgetti A and Win MZ (2009) Ranging with ultrawide bandwidth signals in multipath environments. *Proceedings of the IEEE, Special Issue on UWB Technology & Emerging Applications*, **97**(2), 404–426.

Dardari D, D'Errico R, Roblin C, Sibille A and Win MZ (2010) Ultrawide bandwidth RFID: The next generation? Special Issue on RFID – A Unique Radio Innovation for the 21st Century. *Proceedings of the IEEE*, **98**(9), 1570–1582.

Dardari D, Closas P and Djuric PM (2015) Indoor tracking: Theory, methods, and technologies, *IEEE Transactions in Vehicular Technology*, **64**(4), 1263–1278.

David JS (2015) Indoor wireless RF channels. *JPL's Wireless Communication Reference Website*, http://www.wirelesscommunication.nl/reference/chaptr03/indoor.htm (26 November 2015).

David L (2009) David L. Wilson's GPS accuracy Web page, http://users.erols.com/dlwilson/gps.htm (10 March 2010).

Dhar S and Varshney U (2011) Challenges and business models for mobile location-based services and advertising, *Communications of the ACM*, **54** (5), 121–128.

de Montjoye Y, Hidalgo C, Verleysen M and Blondel V (2013) Unique in the Crowd: The privacy bounds of human mobility. *Nature Srep*, March 25.

Decarli N and Dardari D (2014) Ziv-Zakai bound for time delay estimation of unknown deterministic signals. *Proceedings of the 2014 IEEE International Conference on Acoustics, Speech, and Signal Processing (ICASSP)*, Florence, Italy, pp. 1–5.

Decarli N, Dardari D, Gezici S and D'Amico A (2010) LOS/NLOS detection for UWB signals: a comparative study using experimental data. *Proceedings of the IEEE International Symposium on Wireless Pervasive Computing (ISWPC 2010)*, Modena, Italy, pp. 1–5.

Decarli N, Guidi F and Dardari D (2014) A novel joint RFID and radar sensor network for passive localization: Design and performance bounds. Special Issue on Non-cooperative Localization Networks. *IEEE Journal on Selected Topics in Signal Processing*, **8**(1), 80–95.

Decarli N, Guerra A, Guidi F, Chiani M, Dardari D, Costanzo A, Fantuzzi M, Masotti D, Bartoletti S, Dehkordi J, Conti A, Romani A, Tartagni M, Alesii R, Di Marco P, Santucci F, Roselli L, Virili M, Savazzik P and Bozzik M (2015) The GRETA architecture for energy efficient radio identification and localization. *Proceedings of the International EURASIP Workshop on RFID Tech. (EURFID)*, pp. 1–8.

Decarli N, Guidi F and Dardari D (2016) Passive UWB RFID for tag localization: Architectures and design. *IEEE Sensors Journal*, **16**(5), 1385–1397.

Deffenbaugh M, Bellingham J and Schmidt H (1996) The relationship between spherical and hyperbolic positioning. *Conference Proceedings OCEANS '96, MTS/IEEE, Prospects for the 21st Century*, Fort Lauderdale, FL, Vol. 2, pp. 590–595.

Della Rosa F, Xu L, Leppakoski H and Nurmi J (2011a) Human-Induced Effects on RSS Ranging Measurements for Cooperative Positioning. *International Conference on Indoor Positioning and Indoor Navigation (IPIN)*, 21–23 September, Guimarães, Portugal.

Della Rosa F, Leppakoski H, Ghalib A, Ghazanfari L, Garcia O, Frattasi S and Nurmi J (2011b) Ad hoc networks for cooperative mobile positioning. *IN-TECH*, ISBN- 978-953-7619 2011.

Della Rosa F, Pelosi M and Nurmi J (2012) Human-Induced Effects on RSS Ranging Measurements for Cooperative Positioning. *International Journal of Navigation and Observation*, **2012**, Article ID 959140, 13 pages. doi:10.1155/2012/959140.

Della Rosa F, Paakki T, Pelosi M and Nurmi J (2013) Exploiting RSS Measurements among Neighbouring Devices: a Matter of Trust. *2013 International Conference on Indoor Positioning and Indoor Navigation*, Montbéliard, France.

Deng P and Fan P (2000) An AOA assisted TOA positioning system. *Proceedings of the World Computer Congress/International Conference on Communication Technology (WCC/ICCT)*, Beijing, China, Vol. 3, pp. 1501–1504.

Denis B, Keignart J and Daniele N (2003) Impact of NLOS propagation upon ranging precision in UWB systems. *Proceedings of the IEEE Conference on Ultrawideband Systems Technology (UWBST)*, Reston, VA, pp. 379–383.

Denis B, Pierrot JB and Abou-Rjeily C (2006) Joint distributed synchronization and positioning in UWB ad hoc networks using TOA. *IEEE Transactions on Microwave Theory and Techniques*, **54**(4), 1896–1911.

Dennis JE and Schnabel RB (1983) *Numerical Methods for Unconstrained Optimization and Nonlinear Equations*, Prentice-Hall, Englewood Cliffs, NJ.

Departments of Defense and of Transportation (2001) *2001 Federal Radionavigation Systems*. Department of Defense and Department of Transportation, National Technical Information Service, Springfield, VA.

Dey AK (2000) *Providing architectural support for building context-aware applications*. PhD thesis, College of Computing, Georgia Institute of Technology.

Divis DA (2014) FCC's new E911 location requirements could boost positioning services in general. *INSIDE GNSS News*, http://www.insidegnss.com/node/3930 (18 March 2014).

Divakaran A (2008) *Multimedia content analysis: theory and applications (signals and communication technology)*, Springer.

Dizdarevic V and Witrisal K (2006) On impact of topology and cost function on LSE position determination in wireless networks. *Proceedings of the Workshop on Positioning, Navigation, and Communication (WPNC)*, Hannover, Germany, pp. 129–138.

Dodgeball (2009) http://www.dodgeball.com (13 April 2009).

Doucet A, de Freitas N and Gordon N (2001) *Sequential Monte Carlo Methods in Practice*, Springer-Verlag.

Drane C, Macnaughtan M and Scott C (1998) Positioning GSM telephones. *IEEE Communications Magazine*, **36**(4), 46–59.

Dumont L, Fattouche M and Morrison G (1994) Super-resolution of multipath channels in a spread spectrum location system. *Electronics Letters* **30**(19), 1583–1584.

ECMA-368 (2005) *High rate ultra wideband PHY and MAC standard*, http://www.ecma-international .org/publications/files/ECMA-ST/ECMA-368.pdf.

Eltaher A (2009) *Electro-Optical Ranging for Short Range Applications: Design and Realization Aspects*, Shaker.

EPC Global (2008) *EPC radio-frequency identity protocols class-1 generation 2 UHF RFID protocol for communications at 860 MHz–960 MHz*.

Erceg V (2001) *Channel model for fixed wireless application*. Technical report, IEEE 802.16 Working Group.

Erceg V, Greenstein LJ, Tjandra SY, Parkoff SR, Gupta A, Kulic B, Julius AA and Bianchi R (1999a) An empirically based path loss model for wireless channels in suburban environments. *IEEE Journal on Selected Areas in Communications*, **17**(7), 1205–1211.

Erceg V, Michelson DG, Ghassemzadeh SS, Greenstein LJ, Rustako AJ, Guerlain PB, Dennison MK, Roman RS, Barnickel DJ, Wang SC and Miller RR (1999b) A model for the multipath delay profile of fixed wireless channels. *IEEE Journal on Selected Areas in Communications*, **17**(3), 399–410.

ETRI S1-160014. *Proposed text clean-up for Critical Communications*, ETRI.

ETSI (1998) Universal Mobile Telecommunications System (UMTS): Selection procedures for the choice of radio transmission technologies of the UMTS (UMTS 30.03 version 3.1.0). Technical report.

EUROPA (2003) Commission pushes for rapid deployment of location enhanced 112 emergency services. DN: IP/03/1122.

European Commission Information and Media (2005) *eSafety: Improving road safety using information & communication technologies*. Fact Sheet 48, http://www.esafetysupport.org/download/European_Commission/048-esafety.pdf (10 March 2010).

Falsi C, Dardari D, Mucchi L and Win MZ (2006) Time of arrival estimation for UWB localizers in realistic environments. *EURASIP Journal of Applied Signal Processing*, **2006**, 1–13.

Fantuzzi M, Masotti D and Costanzo A (2015) A novel integrated UWB-UHF one-port antenna for localization and energy harvesting. *IEEE Transactions on Antennas and Propagation*, **63**(9), 3839–3848.

Farhang-Boroujeny B (2011) OFDM Versus Filter Bank Multicarrier. *IEEE Signal Processing Magazine*, **8**(3), May 2011.

Farrell J and Barth M (1999) *The Global Positioning System and Inertial Navigation*, McGraw-Hill.

Fathi H,Shin S, Kobara K and Imai H (2008) Protocols for Authenticated Anonymous Communications in IP Networks. *Journal of Computer Communications*, 3662–3671.

Fathi H,Shin S, Kobara K and Imai H (2010) Purpose-restricted Anonymous Mobile Communications using Anonymous Signatures in Online Credential Systems. *Wireless Personal Communication Journal*, **54**(1), 225–236.

FCC (1999) *FCC acts to promote competition and public safety in enhanced wireless 911 services*. DC: WT Rep. 99-27, FCC.

FCC (2001) FCC wireless 911 requirements. Fact sheet WTB/Policy.

FCC Rules (2015) *Wireless E911 Location Accuracy Requirements*, FCC.

FCC Staff (2012) *Location-Based Services: An Overview of Opportunities and Other Considerations*. FCC Report, 25 May.

FCC (2002) *First Report and Order 02-48*.

FCC (2014) FCC-14-13, *Wireless E911 Location Accuracy Requirements. Proposes new indoor requirements and revisions to existing E911 rules*, FCC.

Feess WA and Stephens SG (1987) Evaluation of GPS ionospheric time-delay model. *IEEE Transactions on Aerospace and Electronic Systems*, **23**, 332–338.

Feldmann S, Kyamakya K, Zapater A and Lue Z (2003) An indoor Bluetooth-based positioning system: Concept, implementation and experimental evaluation. *Proceedings of the International Conference on Wireless Networks*, Las Vegas, NV, pp. 109–113.

Figueiras J (2008) *Accuracy enhancements for positioning of mobile devices in wireless communication networks*. PhD thesis, Aalborg University.

Figueiras J and Schwefel HP (2008) Wireless positioning based on a segment-wise linear approach for modeling the target trajectory. *IEEE Global Communications Conference (IEEE GLOBECOM 2008)*, New Orleans, LA, pp. 4825–4829.

Figueiras J, Schwefel HP and Kovacs I (2005) Accuracy and timing aspects of location information based on signal-strength measurements in Bluetooth. *16th International Symposium on Personal, Indoor, and Mobile Radio Communication (PIMRC 2005)*, Berlin, Germany, Vol. 4, pp. 2685–2690.

Foschini GJ (1996) Layered space–time architecture for wireless communication in a fading environment when using multiple antennas. *Bell Labs Technical Journal*, **6**, 41–59.

Foschini GJ and Gans MJ (1998) On limits of wireless communications in a fading environment when using multiple antennas. *Wireless Personal Communications*, **6**, 311–335.

Fox D, Hightower J, Liao L, Schulz D and Borriello G (2003) Bayesian filtering for location estimation. *IEEE Pervasive Computing*, **2**(3), 24–33.

Frattasi S (2007) *Link layer techniques enabling cooperation in fourth generation (4G) wireless networks*. PhD thesis, Aalborg University.

Frattasi S and Figueiras J (2007) *Ad-Coop Positioning System: Using an Embedded Kalman Filter Data Fusion*, CRC Press, pp. 120–131.

Frattasi S and Monti M (2006) On the use of cooperation to enhance the location estimation accuracy. *Proceedings of the IEEE International Symposium on Wireless Communication Systems (ISWCS)*, Valencia, Spain, pp. 545–549.

Frattasi S and Monti M (2007a) Ad-coop positioning system (ACPS): Positioning for cooperating users in hybrid cellular ad-hoc networks. *European Transactions on Telecommunications*, **19**(8), 929–934.

Frattasi S and Monti M (2007b) Cooperative mobile positioning in 4G wireless networks, *Cognitive Wireless Networks*, Springer, pp. 213–233.

Frattasi S, Monti M and Prasad R (2006) A cooperative localization scheme for 4G wireless communications. *Proceedings of the IEEE Radio & Wireless Symposium (RWS)*, San Diego, CA, pp. 287–290.

Frattasi S and Nielsen L (2015) IP: all-round business asset. *Horizon 2020 Projects: Portal*, **7**, pp.18–19.

Gezici S (2008) A survey on wireless position estimation. *Wireless Personal Communications*, **44**(3), 263–282.

Gezici S and Sahinoglu Z (2004) *UWB geolocation techniques for IEEE 802.15.4a Personal area networks*. MERL technical report.

Gezici S, Kobayashi H and Poor H (2003) Nonparametric non-line-of-sight identification. *Proceedings of the 58th IEEE Vehicular Technology Conference*, Vol. 4, pp. 2544–2548.

Giorgetti A and Chiani M (2013) Time-of-arrival estimation based on information theoretic criteria. *IEEE Transactions on Signal Processing*, **61**(8), 1869–1879.

Goldsmith A (2005) *Wireless Communications*, Cambridge University Press.

Gonzalez-Castano F and Garcia-Reinoso J (2002) Bluetooth location networks. *IEEE Global Telecommunications Conference, GLOBECOM 2002*, Taipei, Taiwan, Vol. 1, pp. 233–237.

Goodwin CJ (2005) *Research in Psychology: Methods and Design*. John Wiley & Sons, Inc.

Google (2006) Review guide. Google Maps for Mobile (beta), http://maps.google.com (10 March 2010).

Google (2007) Google Maps for Mobile with My Location (beta). Online video, http://www.youtube.com/watch?v=v6gqipmbcok (10 March 2010).

Gordon N, Salmond D and Smith A (1993) Novel approach to nonlinear/non-Gaussian Bayesian state estimation. *Radar and Signal Processing, IEE Proceedings F*, **140**(2), 107–113.

Graphisoft R&D (2007) Location-based construction management, http://martasari.files.wordpress.com/2007/09/c2007_print_a4.pdf (10 March 2010).

Green J, Betti D and Davison J (2000) *Mobile location services: Market strategies*. Technical Report, Ovum.

Greenstein L, Yeh VEY and Clark M (1997) A new path-gain/delay-spread propagation model for digital cellular channels. *IEEE Transactions on Vehicular Technology*, **46**(2), 477–485.

Grewal M and Andrews A (2001) *Kalman Filter: Theory and Practice Using MATLAB*, 2nd edn, John Wiley & Sons.

Grewal MS, Weill LR and Andrews AP (2007) *Global Positioning Systems, Inertial Navigation, and Integration*, 2nd edn, John Wiley & Sons.

Grubbs F (1969) Procedures for detecting outlying observations in samples. *Technometrics*, **11**(1), 1–21.

Guan Y, Zhou Y and Law CL (2007) High-resolution UWB ranging based on phase-only correlator. *Proceedings of the IEEE International Conference on Ultra-Wideband (ICUWB)*, Singapore, pp. 100–104.

Gudmundson M (1991) Correlation Model for Shadow Fading in Mobile Radio Systems. *Electronic Letters*, 2145–2146.

Guidi F, Decarli N, Bartoletti S, Conti A and Dardari D (2014b) Detection of multiple tags based on impulsive backscattered signals. *IEEE Transactions on Antennas and Propagation*, **62**(11), 3918–3930.

Guidi F, Sibille A, Roblin C, Casadei V and Dardari D (2014a) Analysis of UWB tag backscattering and its impact on the detection coverage. *IEEE Transactions on Antennas and Propagation*, **62**(8), 4292–4303.

Guidi F, Guerra A and Dardari D (2015) Personal mobile radars with millimeter-wave massive arrays for indoor mapping. *IEEE Transactions in Mobile Computing*, **15**(6), 1471–1484.

Guidi F, Decarli N, Dardari D, Natali F, Savioli E and Bottazzi M (2016) A low complexity scheme for passive UWB-RFID: Proof of concept. *IEEE Commication Letters*, **24** (4), 676–679.

Gustafsson F and Gunnarsson F (2005) Mobile positioning using wireless networks: Possibilites and fundamental limitations based on available wireless network measurements. *IEEE Signal Processing Magazine*, **22**(4), 41–53.

Guvenc I and Arslan H (2006) Comparison of two searchback schemes for non-coherent TOA estimation in IR-UWB systems. *Proceedings of the IEEE Sarnoff Symposium*, Princeton, NJ.

Guvenc I and Chong CC (2009) A survey on TOA based wireless localization and NLOS mitigation techniques. *IEEE Communications Surveys and Tutorials*, **11**(3), 107–124.

Guvenc I and Sahinoglu Z (2005a) Multiscale energy products for TOA estimation in IR-UWB systems. *Proceedings of the IEEE Global Telecommunications Conference (GLOBECOM)*, St Louis, MO, Vol. 1, pp. 209–213.

Guvenc I and Sahinoglu Z (2005b) Threshold-based TOA estimation for impulse radio UWB systems. *Proceedings of the IEEE International Conference on Ultra-Wideband*, Zurich, Switzerland, pp. 420–425.

Guvenc I and Sahinoglu Z (2005c) Threshold selection for UWB TOA estimation based on kurtosis analysis. *IEEE Communication Letters*, **9**(12), 1025–1027.

Guvenc I and Sahinoglu Z (2005d) TOA estimation with different IR-UWB transceiver types. *Proceedings of the IEEE International Conference on UWB (ICU)*, Zurich, Switzerland, pp. 426–431.

Guvenc I, Sahinoglu Z, Molisch AF and Orlik P (2005) Non-coherent TOA estimation in IR-UWB systems with different signal waveforms. *Proceedings of the IEEE International Workshop on Ultra-wideband Networks (UWBNETS)*, Boston, MA, pp. 245–251.

Guvenc I, Sahinoglu Z and Orlik P (2006) TOA estimation for IR-UWB systems with different transceiver types. *IEEE Transactions on Microwave Theory and Techniques (Special Issue on Ultrawideband)*, **54**(4), 1876–1886.

Guvenc I, Chong CC and Watanabe F (2007) NLOS identification and mitigation for UWB localization systems. *Proceedings of the IEEE International Conference on Wireless Communication and Networking (WCNC)*, Hong Kong, pp. 1571–1576.

Guvenc I, Chong CC, Watanabe F and Inamura H (2008) NLOS identification and weighted least squares localization for UWB systems using multipath channel statistics. *EURASIP Journal of Advances in Signal Processing (Special Issue on Signal Processing for Location Estimation and Tracking in Wireless Environments)*, **1**, 1–14.

Hach R (2005) Symmetric double-sided two-way ranging, in *IEEE P802.15 Working Group for Wireless Personal Area Networks (WPANs), June 2005, Berlin, Germany*.

Han S, Lim HS and Lee JM (2007) An efficient localization scheme for a differential-driving mobile robot based on RFID system. *IEEE Transactions on Industrial Electronics*, **54**, 3362–3369.

Hasu V (2007) Radio resource management in wireless communication: Beamforming, transmission power control, and rate allocation. PhD thesis, Helsinki University of Technology.

Heath R and Paulraj A (2000) Switching between multiplexing and diversity based on constellation distance. *Allerton Conference on Communication, Control and Computing*, Morticello, IL, pp. 212–219.

Heidari G (2008) *WiMedia UWB for W-USB and Bluetooth Interpretation of Standards, Regulations and Applications*, John Wiley & Sons.

Heinzelman W, Chandrakasan A and Balakrishnan H (2002) An application-specific protocol architecture for wireless microsensor networks. *IEEE Transactions on Wireless Communications*, **1**(4), 660–670.

Hightower J, Brumitt B and Borriello G (2002) The location stack: A layered model for location in ubiquitous computing. *Proceedings of the Fourth IEEE Workshop on Mobile Computing Systems and Applications (WMCSA)*, Callicoon, NY, pp. 22–28.

Hong D and Rappaport SS (1986) Traffic model and performance analysis for cellular mobile radio telephone system with prioritized and nonprioritized handoff procedures. *IEEE Transactions on Vehicular Technology*, **35**, 77–92.

Hong S, Brand J, Choi J, Jain M, Mehlman J, Katti S and Levis P (2014) Applications of self-interference cancellation in 5G and beyond. *IEEE Communications Magazine*, 114–121, February.

Howard A, Matarić M and Sukhatme G (2003) Cooperative relative localization for mobile robot teams: An ego-centric approach. *Proceedings of the Naval Research Laboratory Workshop on Multi-Robot Systems*, Washington, DC, pp. 65–76.

Hu J and Tirkkonen O (2008) Self-organized synchronization in wireless networks. *2nd IEEE International Conference on Self-Adaptive and Self-Organizing Systems*, Venice, Italy.

Huber PJ (1964) Robust estimation of a location parameter. *Annals of Mathematics and Statistics*, **35**(1), 73–101.

Hunter TE and Nosratinia A (2006) Diversity through coded operations. *IEEE Transactions on Wireless Communications*, **5**(2), 283–289.

IEEE (2002a) *Wireless Medium Access Control (MAC) and Physical Layer (PHY) Specifications for Wireless Personal Area Networks (WPANs)*, IEEE Std. 802.15.1-2002, IEEE Press.

IEEE (2002b) IEEE Standard 803 for Local and Metropolitan Area Networks. *Overview and architecture*. Technical report, IEEE Press.

IEEE (2003) IEEE Standard 802.15.4. *Part 15.4: Wireless medium access control (MAC) and physical layer (PHY) specifications for low-rate wireless personal area networks (LR-WPANS)*, IEEE Press.

IEEE (2005) IEEE Standard 802.15.4. *Wireless medium access control (MAC) and physical layer (PHY) specifications for low-rate wireless personal area networks (LR-WPANs)*, IEEE Press.

IEEE (2007) IEEE Standard for Information Technology. *Telecommunications and information exchange between systems – local and metropolitan area networks. Specific Requirements Part 15.4: Wireless medium access control (MAC) and physical layer (PHY) specifications for low-rate wireless personal area networks (WPANs)*, IEEE Press.

IEEE (2012a) IEEE Standard for Information Technology. *Telecommunications and Information Exchange Between Systems – Local and Metropolitan Area Networks. Specific Requirements Part 15.4: Wireless medium access control (MAC) and physical layer (PHY) specifications for low-rate wireless personal area networks (WPANs). Amendment: Active RFID System PHY*, IEEE Press.

IEEE (2012b) IEEE Standard for Local and Metropolitan Area Networks. *Part 15.6: Wireless Body Area Networks*, IEEE Press.

IF (2009) Issued and pending patents offered to Microsoft Corporation, http://innfundllc.com/press/pr_10.html (7 July 2009).

ISO (2013) ISO/IEC FDIS 24730-62. *Information technology – Real time locating systems (RTLS) Part 62: High rate pulse repetition frequency Ultra Wide Band (UWB) air interface*.

Jacob K (2007) Location-based services. http://www.kenneyjacob.com/2007/05/13/location-based-services (10 March 2010).

Jakes WC (1994) *Microwave Mobile Communications*, John Wiley & Sons.

Jechlitschek C (2006) *A survey paper on radio frequency identification (RFID) trends*, http://www.cse.wustl.edu/~jain/cse574-06/rfid.htm (26 November 2015).

Jiang B, Fishkin KP, Roy S and Philipose M (2006) Unobtrusive long-range detection of passive RFID tag motion. *IEEE Transactions on Instrumentation and Measurement*, **55**(1), 187–196.

Jiang Y and Leung V (2007) An asymmetric double-sided two-way ranging for crystal offset. *Proceedings of the International Symposium on Signals, Systems and Electronics (ISSSE)*, pp. 525–528.

Jiao W, Jiang P, Liu R, Wang W and Ma Y (2008) Providing location service for mobile WiMAX. *IEEE International Conference on Communications, 2008 (ICC '08)*, Beijing, China, pp. 2685–2689.

Jones Q and Grandhi S (2005) P3 systems: Putting the place back into social networks. *IEEE Internet Computing*, **9**(5), 38–46.

Jourdan DB and Roy N (2006) Optimal sensor placement for agent localization. *Proceedings of the IEEE Position, Location, and Navigation Symposium (PLANS)*, San Diego, CA, pp. 128–139.

Jourdan DB, Dardari D and Win MZ (2006) Position error bound for UWB localization in dense cluttered environments. *Proceedings of the IEEE International Conference on Communications (ICC)*, Istanbul, Vol. 8, pp. 3705–3710.

Jourdan DB, Dardari D and Win MZ (2008) Position error bound for UWB localization in dense cluttered environments. *IEEE Transactions in Aerospace Electronic Systems*, **44**(2), 613–628.

Julier S and Uhlmann J (1997) A new extension of the Kalman filter to nonlinear systems, *International Symposium on Aerospace/Defense Sensing, Simulation and Controls*, Orlando, FL.

Jung P (1997) *Analyse und Entwurf digitaler Mobilfunksysteme*, Teubner.

Kailath T, Sayed AH and Hassibi B (2000) *Linear Estimation*, Prentice Hall Information and Systems Science Series, Prentice Hall, Upper Saddle River, NJ.

Kalman R (1960) A new approach to linear filtering and prediction problems. *Journal of Basic Engineering*, **82**(1), 35–45.

Kaplan E and Hegarty C (2006) *Understanding GPS: Principles and Application*, Artech House.

Keeble E (2014) Passive WiFi Tracking. http://edwardkeeble.com/2014/02/passive-wifi-tracking/ (26 February 2014).

Karthaus U and Fischer M (2003) Fully integrated passive UHF RFID transponder IC with 16.7-μW minimum RF input power. *IEEE Journal of Solid-State Circuits*, **38**, 1602–1608.

Khandani AE, Abounadi J, Modiano E and Zheng L (2007) Cooperative Routing in Static Wireless Networks. *IEEE Transactions on Communications*, **55**(11), 2185–2192.

Kim W, Lee JG and Jee GI (2006) The interior-point method for an optimal treatment of bias in trilateration location. *IEEE Transactions on Vehicular Technology*, **55**(4), 1291–1301.

Kim S, Jeon H, Lee H and Ma JS (2008) Robust transmission power and position estimation in cognitive radio. *Springer Lecture Notes in Computer Science*, **5200**, 719–728.

Kingpin (1998) MAC address cloning, http://www.packetstormsecurity.org/docs/hack/mac_address_cloning.pdf (10 March 2010).

Kivanc D and Liu H (2000) Subcarrier allocation and power control for OFDMA. *IEEE Signals, Systems and Computers*, Pacific Grove, CA, pp. 147–151.

Kivera (2002) Value and applications for location-based services. An overview of opportunities and challenges for deploying state-of-the-art location technologies, http://www.kivera.com/downloads/whitepaper.pdf (10 March 2010).

Kjaergaard MB and Munk CV, (2008) Hyperbolic location fingerprinting: A calibration-free solution for handling differences in signal strength (concise contribution). *Sixth Annual IEEE International Conference on Pervasive Computing and Communications*, 17–21 March 2008, ISBN: 978-0-7695-3113-7.

Klingenbrunn T and Mogensen P (1999) Modelling cross-correlated shadowing in network simulations. *Proceedings of the IEEE Vehicular Technology Conference (VTC-Fall)*, Vol. 3, pp. 1407–1411.

Klobuchar J (1976) Ionospheric time-delay corrections for advanced satellite ranging systems. *NATO AGARD Conference Proceedings*, Paris, France, Vol. 209.

Knapp C and Carter G (1976) The generalized correlation method for estimation of time delay. *IEEE Transactions on Acoustics, Speech and Signal Processing*, **24**(4), 320–327.

Knight NL and Wang J (2009) A comparison of outlier detection procedures and robust estimation methods in GPS positioning. *Journal of Navigation, Royal Institute of Navigation*, **62**(4), 699–709.

Ko Y and Vaidya N (1998) Location-aided routing (LAR) in mobile ad-hoc networks, *Proceedings of the Fourth Annual ACM/IEEE International Conference on Mobile Computing and Networking (MOBICOM)*, Dallas, TX, pp. 66–75.

Koeppel I (2002) What are location services? – from a GIS perspective, http://lbs360.directionsmag.com/LBSArticles/ESRI.What%20are%20LS%20Whitepaper.pdf (10 March 2010)

Küpper A (2005) *Location-Based Services: Fundamentals and Operation*, John Wiley & Sons, Ltd, Chichester.

Kyammaka K and Jobmann K (2005) Location management in cellular networks: Classification of the most important paradigms, realistic simulation framework, and relative performance analysis. *IEEE Transactions on Vehicular Technology*, **54**, 687–708.

Laneman JN and Wornell GW (2003) Distributed space-time-coded protocols for exploiting cooperative diversity in wireless networks. *IEEE Transactions on Information Theory*, **49**(10), 2415–2425.

Laneman JN, Wornell GW and Tse DNC (2001) An efficient protocol for realizing cooperative diversity in wireless networks. *Proceedings of the IEEE ISIT*, Washington, DC, USA.

Larsson EG (2004) Cramer–Rao bound analysis of distributed positioning in sensor networks. *IEEE Signal Processing Letters*, **11**(3), 334–337.

Lathi MJ (2000) IEEE 802.11 wireless LAN. *Proceedings of the Seminar on Internetworking on Ad-Hoc Networking*, Helsinki University of Technology.

LaViola J (2003) Double exponential smoothing: An alternative to Kalman filter-based predictive tracking. *Proceedings of the Immersive Projection Technology and Virtual Environments 2003*, Zurich, Switzerland, pp. 199–206.

Lawton MC, Davies RL and McGeehan JP (1991) An analytical model for indoor multipath propagation in the picocellular environment. *Proceedings of the 6th International Conference on Mobile Radio and Personal Communications*, Coventry, UK, pp. 1–8.

Lazaro A, Ramos A, Villarino R and Girbau D (2013a) Active UWB reflector for RFID and wireless sensor networks. *IEEE Transactions Antennas and Propagation*, **61**(9), 4767–4774.

Lazaro A, Ramos A, Villarino R and Girbau D (2013b) Time-domain UWB RFID tag based on reflection amplifier. *IEEE Antennas and Wireless Propagation Letters*, **12**, 520–523.

Lee JS and Miller LE (1998) *CDMA Systems Engineering Handbook*, Artech House, Boston.

Lee J-Y and Scholtz RA (2002) Ranging in a dense multipath environment using an UWB radio link, *IEEE Journal of Selected Areas Communication*, **20**(9), 1677–1683.

Lei Z, Chin F and Kwok Y-S (2006) UWB ranging with energy detectors using ternary preamble sequences. *Proceedings of the IEEE Wireless Communications Networking Conference (WCNC)*, Las Vegas, NE, pp. 872–877.

Leong KS, Ng ML and Cole PH (2005) The reader collision problem in RFID systems. *Proceedings of the IEEE International Symposium on Microwave, Antenna, Propagation and EMC Technologies for Wireless Communications*, Beijing, China, Vol. 1, pp. 658–661.

Levine DA, Akyildiz IF and Naghshineh M (1997) A resource estimation and call admission algorithm for wireless multimedia networks using the shadow cluster concept. *IEEE/ACM Transactions on Networking*, **5**, 1–12.

LG Electronics Inc. (2008) *Method for controlling display characteristic and television receiver using the same, EP1874048A1.*

Li X (2006) An iterative NLOS mitigation algorithm for location estimation in sensor networks. *Proceedings of the IST Mobile and Wireless Communications Summit*, Mykonos, Greece.

Li X and Pahlavan K (2004) Super-resolution TOA estimation with diversity for indoor geolocation. *IEEE Transactions in Wireless Communication*, **3**(1), 224–234.

Li Z, Trappe W, Zhang Y and Nath B (2005) Robust statistical methods for securing wireless localization in sensor networks. *Proceedings of the IEEE International Symposium on Information Processing in Sensor Networks (IPSN)*, Los Angeles, CA, pp. 91–98.

Li B, Dempster A, Rizos C and Lee HK (2006) A database method to mitigate NLOS error in mobilephone positioning. *Proceedings of the IEEE Position, Location, and Navigation Symposium (PLANS)*, San Diego, CA, pp. 173–178.

Liberti J and Rappaport T (1992) Statistics of shadowing in indoor radio channels at 900 and 1900 MHz. *IEEE Military Communications Conference, 1992, MILCOM '92, Conference Record, Communications – Fusing Command, Control and Intelligence*, Vol. 3, pp. 1066–1070.

Lima P (2007) A Bayesian approach to sensor fusion in autonomous sensor and robot networks. *IEEE Instrumentation and Measurement Magazine*, **10**(3), 22–27.

Liu H, Darabi H, Banerjee P and Liu J (2007) Survey of wireless indoor positioning techniques and systems. *IEEE Transactions on Systems, Man, and Cybernetics, Part C: Applications and Reviews*, **37**(6), 1067–1080.

López-Salcedo J and Vázquez G (2005), Nda maximum-likelihood timing acquisition of UWB signals. *Proceedings of the IEEE Workshop on Signal Processing Advances in Wireless Communications*, New York, USA.

Lottici V, D'Andrea A and Mengali U (2002) Channel estimation for ultra-wideband communications. *IEEE Journal on Selected Areas in Communications*, **20**(9), 1638–1645.

Lu S and Bharghavan V (1996) Adaptive resource management algorithms for indoor mobile computing environment. *ACM SIGCOMM*, Stanford, CA, pp. 231–242.

Luba O, Boyd L, Gower A and Crum J (2005) GPS III system operations concepts. *IEEE Aerospace and Electronic Systems Magazine*, **20**, 10–18.

Maali A, Mimoun H, Baudoin G and Ouldali A (2009) A new low complexity NLOS identification approach based on UWB energy detection. *Proceedings of the IEEE Radio and Wireless Symposium, 2009. RWS '09'*, pp. 675–678.

Madsen K, Nielsen H and Tingleff O (2004) *Methods for Non-linear Least Squares Problems*, 2nd edn, Technical University of Denmark, Lyngby.

Mallat A, Gezici S, Dardari D, Craeye C and Vandendorpe L (2014) Statistics of the MLE and approximate upper and lower bounds – Part I: Application to TOA estimation. *IEEE Transactions in Signal Processing*, **62**(21), 5663–5676.

Malm A (2006) *GPS and Galileo in Mobile Handsets*. LBS Research Series, Technical Report, Berg Insight.

Mao G, Fidan B and Anderson B (2007) Wireless sensor network localization techniques. *Computer Networks: The International Journal of Computer and Telecommunications Networking*, **51**(10), 2529–2553.

MapInfo (2009) http://www.pbinsight.com (10 March 2010).

Maranò S, Gifford WM, Wymeersch H and Win MZ (2009) Nonparametric obstruction detection for UWB localization. *Proceedings of the IEEE Global Telecommunications Conference, 2009. GLOBECOM '09'*.

Maxim (2003) An introduction to spread-spectrum communications, http://www.maxim-ic.com/appnotes.cfm/an_pk/1890/ (10 March 2010).

Mayorga CLF, Della Rosa F, Wardana SA, Simone G, Raynal MCN, Figueiras J and Frattasi S (2007) Cooperative positioning techniques for mobile localization in 4G cellular networks. *Proceedings of the IEEE International Conference on Pervasive Services*, pp. 39–44.

Meel J (1999) Spread spectrum – introduction, De Nayer Instituut, http://sss-mag.com/pdf/Ss_jme_denayer_intro_print.pdf (10 March 2010).

Meissner P and Witrisal K (2012) Multipath-assisted single-anchor indoor localization in an office environment. *Proceedings of the 19th International Conference on Systems, Signals and Image Processing (IWSSIP)*, pp. 22–25.

Merwe R and Rebel E (2006) Recursive Bayesian estimation library, http:choosh.csee.ogi.edu/rebel (10 March 2010).

Mhatre V and Rosenberg C (2004) Design guidelines for wireless sensor networks: Communication, clustering and aggregation. *Ad-Hoc Networks Journal*, **2**(1), 45–63.

Mhatre V and Rosenberg C (2005) Energy and cost optimizations in wireless sensor networks: a survey. *Performance Evaluation and Planning Methods for the Next Generation Internet* (eds Girard A, Sansò B and Vázquez-Abad F), Springer, pp. 227–248.

Miao H (2007) *Channel estimation and positioning for multiple antenna systems*. PhD thesis, University of Oulu.

Michailow N, Gaspar I, Krone S, Lentmaier M and Fettweis G (2012) Generalized frequency division multiplexing: An alternative multi-carrier technique for next-generation cellular systems. *Proceedings of the 9th International Symposium on Wireless Communication Systems*, Paris, France, 2012.

Mirollo RE and Strogatz SH (1990) Synchronization of pulse-coupled biological oscillators. *SIAM Journal of Applied Mathematics*, **50**, 1645–1662.

Microsoft Corp. (2015) *Ambient light context-aware display, WO2015038407A1*.

Mitola J and Maguire GQ (1999) Cognitive radio: Making software radios more personal. *IEEE Personal Communications*, **6**(4), 13–18.

MobileIN (2009) Value-added services, http://www.mobilein.com/what_is_a_VAS.htm (10 March 2010).

MobileRobots (2009) http://www.mobilerobots.com/Mobile_Robots.aspx (10 March 2010).

Mohay G, Anderson A, Collie B, De Vel O and McKemmish R (2003) *Computer and Intrusion Forensics*, Artech House.

Molisch AF (2005) *Wireless Communications*, John Wiley & Sons.

Molisch A, Balakrishnan K, Cassioli D, Chong C-C, Emami S, Fort A, Karedal J, Kunisch J, Schantz H, Schuster U and Siwiak K (2005) IEEE p802.15-04/662r2-tg4a, Technical report, IEEE P802.15 WPAN, "ftp://ftp.802wirelessworld.com/15/04/".

Molisch AF, Cassioli D, Chong C-C, Emami S, Fort A, Kannan B, Karedal J, Kunisch J, Schantz H, Siwiak K and Win MZ (2006) A comprehensive standardized model for ultrawideband propagation channels. *IEEE Transactions in Antenna Propagation*, **54**(11), 3151–3166.

Montalvo UD, Ballon P and de Kar EV (2003) Business models for location-based services. *Proceedings of the 6th AGILE Conference on GIScience*, Lyon, France.

Morelli C, Nicoli M, Rampa V and Spagnolini U (2007) Hidden Markov models for radio localization in mixed LOS/NLOS conditions. *IEEE Transactions in Signal Processing*, **55**(4), 1525–1542.

Mourikis A and Roumeliotis S (2006a) Optimal sensor scheduling for resource-constrained localization of mobile robot formations. *IEEE Transactions on Robotics*, **22**(5), 917–931.

Mourikis A and Roumeliotis S (2006b) Performance analysis of multirobot cooperative localization. *IEEE Transactions on Robotics*, **22**(4), 666–681.

Muhammad W, Grosicki E, Abed-Meraim K, Delmas JP and Desbouvries F (2002) Uplink versus downlink wireless mobile positioning in UMTS cellular radio systems. *Proceedings of the European Signal Processing Conference (EUSIPCO'02)*, Toulouse, France, Vol. 2, pp. 373–376.

Murphy K (1988) *Switching Kalman filters*. Technical report, Department of Computer Science, University of California, Berkeley.

Musolesi M, Hailes S and Mascolo C (2004) An ad hoc mobility model founded on social network theory. *ACM International Conference on Modeling, Analysis and Simulation of Wireless and Mobile Systems 2004, MSWiM'04*, Venice, Italy, pp. 20–24.

Nepper P, Treu G and Küpper A (2008) Adding speech to location-based services. *Wireless Personal Communications, Special Issue, Towards Global & Seamless Personal Navigation*, **44**(3), 245–261.

Nguyen DN and Krunz M (2013) Cooperative MIMO in Wireless Networks: Recent Developments and Challenges. *IEEE Network Magazine*, **27**(4), 48–54.

Niedzwiadek H (2002) Where's the value in location services, http://lbs360.directionsmag.com/LBSArticles/HN.Where's%20the%20Value%20in%20Location%20Services.pdf (10 March 2010).

Nikitin PV,Martinez R,Ramamurthy S,Leland H, Spiess G and Rao KVS (2010) Phase-based spatial identification of UHF RFID tags. *Proceedings of the IEEE International Conference on RFID*, Orlando, FL, pp. 102–109.

NIST/SEMATECH (2007) *e-Handbook of Statistical Methods*, NIST, http://www.itl.nist.gov/div898/handbook/ (10 March 2010).

Nosratinia A, Hunter TE and Hedayat A (2004) Cooperative communication in wireless networks. *IEEE Communications Magazine*, **42**(10), 74–80.

Obayashi S and Zander J (1998) A body-shadowing model for indoor radio communication environments. *IEEE Transactions on Antennas and Propagation*, **46**(6), 920–927.

OGC (2003) Open GIS location services (OPENLS): Core services. OGC 03-006r1, OGC Implementation Specification OGC 03-006r1, Open GIS Consortium Inc.

Ohly P, Lombard DN and Stanton KB (2009) Hardware-assisted precision time protocol. Design and case study. *9th LCI International Conference on High Performance Clustered Computing*, Urbana, IL, USA, 28 April–28 May, 2008.

Olsen RL,Figueiras J, Rasmussen JG and Schwefel HP (2010) *How precise should localization be? A quantitative analysis of the impact of delay and mobility on reliability of location information. Proceedings of the IEEE Global Telecommunications Conference (GLOBECOM 2010)*, Miami, FL, pp. 1–6.

Pantel L and Wolf LC (2002) On the suitability of dead reckoning schemes for games. *Proceedings of the 1st workshop on Network and system support for games (NetGames)*, Braunschweig, Germany, pp. 79–84.

Parker L (1996) Multi-robot team design for real-world applications. *Distributed Autonomous Systems 2*, Springer-Verlag, Tokyo, pp. 91–102.

Parkinson B and Spilker J (1996) *The Global Positioning System: Theory and Applications*, American Institute of Aeronautics and Astronautics.

Parsons J (2000) *The Mobile Radio Propagation Channel*, John Wiley & Sons.

Pathanawongthum N and Chemtanomwong P (2010) RFID-based localization techniques for indoor environment. *Proceedings of the 12th International Conference on Advanced Communication Technology (ICACT)*, Vol. 2, pp. 1418–1421.

Patwari N, Ash J, Kyperountas S, Hero A, Moses R and Correal N (2005) Locating the nodes: Cooperative localization in wireless sensor networks. *IEEE Signal Processing Magazine*, **22**(4), 54–69.

Paulraj A and Kailath T (1994) Increasing capacity in wireless broadcast systems using distributed transmission/directional reception. US Patent 5345599.

Paulraj A, Nabar R and Gore D (2003) *Introduction to Space–Time Wireless Communications*, Cambridge University Press.

Pellerano S, Alvarado Jr, J and Palaskas Y (2010) A mm-wave power-harvesting RFID tag in 90 nm CMOS. *IEEE Journal of Solid-State Circuits*, **45**(8), 1627–1637.

Pels M, Barhorst J, Michels M, Hobo R and Barendse J (2005) *Tracking people using Bluetooth: Implications of enabling Bluetooth discoverable mode*. Technical Report, University of Amsterdam.

Penrose R (1955) A generalized inverse for matrices. *Proceedings of the Cambridge Philosophical Society*, **51**, 406–413.

Peressini AL, Sullivan FE and Uhl JJ (1988) *The Mathematics of Nonlinear Programming*, Springer.

Petrus P (1999) Robust Huber adaptive filter. *IEEE Transactions on Signal Processing*, **47**(4), 1129–1133.

Pierrot JB, Denis B and Abou-Rjeily C (2006) Joint distributed time synchronization and positioning in UWB ad hoc networks using TOA. *IEEE Transactions on MTT, Special Issue on Ultra Wideband*, **54**, 1896–1911.

Prasad R, Mohr W and Konhauser W (2000) *Third Generation Mobile Communication Systems*, Artech House.

Prasad R, Rahman MI, Das SS and Marchetti N (2009) *Single- and Multi-Carrier MIMO Transmission for Broadband Wireless Systems*, River.

Proakis JG (1995) *Digital Communications*, Prentice Hall.

Pursula P, Vähä-Heikkilä T, Müller A, Neculoiu D, Konstantinidis G, Oja A and Tuovinen J (2008) Millimeter-wave identification – a new short-range radio system for low-power high data-rate applications. *IEEE Transactions on Microwave Theory Technology*, **56**(10), 2221–2228.

Qi Y, Kobayashi H and Suda H (2006) Analysis of wireless geolocation in a non-line-of-sight environment. *IEEE Transactions on Wireless Communication*, **5**(3), 672–681.

Rabbachin A, Montillet J, Cheong P, de Abreu G and Oppermann I (2005) Non-coherent energy collection approach for toa estimation in UWB systems. *Proceedings of the IST Mobile & Wireless Communications Summit*, Dresden, Germany.

Rabbachin A, Oppermann I and Denis B (2006) ML time-of-arrival estimation based on low complexity UWB energy detection. *Proceedings of the IEEE International Conference on Ultra-Wideband (ICUWB)*, Waltham, MA, pp. 598–604.

Ramos A, Lazaro A and Girbau D (2013) Semi-passive time-domain UWB RFID system. *IEEE Transactions on Microwave Theory and Technology*, **61**(4), 1700–1708.

Rappaport TS (1996) *Wireless Communications Principles and Practice*, Prentice Hall.

Rauner M (2012) Indoor Positioning Technologies. Doctoral and Habilitation Theses, Department of Civil, Environmental and Geomatic Engineering, Institute of Geodesy and Photogrammetry.

Rentel CH and Kunz T (2008) A mutual network synchronization method for wireless ad hoc and sensor networks. *IEEE Transactions on Mobile Computing*, **8**(5), 633–646.

RFID Handbook (2015) *RFID frequencies*, http://rfid-handbook.de/about-rfid/radio-regulation/19-frequency-ranges.html (26 November 2015).

Rhee W and Cioffi JM (2000) Increase in capacity of multiuser OFDM system using dynamic subchannel allocation. *IEEE Vehicular Technology Conference Spring'00*, Tokyo, pp. 1085–1089.

Riba J and Urruela A (2004) A non-line-of-sight mitigation technique based on ML-detection. *Proceedings of the IEEE International Conference on Acoustics, Speech, and Signal Processing (ICASSP)*, Quebec, Canada, Vol. 2, pp. 153–156.

Rios S (2009) Location based services: Interfacing to a mobile positioning center, http://www.wirelessdevnet.com/channels/lbs/features/lbsinterfacing.html (10 March 2010).

Ripley BD (2004) Robust statistics, http://www.stats.ox.ac.uk/pub/StatMeth (10 March 2010).

Rosa FD (2007) Cooperative mobile positioning and tracking in hybrid WIMAX/WLAN networks. Master's thesis, Aalborg University.

Rosa FD, Simone G, Laurent P, Charaffedine R, Mahmood N, Pietrarca B, Kyritsi P, Marchetti N, Perrucci G, Figueiras J and Frattasi S (2007a) Emerging directions in wireless location: Vista from the COMET project. *Proceedings of the 16th IST Mobile & Wireless Communications Summit*, Budapest, Hungary, pp. 1–5.

Rosa FD, Wardana S, Mayorga C, Naima GSM, Figueiras J and Frattasi S (2007b) Experimental activity on cooperative mobile positioning in indoor environments. *Proceedings of the 2nd IEEE Workshop on Advanced Experimental Activities on Wireless Networks & Systems (EXPONWIRELESS)*, Helsinki, Finland, pp. 1–5.

Roy R and Kailath T (1989) *ESPRIT-Estimation of Signal Parameters via Rotational Invariance Techniques*, **37**(7), 984–995.

Ruixue L and Fang Z (2010) LLA: A New High Precision Mobile Node Localization Algorithm Based on TOA. *Journal of Communication*, **8**, 604–611.

Rybaczyk P (2005) *Expert Network Time Protocol: An Experience in Time with NTP*, Apress.

Sanchez L, Lanza J, Olsen RL, Bauer M and Girod-Genet M (2006) *A generic context management framework for personal networking environments*. *Proceedings of the 3rd IEEE Annual International Conference on Mobile and Ubiquitous Systems – Workshops*, pp. 1–8.

Sahinoglu Z and Guvenc I (2006) Multiuser interference mitigation in noncoherent UWB ranging via nonlinear filtering. *EURASIP Journal of Wireless Communication and Networking*. Article ID 56849, 10 pages.

Sari H and Karam G (1996) Orthogonal frequency-division multiple access for the return channel on CATV networks. *Proceedings of the International Conference on Telecommunications (IEEE ICT)*, Istanbul, Turkey, pp. 602–607.

Sayed A and Yousef NR (2003) Wireless location, *Wiley Encyclopedia of Telecommunications* (ed. Proakis J), John Wiley & Sons, New York.

Sayed A, Tarighat A and Khajehnouri N (2005a) Network-based wireless location: Challenges faced in developing techniques for accurate wireless location information. *IEEE Signal Processing Magazine*, **22**(4), 24–40.

Sayed AH, Tarighat A and Khajehnouri N (2005b) Network-based wireless location. *IEEE Signal Processing Magazine*, **22**(4), 24–40.

Schilit B, Adams N and Want R (1994) *Context-aware computing applications*. *Proceedings of the IEEE Workshop on Mobile Computing Systems and Applications (WMCSA'94)*, Santa Cruz, CA, US, pp. 89–101.

Schiller J and Voisard A (2004) *Location-Based Services*, Morgan Kaufmann.

Schmidt R (1986) Multiple emitter location and signal parameter estimation. *IEEE Transactions on Antennas and Propagation*, **34**(3), 276–280.

Scholtz RA, Pozar DM and Namgoong W (2005) Ultra-wideband radio. *EURASIP Journal of Applied Signal Processing*, **2005**(3), 252–272.

Schroeder J, Galler S, Kyamakya K and Kaiser T (2007) Three-dimensional indoor localization in non line of sight UWB channels. *Proceedings of the IEEE International Conference on Ultra Wideband (ICUWB)*, Singapore.

SDMA Toolbox for IT (2009) http://it.toolbox.com/wiki/index.php/SDMA (10 March 2010).

Second Life (2009) http://secondlife.com/ (10 March 2010).

Seidel SY and Rappaport TS (1992) 914 MHz path loss prediction models for indoor wireless communications in multi-floored buildings. *IEEE Transactions on Antennas and Propagation*, **40**(2), 207–217.

Senarath G (2006) *Multi-hop relay system evaluation methodology (channel model and performance metric)*, IEEE 802.16J-06/013R3.

Shen Y and Win MZ (2010) Fundamental limits of wideband localization – Part I: A general framework. *IEEE Transactions on Information Theory*, **56**(10), 4956–4980.

Shiode N, Li C, Batty M, Logley P and Maguire D (2004) The impact and penetration of location-based services. *Telegeoinformatics: Location-Based Computing and Services*, CRC Press, pp. 349–366.

Shiraishi T,Komuro N,Ueda H, Kasai H and Tsuboi T (2008) *Indoor location estimation technique using UHF band RFID. Proceedings of the International Conference on Information Networking (ICOIN)*, pp. 1–5.

Sivrikaya F and Yener B (2004) Time synchronization in sensor networks: A survey. *IEEE Network*, **18**(4), 45–50.

Skyhook wireless (2007) http://www.businesswire.com/news/home/20070529005197/en/Skyhook-Wireless-Launches-Loki-2.0---Location-Aware (6 April 2017).

Small Cell Forum (2012) Femtocells synchronization and location. A Small Cell Forum white paper.

Sobhani B, Paolini E, Giorgetti A, Mazzotti M and Chiani M (2014) Target tracking for UWB multistatic radar sensor networks. *IEEE Journal of Selected Topics in Signal Processing*, **8**(1), 1932–4553.

Song SH and Zhang QT (2008) Multi-dimensional detector for UWB ranging systems in dense multipath environments. *IEEE Transactions on Wireless Communications*, **7**(1), 175–183.

Soork AS, Saadat R and Tadaion AA (2008) Cooperative mobile positioning based on received signal strength. *International Symposium on Telecommunications, 2008 (IST 2008)*, Tehran, Iran, pp. 273–277.

Steele R (1994) *Mobile Radio Communications*, IEEE Press.

Steinfield C (2004) *The Development of Location Based Services in Mobile Commerce*. Springer, pp. 177–197.

Steininger S, Neun M and Edwardes A (2006) *Foundations of location-based services*. Technical Report, University of Zurich.

Stoica L, Rabbachin A and Oppermann I (2006) A low-complexity noncoherent IR-UWB transceiver architecture with TOA estimation. *IEEE Transactions on Microwave Theory and Techniques*, **54**(4), 1637–1646.

Sun GL and Guo W (2004) Bootstrapping M-estimators for reducing errors due to non-line-of-sight (NLOS) propagation. *IEEE Communications Letters*, **8**(8), 509–510.

Talukdar AK, Badrinath BR and Acharya A (1997) On accommodating mobile hosts in an integrated services packet network. *Sixteenth Annual Joint Conference of the IEEE Computer and Communication Societies (IEEE INFOCOM'97)*, Kobe, Japan, pp. 1048–1055.

Tarokh V, Seshadri N and Calderbank AR (1998) Space–time codes for high data rate wireless communications: Performance criterion and code construction. *IEEE Transactions on Information Theory*, **44**, 744–765.

Taylor J (2002) Robust modelling using the t-distribution. *Proceedings of the Australasian Genstat Conference*, Busselton, Australia.

Tejavanija K, Williamson K and Kang J (2003) Development of location-based facilities management system for mobile devices. *Proceedings of the 1st China–US Relations Conference*, College Station, TX.

Telatar I (1999) Capacity of multi-antenna Gaussian channels. *European Transactions on Telecommunications*, **10**(6), 585–595.

Telenity (2009) http://www.telenity.com/ (10 March 2010).

Todling R, Cohn S and Sivakumaran N (1997) Suboptimal schemes for retrospective data assimilation based on the fixed-lag Kalman smoother. *Monthly Weather Review*, **126**(8), 2274–2286.

Trees HLV (2001) *Detection, Estimation, and Modulation Theory: Part I*, 2nd edn, John Wiley & Sons, Inc., New York.

Tripathi ND, Reed JH and VanLandingham HF (1998) Handoff in cellular systems. *IEEE Personal Communications*, **5**(6), 26–37.

TruePosition (2009) http://www.trueposition.com (10 March 2010).

Tyrrell A, Auer G and Bettstetter C (2006) Firefly synchronization in ad hoc networks. *Proceeding MiNEMA Workshop*, Sintra, Portugal.

Vaidya N and Das SR (2008) RFID-based networks: exploiting diversity and redundancy. *ACM SIGMOBILE Mobile Computing and Communications Review*, **12**(1), 2–14.

Vakilian V, Wild T, Schaich F, ten Brink S and Frigon J-F (2013) Universal-filtered multi-carrier technique for wireless systems beyond LTE. *9th IEEE Broadband Wireless Access Workshop, Globecom*, Atlanta, GA, USA.

Van de Kar EAM (2004) Designing mobile information services: An approach for organisations in a value network. PhD thesis, Delft University of Technology.

Vanderveen M, Papadias C and Paulraj A (1997) Joint angle and delay estimation (JADE) for multipath signals arriving at an antenna array. *IEEE Communications Letters*, **1**(1), 12–14.

Venkatesh S and Buehrer RM (2006) A linear programming approach to NLOS error mitigation in sensor networks. *Proceedings of the IEEE International Symposium on Information Processing in Sensor Networks (IPSN)*, Nashville, TN, pp. 301–308.

Venkatesh S and Buehrer RM (2007) NLOS mitigation using linear programming in ultrawideband location-aware networks. *IEEE Transactions on Vehicular Technology*, **56**(5), 3182–3198.

Venkatraman S and Caffery J (2004) Hybrid TOA/AOA techniques for mobile location in non-line-of-sight environments. *Proceedings of the Wireless Communications and Networking Conference (WCNC)*, Atlanta, GA, Vol. 1, pp. 274–278.

Venkatraman S, Caffery JJ and You H-R (2004) A novel ToA location algorithm using LoS range estimation for NLoS environments. *IEEE Transactions on Vehicular Technology*, **53**(5), 1515–1524.

Verdone R, Dardari D, Mazzini G and Conti A (2008) *Wireless Sensor and Actuator Networks: technologies, analysis and design*, Elsevier.

Virrantaus K, Veijalainen J, Markkula J, Katanosov A, Garmash A, Tirri H and Terziyan V (2001) Developing GIS-supported location-based services. *Proceedings of the 1st International Workshop on Web Geographical Information Systems (WGIS 2001)*, Kyoto, Japan, Vl. 2, pp. 66–75.

Vossiek M, Wiebking L, Gulden P, Wieghardt J, Hoffmann C and Heide P (2003) Wireless local positioning. *IEEE Microwave Magazine*, **4**, 77–86.

Waadt A, Hessamian-Alinejad A, Wang S, Statnikov K, Bruck G and Jung P (2008) Mobile-assisted positioning in GSM networks. *Proceedings of the International Workshop on Signal Processing and its Applications (WoSPA2008)*, Sharjah, UAE.

Wang X, Wang Z and Dea BO (2003) A TOA based location algorithm reducing the errors due to non-line-of-sight (NLOS) propagation. *IEEE Transactions on Vehicular Technology*, **52**(1), 112–116.

Wang Z, Tameh E and Nix A (2004) Statistical peer-to-peer channel models for outdoor urban environments at 2 GHz and 5 GHz. *Proceedings of the 60th IEEE Vehicular Technology Conference (VTC-Fall)*, Los Angeles, CA, Vol. 7, pp. 5101–5105.

Ward A, Jones A and Hopper A (1997) A new location technique for the active office. *IEEE Personal Communications*, **4A**(5), 42–47.

Wei D and Chan A (2005) A survey on cluster schemes in ad-hoc wireless networks. *Proceedings of the 2nd International Conference on Mobile Technology, Applications and Systems*, Guangzou, China, pp. 1–8.

Wengerter C, Ohlhorst J and Elbwart AV (2005) Fairness and throughput analysis for generalized proportional fair frequency scheduling in OFDMA. *Proceedings of the 61st IEEE Vehicular Technology Conference (VTC)*, Stockholm, Sweden, Vol. 3, pp. 1903–1907.

Werner-Allen G, Tewari G, Patel A, Welsh M and Nagpal R (2005) Firefly-inspired sensor network synchronicity with realistic radio effects. *Proceeding SenSys*. Proceedings of the 3rd international conference on Embedded networked sensor systems, San Diego, CA, USA, pp. 142–153.

Win M and Scholtz R (1998a) Impulse radio: How it works. *IEEE Communication Letters*, **2** (1), 10–12.

Win MZ and Scholtz RA (1998b) On the energy capture of ultra-wide bandwidth signals in dense multipath environments. *IEEE Communication Letters*, **2**(9), 245–247.

Win MZ and Scholtz RA (2002) Characterization of ultra -wide bandwidth wireless indoor communications channel: A communication theoretic view. *IEEE Journal on Selected Areas in Communications*, **20**(9), 1613–1627.

Win MZ, Dardari D, Molisch AF, Wiesbeck W and Jinyun Z (2009) History and applications of UWB (scanning the issue). Special Issue on UWB Technology & Emerging Applications. *Proceedings of the IEEE*, **97**(2), 198–204.

Witrisal K (2002) *OFDM air interface design for multimedia communications*. PhD thesis, Delft University of Technology.

Witrisal K, Meissner P, Leitinger E, Shen Y, Gustafson C, Tufvesson F, Haneda K, Dardari D, Molisch AF, Conti A and Win MZ (2016) High-accuracy localization for assisted living: 5G systems will turn multipath channels from foe to friend. *IEEE Signal Processing Magazine*, **33**, (2), 59–70,.

Wong CY, Cheng RS, Letaief KB and Murch RD (1999a) Multiuser OFDM with adaptive subcarrier, bit, and power allocation. *IEEE Journal on Selected Areas in Communications*, **17**(10), 1747–1758.

Wong CY, Tsui CY, Cheng RS and Letaief KB (1999b) A real-time subcarrier allocation scheme for multiple access downlink OFDM transmission. *IEEE VTC*, Amsterdam, The Netherlands, Vol. 2, pp. 1124–1128.

Wormell D, Foxlin E and Katzman P (2007) Improved 3d interactive devices for passive and active stereo virtual environments. *Proceedings of the 13th Eurographics Workshop on Virtual Environments*, Weimar, Germany.

Wymeersch H, Lien J and Win M (2009) Cooperative localization in wireless networks. *Proceedings of the IEEE*, **97**(2), 427–450.

Xiong L (1998) A selective model to suppress NLOS signals in angle-of-arrival AOA location estimation. *Proceedings of the IEEE International Symposium on Personal, Indoor, and Mobile Radio Communications (PIMRC)*, Boston, MA, Vol. 1, pp. 461–465.

Xu G (2003) *GPS: Theory, Algorithms and Applications*, Springer.

Yoon J, Liu M and Noble B (2003) Random waypoint considered harmful. *Twenty-Second Annual Joint Conference of the IEEE Computer and Communications Societies, IEEE INFOCOM 2003*, San Francisco, CA, Vol. 2, pp. 1312–1321.

Zander J and Erikkson H (1993) Asymptotic bounds on the performance of a class of dynamic channel assignment algorithms. *IEEE Journal on Selected Areas in Communications*, **11**, 926–933.

Zeisberg S and Schreiber V (2008) EUWB – coexisting short range radio by advanced ultra-wideband radiotechnology. *Proceedings of the ICT Mobile Summit 2008*, Stockholm, Sweden.

Zhan H, Ayadi J, Farserotu J and Le Boudec J-Y (2007) High-resolution impulse radio ultra wideband ranging. *IEEE International Conference on Ultra-Wideband (ICUWB)*, pp. 568–573.

Zhan H, Ayadi J, Farserotu J and Le Boudec J-Y (2007b) A novel maximum likelihood estimation of superimposed exponential signals in noise and ultra-wideband application. *IEEE Personal and Indoors Mobile*, pp. 1–5.

Zhen B, Li H-B and Kohno R (2007) Clock management in ultra-wideband ranging. *Proceedings of the Mobile and Wireless Communications Summit*, pp. 1–5.

Zimmermann T (1991) *Elektrooptische Entfernungsmessung mit Bandspreizverfahren*, VDI.

Ziv N and Mulloth B (2006) An exploration on mobile social networking: Dodgeball as a case in point. *Proceedings of the 5th IEEE International Conference on Mobile Business (ICMB 2006)*, Copenhagen, Denmark, p. 21.

Zou Z, Shao B, Zhou Q, Zhai C, Mao J, Baghaei-Nejad M, Chen Q and Zheng L (2012) Design and demonstration of passive UWB RFIDs: Chipless versus chip solutions. *Proceedings of the IEEE International Conference on RFID-Technologies and Applications (RFID-TA)*, Nice, France, pp. 6–11.

Zuendt M, Dornbusch P, Schafer T, Jacobi P and Flade D (2005) Integration of indoor positioning into a global location platform. *First Workshop on Positioning, Navigation and Communication (WPNC 2004)*, Hanover, Germany, pp. 55–56.

Index

Mobile Positioning and Tracking: From Conventional to Cooperative Techniques, Second Edition.
Simone Frattasi and Francescantonio Della Rosa.